T0134872

Biofuel and Biorefinery Technologies

Volume 11

This book series provides detailed information on recent developments in biofuels & bioenergy and related research. The individual volumes highlight all relevant biofuel production technologies and integrated biorefinery methods, describing the merits and shortcomings of each, including cost-efficiency. All volumes are written and edited by international experts, academics and researchers in the respective research areas.

Biofuel and Biorefinery Technologies will appeal to researchers and post-graduates in the fields of biofuels & bioenergy technology and applications, offering not only an overview of these specific fields of research, but also a wealth of detailed information.

More information about this series at http://www.springer.com/series/11833

Ajar Nath Yadav · Ali Asghar Rastegari ·
Neelam Yadav · Rajeeva Gaur
Editors

Biofuels Production – Sustainability and Advances in Microbial Bioresources

Editors
Ajar Nath Yadav
Department of Biotechnology
Eternal University
Sirmour, Himachal Pradesh, India

Neelam Yadav
Veer Bahadur Singh Purvanchal University
Jaunpur, Uttar Pradesh, India

Ali Asghar Rastegari
Department of Molecular and Cell Biochemistry
Islamic Azad University
Falavarjan Branch
Isfahan, Iran

Rajeeva Gaur
Department of Microbiology
Dr. Ram Manohar Lohia Avadh University
Faizabad, India

ISSN 2363-7609 ISSN 2363-7617 (electronic)
Biofuel and Biorefinery Technologies
ISBN 978-3-030-53935-1 ISBN 978-3-030-53933-7 (eBook)
https://doi.org/10.1007/978-3-030-53933-7

This Springer imprint is published by the registered company Springer Nature Switzerland AG
The registered company address is: Gewerbestrasse 11, 6330 Cham, Switzerland

Preface

Bioenergy represents a major type of renewable energy. Bioenergy is part of a larger bioeconomy, including agriculture, forestry and manufacturing. Bioenergy typically enhances regional energy access and reduces reliance on fossil fuels. It can vitalize the forestry and agriculture sectors and support increased use of renewable resources as feedstocks for a range of industrial processes. Biofuels are the potential and sustainable alternative sources of fossil fuels. Efforts are continuously being made to develop economically competitive biofuels and bioenergy. Microbes play an important role in the production of biofuels from different bioresources. There are different microbial technologies for the sustainable development of energy. The beneficial microbes also help to reduce climate change for Sustainable developments.

The present book on *Biofuels Production – Sustainability and Advances in Microbial Bioresources* covers biodiversity of plant-associated fungal communities and their role in plant growth promotion, mitigation of abiotic stress and soil fertility for sustainable agriculture. This book will be immensely useful to biological sciences, especially to microbiologists, microbial biotechnologists, biochemists, researchers and scientists of fungal biotechnology. We have honoured that the leading scientists who have extensive, in-depth experience and expertise in plant–microbes interaction and fungal biotechnology took the time and effort to develop these outstanding chapters. Each chapter is written by internationally recognized researchers/scientists so the reader is given an up-to-date and detailed account of our knowledge of the fungal biotechnology and innumerable agricultural applications of fungal communities.

Baru Sahib, Himachal Pradesh, India — Ajar Nath Yadav
Isfahan, Iran — Ali Asghar Rastegari
Mau, Uttar Pradesh, India — Neelam Yadav
Faizabad, Uttar Pradesh, India — Rajeeva Gaur

Acknowledgments

All the authors are sincerely acknowledged for contributing up-to-date information on the advances in microbial bioresources for biofuels production for sustainability. The editors are thankful to all the authors for their valuable contributions.

All editors would like to thank their families who were very patient and supportive during this journey. Our sincere thanks to the whole Springer team who was directly or indirectly involved in the compilation of this book. We are grateful to the many people who helped to bring this book to light. Editors wish to thank Dr. Miriam Sturm, Editor, Springer; Dr. Vijai Kumar Gupta, and Prof. Maria G. Tuohy, Series editor, Biofuel and Biorefinery Technologies; Arumugam Deivasigamani, Project Coordinator, Springer for generous assistance, constant support and patience in initializing the volume.

The editor, Dr. Ajar Nath Yadav is grateful to his Ph.D. research scholars Tanvir Kaur, Rubee Devi, Divjot Kour, Kusam Lata Rana and colleagues for their support, love and motivation in all his efforts during this project.

We are very sure that this book will interest scientists, graduates, undergraduates and postdocs who are investigating "Biofuel and Biorefinery Technologies".

Acknowledgments

Contents

Editors and Contributors

About the Editors

Ajar Nath Yadav is an Assistant Professor (Senior Scale) and Assistant Controller of Examination at Eternal University, Baru Sahib, Himachal Pradesh, India. He has 5 years of teaching and 11 years of research experience in the field of Microbial Biotechnology, Microbial Diversity and Plant–Microbe Interactions. Dr. Yadav obtained his doctorate degree in Microbial Biotechnology, jointly from Indian Agricultural Research Institute, New Delhi and Birla Institute of Technology, Mesra, Ranchi, India. Dr. Yadav has 196 publications, with h-index of 45, i10-index of 111 and 5151 citations (Google Scholar—on 29/09/2020). Dr. Yadav is the editor of 14 books in Springer, 02 books in Elsevier and 05 books in CRC Press Taylor & Francis. Dr. Yadav has published 115 research communications in different international and national conferences. In his credit one granted patent "Insecticidal formulation of novel strain of *Bacillus thuringiensis* AK 47". Dr. Yadav has got 12 Best Paper Presentation Awards, and 01 Young Scientist Award (NASI-Swarna Jyanti Purskar). Dr. Yadav had received the "Outstanding Teacher Award" in the 6th Annual Convocation, 2018, by Eternal University, Baru Sahib, Himachal Pradesh. Dr. Yadav has a long-standing interest in teaching at the UG, PG and Ph.D. level and is involved in taking courses in microbiology and microbial biotechnology. Dr. Yadav is currently handling two projects. Dr. Yadav has guided 01 Ph.D. and 01 M.Sc. Scholar and presently he is guiding 05 scholars for

Ph.D. degree. He has been serving as an editor/editorial board member and reviewer for different national and international peer-reviewed journals. He has lifetime membership in the Association of Microbiologists, India, and the Indian Science Congress Council, India. Please visit https://sites.google.com/site/ajarbiotech/ for more details.

Ali Asghar Rastegari is currently working as an assistant professor in the Faculty of Biological Science, Department of Molecular and Cellular Biochemistry, Falavarjan Branch, Islamic Azad University, Isfahan, I.R. Iran. He has 13 years of experience in the field of Enzyme Biotechnology, Nanobiotechnology, Biophysical Chemistry, Computational Biology and Biomedicine. Dr. Rastegari gained a Ph.D. in Molecular Biophysics in 2009, the University of Science and Research, Tehran Branch, Iran; M.Sc. (Biophysics), in 1994, from the Institute of Biochemistry and Biophysics, the University of Tehran and B.Sc. (Microbiology) in 1990, from the University of Isfahan, Iran. He has in his honour 39 Publications [21 Research papers, 02 Book, 16 Book chapters] in various supposed international, national journals and publishers. He is the editor of 02 books in Elsevier, 02 in CRC press, Taylor & Francis and 06 in Springer, under the process of publication. He has issued 12 abstracts in different conferences/symposiums/workshops. He has presented 02 papers at national and international conferences/symposiums. Dr. Rastegari is a reviewer of different national and international journals. He has a lifetime membership of Iranian Society for Trace Elements Research (ISTER), The Biochemical Society of I.R. IRAN, Member of Society for Bioinformatics in Northern Europe (SocBiN), Membership of Boston Area Molecular Biology Computer Types (BAMBCT), Bioinformatics/Computational Biology Student Society (BIMATICS Membership), Ensemble genome database and Neuroimaging Informatics Tools and Resources Clearinghouse (NITRC).

Neelam Yadav is currently working on microbial diversity from diverse sources and their biotechnological applications in agriculture and allied sectors. She obtained her post-graduation degree from Veer Bahadur Singh Purvanchal University, Uttar Pradesh, India. She has research interest in the area of beneficial microbiomes and their biotechnological application in agriculture, medicine environment and allied sectors. She has to her credit 65 publications in different reputed international, national journals and publishers with h-index of 20, i10-index of 35, and 1216 citations (Google Scholar—on 29/09/2020). She is editor of 08 books in Springer, 02 Books in Elsevier and 02 books in CRC Press, Taylor & Francis. She is Editor/associate editor/reviewer of different international and national journals including Plos One, Extremophiles, Annals of Microbiology, Journal of Basic Microbiology, Advance in Microbiology and Biotechnology. She has the lifetime membership in the Association of Microbiologists, India, Indian Science Congress Council, India and National Academy of Sciences, India. Please visit https://sites.google.com/site/neelamanyadav/ for more details.

Rajeeva Gaur is a Professor in the Department of Microbiology, Dr. Ram Manohar Lohia Avadh University, Faizabad, India. Dr. Gaur received his B.Sc. (CBZ) from D. D. U. Gorakhpur University, Gorakhpur, M.Sc. in Microbiology from G. B. Pant University of Agriculture and Technology, Pantnagar and Ph.D. in Microbiology from Banaras Hindu University, Varanasi, Uttar Pradesh, India. Dr. Gaur has published 62 National and International research papers, 06 Books and 21 Book chapters. Dr. Gaur completed four projects funded by UPCST, Lucknow and UGC New Delhi. Dr. Gaur has a long-standing interest in teaching at the UG, PG and Ph.D. level and is involved in taking courses in microbiology. He is an Editor/associate editor/reviewer of different international and national journals. Dr. Gaur has lifetime membership in the Association of Microbiologists of India and Association of Biochemistry of India.

Contributors

I. Abernaebenezer Selvakumari Department of Chemical Engineering, Vel Tech High Tech Dr. Rangarajan Dr. Sakunthala Engineering College, Chennai, India

M. A. Amer Bio-System Engineering Department, Agricultural Engineering Research Institute (AEnRI), Agricultural Research Center (ARC), Giza, Egypt

R. R. L. Araújo Federal Institute of Pernambuco, Campus Caruaru, Recife, Brazil

Cristian B. Arenas Chemical and Environmental Bioprocess Engineering Group, Natural Resources Institute (IRENA), University of León, Leon, Spain

Richa Arora Department of Microbiology, Punjab Agricultural University, Ludhiana, Punjab, India

Shailendra Kumar Arya Department of Biotechnology, University Institute of Engineering and Technology (UIET), Panjab University (PU), Chandigarh, Punjab, India

Mehwish Aslam School of Biological Sciences, University of the Punjab, Lahore, Pakistan

Hamideh Bakhshayeshan-Agdam Department of Plant Sciences, Faculty of Natural Sciences, University of Tabriz, Tabriz, Iran

Sangita Banga Department of Chemistry, Manav Rachna University, Faridabad, Haryana, India

B. Bharathiraja Department of Chemical Engineering, Vel Tech High Tech Dr. Rangarajan Dr. Sakunthala Engineering College, Chennai, India

Sonali Bhardwaj School of Bioengineering and Biosciences, Lovely Professional University, Phagwara, Punjab, India

Prem Kumar Dantu Department of Botany, Dayalbagh Educational Institute, Agra, Uttar Pradesh, India

F. P. De Andrade Chemical Engineering, Federal University of Alagoas, Maceió, Brazil

C. E. De Farias Silva Chemical Engineering, Federal University of Alagoas, Maceió, Brazil

M. L. F. De Sá Filho Chemical Engineering, Federal University of Alagoas, Maceió, Brazil

Esteffany de Souza Candeo Department of Bioprocess Engineering and Biotechnology, Universidade Tecnológica Federal do Paraná (UTFPR), Ponta Grossa, Paraná, Brazil

Rubee Devi Department of Biotechnology, Dr. KSG Akal College of Agriculture, Eternal University, Baru Sahib, Sirmour, Himachal Pradesh, India

M. E. Egela Bio-System Engineering Department, Agricultural Engineering Research Institute (AEnRI), Agricultural Research Center (ARC), Giza, Egypt

Xiomar Gómez Chemical and Environmental Bioprocess Engineering Group, Natural Resources Institute (IRENA), University of León, Leon, Spain

José Francisco González-Álvarez School of Industrial, Informatics and Aerospace Engineering, University of León, Leon, Spain

Judith González-Arias Chemical and Environmental Bioprocess Engineering Group, Natural Resources Institute (IRENA), University of León, Leon, Spain

Deepika Goyal Department of Botany, Dayalbagh Educational Institute, Agra, Uttar Pradesh, India

Elisabete Hiromi Hashimoto Department of Bioprocess Engineering and Biotechnology, Universidade Tecnológica Federal do Paraná (UTFPR), Ponta Grossa, Paraná, Brazil

N. K. Ismail Bio-System Engineering Department, Agricultural Engineering Research Institute (AEnRI), Agricultural Research Center (ARC), Giza, Egypt

J. Jayamuthunagai Centre for Biotechnology, Anna University, Chennai, India

Yachana Jha N. V. Patel College of Pure and Applied Sciences, S. P. University, Anand, Gujarat, India

Meena Kapahi Department of Chemistry, Manav Rachna University, Faridabad, Haryana, India

Tanvir Kaur Department of Biotechnology, Dr. KSG Akal College of Agriculture, Eternal University, Baru Sahib, Sirmour, Himachal Pradesh, India

Divjot Kour Department of Biotechnology, Dr. KSG Akal College of Agriculture, Eternal University, Baru Sahib, Sirmour, Himachal Pradesh, India

Arvind Kumar Environmental Pollution Abatement Laboratory, Department of Chemical Engineering, National Institute of Technology, Rourkela, India

Sachin Kumar Biochemical Conversion Division, Sardar Swaran Singh National Institute of Bio-Energy, Kapurthala, India

Manoj Kumar Mahapatra Environmental Pollution Abatement Laboratory, Department of Chemical Engineering, National Institute of Technology, Rourkela, India

Sushma Mishra Department of Botany, Dayalbagh Educational Institute, Agra, Uttar Pradesh, India

Salma Mukhtar School of Biological Sciences, University of the Punjab, Lahore, Pakistan

Puneet Negi Department of Physics, Akal College of Basic Sciences, Eternal University, Baru Sahib, Sirmour, Himachal Pradesh, India

Versha Pandey Agronomy and Soil Science, Central Institute of Medicinal and Aromatic Plants, Lucknow, Uttar Pradesh, India

Vinayak Vandan Pathak Department of Chemistry, Manav Rachna University, Faridabad, Haryana, India

J. Tony Pembroke Laboratory of Structural and Molecular Biochemistry, Department of Chemical Sciences, School of Natural Sciences and Bernal Institute, University of Limerick, Limerick, Ireland

Shiv Prasad Centre for Environment Science & Climate Resilient Agriculture, ICAR-Indian Agricultural Research Institute, New Delhi, India

Roopa Rani Department of Chemistry, Manav Rachna University, Faridabad, Haryana, India

T. R. M. Ribeiro Chemical Engineering, Federal University of Sergipe, São Cristóvão, Brazil

Michael P. Ryan Laboratory of Structural and Molecular Biochemistry, Department of Chemical Sciences, School of Natural Sciences and Bernal Institute, University of Limerick, Limerick, Ireland

A. G. Saad Bio-System Engineering Department, Agricultural Engineering Research Institute (AEnRI), Agricultural Research Center (ARC), Giza, Egypt

Seyed Yahya Salehi-Lisar Department of Plant Sciences, Faculty of Natural Sciences, University of Tabriz, Tabriz, Iran

K. Senthilkumar Kongu Engineering College, Perundurai, Erode, India

Archita Sharma Department of Biotechnology, University Institute of Engineering and Technology (UIET), Panjab University (PU), Chandigarh, Punjab, India

A. E. Silva Chemical Engineering, Federal University of Alagoas, Maceió, Brazil

Anoop Singh Department of Scientific and Industrial Research (DSIR), Ministry of Science and Technology, Government of India, Technology Bhawan, New Delhi, India

Carlos Ricardo Soccol Department of Bioprocess Engineering and Biotechnology, Federal University of Paraná, Curitiba, Paraná, Brazil

Alessandra Cristine Novak Sydney Department of Bioprocess Engineering and Biotechnology, Universidade Tecnológica Federal do Paraná (UTFPR), Ponta Grossa, Paraná, Brazil

Eduardo Bittencourt Sydney Department of Bioprocess Engineering and Biotechnology, Universidade Tecnológica Federal do Paraná (UTFPR), Ponta Grossa, Paraná, Brazil

Jaya Tuteja Department of Chemistry, Manav Rachna University, Faridabad, Haryana, India

Ajar Nath Yadav Department of Genetics, Plant Breeding and Biotechnology, Dr. Khem Singh Gill Akal College of Agriculture, Eternal University Baru Sahib, Sirmour, Himachal Pradesh, India

Neelam Yadav Gopi Nath P.G. College, Veer Bahadur Singh Purvanchal University, Ghazipur, Uttar Pradesh, India

Gholamreza Zarrini Department of Animal Sciences, Faculty of Natural Sciences, University of Tabriz, Tabriz, Iran

Chapter 1
Microbial Bioresources for Biofuels Production: Fundamentals and Applications

Esteffany de Souza Candeo, Alessandra Cristine Novak Sydney, Elisabete Hiromi Hashimoto, Carlos Ricardo Soccol, and Eduardo Bittencourt Sydney

Abstract Biofuel production is increasingly aroused by new market demands, increased societal pressure on sustainability, and increasingly restrictive action by environmental agencies and legislations. Within the already known and consolidated technologies to produce biofuels from plant biomass (first-generation), there is growing interest in the use of microbial biomass, whose use through different techniques gives rise to second, third, and fourth-generation biofuels. The development of biofuel production technologies from microbial resources play an important economic and environmental role, specifically regarding the bioeconomy and the concepts of biorefinery and circular economy, due to its easy integration into existing industrial systems. Considering the use of different microorganisms (bacteria, yeast, fungi, and microalgae) and classes of molecules produced by them for production of liquid and gaseous biofuels, it is presented the fundamentals of biological, chemical, and thermochemical conversion of microbial-derived molecules such as carbohydrates, lipids, proteins, and/or microbial biomasses to ethanol (2nd, 3rd, and 4th generation), biodiesel, bio-oil, biogas, and biohydrogen.

1.1 Introduction

Renewable energy has gained global prestige in response to environmental problems caused by the overuse of nonrenewable energy sources (coal, oil, and natural gas). Reducing greenhouse gas emissions is the main environmental contribution of renewable energy. In addition, sustainable options (biomass availability, sunlight,

E. de Souza Candeo · A. C. N. Sydney · E. H. Hashimoto · E. B. Sydney (✉)
Department of Bioprocess Engineering and Biotechnology, Universidade Tecnológica Federal do Paraná (UTFPR), Ponta Grossa, Paraná, Brazil
e-mail: eduardosydney@utfpr.edu.br

C. R. Soccol
Department of Bioprocess Engineering and Biotechnology, Federal University of Paraná, Curitiba, Paraná, Brazil

A. N. Yadav et al. (eds.), *Biofuels Production – Sustainability and Advances in Microbial Bioresources*, Biofuel and Biorefinery Technologies 11,
https://doi.org/10.1007/978-3-030-53933-7_1

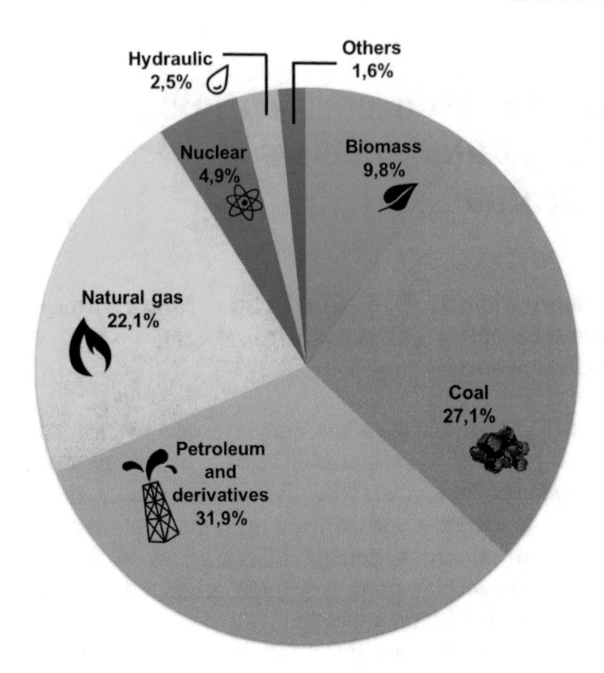

Fig. 1.1 World energy matrix in 2016. Adapted from EIA (2019)

winds, tides, etc.) value regional production chains and allow diversification of the predominantly (approximately 80%) nonrenewable global energy matrix (Rasool and Hemalatha 2016). Renewable energy sources can be classified in traditional, conventional, and new (Fig. 1.1). Traditional renewable energy sources are based on burning deforestation wood for noncommercial purposes (Rastegari et al. 2020). Conventional sources are based on commercially disseminated processes such as medium and large hydroelectricity (Schutz et al. 2013). The new ones encompass all of the recent renewable energy sources in the market, such as biomass conversion processes and production of second-generation biofuels by microbial resources (Table 1.1).

As a result of population growth and the improvement in quality of life for part of the people, energy demand is expected to increase over the next decades. Meeting this demand will require increasing energy efficiency (reducing transport losses, increasing motor's and generator's efficiency, etc.); diversify the energy matrix; and innovate the development of integrated biomass harvesting processes (Chisti 2008; EIA 2018).

In this context, the conversion of microbial resources into energy gains importance due to the diversity of biofuels that can be generated. Microbial resources are advantageous regardless of the availability of sun, wind, and fertile soils; they present stability in their production and promote the valorization of local productive chains. The raw material used in the microbial biomass/energy generation may come from

Table 1.1 Classification of primary and secondary energy sources according to Goldemberg and Lucon 2007

Sources of energy		Primary energy		Second energy
Nonrenewable	Fossil	Mineral coal		Thermoelectricity, heat, transport fuel
		Petroleum and derivatives		
		Natural gas		
	Nuclear	Fissile materials		
Renewable	"Traditional"	Primitive biomass		Heat
	"Conventional"	Hydroelectric		Hydroelectricity
	"New"	Hydroelectric		
		Modern biomass		Biofuel, heat, thermoelectricity
		Others	Solar energy	Heat, photovoltaic electricity
			Geothermal	Heat, electricity
			Wind	Electricity
			Tidal wave	

natural processes (lignocellulosic materials, aquatic biomass) or as a result of anthropogenic processes (industrial organic waste, agricultural waste, urban solid waste, among others) (Fig. 1.2). These raw materials are abundant, inexpensive, and seasonally free and they can be converted by microorganisms to energy such as bioethanol, biogas, biodiesel, and biohydrogen (Kour et al. 2019a; Yadav et al. 2020; Vassilev et al. 2012).

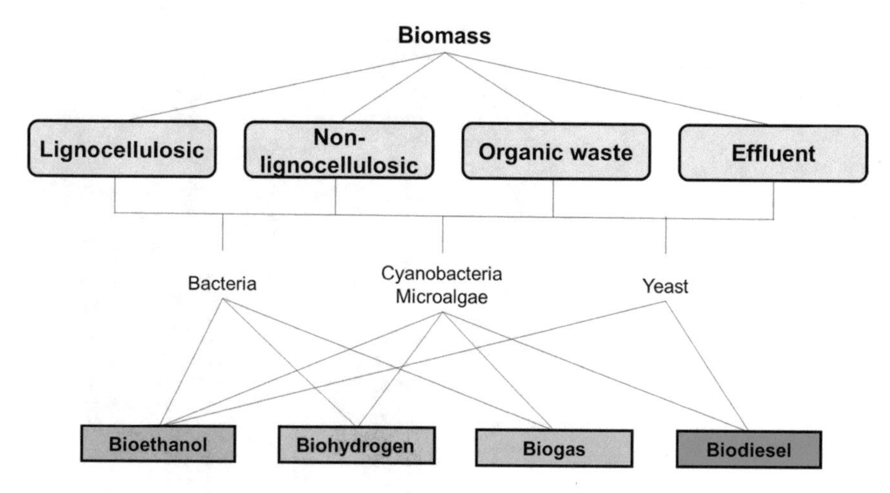

Fig. 1.2 Routes of biomass conversion to biofuels

1.2 Advantages of Biofuels

Biofuels are solid, liquid, and gaseous products with a high concentration of energy from renewable sources. Biofuels can be divided according to raw material and production process into first, second, third or fourth-generation (Fig. 1.3) (Naik et al. 2010).

First-generation biofuels are those made from raw materials composed primarily of sugars or oils. They are often available in plant crops that can also be used for food and may compete negatively for arable fertile soils. In the second-generation, modern and sustainable bioprocesses are used from residual biomass, such as ligno-cellulosic materials from the industrial processing of plant material. Third-generation biofuels are characterized by the direct conversion of solar energy into energy molecules (mainly carbohydrates and lipids) through the photosynthesis of algae and microalgae. The fourth-generation is based on the production of biofuels from engineering-designed microbial sources and synthetic biology for better energy yields and lower environmental impacts (Aro 2016; Dutta and Davereya 2014; Elegbede and Guerrero 2016; Liew et al. 2014; Naik et al. 2010).

Among the main advantages of replacing fossil fuels with biofuels is the reduction of greenhouse gas emissions, such as carbon dioxide (CO_2). The life cycle of these biofuels allows CO_2 to be recycled because the gas emitted by burning biofuel is approximately equivalent to that used in the biosynthesis of renewable sources, maintaining a supposed CO_2 consumption-release balance (Fig. 1.4). In the case of fourth-generation biofuels, the removal of gas from the atmosphere results in positive

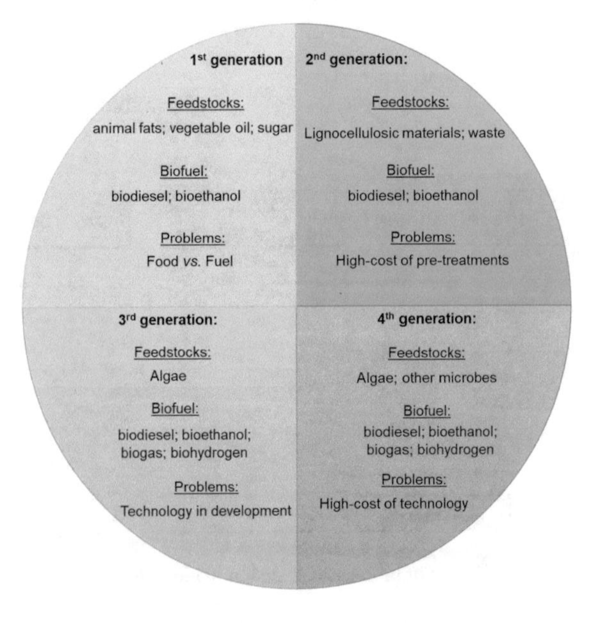

Fig. 1.3 Different biofuels generation: general characteristics of the process. Adapted from Dutta and Davereya (2014), Naik et al. (2010)

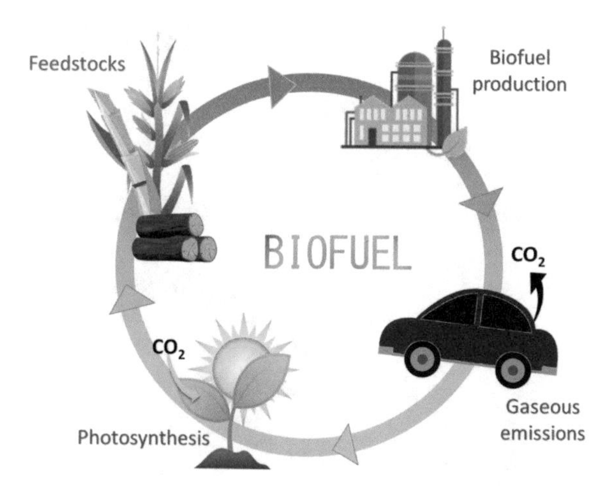

Fig. 1.4 Life cycle of a biofuel: CO_2 absorption for biomass biosynthesis; generation of renewable raw materials; biofuel processing; fuels burning and gas emissions

credits (Cuellar-Bermudez et al. 2015; Dutta and Davereya 2014; Lora and Venturini 2012; Porqueras et al. 2012).

In the political-economic context, biofuel production results in local energy independence and establishment of a biobased economy. Possible impacts of fluctuations in the fossil fuel market are minimized by the consumption of biofuels. The stimulation of the agro-industrial sector is also relevant. Given the large availability of biomass, it is possible to integrate processes in biorefineries and reuse solid, liquid, and gaseous waste (Demirbas 2009; Machado and Atsumi 2012). Moreover, the use of waste (agricultural, timber, food, industrial effluents, saline and wastewater, domestic sewage, carbon dioxide, etc.) adds value to previously unusable sources, associating the reduction of organic load with obtaining biomass with high energy content (Kour et al. 2019b; Larkum et al. 2012).

1.3 Microbial Cell Composition

Biofuel production can be carried out from microbial resources such as bacteria, fungi, yeast, and microalgae. Different raw materials can be converted into chemical products of high energy by fermentation or directly by thermochemical processing of cells. Through cell fractionation, it is possible to separate carbohydrates, lipids, and proteins present in all prokaryotic or eukaryotic cells (Fig. 1.5).

1.3.1 Microbial Carbohydrates

A carbohydrate molecule has an empirical formula $(CH_2O)_n$ with $n \geq 3$, and it can contain phosphorus, nitrogen or sulfur content on its composition. The monomer, the

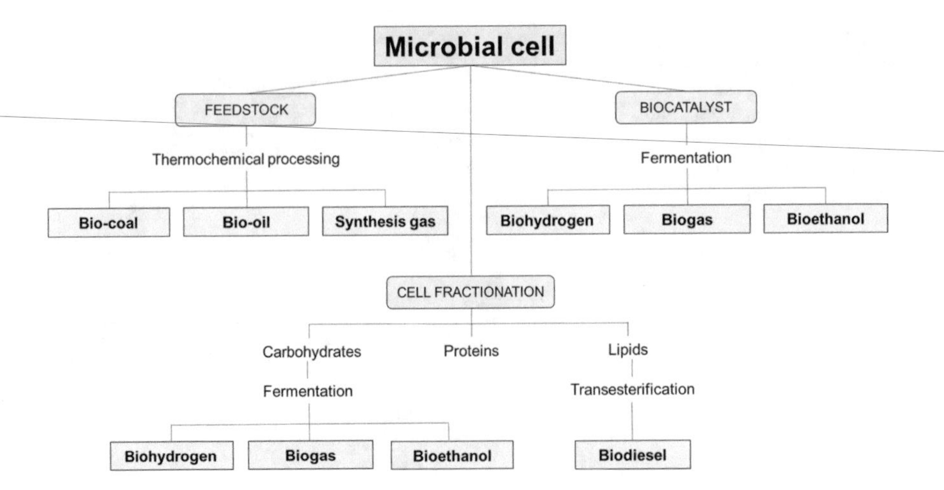

Fig. 1.5 Conversion of biomass to biofuels by technological processes using microbial resources as biocatalysts or direct source of raw material

basic unit of a carbohydrate, may be a polyhydroxy ketone or a polyhydroxy aldehyde, with two or more hydroxyl groups. They have a common characteristic of high hydro solubility, and consequently, they are soluble in the cytoplasm or polymerized and conjugated with other organic molecules of the cell. Carbohydrates are classified into four classes according to the number of monomers. The monosaccharide comprises a single monomer, which may be an aldose or ketosis, depending on the organic class (Nelson and Cox 2014). D-glucose is the most common representative of aldoses and D-fructose is an example of ketose. Disaccharides are composed of two monosaccharides linked by a covalent glycosidic bond. Sucrose, lactose, and maltose are common examples of disaccharides. Oligosaccharides are short-chain carbohydrates with two or more monomers, such as dextrins.

Polysaccharides are polymers with more than 20 monosaccharide units. Polysaccharides have high molecular mass (MM > 20,000). Depending on the type of glycosidic bond, the polysaccharide chain is branched or linear, resulting in different properties and functions. Polysaccharides may be classified as homopolymers or heteropolymers. When they are made up of a single monomeric species, they are called homopolysaccharides, such as starch and glycogen (made up of D-glucose molecules). When they are made up of two or more different monomers, they are called heteropolysaccharides (Nelson and Cox 2014). Polysaccharides are molecules that store the energy in the cells. Algae and vegetables store starch; bacterial and animal cells store glycogen. These D-glucose polymers can be extracted and converted into biofuels (Cheng et al. 2011; Kurita 2006; Nelson and Cox 2014; Scott et al. 2010).

Carbohydrates can be absorbed from the extracellular medium or synthesized inside microbial cells. Carbohydrates are metabolized by common pathways generating energy for reproduction, development, and growth of the cells. Carbohydrates can also be converted to products by fermentation, such as ethanol (Costa and Morais 2011; Nelson and Cox 2014). First-generation ethanol production is characterized by fermentation of saccharide or starch sources (energy reserves of higher plant species). While sucrose can be quickly converted to ethanol, starch must be previously hydrolysed to D-glucose molecules, which can be converted to ethanol by a fermentation agent (Bai et al. 2008; Wang et al. 2007).

In the second-generation ethanol production, the raw material is usually lignocellulosic biomass, such as agro-industry residues (bagasse and bark), cellulose paper (scraps), and timber (thin chips). These materials are abundant in sugars, but they are polymerized in cellulose and hemicellulose chains and protected by recalcitrant lignin chains (Aditiya et al. 2016; Cheng and Timilsina 2011). Pretreatment steps are required due to the structural arrangement of the lignocellulosic biomass matrix. This step disorganizes the lignocellulosic complex and increases the susceptibility of cellulose chains to the enzymatic hydrolysis. After this, simple sugars such as L-arabinose, D-xylose, D-glucose, D-mannose, and D-galactose are released from the hemicellulose chains and the lignin is removed; the cellulose fraction remains. The hydrolysis process (acid, basic or enzymatic) degrades cellulose polymers and hemicellulose residues, resulting in a fermentable carbohydrate broth. In the end, the broth rich in pentoses and hexoses can be converted anaerobically to ethanol (Aditiya et al. 2016; Nigam and Singh 2011; Pereira et al. 2008; Rosgaard and Meyer 2007; Schädel et al. 2010). However, expensive biomass pretreatment processes and low efficiency of the fermentation restrict the viability and commercial competitiveness of second-generation ethanol production technologies from cellulose and hemicellulose biomass (Cheng and Timilsina 2011; Nigam and Singh 2011).

The ethanol production by third-generation technology is based on the conversion of solar energy into biofuel. This process is possible due to some algae ability to store intracellular carbohydrates, especially starch by *Euglena* spp., *Chlorella* spp., *Chlamydomonas* spp., and *Scenedesmus* spp (Kumar et al. 2019; Yadav et al. 2017; Yadav et al. 2019b). Different species can accumulate about 8–64% of carbohydrates on a dry basis (Table 1.2). These algae are photosynthetic organisms and they grow absorbing light and assimilating CO_2 (industrial emissions, for example) and some inorganic nutrients. Some microalgae can use both: CO_2 in photosynthesis and organic carbon in the respiration process, obtaining better growth rates (John et al. 2011; Nigam and Singh 2011; Subashchandrabose et al. 2013), but null or reduced CO_2 balance.

For the third-generation bioethanol production, microalgae grow in open ponds or photobioreactors until they have a high carbohydrate intracellular content, usually in the stationary phase. Then, they are harvested and processed to disrupt the cell wall and release intracellular macromolecules. Starch is hydrolyzed and the resulting simple sugars are fermented (Günerken et al. 2015; Kose and Oncels 2016; Mata et al. 2010). With advances and improvements, the third-generation ethanol production is becoming economically and environmentally profitable: microalgae can grow in nonagricultural areas, avoiding competition with the food industry. The source of water for microalgae cultivation may be brackish, saline or even residual, and contaminants from wastewater treatment can be used as nutrients. The bio-fixation

Table 1.2 Percentage of carbohydrate accumulation in microorganisms

Microrganism	Carbohydrates (% dry matter basis)
Porphyridium cruentum	40–57
Spirogyna sp.	33–64
Dunaliella salina	32
Chlorella pyrenoidosa	26
Prymnesium parvum	25–33
Anabaena cylindrical	25–30
Scenedesmus dimorphus	21–52
Chlamydomonas rheinhardii	17
Synechoccus sp.	15
Tetraselmis maculate	15
Euglena gracilis	14–18
Spirulina maxima	13–16
Chlorella vulgaris	12–17
Scenedesmus obliquus	10–17
Spirulina platensis	8–14

Source Adapted from Demirbas (2011)

of CO_2 from the atmosphere during algal growth is efficient and high microalgal growth rates and bioethanol yields achieve high efficiency of energy (Nigam and Singh 2011; Subhadra and Edwards 2010).

The challenges of the third-generation ethanol production depend on the full development of microalgae studies by genetic engineering. The application of this technology can optimize growth rates and cell potential against adverse growth conditions, increase chloroplast expression levels, and direct metabolism to carbohydrate accumulation (John et al. 2011; Machado and Atsumi 2012).

1.3.2 Microbial Lipids

Lipid macromolecules are a diversified group of chemical compounds, which have the common characteristics of insolubility in water and solubility in nonpolar organic solvents. The lipid classes have various functions, as energy storage in oil and fats, composition of structural components of biological membranes (phospholipids and sterols), enzymatic cofactors, hormones, photosensitive pigments, protein hydrophobic anchors, intracellular messengers, and others (Nelson and Cox 2014). Oils and fats are derived from saturated or unsaturated fatty acids, with 4–36 carbons. Triacylglycerol molecule is composed of glycerol linked by bonds to three fatty acids. The function of triglycerides in the cells are the reserve of energy. The advantages of energy storage in lipids are that they occupy less volume; the oxidation of one gram of triglycerides releases more than two times the energy released by the oxidation of one gram of carbohydrates; and the hydrophobicity of molecules, without hydration water (Nelson and Cox 2014).

In the first-generation technology, biodiesel is the biofuel produced from transesterification reactions of triacylglycerols that proceeded from vegetable oils (edible or not) or animal fats. However, the use of vegetable oils for fuel production competes with the food industry and impacts on the economy of both sectors. Thus, lipids contained in microalgae, bacteria, yeast, and mold (Table 1.3), are potential raw materials for biofuel production by second and third-generation technologies. The similarity of microbial oils with those of vegetable origin reinforces the advantageous production of biodiesel from this raw material (Béligon et al. 2015; Mata et al. 2010; Poli et al. 2014).

Microbial oils that have intracellular lipid accumulation above 20% of their biomass weight can be used as biomass for biodiesel production. These lipid-storing microorganisms are usually referred to as single cell oils (SCO). The microalgae and yeasts represent the classes with higher lipid content, accumulating up to 75 and

Table 1.3 Percentage of oil accumulation in different species of microorganisms

Microrganisms	Oil content (% dry wt)
Microalgae	
Schizochytrium sp.	50–77
Nitzchia sp.	45–47
Nannochloropsis sp.	31–68
Botryococcus braunii	25–75
Cylindrotheca sp.	16–37
Bacteria	
Arthrobacter sp.	>40
Acinetobacter calcoaceticus	27–38
Rhodococcus opacus	24–25
Bacillus alcalophilus	18–24
Yeast	
Rhodotorula glutinis	72
Cryptococcus albidus	65
Lipomyces starkeyi	64
Candida curvata	58
Fungi	
Mortierella isabelline	86
Humicola lanuginose	75
Mortierella vinacea	66
Aspergillus oryzae	57

Source Adapted from Demirbas (2011), Meng (2009)

70% of their dry weight, respectively (Alvarez and Steinbüchel 2002; Busic et al. 2018; Demirbas 2011; Dutta and Davereya 2014; Galafassi et al. 2012; Meng 2009; Ratledge and Cohen 2008; Saenge et al. 2011).

The production of SCO for biodiesel generation consists in the cultivation of pure strains in bioreactors, under controlled conditions and medium with high C/N ration (reduced concentration of nitrogen and high content of carbon). After cultivation, the biomass is processed to cells lysis and the triglycerides are released. Then, the extracted oils are recovered and the extracted biomass can be processed to the production of bioethanol or biogas, depending on the carbohydrate's availability in the cells. The oily fraction is submitted to the transesterification step with short-chain alcohols (methanol or ethanol), generating biodiesel, and glycerol as a high-value co-product. The glycerol can be fed back into the bioreactor as a carbon source, generating economy in the process (Brennan and Owende 2010; Cuellar-Bermudez et al. 2015; Nigam and Singh 2011; Scott et al. 2010).

Opportunities for improving the process of obtaining biodiesel from single-cell oils consist in reducing the costs: increasing the levels of triacylglycerol in the cells and improving the biomass processing and the lipid extraction. In addition, the possibility of other high value compounds production can help the process viability, minimizing costs, and waste (Mata et al. 2010; Scott et al. 2010).

1.3.3 Microbial Proteins

Usually, microorganisms have high protein content. Some microalgae species, for example, can accumulate about 70% of the protein in dry base. Proteins extracted from microalgae are intended for supplementation in human food, animal feed, as biofertilizers, and sources of nitrogen to the soil. However, due to the technological and economic barriers, microbial proteins have not yet been applied to biofuels generation, except for recent theoretical studies with genetically modified microorganisms (Boland et al. 2013; Brennan and Owende 2010; Choi et al. 2014; Liew et al. 2014).

Although proteins are not directly converted to biofuels, these macromolecules can serve as raw materials to produce microbial species capable of synthesizing high energy compounds. The Proteolysis, the process of proteins digestion into amino acids, is the first step to use these macromolecules as nutrients. An example of this process was performed by genetically modified *Bacillus subtilis,* which produced proteases and advanced biofuels such as higher alcohols from amino acids consumption (Choi et al. 2014).

1.4 Conversion of Whole Cells to Biofuels

In order to take full advantage of microbial biomass and reduce costs with specific components separation, some biofuels allow the use of whole cells in the production process.

1.4.1 Microalgae Biomass as Substrates for Biogas Production

Biogas is composed mainly of methane (CH_4) and carbon dioxide (CO_2), but it contains water vapor, trace quantities of H_2S, NH_3, N_2, CO, particulate compounds, and siloxanes. It is produced through an anaerobic process where organic residues such as sewage sludge, municipal solid waste, lignocellulosic biomass, slaughterhouse waste, animal waste, and others are degraded into CH_4 and CO_2 by a microbial community (Porpatham et al. 2008; Rasi et al. 2011). *Clostridia* sp., *Bacteriocides* sp., *Bifidobacteria* sp., *Streptococcus* sp., *Acetobacterium woodii, Clostridium aceticum, Methanosarcina barkeri, Methanococcus mazei, Methanotrix soehngenii* are examples of microbes included in this microbial community (Busic et al. 2018; Rana et al. 2019; Yadav et al. 2019a).

To achieve high efficiency in biogas production systems, the substrate should be submitted to pretreatments, such as size reduction, in order to optimize the action of microorganisms and enzymes in the process. The initial digestion phase of organic matter is called hydrolysis. In this phase, the enzymes secreted by the microorganisms promote the decomposition of the organic polymers in their respective monomers, in order to facilitate the absorption of the monomers. Then, in the acidogenic phase, different microorganisms hydrolyze the monomers into various volatile fatty acids, CO_2, hydrogen (H_2), and other simpler organic compounds. In the next phase, called acetogenesis, the simple organic molecules are metabolized into acetate, CO_2, and H_2. Then, in the methanogenic phase, the acetate, CO_2, and H_2 molecules are converted into CH_4 (Busic et al. 2018; Kiran et al. 2016).

To increase the efficiency of biogas production from waste biomass, it is common to use the addition of other substrates that will be co-digested, supplementing possible nutrient imbalances or adverse conditions of production. Domestic wastewaters, for example, have low solids content and consequently result in low CH_4 production. Manure and slaughter effluents, on the other hand, have such high concentrations of nitrogen that they can inhibit the methanogenic bacteria. Combining domestic wastewater and manure in an adequate proportion can be a rational solution in optimizing the production of methane. Thus, the co-digestion of different substrates may increase the process stability due to the nutrient balance, as well as reduce the inhibitory effects of ammonia and sulfides, resulting in higher CH_4 content. (Koch et al. 2015; Montingelli et al. 2015; Nayono et al. 2010; Zhang et al. 2013).

The produced biogas can be purified by condensation, activated carbon adsorption, cryogenic separation, and other techniques. Biomethane generated at high concentrations can be converted into electricity, heat or even applied as a gaseous fuel in the transport sector. The liquid effluent from the biogas production process is called digestate and it can be applied to arable soil as a biofertilizer and a nitrogen source (Busic et al. 2018; Kour et al. 2020; Rastegari et al. 2019b). An alternative for biogas production is the direct anaerobic digestion of microalgae biomass, especially those produced in substrates that may not be used as food/feed, such as domestic wastewaters. Basically, after cultivation in ponds, microalgae are recovered by sedimentation and centrifugation. In the next step, they are added to anaerobic digestion systems for biogas generation, and then they can undergo the same steps of purification of conventional production with organic waste. Considering that microalgae do not have lignin in its composition, pretreatment steps are unnecessary. The process, thus, is independent of separation and concentration of carbohydrate or lipid fractions of the cells. Circular systems, where microalgae cultures are used for CO_2 capture and biogas purification, resulting in increased biomass production that feeds the anaerobic reactor, were proposed (Collet et al. 2011; Montingelli et al. 2015).

1.4.2 Biohydrogen Production

Another gaseous product of microbial metabolism is biohydrogen, which can be generated by biophotolysis or dark fermentation (Fig. 1.6). Anaerobic digestion of organic matter results in high concentrations of H_2 during the anaerobic digestion phase called acidogenesis (Chandrasekhar et al. 2015; Mohan and Pandey 2013). The acidogenic bacteria grow faster and resist to lower pH when compared to methanogenic bacteria (Rastegari et al. 2019a). Thus, substrate feeding rate and pH can be used to select acidogenic rather than methanogenic in an anaerobic bioreactor. Biohydrogen, as well as biogas, can be produced by a microbial community capable to process complex macromolecules and produce H_2.

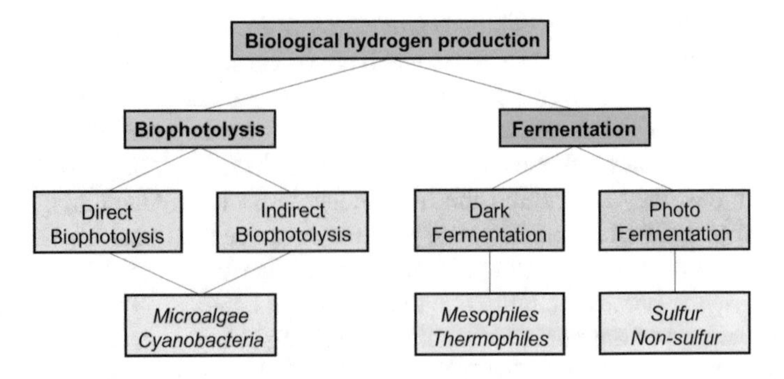

Fig. 1.6 Biohydrogen production routes. Modified from Gopalakrishnan et al. (2019)

Biohydrogen can also be obtained by cultivating microalgae and cyanobacteria by water biophotolysis using inorganic CO_2 in the presence of sunlight, as well as by photosynthetic bacteria, by photo fermentation processes of various organic acids in the presence of light (Gopalakrishnan et al. 2019). The generated hydrogen can be captured from the production system, purified, and then it can be used as biofuels and for the production of electricity in fuel cells (Show et al. 2012).

1.4.3 Thermochemical Routes

Another possibility for the destination and uses of microorganisms in the production of biofuels by thermochemical routes, mainly pyrolysis, gasification, torrefaction, and liquefaction. Each of the thermochemical routes generates energy products with different characteristics and applications (Chen et al. 2015; Rastegari et al. 2019c).

In the pyrolysis production system, the biomass is subjected to high temperatures (above 673 K) in the absence of oxygen, resulting in the formation of two fractions: (i) gaseous, which is the fraction that the bio-oil (liquid) is extracted by condensation, and (ii) solid, called biochar. Gasification process involves heating the biomass above 873 K, thus maximizing the fraction of combustible gasses (e.g. H_2, CH_4, CO_2, and ammonia) and minimizing the generation of bio-oil. The thermal processing of biomass in an inert atmosphere at temperatures above 473 K is called torrefaction, which results in only a solid fraction of high energy content. Finally, liquefaction is carried at critical conditions of temperature and pressure (above 573 K and 5 MPa), which are applied to the wet biomass, generating bio-oil, CO_2, H_2, CH_4, and hydrocarbon gas mixture, as well as a solid fraction of biochar (Brennan and Owende 2010; Bridgwater 2012; Chen et al. 2015; Demirbas 2011; Liew et al. 2014).

1.5 Conclusion and Future Vision

Fuel production is of enormous interest because large amounts of energy are required in industrial processes. This energy can come from fossil, renewable sources or microbial resources. The advantage of biological processes is that they promote the use of industrial waste and by-products for bioenergy production, which can be used in situ, resulting in autonomy and economy of energy, and increased environmental sustainability. Biological processes may involve (i) the use of microorganisms as catalysts, transforming raw materials into high energy bioproducts (ethanol, bio-oil, biogas, biohydrogen); (ii) the use of cellular components, especially carbohydrates and lipids from bacteria, fungi, and microalgae for biofuel production; and (iii) utilization of intact microbial biomass in biofuels generation through pyrolysis, gasification, among others. The development of biofuel production technologies from microbial resources play an important economic and environmental role, especially regarding the bioeconomy and the concepts of biorefinery and circular economy, due to its easy integration into existing industrial facilities.

References

Aditiya HB, Mahlia TMI, Chong WT, Nur H, Sebayang AH (2016) Second generation bioethanol production: a critical review. Renew Sust Energ Rev 66:631–653

Alvarez H, Steinbüchel A (2002) Triacylglycerols in prokaryotic microorganisms. Appl Microbiol Biotechnol 60:367–376

Aro E-M (2016) From first generation biofuels to advanced solar biofuels. Ambio 45:24–31

Bai FW, Anderson WA, Moo-Young M (2008) Ethanol fermentation technologies from sugar and starch feedstocks. Biotechnol Adv 26:89–105

Béligon V, Poughon L, Christophe G, Lebert A, Larroche C, Fontanille P (2015) Improvement and modeling of culture parameters to enhance biomass and lipid production by the oleaginous yeast *Cryptococcus curvatus* grown on acetate. Bioresour Technol 192:582–591

Boland MJ, Rae AN, Vereijken JM, Meuwissen MP, Fischer AR, Van Boekel MA et al (2013) The future supply of animal-derived protein for human consumption. Trends Food Sci Technol 29:62–73

Brennan L, Owende P (2010) Biofuels from microalgae A review of technologies for production, processing, and extractions of biofuels and co-products. Renew Sust Energ Rev 14:557–577

Bridgwater AV (2012) Review of fast pyrolysis of biomass and product upgrading. Biomass Bioenergy 38:68–94

Busic A, Kundas S, Morzak G, Belskaya H, Santek MI, Komes D et al (2018) Recent trends in biodiesel and biogas production. Food Technol Biotechnol 56:152–173

Capuano L (2018) International energy outlook 2018 (IEO2018) US Energy Information Administration (EIA): Washington, DC. USA 2018:21

Chandrasekhar K, Lee Y, Lee D (2015) Biohydrogen production: strategies to improve process efficiency through microbial routes. Int J Mol Sci 16:8266–8293

Chen WH, Lin BJ, Huang MY, Chang JS (2015) Thermochemical conversion of microalgal biomass into biofuels: a review. Biores Technol 184:314–327

Cheng G, Varanasi P, Li C, Liu H, Melnichenko YB, Simmons BA et al (2011) Transition of cellulose crystalline structure and surface morphology of biomass as a function of ionic liquid pretreatment and its relation to enzymatic hydrolysis. Biomacromol 12:933–941

Cheng JJ, Timilsina GR (2011) Status and barriers of advanced biofuel technologies: a review. Renew Energy 36:3541–3549

Chisti Y (2008) Biodiesel from microalgae beats bioethanol. Trends Biotechnol 26:126–131

Choi KY, Wernick DG, Tat CA, Liao JC (2014) Consolidated conversion of protein waste into biofuels and ammonia using *Bacillus subtilis*. Metab Eng 23:53–61

Collet P, Hélias A, Lardon L, Ras M, Goy RA, Steyer JP (2011) Life-cycle assessment of microalgae culture coupled to biogas production. Bioresour Technol 102:207–214

Costa JAV, Morais MG (2011) the role of biochemical engineering in the production of biofuels from microalgae. Bioresour Technol 102:2–9

Cuellar-Bermudez SP, Garcia-Perez JS, Rittmann BE, Parra-Saldivar R (2015) Photosynthetic bioenergy utilizing CO_2: an approach on flue gases utilization for third generation biofuels. J Clean Prod 98:53–65

Demirbas MF (2009) Biorefineries for biofuel upgrading: a critical Rev. Appl Energy 86:151–161

Demirbas MF (2011) Biofuels from algae for sustainable development. Appl Energy 88:3473–3480

Dutta K, Davereya Lin J (2014) Evolution retrospective for alternative fuels: first to fourth generation. Renew Energy 69:114–122

Elegbede I, Guerrero C (2016) Algae biofuel in the Nigerian energy context. Environ Climate Technol 17:44–60

Galafassi S, Cucchetti D, Pizza F, Franzosi G, Bianchi D, Compagno C (2012) Lipid production for second generation biodiesel by the oleaginous yeast *Rhodotorula graminis*. Bioresour Technol 111:398–403

Goldemberg J, Lucon O (2007) Energias renováveis: um futuro sustentável. Revista USP72: 6-15

Gopalakrishnan B, Khanna N, Das D (2019) Dark-fermentative biohydrogen production. Biohydrogen 2:79–122

Günerken E, D'hondt E, Eppink MHM, Garcia-Gonzalez L, Elst K, Wijffels RH (2015) Cell disruption for microalgae biorefineries. Biotechnol Adv 33: 243–260

John RP, Anisha GS, Nampoothiri KM, Pandey A (2011) Micro and macroalgal biomass: a renewable source for bioethanol. Bioresour Technol 102:186–193

Kiran EU, Stamatelatou K, Antonopoulou G, Lyberatos G (2016) Production of biogas via anaerobic digestion. Handb Biofuels Prod 1:259–301

Koch K, Helmreich B, Drewes J (2015) Co-digestion of food waste in municipal wastewater treatment plants: effect of different mixtures on methane yield and hydrolysis rate constant. Appl Energy 137:250–255

Kose A, Oncels S (2016) Algae as a promising resource for biofuel industry: facts and challenges. Intl J Energy Res 41:924–951

Kour D, Rana KL, Yadav AN, Yadav N, Kumar M, Kumar V et al (2020) Microbial biofertilizers: bioresources and eco-friendly technologies for agricultural and environmental sustainability. Biocatal Agric Biotechnol 23:101487. https://doi.org/10.1016/j.bcab.2019.101487

Kour D, Rana KL, Yadav N, Yadav AN, Rastegari AA, Singh C et al. (2019a) Technologies for biofuel production: current development, challenges, and future prospects. In: Rastegari AA, Yadav AN, Gupta A (eds) Prospects of renewable bioprocessing in future energy systems. Springer International Publishing, Cham, pp 1–50. https://doi.org/10.1007/978-3-030-14463-0_1

Kour D, Rana KL, Yadav N, Yadav AN, Singh J, Rastegari AA et al. (2019b) Agriculturally and industrially important fungi: current developments and potential biotechnological applications. In: Yadav AN, Singh S, Mishra S, Gupta A (eds) Recent advancement in white biotechnology through fungi, Volume 2: Perspective for value-added products and environments. Springer International Publishing, Cham, pp 1–64. https://doi.org/10.1007/978-3-030-14846-1_1

Kumar S, Sharma S, Thakur S, Mishra T, Negi P, Mishra S et al (2019) Bioprospecting of microbes for biohydrogen production: current status and future challenges. In: Molina G, Gupta VK, Singh BN, Gathergood N (eds) Bioprocessing for biomolecules production. Wiley, USA, pp 443–471

Kurita K (2006) Chitin and chitosan: functional biopolymers from marine crustaceans. Mar Biotechnol 8:203–226

Larkum AWD, Ross IL, Kruse O, Hankamer B (2012) Selection, breeding and engineering of microalgae for bioenergy and biofuel production. Trends Biotechnol 30:198–205

Liew WH, Hassim MH, Ng DKS (2014) Review of evolution, technology and sustainability assessments of biofuel production. J Clean Prod 71:11–29

Lora EES, Venturini OJ (2012) Biocombustíveis. Interciência 1:1–585

Machado IMP, Atsumi S (2012) Cyanobacterial biofuel production. J Biotechnol 162:50–56

Mata TM, Martins AA, Caetanons (2010) Microalgae for biodiesel production and other applications: a review. Renew Sustain Energy Rev 14:217–232

Meng X (2009) Biodiesel production from oleaginous microorganisms. Renew Energy 34:1–5

Mohan SV, Pandey A (2013) Biohydrogen production. Biohydrogen 2:1–24

Montingelli ME, Tedesco S, Olabi AG (2015) Biogas production from algal biomass: A review. Renew Sustain Energy Rev 43:961–972

Naik SN, Goud VV, Rout PK, Dalai AK (2010) Production of first and second generation biofuels: a comprehensive review. Renew Sustain Energy Rev 14:578–597

Nayono SE, Gallert C, Winter J (2010) Co-digestion of press water and food waste in a biowaste digester for improvement of biogas production. Bioresour Technol 101:6987–6993

Nelson D, Cox MM (2014) Princípios de bioquímica de Lehninger

Nigam PS, Singh A (2011) Production of liquid biofuels from renewable resources. Prog Energ Combust Sci 37:52–68

Pereira JRN, Couto M, Santa Anna LM (2008) Biomass of lignocellulosic composition for fuel ethanol production and the context of biorefinery. Series Biotechnol 2:2–45

Poli JS, Silva MAN, Siqueira EP, Pasa VMD, Rosa CA, Valente P (2014) Microbial lipid produced by *Yarrowia lipolytica* QU21 using industrial waste: a potential feedstock for biodiesel production. Bioresour Technol 16:320–326

Porpatham E, Ramesh A, Nagalingam B (2008) Investigation on the effect of concentration of methane in biogas when used as a fuel for a spark ignition engine. Fuel 87:1651–1659

Porqueras EM, Rittmann S, Herwig C (2012) Biofuels and CO_2 neutrality: an opportunity. Biofuels 3:413–426

Rana KL, Kour D, Sheikh I, Yadav N, Yadav AN, Kumar V et al (2019) Biodiversity of endophytic fungi from diverse niches and their biotechnological applications. In: Singh BP (ed) Advances in endophytic fungal research: present status and future challenges. Springer International Publishing, Cham, pp 105–144. https://doi.org/10.1007/978-3-030-03589-1_6

Rasi S, Läntelä J, Rintala J (2011) Trace compounds affecting biogas energy utilisation—A review. Energy Convers Manag 52:3369–3375

Rastegari AA, Yadav AN, Yadav N (2020) New and future developments in microbial biotechnology and bioengineering. Trends of microbial biotechnology for sustainable agriculture and biomedicine systems: perspectives for human health. Elsevier, Amsterdam

Rastegari AA, Yadav AN, Gupta A (2019a) Prospects of renewable bioprocessing in future energy systems. Springer International Publishing, Cham

Rastegari AA, Yadav AN, Yadav N (2019b) Genetic manipulation of secondary metabolites producers. In: Gupta VK, Pandey A (eds) New and future developments in microbial biotechnology and bioengineering. Elsevier, Amsterdam, pp 13–29. https://doi.org/10.1016/B978-0-444-63504-4.00002-5

Rastegari AA, Yadav AN, Yadav N, Tataei Sarshari N (2019c) Bioengineering of secondary metabolites. In: Gupta VK, Pandey A (eds) New and future developments in microbial biotechnology and bioengineering. Elsevier, Amsterdam, pp 55–68. https://doi.org/10.1016/B978-0-444-63504-4.00004-9

Ratledge C, Cohen Z (2008) Microbial and algal oils: do they have a future for biodiesel or as commodity oils? Lipid Technol 20:155–160

Rosgaard L, Meyer AS (2007) Enzymatic hydrolysis of pretreated barley and wheat straw. Chem Eng Tech Univ Denmark 1:1–124

Saenge C, Cheirsilp B, Suksaroge TT, Bourtoom T (2011) Efficient concomitant production of lipids and carotenoids by oleaginous red yeast Rhodotorula glutinis cultured in palm oil mill effluent and application of lipids for biodiesel production. Biotechnol Bioprocess Eng 16(1):23–33

Schädel C, Blöchl A, Richter A, Hoch G (2010) Quantification and monosaccharide composition of hemicelluloses from different plant functional types. Plant Physiol Biochem 48:1–8

Schutz F, Massuquetti A, Alves TW (2013) Demanda e oferta energética: uma perspectiva mundial e nacional para o etanol. Revista Eletrônica em Gestão, Educação e Tecnologia Ambiental 16:3167–3186

Scott SA, Davey MP, Dennis JS, Horst I, Howe CJ, Lea-Smith DJ et al (2010) Biodiesel from algae: challenges and prospects. Curr Opin Biotechnol 21:277–286

Show KY, Lee DJ, Tay JH, Lin CY, Chang JS (2012) Biohydrogen production: current perspectives and the way forward. Intl J Hydrog Energy 37:15616–15631

Subashchandrabose SR, Ramakrishnan B, Megharaj M, Venkateswarlu K, Naidu R (2013) *Mixotrophic cyanobacteria* and microalgae as distinctive biological agents for organic pollutant degradation. Environ Int 51:59–72

Subhadra B, Edwards M (2010) An integrated renewable energy park approach for algal biofuel production in United States. Energ Policy 38(9):4897–4902

Vassilev SV, Baxter D, Andersen LK, Vassileva CG, Morgan TJ (2012) An overview of the organic and inorganic phase composition of biomass. Fuel 94:1–33

Wang R, Ji Y, Melikoglu M, Koutinas A, Webb C (2007) Optimization of innovative ethanol production from wheat by response surface methodology. Process Saf Environ Prot 85:404–412
Yadav AN, Kumar R, Kumar S, Kumar V, Sugitha T, Singh B et al (2017) Beneficial microbiomes: biodiversity and potential biotechnological applications for sustainable agriculture and human health. J Appl Biol Biotechnol 5:45–57
Yadav AN, Rastegari AA, Yadav N (2020) Microbiomes of extreme environments: biodiversity and biotechnological applications. CRC Press, Taylor & Francis, Boca Raton, USA
Yadav AN, Singh S, Mishra S, Gupta A (2019a) Recent advancement in white biotechnology through fungi. Volume 2: perspective for value-added products and environments. Springer International Publishing, Cham
Yadav AN, Yadav N, Sachan SG, Saxena AK (2019b) Biodiversity of psychrotrophic microbes and their biotechnological applications. J Appl Biol Biotechnol 7:99–108
Zhang C, Xiao G, Peng L, Su H, Tan T (2013) the anaerobic co-digestion of food waste and cattle manure. Bioresour Technol 129:170–176

Chapter 2
Bioprospecting of Microorganisms for Biofuel Production

Sonali Bhardwaj, Sachin Kumar, and Richa Arora

Abstract Microbial biofuels have captured considerable scientific attention as they can reduce the world's reliance on fossil energy sources by meeting the rising energy demands, reducing the emission of greenhouse gases and overcoming other environmental challenges. It acts as a clean alternative energy source, therefore ensuring energy security and combating the drastic climate change. Extensive research is being done to develop microbial biofuels having high yield and social stability which can be used as affordable energy. This chapter emphasizes a range of microbes used in the valorization of multifarious lignocellulosic biomass into sustainable and economically significant biofuel. Although many microorganisms are reported to be involved in biofuel production, efficient bioconversion of complex sugars into simple sugars still remains a challenge. Great strides have been made in recent years towards genetic engineering of microorganisms to enhance saccharification of lignocellulosic biomass, lessen the production of inhibitory sugars and enhance the tolerance of desirable end product towards fermenting microbes. *Saccharomyces* sp., *Kluyveromyces* sp., *Clostridium* sp., and *Trichoderma* sp. have been extensively exploited to obtain a high yield of simpler sugars, less amount of inhibitory compounds and high yield of biofuel. This chapter aims to review the important fermenting microbes being used in the production of different types of biofuels.

The original version of this chapter was revised: The authors' names have been corrected to "Sonali Bhardwaj, Sachin Kumar and Richa Arora". The correction to this chapter is available at https://doi.org/10.1007/978-3-030-53933-7_18

S. Bhardwaj
School of Bioengineering and Biosciences, Lovely Professional University, Phagwara, Punjab, India

S. Kumar
Biochemical Conversion Division, Sardar Swaran Singh National Institute of Bio-Energy, Kapurthala, India

R. Arora (✉)
Department of Microbiology, Punjab Agricultural University, Ludhiana, Punjab, India
e-mail: richaarora@pau.edu

A. N. Yadav et al. (eds.), *Biofuels Production – Sustainability and Advances in Microbial Bioresources*, Biofuel and Biorefinery Technologies 11,
https://doi.org/10.1007/978-3-030-53933-7_2

2.1 Introduction

Energy is essential for sustainable and economic development and major fraction of energy being used is derived from fossil fuels (Perumal Saravanan et al. 2018). Our planet is a limited reservoir of fossil fuels. The accelerated depletion of environmental-unfriendly fossil fuels has compelled researchers all over the globe to hunt for viable alternative renewable energy sources to satiate the increasing energy demands of global population, industrial sector and transportation sector, also reducing the emission of greenhouse gases and overcoming other environmental challenges. When the extraction of fossil fuels reaches its peak, the production will enter the terminal decline phase resulting in geopolitical instability and threatening the international energy security (Sarmiento et al. 2017). In underdeveloped countries, the hikes in the prices of fossil fuels are also a matter of great concern. Biofuel is a minimally toxic renewable fuel that can be derived from a plethora of biomass feedstock such as food crops, lignocellulosic biomass, algal biomass, etc. (Rastegari et al. 2019).

Biofuels can be an auspicious substitute for unsafe fossil fuels as it can overcome the problems encumbered by the use of fossil fuels, therefore ensuring energy security by reducing world's reliance on fossil energy sources, meeting the rising energy demands and combating the drastic climate change (Perumal Saravanan et al. 2018). Research has unveiled the potential of microorganisms for the production of biofuel beyond their conventional uses such as wastewater treatment process; production of antibiotics, vaccines, probiotics, fermented food products, etc. Thereby, microbial biofuels have captured considerable scientific attention as they can act as a cleaner alternative energy source and leave a positive impact on the environment. Microbes can produce a range of biofuels such as bioethanol, biobutanol, biodiesel, biohydrogen and biogas. Microbes have innate pathways by which they can utilize comprehensive substrates for biofuel production. Microbes play a crucial role in pretreatment, saccharification and fermentation of biomass. Prospective role of ligninolytic enzymes for pretreatment of lignocellulosic biomass has captured significant attention. Extensive research is being done to develop microbial biofuels having high yield and social stability which can be used as affordable energy.

2.2 Feedstock for Biofuels Production

Microbial biofuels, depending on the biomass or feedstock used can be broadly grouped as first, second, third and fourth generation biofuels. All edible food crops such as sugar crops, starch crops, oil crops and even animal fat are included in first-generation biofuels. Utilization of food crops as biofuel feedstock directly competes for food consumption, agricultural land as well as resources, endangering the food security (Bhatia et al. 2017). Non-food crops such as lignocellulosic biomass and biowaste such as agricultural waste and municipal waste consolidate as second-generation biofuels. Improper disposal of agricultural waste and other lignocellulosic biomass poses jeopardy to the global economy as well as the environment; therefore, its valorization to the alternative energy source is a sustainable approach also ensuring food safety (Rastogi and Shrivastava 2017). Lignocellulosic biomass is abundant in environment constituting of polymers like lignin, cellulose and hemicellulose (Bhatia et al. 2017). This generation, however, overcomes the limitations posed by first-generation biofuel feedstock; despite challenges still exist owing to unique chemical composition, recalcitrance to degradation due to lignin sheath, efficient hydrolysis of cellulose, heterogeneity of hemicellulose (Rastogi and Shrivastava 2017). Lignolytic microbes can potentially achieve 80% delignification of LCB by the action of enzyme laccase (Avanthi and Banerjee 2016).

Biological pretreatment using enzymes such as lignin peroxide, manganese peroxide, laccase etc. is efficient but consumes a lot of time and increases the cost of production process. Chemical or physical pretreatment of biomass is often required to reduce crystallinity of cellulose to make it more vulnerable towards enzymatic hydrolysis, minimize the formation of inhibitory sugars, avert degradation of sugars and maximize recovery of lignin for its valorization into valuable products and reduction in process costs. The third-generation biofuels make the use of photosynthetic algal biomass as feedstock, which can grow on nonarable land neutralize greenhouse gas emissions and yield high amount of lipids. Algal biomass has high growth rates and short harvesting cycles also overcoming limitations of both first as well as second generation. However, the use of algae as feedstock requires extensive modification of process equipment which large-scale biomass production making it an expensive process. Fourth-generation biofuels use genetically modified algal biomass with altered properties such as increased carbohydrate and lipid content for enhanced biofuel production. Engineering of algae for dwindled photoinhibition and escalated light penetration helps boost photosynthetic efficiency of algae (Tandon and Jin 2017).

2.3 Bioethanol-Producing Microorganisms

Bioethanol, a 'drop in fuel' is chemically ethyl alcohol (C_2H_5OH), a metabolite of the biochemical pathway of alcohol fermenting microorganisms. Global bioethanol production has escalated from 25 billion gallons in 2014 to 28.5 billion gallons in 2018, with the United States and Brazil being the leading producers, accounting for 56% and 28% of 2018 global ethanol production and utilizing first-generation biomass, corn and sugarcane, respectively. India accounts for only 1.0% of 2018 global ethanol production despite being the second largest producer of sugarcane in the world. India's ethanol production raised from 85 million gallons in 2014 to 400 million gallons in 2018 (www.ethanolrfa.org). Currently, India accounts for 2.0% bioethanol blending in gasoline and has proposed to achieve

20% blending of bioethanol in petrol by 2030 as per National Policy on Biofuels-2018 (petroleum.nic.in). Minimal generation of particulate matter and greenhouse gases favours the blending. There are various blends of gasoline with bioethanol such as E100, E85, E25, E15, E10 and E5. E5 with 5% blending in gasoline requires almost no modification in machinery but upon the increase in bioethanol blending concentration modifications become necessary (Di Donato et al. 2019).

Since ages, *Saccharomyces cerevisiae* has been used in alcohol fermentation owing to its high fermentative power, low pH values, scarce oxygen availability, ability to tolerate high levels of ethanol and organic acids (Albergaria and Arneborg 2016). Till date, a plethora of yeasts and other microbes have been explored having the inherent potential to produce bioethanol apart from brewer's yeast. Bioethanol production process starts with the pretreatment of biomass followed by saccharification and fermentation. Pretreatment process can be physical, chemical, thermal, biological or combination of any of these. An enzymatic pretreatment using lignocellulolytic enzyme is gaining interest due to its effectiveness in degrading recalcitrant lignin and aiding separation of cellulose and hemicellulose. Fungal species of genus *Trichoderma* have been immensely exploited to obtain hypercellulase-secreting strains by strategies like mutagenesis and chemical irradiation (Kour et al. 2019b). Nowadays, research focus is on the production of thermostable hypercellulases with hyperactivity as well as high specificity which can act synergistically to harness the locked-up potential of cellulosic conversion to bioethanol and make the enzymatic hydrolytic process economically feasible.

Thermostable cellulases allow complete hydrolysis of substrate by enhancing the reaction rate, the bioavailability of substrate, the diffusion coefficient and the substrate solubility meanwhile, decreasing the viscosity of medium, the risk of microbial contamination even after long storage period and the cooling costs during the fermentation process. Thermophilic fungi, *Sporotrichum thermophile* is capable of producing cellulase 2–3 folds higher in comparison to mesophilic fungi (Acharya and Chaudhary 2012). *Saccharomyces cerevisiae* is able to ferment a wide range of substrates such as sugarcane bagasse (Jugwanth, et al. 2019), pine slurry (Dong et al. 2017), corn cobs (Sewsynker-Sukai and Kana 2018), potato peels (Chohan et al. 2019), etc. to produce bioethanol but is unable to utilize pentose sugars (Kumar et al. 2009). Yeasts from the genera *Pachylosen*, *Pichia*, *Candida* and *Schizosaccharomyces* have been reported to ferment pentose sugar to bioethanol. Moreover, the yeast strain is rendered non-viable due to stressful conditions during industrial production process like osmotic stress, production of inhibitory sugars or compounds, contamination by other microorganisms, rise in temperature (35–45 °C) and ethanol concentration (>20%). To overcome these problems, research focus is shifted towards thermotolerant and ethanol-tolerant strains with the ability to sustain growth in the presence of inhibitory compounds. *Kluyveromyces marxianus* NIRE-K3 is a thermotolerant yeast capable of producing 19.01 g/L of ethanol after 16 h of fermentation at 45 °C with a high productivity of 3.17 g/L/h and can also ferment pentose sugar like xylose (Arora et al. 2017) (Table 2.1).

Table 2.1 List of microorganisms producing bioethanol

Microorganisms	Substrate	Conc	Productivity (g/l/h)	Fermentation condition	References
FUNGI					
Saccharomyces cerevisiae	Banana frond juice	45.75 g/L	–	30 °C, 100 rpm, 57 h	Tan et al. (2019)
Saccharomyces cerevisiae strain BY4743	Corn cobs	36.92 g/L	–	30 °C, 120 rpm, 18 h	Sewsynker-Sukai and Kana (2018)
Saccharomyces cerevisiae BY4743	Potato peels	15.475 g/L	1.513	40 °C, 120 rpm, 24 h	Chohan et al. (2019)
Saccharomyces cerevisiae BY4743	Waste sorghum leaves	17.15 g/L	0.52	30 °C, 120 rpm, 24 h	Rorke and Kana (2017)
Saccharomyces cerevisiae	*Hindakia tetrachotoma* ME03 biomass (microalgae)	11.2 g/L	–	30 °C, 150 rpm, 36 h	Onay (2019)
Saccharomyces cerevisiae	Momentary pine slurry	82.1 g/L	–	35 °C, 24 h	Dong et al. (2017)
Saccharomyces cerevisiae	Sugarcane bagasse	3.12 g/L	0.29	39 °C, 120 rpm, 60 h	Jugwanth et al. (2019)
Kluveromyces marxianus MTCC 1389	Palmwood	22.9 g/L	–	45 °C, 156 rpm 84 h	Sathendra et al. (2019)
Kluveromyces marxianus MTCC 4136	Wheat straw	21.6 g/L	–	30 °C, 150 rpm, 48 h	Singhania et al. (2014)
Kluveromyces marxianus	Pomegranate peel waste	7.20 g/L	–	30 °C, 100 rpm, 12 h	Demiray et al. (2019)
Pichia stipitis	Pomegranate peel waste	2.95 g/L	–	30 °C, 100 rpm, 12 h	Demiray et al. (2018)

(continued)

Table 2.1 (continued)

Microorganisms	Substrate	Conc	Productivity (g/l/h)	Fermentation condition	References
Kluveromyces marxianus K21	Taro waste	48.98 g/L	2.23	40 °C, 150 rpm, 20 h	Wu et al. (2015)
Kluyveromyces lactis CBS2359	Cheese whey permeate	15.0 g/L	0.31	30 °C, 72 h	Sampaio et al. (2019)
Candida glabrata	Corncob	31.32 g/L	0.33	40 °C, 150 rpm, 120 h	Boonchuay et al. (2018)
Saccharomyces cerevisiae TISTR 5339	Fresh vetiver grass	5.85 g/L	–	30 °C, 36 h	Subsamran et al. (2018)
Saccharomyces cerevisiae	Sugarcorn juice	45.6 g/L	–	30 °C, 200 rpm, 72 h	Gomez-Flores et al. (2018)
Pachysolen tannophilus	Ulva rigida (green seaweed)	11.92 g/L	0.5	30 °C, 120 rpm, 96 h	El Harchi et al. (2018)
K. marxianus NIRE-K1	Glucose	17.73 g/L	2.22	45 °C, 150 rpm, 16 h	Arora et al. (2017)
K. marxianus NIRE-K3	Glucose	19.01 g/L	3.17	45 °C, 150 rpm, 16 h	Arora et al. (2017)
BACTERIA					
Escherichia coli strain MS04	Corn stover	24.5 g/L	–	37 °C, 60 rpm, 24 h	Vargas-Tah et al. (2015)
Thermoanaerobacter sp. DBT-IOC-X2	Rice straw	0.29 g/g	–	70 °C, 48 h	Singh et al. (2018a, b)

2.4 Biodiesel Producing Microorganisms

Biodiesel, a petroleum-free alternative to petro-diesel is a blend of fatty acid alkyl esters produced by transesterification reaction in which triglycerides, irrespective of their origin, react with short-chain alcohols. It is a sustainable fuel owing to its biodegradability, renewability, non-toxicity and negligible carbon dioxide emissions (Kour et al. 2019a). Use of biodiesel in pure form without any modifications in engines make it favourable for using as a transport fuel in diesel operated vehicles. Typically, it is produced by transesterification of edible vegetable oils, a first-generation feedstock which compromises global food security. However, use of second-generation biodiesel feedstock such as (non-edible crops like *Jatropha* overcomes the problem of food consumption and security but is not yet fruitful as it is not enough to meet global transportation demand.

Deforestation and requirement of huge land area for cultivation are limitations of plant-based biodiesel feedstock. Recent emphasis has shifted towards the cultivation of oligeanous microorganisms capable of growing on low-cost substrates with enormous lipid yield to be used as a source of oil (Ananthi et al. 2019). Oligeanous microbes such as bacteria, microalgae, filamentous fungi and yeasts, produce single cell oils which have a fatty acid composition similar to that of vegetable oil making them a promising feedstock for biodiesel production. Microorganisms belonging to genera *Anabaena, Chlorella, Nanochloropsis, Chlamydomonas, Scenedesmus, Schizochytrium and Botryococcus* produce accumulate lipids but lipid content is usually lower than its theoretical maximum. Certain yeasts like *Rhodosporidium, Rhodotorula, Myerozyma, Cryptococcus, Yarrowia* and *Pichia* can accumulate high amount of lipids in response to environmental stress and some of them form intracellular lipid bodies. *Rhodococcus* sp. is able to utilize lignin-derived various aromatic compounds such as p-coumaric acids, cresol, 2,6 dimethoxyphenol, etc., *to accumulate lipids.* Among bacteria, *Mycobacterium, Lentibacillus, Rhodococcus, Serratia,* etc., have been reported to accumulate lipids (Table 2.2).

Table 2.2 List of microorganisms producing biodiesel

Microorganisms	Feedstock	Lipid content	References
FUNGI			
Naganishia liquefaciens NITTS2	Municipal waste activated sludge	11.68 g/L	Selvakumar et al. (2019)
Cryptococcus psychrotolerans IITRFD	Groundnut shell hydrolysate	6.33 g/L	Deeba et al. (2017)
Meyerozyma caribbica MH267795	Crude glycerol	9.40 g/L	Chebbi et al. (2019)
Cryptococcus curvatus NRRL Y-1511	Ricotta cheese whey	6.8 g/L	Carota et al. (2017)
Cryptococcus laurentii UCD 68-201	Ricotta cheese whey	5.10 g/L	Carota et al. (2017)
Meyerozyma guilliermondii	Sugarcane bagasse hydrolysate	2.33 g/L	Ananthi et al. (2019)
Rhodotorula mucilaginosa	Sugarcane bagasse hydrolysate	1.99 g/L	Ananthi et al. (2019)
Meyerozyma guilliermondii	Rice husk	2.37 g/L	Ananthi et al. (2019)
Pichia kudriavzevii	Rice husk	2.39 g/L	Ananthi et al. (2019)
Rhodotorula mucilaginosa	Rice husk	1.96 g/L	Ananthi et al. (2019)
Cryptococcus sp.	Banana peel	1.12 g/L	Han et al. (2019)
Aspergillus awamori	Laccase-treated waste liquor of Ricinus communis	35.0% w/w	Gujjala et al. (2019)
BACTERIA			
Mycobacterium smegmatis LZ-K2	Corn straw	0.0715 g/L	Zhang et al. (2019a, b)
Lentibacillus salarius NS12IITR	Wheat bran hydrolysate	0.70 g/L	Singh and Choudhury (2019)
Rhodococcus sp. YHY01	barley straw lignin	39% w/w	Bhatia et al. (2019)
Serratia sp. ISTD04	MSS(15 g L-1 SS) + 50 mM $NaHCO_3$	1.96 g/L	Kumar and Thakur (2018)

2.5 Biohydrogen-Producing Microorganisms

In the current scenario, hydrogen is a promising energy source with high energy content (143 MJ/kg) and is considered as the cleanest fuel among other biofuels as its combustion generates only water as a by-product (Abubackar et al. 2019; Kumar et al. 2019). Using Microbial Fuel Cells (MFC's), hydrogen can be converted to electricity to meet energy demands. Deficient in nitrogen and rich in carbohydrate are basic requirements of an ideal hydrogen-producing feedstock. Hydrogen production employs various processes among which dark fermentation is most promising in large-scale production due to utilization of a wide range of low-cost substrates and continuous production of hydrogen with higher rates (Hu et al. 2017). In a study, the photofermentative production of hydrogen carried at 30 ± 2.0 °C yielded 1.96 mol H2/mole of sugar. Co-cultures of microorganisms can potentially lead to increased process performance. For example, the biohydrogen production of 119.7 mM was achieved using co-culture of *Clostridum thermocellum* and *Thermoanaerobacterium thermosaccharolyticum*, which was 2.7 folds higher than using monocultures (Hu et al. 2018). In general, primary hydrogen-producing bacteria are strict anaerobes such as *Clostridium,* which require maintenance of anaerobic conditions during fermentation. This is less of consideration when using facultative anaerobes like *Enterobacter aerogenes* and *Escherichia coli* (Table 2.3).

2.6 Conclusion

Microbial biofuels have captured considerable scientific attention as they can reduce the world's reliance on fossil energy sources by meeting the rising energy demands, reducing the emission of greenhouse gases and overcoming other environmental challenges. It acts as a clean alternative energy source, therefore ensuring energy security and combating the drastic climate change. Extensive research is being done to develop microbial biofuels having high yield and social stability which can be used as affordable energy.

Table 2.3 List of biohydrogen producing microorganisms

Inoculum	Feedstock	Conc./yield/productivity	Fermentation conditions	References
Anaerobic sludge (*Clostridium, Bacillus* and *Rummeliibacillus*)	Sewage sludge and fallen leaves	37.8 mL/g-VS	37 °C, 100 rpm, 48 h	Yang et al. (2019)
Clostridium thermocellum ATCC 27405	Sugarcane bagasse	187.44 mmol/L	50 °C, 150 rpm, 24 h	Swathy et al. (2019)
Anaerobic mixed consortium	Fruit and vegetables wastes	258 mmol/L	–	Keskin et al. (2018)
Rhodobacter M 19 and *Enterobacter aerogenes* co-culture	Brewery effluent	1.96 mol/mol of sugar	31 °C, 70 h	Veeramalini et al. (2019)
Co-culture of *Clostridum thermocellum* and *Thermoanaerobacterium thermosaccharolyticum*	Sugarcane bagasse	119.7 mM	55 °C, 150 rpm, 120 h	Hu et al. (2018)
Clostridium thermocellum ATCC 27405	Waste date seeds	146.19 mmol/L	50 °C, 150 rpm, 168 h	Rambabu et al. (2019)
Domestic sewage treatment plant	Glucose	100.2 mol/m^3-d	35 °C, 48 h	Lu et al. (2019)
Rhodobacter capsulatus-PK	Sugarcane bagasse	357.3 ml/L	40 °C, 48 h	Mirza et al. (2019)
Enterobacter aerogenes	Sago wastewater	7.42 mmol/mol glucose	31 °C, 150 rpm, 36 h	Ulhiza et al. (2018)

(continued)

Table 2.3 (continued)

Inoculum	Feedstock	Conc./yield/productivity	Fermentation conditions	References
N.d	Palm oil mill effluent	3479 ml/L	60 °C, 150 rpm, 24 h	Srirugsa et al.(2019)
Heat-treated anaerobic sludge	Waste wheat powder starch	1.22 mol/mol glucose	37 °C, 3 h	Gorgec and Karapinar (2019)
Anaerobic sludge	Rice straw	129 mL/g COD	10 d	Kannah et al. (2018)
Anaerobically digested sludge	Sea eelgrass biomass	23.2 mL/g-VS	37 °C, 120 rpm	Banu et al. (2019)
Anaerobic digested sludge	*Chlorella vulgaris* (microalgae)	190.90 mL/g-VS	35 °C	Stanislaus et al. (2018)
Thermoanaerobacterium thermosaccharolyticum	Sugarcane bagasse	277.4 mM/L	120 h	Hu et al. (2017)
Enterobacter aerogenes AS1.48	*Humulus scandens*	64.08 ml/g raw material	50 °C, 150 rpm	Zhang et al. (2019a, b)
Anaerobic digested sludge	Waste biological sludge	41 mmol/g-TS	37 °C	Ilgi and Onur (2019)

References

Abubackar HN, KeskinT YO, Gunay B, Arslan K, Azbar N (2019) Biohydrogen production from autoclaved fruit and vegetable wastes by dry fermentation under thermophilic condition. Int J Hydrogen Energ 44:18776–18784. https://doi.org/10.1016/j.ijhydene.2018.12.068

Acharya S, Chaudhary A (2012) Bioprospecting thermophiles for cellulase production: a review. Braz J Microbiol 43:844–856

Albergaria H, Arneborg N (2016) Dominance of Saccharomyces cerevisiae in alcoholic fermentation processes: role of physiological fitness and microbial interactions. Appl Microbiol Biotechnol 100(5):2035–2046

Ananthi V, Siva Prakash G, Chang SW, Ravindran B, Nguyen DD, Vo D-VN, Arun A (2019) Enhanced microbial biodiesel production from lignocellulosic hydrolysates using yeast isolates. Fuel 256:115932

Arora R, Behera S, Sharma NK, Kumar S (2017) Augmentation of ethanol production through statistically designed growth and fermentation medium using novel thermotolerant yeast isolates. Renew Energy 109:406–421. https://doi.org/10.1016/j.renene.2017.03.059

Avanthi A, Banerjee R (2016) A strategic laccase mediated lignin degradation of lignocellulosic feedstocks for ethanol production. Ind Crop Prod 92:174–185

Banu JR, Tamilarasan T, Kavitha S, Gunasekaran M, Kumar G, Al-Muhtaseb AH (2019) Energetically feasible biohydrogen production from sea eelgrass via homogenization through a surfactant, sodium tripolyphosphate. Int J Hydrogen Energy. https://doi.org/10.1016/j.ijhydene.2019.03.206

Bhatia SK, Gurav R, Choi T-R, Han YH, Park Y-L, Park JY, Jung H-R, Yang S-Y, Song H-S, Kim S-H, Choi K-Y, Yang Y-H (2019) Bioconversion of barley straw lignin into biodiesel using Rhodococcus sp.YHY01. Bioresour Technol 289: 121704

Bhatia SK, Kim S-H, Yoon J-J, Yang Y-H (2017) Current status and strategies for second generation biofuel production using microbial systems. Energy Convers 148:1142–1156

Boonchuay P, Techapun C, Leksawasdi N, Seesuriyachan P, Hanmoungjai P, Watanabe M, Chaiyaso T (2018) An integrated process for xylooligosaccharide and bioethanol production from corncob. Bioresour Technol 256:399–407. https://doi.org/10.1016/j.biortech.2018.02.004

Carota E, Crognale S, D'Annibale A, Gallo AM, Stazi SR, Petruccioli M (2017) A sustainable use of Ricotta cheese whey for microbial biodiesel production. Sci Total Environ 584–585:554–560. https://doi.org/10.1016/j.scitotenv.2017.01.068

Chebbi H, Leiva-Candia D, Carmona-Cabello M, Jaouania A, Dorado MP (2019) Biodiesel production from microbial oil provided by oleaginous yeasts from olive oil mill wastewater growing on industrial glycerol. Ind Crops Prod 139:111535

Chohan NA, Aruwajoye GS, Sewsynker-Sukai Y, Kana EBG (2019) Valorisation of potato peel wastes for bioethanol production using simultaneous saccharification and fermentation: process optimization and kinetic assessment. Renew Energy 146:1031–1040

Deeba F, Pruthi V, Negi YS (2017) Fostering triacylglycerol accumulation in novel oleaginous yeast Cryptococcus psychrotolerans IITRFD utilizing groundnut shell for improved biodiesel production. Bioresour Technol 242:113–120. https://doi.org/10.1016/j.biortech.2017.04.001

Demiray E, Karatay SE, Dönmez G (2018) Evaluation of pomegranate peel in ethanol production by Saccharomyces cerevisiae and Pichia stipitis. Energy 159:988–994. https://doi.org/10.1016/j.energy.2018.06.200

Demiray E, Karatay SE, Dönmez G (2019) Efficient bioethanol production from pomegranate peels by newly isolated Kluyveromyces marxianus. Energy Sources, Part A: Recovery, Utilization, and Environmental Effects

Di Donato P, Finore I, Poli A, Nicolaus B, Lama L (2019) The production of second generation bioethanol: the biotechnology potential of thermophilic bacteria. J Clean Prod 233:1410–1417

Dong C, Wang Y, Zhang H, Leu S-Y (2017) Feasibility of high-concentration cellulosic bioethanol production from undetoxified whole Monterey pine slurry. Bioresour Technol 250:102–109

El Harchi M, Fakihi Kachkach FZ, El Mtili N (2018) Optimization of thermal acid hydrolysis for bioethanol production from Ulva rigida with yeast Pachysolen tannophilus. Botany 115:161–169. https://doi.org/10.1016/j.sajb.2018.01.021

Gomez-Flores R, Thiruvengadathan TN, Nicol R, Gilroyed B, Morrison M, Reid LM, Margaritis A (2018) Bioethanol and biobutanol production from sugarcorn juice. Biomass Bioenergy 108:455–463. https://doi.org/10.1016/j.biombioe.2017.10.038

Gorgec FK, Karapinar I (2019) Production of biohydrogen from waste wheat in continuously operated UPBR: the effect of influent substrate concentration. Int J Hydrog Energy 44:17323–17333. https://doi.org/10.1016/j.ijhydene.2018.12.213

Gujjala LKS, Bandyopadhyay TK, Banerjee R (2019) Production of biodiesel utilizing laccase pretreated lignocellulosic waste liquor: An attempt towards cleaner production process. Energy Convers 196:979–987

Han S, Kim G-Y, Han J-I (2019) Biodiesel production from oleaginous yeast, Cryptococcus sp. by using banana peel as carbon source. Energy Reps 5:1077–1081

Hu B-B, Li M-Y, Wang Y-T, Zhu M-J (2017) High-yield biohydrogen production from nondetoxified sugarcane bagasse: Fermentation strategy and mechanism. Chem Eng J. https://doi.org/10.1016/j.cej.2017.10.157

Hu B-B, Li M-Y, Wang Y-T, Zhu M-J (2018) Enhanced biohydrogen production from dilute acid pretreated sugarcane bagasse by detoxification and fermentation strategy. Int J Hydrog Energy. https://doi.org/10.1016/j.ijhydene.2018.08.164

Ilgi K, Onur B (2019) Biohydrogen production from acid hydrolyzed wastewater treatment sludge by dark fermentation. Int J Hydrog Energy. https://doi.org/10.1016/j.ijhydene.2019.03.230

Jugwanth Y, Sewsynker-Sukai Y, Kana EBG (2019) Valorization of sugarcane bagasse for bioethanol production through simultaneous saccharification and fermentation: optimization and kinetic studies. Fuel 116552

Kannah RY, Kavitha S, Sivashanmugham P, Kumar G, Nguyen DD, Chang SW, Banu JR (2018) Biohydrogen production from rice straw: Effect of combinative pretreatment, modelling assessment and energy balance consideration. Int J Hydrog Energy. https://doi.org/10.1016/j.ijhydene.2018.07.201

Keskin T, Abubackar HN, Yazgin O, Gunay B, Azbar (2018) N Effect of percolation frequency on biohydrogen production from fruit and vegetable wastes by dry fermentation. Int J Hydrog Energy. https://doi.org/10.1016/j.ijhydene.2018.12.099

Kour D, Rana KL, Yadav N, Yadav AN, Rastegari AA, Singh C et al (2019a) Technologies for biofuel production: current development, challenges, and future prospects. In: Rastegari AA, Yadav AN, Gupta A (eds) Prospects of renewable bioprocessing in future energy systems. Springer International Publishing, Cham, pp 1–50. https://doi.org/10.1007/978-3-030-14463-0_1

Kour D, Rana KL, Yadav N, Yadav AN, Singh J, Rastegari AA et al (2019b) Agriculturally and industrially important fungi: current developments and potential biotechnological applications. In: Yadav AN, Singh S, Mishra S, Gupta A (eds) Recent advancement in white biotechnology through fungi, Volume 2: Perspective for value-added products and environments. Springer International Publishing, Cham, pp 1–64. https://doi.org/10.1007/978-3-030-14846-1_1

Kumar A, Singh LK, Ghosh S (2009) Bioconversion of lignocellulosic fraction of water-hyacinth (Eichhornia crassipes) hemicellulose acid hydrolysate to ethanol by Pichia stipitis. Bioresour Technol 100(13):3293–3297. https://doi.org/10.1016/j.biortech.2009.02.023

Kumar M, Thakur IS (2018) Municipal secondary sludge as carbon source for production and characterization of biodiesel from oleaginous bacteria. Bioresour Technol Rep. https://doi.org/10.1016/j.biteb.2018.09.011

Kumar S, Sharma S, Thakur S, Mishra T, Negi P, Mishra S et al (2019) Bioprospecting of microbes for biohydrogen production: Current status and future challenges. In: Molina G, Gupta VK, Singh BN, Gathergood N (eds) Bioprocessing for biomolecules production. Wiley, USA, pp 443–471

Lu C, Wang Y, Lee, Zhang Q, Zhang H, Tahir N, Jing Y, Liu H, Zhang K (2019) Biohydrogen production in pilot-scale fermenter: effects of hydraulic retention time and substrate concentration. J Clean Prod 229:751e760

Mirza SS, QaziJI LiangY, Chen S (2019) Growth characteristics and photofermentative biohydrogen production potential of purple non sulfur bacteria from sugar cane bagasse. Fuel 255:115805

Onay M (2019) Bioethanol production via different saccharification strategies from *H. tetrachotoma* ME03 grown at various concentrations of municipal wastewater in a flat-photobioreactor. Fuel 239:1315–1323

Perumal Saravanan AP, Mathimani T, Deviram G, Rajendran K, Pugazhendhi A (2018) Biofuel policy in India: a review of policy barriers in sustainable marketing of biofuel. J Clean Prod. https://doi.org/10.1016/j.jclepro.2018.05.033

Rambabu K, Show P-L, Bharath G, Banat F, Naushad M, Jo-Shu Chang J-S (2019) Enhanced biohydrogen production from date seeds by Clostridium thermocellum ATCC 27405. Int J Hydrog Energy. https://doi.org/10.1016/j.ijhydene.2019.06.133

Rastegari AA, Yadav AN, Gupta A (2019) Prospects of renewable bioprocessing in future energy systems. Springer International Publishing, Cham

Rastogi M, Shrivastava S (2017) Recent advances in second generation bioethanol production: an insight to pretreatment, saccharification and fermentation processes. Renew Sustain Energy Rev 80:330–340

Rorke DCS, Kana EBG (2017) Kinetics of bioethanol production from waste sorghum leaves using *Saccharomyces cerevisiae* BY4743. Fermentation 3(2):19

Sampaio FC, de Faria JT, da Silva MF, de Souza Oliveira RP, Converti A (2019) Cheese whey permeate fermentation by kluyveromyces lactis: a combined approach to wastewater treatment and bioethanol production. Environ Technol. https://doi.org/10.1080/09593330.2019.1604813

Sarmiento F, Espina G, Boehmwald F, Peralta R, Blamey JM (2017) Bioprospection of extremozymes for conversion of lignocellulosic feedstocks to bioethanol and other biochemicals. Extremophilic enzymatic processing of lignocellulosic feedstocks to bioenergy. 271–297. https://doi.org/10.1007/978-3-319-54684-1_14

Sathendra ER, Baskar G, Praveenkumar R, Gnansounou E (2019) Bioethanol production from palm wood using *Trichoderma reesei* and *Kluveromyces marxianus*. Bioresour Technol 271:345–352. https://doi.org/10.1016/j.biortech.2018.09.134

Selvakumar P, Arunagiri A, Sivashanmugam P (2019) Thermo-sonic assisted enzymatic pretreatment of sludge biomass as potential feedstock for oleaginous yeast cultivation to produce biodiesel. Renew Energy 139:1400–1411. https://doi.org/10.1016/j.renene.2019.03.040

Sewsynker-Sukai Y, Kana EBG (2018) Simultaneous saccharification and bioethanol production from corn cobs: process optimization and kinetic studies. Bioresour Technol 262:32–41

Singh N, Choudhury B (2019) Valorization of food-waste hydrolysate by *Lentibacillus salarius* NS12IITR for the production of branched chain fatty acid enriched lipid with potential application as a feedstock for improved biodiesel. Waste Manage 94:1–9

Singh N, Puri M, Tuli DK, Gupta RP, Barrow CJ, Mathur AS (2018a) Bioethanol production by a xylan fermenting thermophilic isolate Clostridium strain DBT-IOC-DC21. Anaerobe 51:89–98. https://doi.org/10.1016/j.anaerobe.2018.04.014

Singh N, Puri M, Tuli DK, Gupta RP, Barrow CJ, Mathur AS (2018b) Bioethanol production potential of a novel thermophilic isolate Thermoanaerobacter sp. DBT-IOC-X2 isolated from Chumathang hot spring. Biomass Bioenergy 116:122–130. https://doi.org/10.1016/j.biombioe.2018.05.009

Singhania RR, Saini JK, Saini R, Adsul M, Mathur A, Gupta R, Tuli DK (2014) Bioethanol production from wheat straw via enzymatic route employing Penicillium janthinellum cellulases. Bioresour Technol 169:490–495. https://doi.org/10.1016/j.biortech.2014.07.011

Srirugsa T, Prasertsan S, Theppaya T, Leevijit T, Prasertsan P (2019) Appropriate mixing speeds of Rushton turbine for biohydrogen production from palm oil mill effluent in a continuous stirred tank reactor. Energy 179:823e830

Stanislaus MS, Zhang N, Yuan Y, Zheng H, Zhao C, Hu X, Zhu Q, Yang Y (2018) Improvement of biohydrogen production by optimization of pretreatment method and substrate to inoculum ratio from microalgal biomass and digested sludge. Renew Energy 127:670e677

Subsamran K, Mahakhan P, Vichitphan K, Vichitphan S, Sawaengkaew J (2018) Potential use of vetiver grass for cellulolytic enzyme production and bioethanol production. Biocatal Agric Biotechnol. https://doi.org/10.1016/j.bcab.2018.11.023

Swathy R, Rambabu K, Banat F, Ho S-H, Chu D-T, Pau Loke Show PL (2019) Production and optimization of high grade cellulase from waste date seeds by Cellulomonas uda NCIM 2353 for biohydrogen production. Int J Hydrog Energy. https://doi.org/10.1016/j.ijhydene.2019.06.171

Tan JS, Phapugrangkul P, Lee CK, Lai Z-W, Abu Bakar MH, Murugan P (2019) Banana frond juice as novel fermentation substrate for bioethanol production by *Saccharomyces cerevisiae*. Biocatal Agric Biotechnol 21:101293

Tandon P, Jin Q (2017) Microalgae culture enhancement through key microbial approaches. Renew Sustain Energy Rev 80:1089–1099

Ulhiza TA, Puad NIM, Azmi AS (2018) Optimization of culture conditions for biohydrogen production from sago wastewater by Enterobacter aerogenes using response surface methodology. Int J Hydrog Energy https://doi.org/10.1016/j.ijhydene.2018.10.057

Vargas-Tah A, Moss-Acosta CL, Trujillo-Martinez B, Tiessen A, Lozoya-Gloria E, Orencio-Trejo M, Martinez A (2015) Non-severe thermochemical hydrolysis of stover from white corn and sequential enzymatic saccharification and fermentation to ethanol. Bioresour Technol 198:611–618. https://doi.org/10.1016/j.biortech.2015.09.036

Veeramalini JB, Selvakumari IAE, Park S, Jayamuthunagai J, Bharathiraja B (2019) Continuous production of biohydrogen from brewery effluent using co-culture of mutated Rhodobacter M 19 and enterobacter aerogenes. Bioresour Technol

Wu W-H, Hung W-C, Lo K-Y, Chen Y-H, Wan H-P, Cheng K-C (2015) Bioethanol production from taro waste using thermo-tolerant yeast Kluyveromyces marxianus K21. Bioresour Technol. https://doi.org/10.1016/j.biortech.2015.11.015

Yang G, Hu Y, Wang J (2019) Biohydrogen production from co-fermentation of fallen leaves and sewage sludge, Bioresour. Technol. https://doi.org/10.1016/j.biortech.2019.121342

Zhang K, Xu R, Abomohra AE-F, Xie S, Yu Z, Guo Q, Li X (2019) A sustainable approach for efficient conversion of lignin into biodiesel accompanied by biological pretreatment of corn straw. Energy Convers 199:111928. https://doi.org/10.1016/j.enconman.2019.111928

Zhang L, Holle MJ, Kim J-S, Daum MA, Miller MJ (2019) Nisin incorporation enhances the inactivation of lactic acid bacteria during the acid wash step of bioethanol production from sugarcane juice. Lett Appl Microbiol. https://doi.org/10.1111/lam.13165

Chapter 3
Cyanobacterial Biofuel Production: Current Development, Challenges and Future Needs

J. Tony Pembroke and Michael P. Ryan

Abstract As the need to replace fossil fuels increases and global energy needs expand the drive to find alternative, sustainable sources of fuels have accelerated. Microbial sources are attractive because of the rapid growth rates of microorganisms and their potential techno-economic advantages. Cyanobacteria are prokaryotic photoautotrophs (utilise photosynthesis and CO_2 for energy and carbon needs), which have emerged as potentially ideal candidates as sources of sustainable biofuel producers once metabolically engineered to do so. Over the past decade, there has been much interest in utilising cyanobacterial model species as proof of concept to produce and overexpress a range of biofuel candidates ranging from ethanol, butanol and other compounds ranging from hydrogen to fatty acids. Research on model biofuel candidates has revealed the potential for biofuel production but also revealed a number of challenges to future development. These challenges range from (1) biological, concerning the genetic constructs, their expression, stability and tolerance to the recombinant biofuel product, (2) production efficiency and biofuel recovery strategies and (3) economic, concerning the viability of production at a scale relative to the market price of the biofuel. Here, various technical challenges will be addressed based on experience and insights gained from the production of ethanol in model cyanobacteria, where many of these challenges are identified and strategies for future development discussed based on the current state of the art.

J. T. Pembroke (✉) · M. P. Ryan
Laboratory of Structural and Molecular Biochemistry, Department of Chemical Sciences, School of Natural Sciences and Bernal Institute, University of Limerick, Limerick, Ireland
e-mail: tony.pembroke@ul.ie

M. P. Ryan
e-mail: michaelpryan1983@gmail.com

A. N. Yadav et al. (eds.), *Biofuels Production – Sustainability and Advances in Microbial Bioresources*, Biofuel and Biorefinery Technologies 11,
https://doi.org/10.1007/978-3-030-53933-7_3

3.1 Introduction

The need to replace fossil fuels in an expanding energy market with sustainable fuel sources has never been greater. Much attention has focused on biofuels from microbial sources that are potentially sustainable; do not compete with land usage for food crops and which can potentially be genetically manipulated for high yield in an industrial production-type system. Attention has focused particularly on cyanobacterial species, which are prokaryotic photoautotrophs and derive their energy and carbon needs from sunlight and CO_2, respectively. This not only holds out the possibility of phototrophic biofuel production but also environmentally could impact carbon capture strategies by recycling CO_2.

Initial attention with cyanobacteria has focused on a broad range of compounds for metabolic engineering. These include 3-hydroxybutyrate (Wang et al. 2013), 1,2-propanediol (Li and Liao 2013), isobutanol, isobutyraldehyde (Atsumi et al. 2009; Varman et al. 2013a), 2,3-butanediol (Oliver et al. 2013; Savakis et al. 2013), isopropanol (Kusakabe et al. 2013), free fatty acids (Gao et al. 2012a; Kaiser et al. 2013), fatty alcohols (Tan et al. 2011; Yao et al. 2014), endogenously produced alka(e)nes (Schirmer et al. 2010; Wang et al. 2013), carotinoids (Lagarde et al. 2000), squalene (Englund et al. 2014), sesquiterpene β-caryophyllene (Reinsvold et al. 2011), isoprene (Bentley et al. 2014), terpenoids (Lin and Pakrasi 2019), limonene (Kiyota et al. 2014), heptadecane (Yoshino et al. 2015), ethylene (Takahama et al. 2003; Guerrero et al. 2012), β-Phellandrene (Formighieri and Melis 2014), poly-β-hydroxybutyrate (PHB) (Wu et al. 2002), heparasan (Sarnaik et al. 2019) polyhydroxyalkanoate (Lau et al. 2014), plant essential oils (Formighieri and Melis 2018), hydroxyl propionic acid (Lan et al. 2015) cellulose (Nobles and Brown 2008), sucrose and glucose/fructose carbon substrates (Niederholtmeyer et al. 2010; Ducat et al. 2012), farnesene (Halfmann et al. 2014), mannitol (Jacobsen and Frigaard 2014), lactic acid (Niederholtmeyer et al. 2010; Angermayr et al. 2012; Joseph et al. 2013; Varman et al. 2013b), acetone (Zhou et al. 2012), H_2 production—both directly (Khetkorn et al. 2017) and in microbial electrolytic cells (McCormick et al. 2013) and ethanol (Deng and Coleman 1999; Dexter and Fu 2009; Dexter et al. 2015). The strategies utilised in metabolically engineering and assessing production levels with these various systems have proven useful in providing optimisation strategies towards particular biofuel production systems in cyanobacteria.

Model cyanobacterial systems have been extensively utilised to provide proof of concept for biofuel production systems. Within the cyanobacteria, the model organisms *Synechocystis sp* PCC6803 and *Synechococcus sp* PCC7942, have been utilised most extensively. In both cases, key attributes for choosing these species were based on ease of genetic manipulation, availability of genetic systems for metabolic engineering, nucleotide sequence data for the organisms, mutant strain availability and knowledge of the organisms through extensive usage. There is currently little consensus as to what type of cyanobacteria would make an ideal 'production' candidate, and indeed this may differ depending on the end product. An industrial 'producer' would have to possess many additional traits that are not necessarily optimal in a model strain. These might include competitiveness in an open reactor system necessary for low-value biofuel products, fast growth rates, better partitioning of biofuel to product, high production rates and tolerance to the end product to mention just a few. Then given that these traits may be found, the necessary knowledge on genetics and metabolic engineering within such a candidate may take time to develop. Thus, although there are many challenges, lessons are being learned from model organism manipulation and with single biofuels such as ethanol. These can inform strategies for other organisms and indeed other biofuel products.

3.2 Manipulation Strategies with Model Organisms for Ethanol as a Biofuel

Many potential cyanobacterial niches allow photosynthetic metabolism for only part of the life cycle and many cyanobacteria utilise energy and storage compounds generated during the light phase to promote dark metabolism (Rastegari et al. 2019). Equally because of the niche, for example in microbial mats or lake sediments, the environment may become anoxic requiring inhabiting cyanobacteria to adapt quickly to different metabolic situations. Many cyanobacteria possess fermentative pathways that allow rapid adaptive change to environmental conditions somewhat like a survival tool (Stal and Moezelaar 1997). Some cyanobacteria are therefore capable of producing ethanol principally via heterotrophic anaerobic metabolism and surveys have shown that many strains inhabiting mats are capable of ethanol production (Heyer and Krumbein 1991; Stal and Moezelaar 1997) but at levels far too low and under non-photoautotrophic conditions to be of any practical use. Hence, realistically the only way forward is via metabolic engineering for the production of most cyanobacterial biofuels, including ethanol. A key start point to metabolically engineer biofuel production such as ethanol is to understand the pathways of central metabolism and how key intermediates can be diverted or manipulated (Fig. 3.1).

Synechocystis sp. PCC6803 was one of the first cyanobacteria to be fully sequenced (Kaneko et al. 1996). There are a variety of sub-strains of this model organism but these have in general been derived from the original strain although have acquired a number of mutations as a result of laboratory passage. Currently, there are a number of sub-strain genome sequences available with many single nucleotide polymorphisms that can affect growth and productivity on occasion. The availability of the genome sequence of model organisms such as *Synechocystis* sp. PCC6803 allows the sequence to be analysed via the Kyoto Encyclopaedia of Genes and Genomes (KEGG) database (Ogata et al. 1999). This allows prediction of the metabolic pathways encoded by the organism and allows rational strategies to be developed to examine where metabolic engineering can start (from what intermediate) and what the consequences may be for perturbation of that intermediate. In the case of bioethanol production as a biofuel candidate, the start point is pyruvate

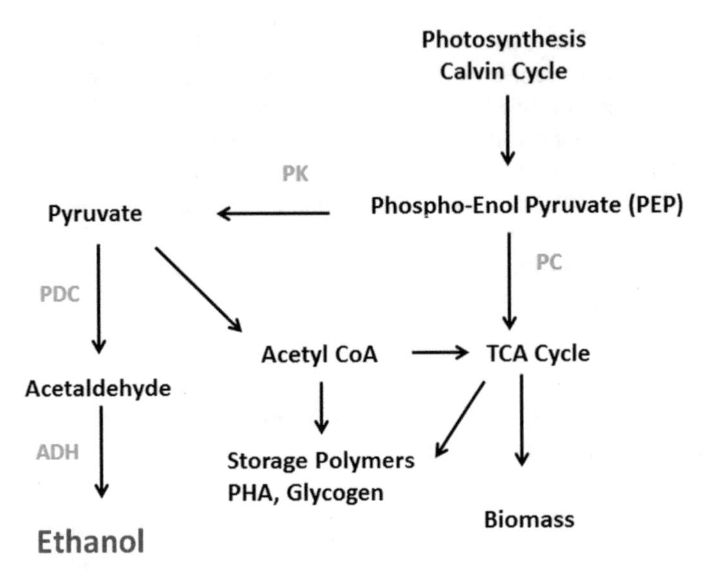

Fig. 3.1 Strategies for manipulating ethanol production in the model cyanobacterium, *Synechocystis* sp. PCC6803. The metabolic intermediate pyruvate can be converted to ethanol via the introduction of a genetic cassette expressing the *Zymomonas mobilis* pyruvate decarboxylase activity (*Pdc*), which channels pyruvate to acetaldehyde. This is in turn converted to ethanol by either the *Zymomonas mobilis* alcohol dehydrogenase II activity or by the native *Synechocystis Adh*A activity. Metabolic intermediates are channelled in the light to storage polymers and to the TCA cycle for biomass needs. Further manipulation of the pathways could be achieved by expressing pyruvate kinase (PK) to increase pyruvate concentrations or by knocking down PEP carboxylase (PC) to channel more PEP to pyruvate

and KEGG will then allow a view of where energy savings can be achieved or where problems may lie should too much intermediate be removed. KEGG may also allow the development of metabolic flux models and the identification of key bottlenecks. For example, in the case of bioethanol production, a key element is the diversion of the metabolic intermediate pyruvate to ethanol via acetaldehyde.

The enzyme pyruvate decarboxylase (*Pdc*) used to convert pyruvate to acetaldehyde (not encoded by cyanobacteria) has a requirement for two co-factors, thiamine diphosphate/pyrophosphate (ThDP) and Mg^{2+}. ThDP is also required for a number of other enzyme activities. Thus, one will be aware at an early stage that overexpression of a *pdc* gene to produce ethanol will give rise to a shortage of ThDP if not Mg^{2+}. Analysis of the *Synechocystis sp* PCC6803 genome and its KEGG database has in this case revealed that no ThDP transporters are present based on genome and KEGG data. Hence, a strategy of increasing expression of pathways to synthesise ThDP will be required or the provision of a heterologous ThDP transporter may need to be developed in addition to the production cassette containing the diverting metabolically engineered enzymes. It is this type of invaluable information that can be obtained from interrogation and analysis of model organism genome data, which

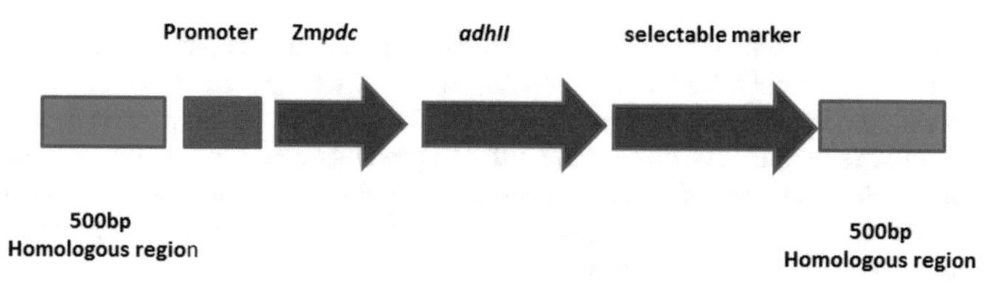

Fig. 3.2 Typical construction of an ethanol cassette utilised to divert flux to ethanol in *Synechocystis sp* PCC6803. The construct consists of two regions, typically of 500 bp with homology to the integration site within the host. The expression is controlled by a functioning cyanobacterial promoter which is linked to the *Zymomonas mobilis* pyruvate decarboxylase Zm*pdc* gene. This is linked to an alcohol dehydrogenase II gene (*adh*II) which is in turn linked to a selectable marker such as a kanamycin resistance determinant for host selection and maintenance of the cassette (Lopes da Silva et al. 2018)

can aid in developing systems for optimal biofuel production. Understanding limitations such as co-factor availability will be key to optimisation of cyanobacterial systems for any biofuel.

One of the first cyanobacterial species to be metabolically engineered to produce biofuel, ethanol, was *Synechococcus sp.* PCC7942 (formally called *Synechococcus elongatus* PCC7942) (Deng and Coleman 1999) by transforming the genes for pyruvate decarboxylase (*pdc*) and alcohol dehydrogenase II (*adh*II) from *Zymomonas mobilis*, into *Synechococcus sp* PCC7942. *Zymomonas mobilis*, an obligate fermenter, is one of the few prokaryotes that produce large amounts of ethanol and whose genetic system has been extensively studied (Hoppner and Doelle 1983; Montenecourt 1985; Neale et al. 1986). These genes were expressed in the initial studies under the control of the cyanobacterial *rbc*LS promoter, alone and in combination with the *Escherichia coli lac* promoter. The reported yields of ethanol produced by the transformed strain reached 54 nmol. OD 730 nm unit^{-1} liter^{-1} day^{-1}(Deng and Coleman 1999).

This set the scene for further manipulation, which provided the initial proof of concept for photoautotrophic biofuel production. There followed a series of attempts to improve yields in *Synechococcus* PCC7942. Initially, (Woods et al. 2004) modified the promoter using the *p*L promoter of the coliphage lambda. This was followed by a number of industrial patents from Joule Unlimited and Algenol Biofuels (US patents US8163516B2 and WO2013098267A1), which focussed on different promoters to drive cassette expression and some construct modifications. Using the same *Zymomonas mobilis* genes for *pdc* and *adh*II Dexter and Fu (2009) reported expression in *Synechocystis* sp. PCC6803 under the control of the light-driven *psb*A2 promoter, reaching a production level of 5.2 mmol. OD 730 nm unit^{-1} litre^{-1} day^{-1}. Attempts to use other *pdc* genes with lower Km values, such as that from *Zymobacter palmae,* have also been undertaken but these have proven to be unsuccessful (Quinn et al. 2019).

Figure 3.2 illustrates the overall construction of a biofuel cassette for metabolic engineering ethanol production in *Synechocystis sp* PCC6803 for chromosomal integration. The homologous regions allow recombination with the appropriate neutral integration site. The cassette can have different promoter constructs to drive expression. The key genes are *pdc* and *adh* genes to convert pyruvate to ethanol via acetaldehyde and the final drug resistance gene allows selection and again can encode various determinants suitable for the model host.

Using two different engineered strains of *Synechocystis sp* PCC6803, one engineered with the ethanol cassette (*pdc* and *adh*) and co-cultured with another PCC6803 strain deleted in the production of glucose-1-phosphate adenylyltransferase and PHA-specific β-ketothiolase (Δ*apx* and Δ*gbk* unable to synthesise glycogen and PHB storage polymers), it was demonstrated that the deleted strain released various metabolites that were utilised by the ethanol producer (Velmurugan and Incharoensakdi 2020). This strategy resulted in an overall increased ethanol yield due to co-culture and may point to innovative co-culture strategies in the future.

In addition to ethanol, several other candidate biofuel products have been investigated in model cyanobacterial species. 1-Butanol is considered as a fuel substitute to displace gasoline which has an energy density of 27 MJ L^{-1}, whereas ethanol is 21 MJ L^{-1}. 1-Butanol can be synthesised by converting butyryl-CoA from acetyl-CoA where butyryl-CoA is then reduced to 1-butanol (Atsumi et al. 2008) To produce 1-butanol, a CoA-dependent 1-butanol pathway was transferred from *Clostridium* and expressed in *Synechococcus elongatus* PCC794 (Lan and Liao 2011). To overcome the need for *Clostridium* ferrodoxins, the *Clostridium acetobutylicum* butyryl-CoA reductase, which requires ferrodoxin, was replaced by a *Treponema denticola* trans-enyl-CoA reductase, which utilises NADH as a co-factor. In addition, the *Ato*B, encoding acetoacetyl CoA thiolase, was derived from *E.coli* as it possesses a higher specific activity and was used to generate a heterologous construct that was integrated in two parts into two separate neutral chromosomal sites (Lan and Liao 2011). An initial worry was the possibility that since these activities were derived from a strict anaerobe then enzyme activity might be oxygen-sensitive, however, this construct demonstrated 1-butanol production, suggesting no or little oxygen sensitivity.

Isobutanol is another biofuel candidate that has been investigated as it contains 98% of the energy content of gasoline, has a lower solubility than ethanol (potentially aiding recovery from metabolically engineered organisms), meaning that it can be blended perhaps more easily with gasoline (Miao et al. 2017). To produce isobutanol, an α-ketoisovalerate decarboxylase, termed *kivd,* from *Lactococcus lactis* was expressed in *Synechocystis sp* PCC6803 (Miao et al. 2017) and strains were shown to produce 3 mg L^{-1} OD 750^{-1} isobutanol in a 6-day growth period. Supplementation with isobutyraldehyde increased yields to 60.8 mg L^{-1} day^{-1}. Miao et al. (2017) demonstrated that overexpressing *kivd* via self-replicating vectors under the control of the strong P*trc* promoter resulted in even higher yields than chromosomal integration.

To aid construct development for metabolic engineering in cyanobacteria a standardised cloning and assembly system has been developed, termed CyanoGate. This has involved the development of a suite of parts and acceptor vectors to generate knockouts, multigene expression and repression systems, for use via replicative vectors initially in *Synechococcus* and *Synechocystis* (Vasudevan et al. 2019). Such systems will aid with the development of synthetic biology approaches in cyanobacteria particularly towards biofuel production into the future.

Given that cyanobacteria have not evolved strategies to produce biofuels and therefore must be metabolically engineered to do so the process of biofuel production, optimisation and scale-up in cyanobacteria are not without their challenges. Such challenges can be considered as a mixture of biological, production and economic issues, which need to be considered in detail to ensure progress with biofuel production. Lessons have been learned from developments principally with ethanol production in model organisms but these lessons have implications for many of the biofuel candidates and merit further consideration and discussion.

3.3 Polyploidy and Cassette Stability

Many cyanobacteria are polyploid, possessing multiple chromosomal copies. Using real-time PCR, it has been shown that the 'Kasusa' strain of *Synechocystis* sp PCC6803 (one of the original *Synechocystis* sub-strains) (Pembroke et al. 2017) could have up to several hundred copies of its chromosome in exponential phase and an average of 58 copies in stationary phase (Griese et al. 2011). Ploidy appears to be a common trait in cyanobacteria, which has been proposed to be a response to life in a high UV environment with multiple copies aiding recombinational repair and chromosome maintenances. This high polyploid chromosomal number poses a number of challenges when using and manipulating polyploid hosts. The carbon, nitrogen and phosphate flux needed to maintain such a high chromosome copy number, which appears to vary with the growth phase is high and hence there is a diversion from using these resources for biofuel production. Copy number mutants with low ploidy have so far not been isolated but this could aid in enhancing biofuel production by lowering carbon flux to maintain chromosome number. A second key issue that emerges from the polyploid nature of cyanobacteria is related to genetically engineering them and maintaining the stability of any metabolically engineered cassette. In most cases, constructs are integrated directly into the host chromosome via recombination of the constructed cassette (Dexter and Fu 2009; Lopes da Silva et al. 2018; Pembroke et al. 2019). To develop stable and expressing recombinants, constructs have to undergo extensive selection. Initially, during selection, recombinant clones will be only partly segregated in the polyploid genome with some chromosomes possessing the insert, while others will not. This can be tracked using PCR primers across the integration site.

The absence of the cloned cassette will manifest with PCR amplicons of a size representing just the integration area, while an integrated cassette will give rise to larger PCR amplicons. In practice, during the early phases of uptake, there will be a mixture of amplicons that will stabilise as each chromosome integrates the cassette or the integrated chromosome is copied. Polyploidy also means that there are gene dosage effects. An integrated cassette with a chromosome copy number of 58 can lead to high expression levels. This coupled to an inducible promoter, such as the *psb*A2 light-inducible promoter, can result in significant heterologous enzyme expression, during the light phase. If selection is not maintained, then the inserted cassette can be lost as there is a biological prerogative to remove the insert as usually its expression, in the case of a cassette to produce biofuel, puts a metabolic burden on the host. Thus, in the case of bioethanol, the cassette needs to be maintained and, even with selection, stability may be affected by mutagenesis to knock out the heterologous genes or delete them entirely.

The variation in polyploid copy numbers can also be problematic for the stability of introduced cassettes. As the chromosome copy number reduces during stationary and late stationary phases (Griese et al. 2011), mutations or deletions picked up at this stage may be advantageous as it may relieve the pressure of biofuel production and allow more rapid selection of non-producers, which by virtue of this relief will be faster growers and may lead to a population of non-biofuel producers. Thus, polyploidy can have major effects on engineered producers and needs further assessment.

3.4 Copy Number and Insert Site Selection

One of the strategies utilised to maximise the expression of constructs to divert flux to biofuels in cyanobacterial model species has been to utilise gene dosage as a tool. By inserting into the chromosome and by virtue of the polyploid nature of model cyanobacteria, there will be a gene dosage effect as the chromosome number increases. Initial studies on bioethanol production in model cyanobacteria were carried out using one copy of an ethanol cassette. The original bioethanol work (Deng and Coleman 1999) utilised replicative plasmids based on the multi-copy *Synechococcus* sp. plasmid pCB4 to generate a series of constructs, pCB4-LRpa and pCB4-LR (TF), utilising ampicillin resistance as a selectable marker. Later, (Dexter and Fu 2009) utilised an integration system in *Synechocystis,* which resulted in higher yields with reported greater stability. Based on such studies, there has thus been the tendency to utilise integration as the engineering technique, particularly with the model organism *Synechocystis sp* PCC6803. A natural progression towards increasing productivity would be to engineer strains with more than one copy of a cassette and, to this end, Gao et al. (2012b) reported the use of a two-cassette system where two copies of the ethanol cassette were integrated in *Synechocystis* sp PCC6803 with a yield of 5.50 g L^{-1} after 26-day growth (or 212 mg L^{-1} day^{-1}). This two-cassette strategy was also utilised towards pilot-scale production (Lopes da Silva et al. 2018), but thus far, a three-cassette model has proven difficult to isolate and with ploidy and gene dosage two copies may be close to the maximum that can be tolerated by an engineered organism (at least in the case of ethanol cassettes).

Indeed, the generation of modified engineered strains, with multiple inserts can be problematic. The modifications themselves can be time-consuming with multiple rounds of transformation, cloning and selection particularly if marker-less clones are needed for production strains where the antibiotic selection may not be an option. In addition with multiple inserts, many different neutral sites may be required for insertion and there may be differential expression and stability at different integration sites. Another option would be to utilise native stable, multi-copy plasmids as integration sites. *Synechocystis sp* PCC6803 contains at least three small plasmids pCA2.4, pCB2.4 and pCC5.2 and in addition at least four large plasmids pSYSM (125 kb), pSYA (119 kb), pSYSG (45 kb) and pSYSX (106 kb) (Kaneko et al 2003). These appear to be highly stable and to have been maintained within the strain since the strain was originally isolated more than 50 years ago. Other cyanobacteria also appear to contain multiple plasmids. Genome sequencing of *Anabaena* sp. PCC7120 has revealed six plasmids ranging from 5.6 to 408 kb (Kaneko et al. 2003) suggesting that possessing native plasmids may be common in cyanobacteria. As a proof of concept, the yellow fluorescent protein gene (*yfp*) was cloned into the small *Synechocystis* sp. PCC6803 pCA2.4 plasmid (Armshaw et al. 2015). This plasmid, of 2.4 kb, is consistently maintained at seven times the chromosomal polyploid copy number and is extremely stable. When the comparison between chromosomal integration and pCA2.4 integration of this YFP cassette was determined it was demonstrated that a 20-fold higher fluorescence could be detected (Armshaw et al. 2015) upon the integration of yfp into pCA2.4. This level of heterologous protein production illustrates interesting possibilities for future integration of biofuel cassettes into multi-copy, stable native plasmids as a strategy towards enhanced gene dosage and production.

As integration strategies for biofuel cassettes are common when manipulating cyanobacteria, it is important to integrate at neutral sites (locations where integration does not impact cell viability or phenotype) (Ng et al. 2015; Pinto et al. 2015). A comprehensive analysis of such sites was carried out for *Synechocystis sp* PCC6803 by analysing insertion and deletion mutations at these sites (Pinto et al. 2015). This type of analysis is essential, as to build metabolic pathways in cyanobacteria many different locations may be necessary and key issues such as stability and expression differences at these sites will ultimately affect production. Potential neutral sites need to be identified and characterised for the stability of integrated cassette while analysing proteome changes that might result from the insertion can also aid choice. There is in addition the possibility that a potential neutral site might be part of a cis-regulatory sequence or be essential during a particular but limited growth phase. Thus, the identification of real neutral sites (Pinto et al. 2015), which show identical growth patterns to wild type may be complex but feasible. Hence, in the future, developing integration-cloning strategies for potentially faster growing, more productive, more resistant, more competitive biofuel strains lessons gained from strategies applied to model cyanobacteria can prove hugely informative.

With any gene expression effect, whether it be promoter-enhanced expression or increased gene dosage, one must reach a point of diminished return. A point, where any increase in a protein expressed from a cassette will not enhance production further. This point may be related to the maximum flux that can be reached and, beyond this point, no extra enzyme level will have a beneficial effect. Producing organisms will have a minimum metabolic intermediate need for maintenance of cell structures, metabolism and other metabolic intermediates. Thus, the flux towards biofuel can only go so far before no further enhancement can occur and diversion towards biofuel production will result in a major decrease in biomass.

3.5 Promoter-Driven Expression and Vector Systems

One of the challenges to the expression of engineered biofuel cassettes has been finding suitable promoter systems to optimise expression which is seen as a key element of the manipulative toolbox. As many inducible systems that operate in *E.coli* do not function well or similarly, in cyanobacteria, native promoters need to be utilised. One of the early promoters investigated was the *psb*A2 light-driven promoter (Dexter and Fu 2009). Use of the *psb*A2 promoter was based on observations that expression patterns of chimeric genes containing the promoter regions of the *psb*A2, gene fused to the firefly luciferase (*luc*) reporter gene indicated that transcription of *psbA2/luc* transgenes was elevated, similarly to that of the endogenous *psbA* gene (Máté et al. 1998). The *psb*A2 promoter has a number of unique characteristics (Asayama et al. 2002; Shibato et al. 2002) involving cis-acting sequences, which are involved in circadian expression and light-driven promotion of fused genes. Thus, this promoter-driven system has been popular in many studies (Dexter and Fu 2009; Gao et al. 2012b; Lopes da Silva et al. 2018).

A systematic analysis of promoters and ribosome binding sites has been carried out specifically for *Synechocystis* (Englund et al. 2016). Comparison with metal inducible, light-inducible and constitutive promoters revealed that the P*nrs*B could be induced some 40 folds by nickel or cobalt addition up to the level of the strong light-inducible promoter *psb*A2 (Englund et al. 2016). Inducible promoters such as P*nrs*B may have an application in decoupling growth from production. On other occasions, obtaining large quantities of enzyme activity may be the goal and, in such cases, a strong promoter will be required. Using the promoter for the *cpc*B gene, P*cpc*560, functional proteins were produced at a level of up to 15% of total soluble protein in *Synechocystis* sp PCC6803, a level comparable to that produced in *E. coli* (Zhou et al. 2014).

There is somewhat of a scarcity of well-characterised replicative vectors for cyanobacterial model systems (Huang et al. 2010; Taton et al. 2014). This has led to the development of new, more functionally designed vector systems with model organism functionality considerations being to the fore. Three plasmids pSEVA251 (KM^R) pSEVA351 (CM^R) and pSEVA451 (SP/SM^R) have been developed to add to the toolbox of existing integrative systems (Ferreira et al. 2018). These vectors now carry a range of promoters based on PT7opol and P*trc* giving up to 41-fold enhanced expression and containing repression systems. Detailed analysis of these vector systems demonstrated that the presence of the plasmid does not lead to an evident phenotype effect on *Synechocystis* growth, with the majority of the cells able to retain the replicative plasmid even in the absence of selective pressure (Ferreira et al. 2018).

Self-replicating shuttle vectors have also been developed based on *pANS* of *Synechococcus elongates* PCC7942. This vector constructs involved the introduction of a heterologous toxin–antitoxin cassette into the shuttle vector for stable plasmid maintenance in the absence of antibiotic selection (Chen et al. 2016). The vector was shown to be stable in *Anabaena* and in cured *Synechococcus* cells and in *E.coli*. It was shown to give rise to ten copies per cell and reporter genes were expressed some 2.5 folds compared to chromosomal integration. Such shuttle systems will add to the toolbox to aid construction in well-developed backgrounds, such as *E.coli,* and allow rapid shuttle into a variety of cyanobacterial strains.

There are sometimes alternative needs and requirements for controlled expression of biofuel cassettes such as decoupling production from growth where it may be useful to allow biomass production firstly and only induce biofuel production when the biomass resources are available. This is analogous to how ethanol is produced in the yeast *Saccharomyces* with ethanol production triggered on shift to anaerobic metabolism.

Several P*trc* riboswitches have been characterised in cyanobacteria and tightly regulated expression shown to be possible using theophylline as the switch inducer (Nakahira et al. 2013). Using a riboswitch technique in *Synechocystis sp* PCC6803, where the ethanol cassette was fused to a riboswitch, has allowed the decoupling of ethanol production from biomass (Armshaw et al. 2018). Here biomass is allowed until ethanol production is induced by theophylline at set points in the exponential or stationary phases; however, there was no increase in overall ethanol productivity reported via such constructs although higher biomass yields were reported. Thus, the toolkit to obtain time-dependent, via an inducible system, or high-level expression, via a strong promoter is available and will be key elements for future genetic manipulation strategies in cyanobacteria.

3.6 Knockouts, Rerouting Carbon Flux and Flux Analysis

Rerouting carbon flux has been utilised in attempts to improve carbon flow to biofuel products in many cases (Dexter et al. 2015; Hendry et al. 2017). The principle underlying this being that if there are two competing pathways, then knocking out one or more may alter the flux towards the preferred product (Fig. 3.1). In cyanobacterial cells, there are many examples of storage compounds such as glycogen or PHA, which are used for energy storage during high levels of photosynthesis, effectively acting as a carbon sink. Theoretically, manipulating or deleting such pathways should alter the flux to product (Dexter et al. 2015). Using ^{13}C metabolic flux analysis, the rerouting of carbon was examined in a glycogen synthase-deficient mutant (*glg*A-I *glg*A-II) strain of *Synechococcus* sp. PCC 7002 (Hendry et al. 2017). Normally, between 10 and 20% of the fixed carbon is stored in the form of glycogen in many cyanobacterial strains during balanced photoautotrophic growth (Hendry et al. 2017). In the *glg*A-I and *glg*A-II mutants, a redistribution of carbon flux occurs, some to other storage compounds such as glucosyl glycerol and sucrose while the rest partitions to other metabolic networks such as glycolysis and the TCA cycle. In this respect, Monshupanee et al (2019) disrupted the γ-aminobutyric acid (GABA) shunt, one of the metabolic pathways for completing the TCA cycle in *Synechocystis*, by inactivating the glutamate decarboxylase (*gdc*) gene. This resulted in an increase in pyruvate levels (1.23 folds) and a 2.5-fold increase in poly(3-hydroxybutyrate) (PHB) production while reducing TCA cycle intermediates. Such a knockout is potentially one of many strategies that could divert flux to pyruvate and hence ethanol in a cassette containing host.

Changing the carbon sink by genetically engineering alternative pathways can also have a major effect on the flux as it redirects the sink within the host organism. Two engineered strains of *Synechocystis* sp. PCC 6803 with altered carbon sink capacity were assayed for their photosynthetic and CO_2 concentrating properties (Holland et al. 2016). A comparison of knocking out and adding a sink via analysis of a Δ*glg*C mutant, where a carbon sink was removed (unable to synthesise glycogen as a storage compound) and strain JU547 (engineered to produce ethylene, a new sink) revealed that the Δ*glg*C mutant displayed a diminished photochemical efficiency, a more reduced NADPH pool, delayed initiation of the Calvin–Benson–Bassham cycle, impairment of linear and cyclic electron flows and a reduced PQ pool, and an undefined dissipative mechanism to spill excess energy. In the case of JU547, more oxidised PQ and NADPH pools were observed with increased rates of cyclic electron flow and enhanced demand for inorganic carbon was observed as suggested by increased expression of the bicarbonate transporter, *Sbt*A (Holland et al. 2016). This study identified that subtle changes in pathways and flux can affect many areas of photosynthetic metabolism, which can ultimately affect the production of metabolically engineered products.

Since cofactors, such as NADPH, are essential for *Adh*II activities (catalyses the reversible oxidation of alcohols to aldehydes or ketones in a NAD(P)-dependent manner) one can see that changes that alter one sink may have unplanned consequences on another pathway and ultimately affect the production of metabolically engineered strains as with ethanol in this case. It has been shown that *Zymomonas Adh*II utilises NAD(H), while the native *Synechocystis Adh*A utilises NADP(H). A number of *adh*II genes have been trialled in conjunction with the *Zymomonas mobilis pdc,* but it has been shown that the *Adh*A (slr1192) of *Synechocystis*, with its NADP(H) co-factor preference, is the most efficient (Gao et al. 2012b). In practice in many engineered hosts both activities, the *Zymomonas* and native, are present (Gao et al. 2012b; Pembroke et al. 2017) however for maximal activity and coupling with the *Pdc* activity any reduction in the NADP(H) pools will affect production.

Flux distribution studies can provide a quantitative view of the way carbon is partitioned in cyanobacterial hosts and much information can be obtained by flux balance analysis, which requires flux values (Baroukh et al. 2015) and metabolic flux analysis data, generally based on isotopic labelling. Such data allow the construction of useful metabolic models (Mueller et al. 2013), which can aid in determining strategies for useful knockout or pathway enhancement protocols. How such adjustments might affect biofuel production or diversion of key intermediates can also be examined by modelling strategies such as Minimisation of Metabolic Adjustments

(MOMA). MOMA strategises that any perturbation will result in minimal adjustment (Segre et al. 2002) and such models have been applied to biofuel producers to predict knockout strategies towards higher yields (Hendry et al. 2016). While modelling can provide useful data, the availability of real data for metabolic flux analysis based on isotopic labelling (Hendry et al. 2017) will in future dramatically aid knockout and enhancement strategies.

In generating designer microbes for biofuel production, Angermayr et al. (2014) has proposed a number of 'design' principles for model organisms such as *Synechocystis sp* PCC6803. Although they focused on the increased production of lactic acid, the principles are valid to other cyanobacterial production systems. The principles include increased expression of the product forming enzymes, co-expression of heterologous pyruvate kinase to increase flux to pyruvate, knockdown of PEP carboxylase to decrease flux to competing pathways (Fig. 3.1) and optimising the production enzymes via mutagenesis to improve the kinetics or co-factor affinity (Angermayr et al. 2014). Many of these principles hold fast for ethanol production and indeed other potential biofuel candidates where pyruvate is also the key cellular intermediate.

3.7 Increasing Carbon Uptake

As engineered strains of cyanobacteria are pushed to produce biofuel products, there is an inevitable reduction of biomass at a fixed CO_2 uptake. Therefore, a potential strategy to increase flux to biofuel is to manipulate the CO_2 uptake systems. Photosynthetic organisms have evolved different forms of Carbon Concentrating Mechanisms (CCM's) to aid RuBisCO in capturing CO_2 from the aqueous/gaseous environment (Badger et al. 2002). The CCM in cyanobacteria is one of the most effective concentrating mechanisms known, able to concentrate carbon up to 1000 folds within the cell. Many cyanobacteria, including *Synechocystis* sp. PCC6803, use a number of scavenging systems to concentrate bicarbonate and CO_2. CCMs in model cyanobacteria studied thus far involve six functional elements: (1) Passive or energised entry of dissolved inorganic carbon, (2) Increase in HCO_3 concentration in the cell, (3) Entry into carboxysomes, (4) Providing saturation of CO_2 near RuBisCO, (5) Fixation of CO_2 and (6) Prevention of CO_2 leakage from the carboxysomes (Kaplan and Reinhold 2002). To effect carbon uptake and concentration, *Synechocystis* sp. PCC6803 has been shown to contain a number of transporters (Shibata et al. 2001, 2002) including three bicarbonate transporters:

(a) A high-affinity inducible *Bct*1 bicarbonate transporter (slr0040–44)
(b) An inducible medium affinity sodium-dependent *Sbt*A bicarbonate transporter (slr1512)
(c) A medium-affinity *Bic*A transporter (sll0834).

The *Bct*1 and *Sbt*A transporters are regulated by the *Ccm*R transcription factor, which senses intercellular levels of α-ketoglutarate and NADP (Daley et al. 2012), while *Bic*A is constitutively expressed (Badger et al. 2002). In addition, there are also a number of CO_2 uptake systems termed NDH-I_3 and NDH-I_4 (multiple variants may also exist) that involve proton translocating NAD(P)H-dependent oxidoreductases, which have a multi-subunit composition, 6 in the case of *Synechocystis* (Battchikova et al. 2010). Although many of the systems are multi-gene, in principle increased levels of CO_2 could be achieved by cloning or overexpressing one or more of these carbon concentrating systems. Strains of the model organism *Synechocystis sp* PCC6803 have been engineered by installing extra bicarbonate transporters via the introduction of inducible copies of the single gene encoding *Bic*A (Kamennaya et al. 2015). When cultured under atmospheric pressure, the strain expressing *Bic*A grew almost twice as fast and accumulated twice as much biomass as the wild type. Interestingly, an accumulation of increased sugar-rich exopolymeric material was also detected in these strains. This indicated that carbon flux could be redirected in a similar manner to enhance biofuel production and such strategies may offer potential in the future for many biofuel candidates.

3.8 Cellular Tolerance to the Engineered Biofuel

Given that model organisms do not naturally produce more than minute quantities of bioethanol during fermentative growth in the dark, it is of interest to examine the response of engineered strains to ethanol or other biofuels (Kumar et al. 2019). Almost all engineered products and biofuel candidates elicit some form of stress or tolerance response (Nicolaou et al. 2010). Commercial production of biofuels such as ethanol at industrial levels may require production strains to produce up to 20% (v/v) ethanol to be economic. Current production levels are still very far from even 1% (v/v) but the production needed at scale illustrates what potentially is required (Dexter et al. 2015; Pembroke et al. 2017). In a comparison of ethanol tolerance between nine different cyanobacterial strains, it was demonstrated that the growth inhibition GI_{50} values ranged from 3 g L^{-1} (0.4% v/v) to 28 g L^{-1} (3.5% v/v) (Kämäräinen et al. 2018). In this study, the most tolerant strains were *Synechocystis* sp. PCC 6803 and *Synechococcus* sp. PCC 7002, with both model organisms showing little effect on growth below ethanol concentrations 9.2 g L^{-1} (1.2% V/V).

Currently, the highest reported ethanol yields are 5.5 g L^{-1} (Gao et al. 2012b) and 7.1 g L^{-1} (Dehring et al. 2012, a US Patent). Both of these studies were from idealised photobioreactor conditions operating in full light under sterile culture conditions, which could be far from potential industrial conditions. Thus, while current yields appear far from the GI_{50} levels, strategies that would improve yields would eventually lead to some form of cellular toxicity. Qiao et al. (2012) demonstrated that the addition of ethanol at 12 g L^{-1} resulted in a 50% growth reduction of *Synechocystis sp* PCC6803. To analyse the effects of ethanol toxicity on model strains of *Synechocystis,* both transcriptomic and proteomic analyses have been carried out (Wang et al. 2012; Qiao et al. 2012; Dienst et al. 2014; Borirak et al. 2015). A comprehensive transcriptomic analysis, using an RNA-seq library was carried out by supplementing cultures of *Synechocystis sp* PCC6803 with 1.25–2.5% (v/v) of ethanol and sampling between 24 and 72 h post-ethanol addition (Wang et al. 2012) and 1.5% was reported to cause a 50% reduction in growth rate and visible aggregation. Some 274 genes were reported to be up-regulated, associated with photosynthesis, the Calvin cycle, ribosomal subunits, energy metabolism and ATP synthesis.

These transcriptomic observations were similar to proteomic data (Qiao et al. 2012) with both studies reporting up-regulation of a number of unknown transcripts and proteins such as *slr*0144, *slr*0373 and *slr*1470 suggesting some potentially, as yet uncharacterised, response mechanisms. The individual transcripts and proteins up and down-regulated suggest that the effected organism may be triggering the oxidative stress response, enhancing transport mechanisms (12 transporters were up-regulated). Equally, proteins associated with membrane modification were expressed suggesting that membrane modification may be a key tolerance mechanism. Squalene hopene cyclase (*slr*2089) was one such protein with overexpressed activity observed (Wang et al. 2012; Qiao et al. 2012) which has been proposed to be involved in strengthening the membrane in response to ethanol. Cell envelope proteins and genes were also up-regulated as were genes involved in PHA storage and some ten signal transduction activities were also up-regulated. Some 1874 genes were down-regulated of which over 60% were of unknown function, indicating a comprehensive response elicited to ethanol.

In an allied study, Dienst et al. (2014) examined the transcriptomic response via high-density microarrays to the continuous production of ethanol from an integrated ethanol cassette at current low production levels (0.03% v/v). This study confirmed that the production of ethanol in this engineered strain resulted in a 40% reduction of biomass. Although the ethanol environment might be expected to be less toxic than adding 1.5% as in the previous addition studies (Wang et al. 2012; Qiao et al. 2012), there was still a significant response. The *adh*A gene encoding the *Synechocystis*-*sAdh*II enzyme was up-regulated suggesting this may have some sort of detoxifying role. An ABC transporter (*slr*1897) was up-regulated as were a number of ribosomal proteins suggesting an initial limited response by protein synthesis, metabolic conversion and perhaps some transport phenomenon may initially be induced. Several genes were downregulated, but there was no evidence of induction of the stress response (Dienst et al. 2014). A similar study (Borirak et al. 2015) carried out a quantitative proteomic study of ethanol producers examining the proteome in a similar way to the transcriptomic study previously reported (Dienst et al. 2014).

Even with low ethanol levels, some 267 proteins were up or downregulated. These included upregulation of carbon fixation, presumably to compensate for the flux away from biomass to ethanol, and evidence of oxidative stress induction was reported. Interestingly, the enzyme phosphor-methyl-pyrimidine synthase (*thi*C), involved in ThDP biosynthesis, was also up-regulated suggesting that even at early stages in production the Pdc co-factor, ThDP, may be in limited supply and could affect overall flux to ethanol (Borirak et al. 2015). Such transcriptomic and proteomic studies may offer strategies to respond to the apparent ethanol toxicity observed. Some interventions such as cloning ThDP transporters to prevent ThDP co-factor limitation, or overexpressing squalene hopene cyclase (*slr* 2089) to increase membrane solidity may be options going forward but others such as induction of stress responses may need the evolution or use of more tolerant model organisms.

Tolerance to other biofuel candidates has also been examined. Isobutanol tolerant strains of *Synechocystis sp* PCC6803 were obtained by long-term laboratory evolution using media containing 2 g L^{-1} (Matsusako et al. 2017). Mutant strains capable of growing at 5 g L^{-1} were isolated and genetic analysis revealed they had accumulated multiple mutations in *slr*1044, (*mcp*A) encoding a methyl-accepting chemotaxis protein which may affect aggregation, *slr*0369 (*envD*), an efflux transporter allowed growth in 5 g L^{-1} isobutanol or *slr*0322 (hik43), a histidine kinase sensor regulator again controlling cell aggregation. These mutations generally demonstrated stress resistance not only to isobutanol but also to other alcohols including ethanol when examined (Matsusako et al. 2017). This observation of synergism may allow data generated on tolerance evolution to be utilised amongst several biofuel production systems as resistant strains to isobutanol also demonstrated increased isobutanol production when containing production cassettes (Matsusako et al. 2017). In an attempt to observe the effects of these isobutanol resistance mutations alone, strains were constructed with deletions in *mcpA*, *hik43* and *envD* and, while no mutation alone gave high-level resistance, it was the synergistic effect of the combination of these mutations that gave rise to the evolved resistance (Matsusako et al. 2017).

Tolerance to n-butanol, which limits growth in cyanobacterial production strains, was also examined via transcriptome (RNA-seq) sequencing (Anfelt et al. 2013) in *Synechocystis sp* PCC6803. Some 80 transcripts were differentially expressed by the addition of 40 mg L^{-1} of n-butanol, while some 280 were differentially expressed at 1 g L^{-1}. Analysis of data suggested that issues with membrane function, impaired photosynthesis, electron transport, reduced biosynthesis and accumulation of reactive oxygen species were all inferred from the transcriptome data (Anfelt et al. 2013). Using the transcriptome data a number of proteins were overexpressed as informed by the differentially expressed transcriptome, one of which *Hsp*A, a small heat shock protein, improved tolerance to butanol. However, the picture can be complex. Comparative quantitative proteomic analysis of the response to n-butanol led to the identification of 303 differentially regulated proteins in metabolically engineered *Synechocystis* sp PCC6803 (Tian et al. 2013) with data analysis concluding that *Synechocystis* employed multiple and synergistic resistance mechanisms in dealing with butanol stress most notably induction of heat shock proteins, cell membrane modification and transporters. Such studies indicate that transcriptome and proteome data can be useful in informing strategies for enhancing tolerance and production.

3.9 Mutation or Deletion of Production Cassettes

Genetic instability in cyanobacteria is considered somewhat of 'an elephant in the room' (Jones 2014). When model cyanobacteria are manipulated with heterologous DNA for biofuel production this puts a metabolic burden on the host. With the use of selective pressure, the construct can be maintained and its presence as part of the polyploid genome means that its removal by the organism is not simple. However, even though there are multiple copies, clones that put a burden on the host are prime targets for an adaptive mutation to remove the burden. Because biofuel production diverts metabolic intermediates away from biomass it leads to slower growth rates in the producer. Therefore, mutations in the construct will alleviate this burden, remove the bottleneck to growth and allow the development of more competitive faster growing (non-producing) mutants. Although there are few publications in the literature (Jones 2014), it is recognised by those who work in the area that genetic instability exists and is indeed common. It is not unusual to see strains emerge suddenly which show very low production rates and examination reveals that constructs that have taken time to construct, transform and select have suddenly undergone mutation. The literature on the mutation of constructs in cyanobacteria is sparse.

Takahama et al. (2003) reported instability in engineered strains of *Synechococcus elongatus* PCC 7942 modified for ethylene synthesis while similar instability was observed in engineered mannitol-producing *Synechococcus sp.* PCC 7002 strain (Jacobsen and Frigaard 2014). Angermayr et al. (2012) reported revertants in a lactate-engineered construct of *Synechocystis sp* PCC6803, while Kusakabe et al. (2013) reported mutations in *ato*B one of the genes central to a construct synthesising isopropanol in *Synechococcus sp.* PCC 7942. We have also observed mutation accumulation in the *pdc* gene of pUL004 a construct utilised to produce ethanol (Lopes da Silva et al. 2018), although at a relatively low rate. Jones (2014) has suggested that mutations may be inducible in cyanobacteria that undergo stress and indeed other organisms also undergo mutation as a result of stress (Sleight and Sauro 2013). However, the added stress both metabolically and through solvent interaction with cell structures and components during biofuel production may force the selection to negate this stress and ultimately effect production. Thus, stability and instability of biofuel constructs will need to be examined as part of the toolkit for generating industrial-scale biofuel producers moving forward.

3.10 Growth and Product Recovery at the Pilot or Industrial Scale

In the drive towards industrial production of biofuels from cyanobacteria at scale, consideration will be needed as to the growth and production strategies. A key determinant of this strategy will be the cost of the biofuel product itself. Currently, ethanol has a market price of around $0.40 per litre (US Grains Council https://grains.org/) and there are similar constraints on other biofuel candidates. Market price will be the main driving force as to Capital Expenditure (CapEX) on any commercial cyanobacterial biofuel enterprise. At the current price for ethanol, large-scale photobioreactors (used to optimise growth and production at laboratory scale) are not an option. Current production rates at laboratory scale are generally carried out

Fig. 3.3 Tubular reactor system for growing metabolically engineered *Synechocystis sp* PCC6803 viewed at various stages of cyanobacterial growth. The darkening colour indicates growth with time

in optimised bioreactors, in a 24-hr cycle with maintenance of optimal conditions of CO_2 supply, mixing, aeration and pH control. As it is generally assumed that such conditions will not be feasible at scale, other growth systems need to be developed or utilised. With relatively volatile biofuels such as ethanol, open pond systems will also pose problems, as there may be evaporative loss and even in high light environments photosynthetic growth can only be maintained during daylight. Hence, some form of a tubular reactor system that will retain the producing organism and the biofuel product will be sensible and in an environment that maximises daylight for photosynthesis. During the recent EU-funded DEMA project (Direct Ethanol from MicroAlgae) http://www.dema-etoh.eu/en/ the production partner A4f in Portugal utilised tubular reactors (Fig. 3.3).

CO_2 can be supplied by air aeration, while optimisation of tube design can allow mixing and retention of the ethanol produced (Lopes et al. 2019). The nature of the culture conditions at scale is important to consider. Such a tubular system does not give rise to axenic (pure) culture but can be optimised to maintain monoculture for many hours of production with optimised inoculum strategies. To keep costs on track the media will not be sterile, although the inoculum may be. If maintenance of sterile conditions were to be essential then the CapEX and OpEx would be uneconomic due to the energy costs of sterilisation of both media and reactor. Under such non-sterile culturing conditions, the production strain would have to be competitive with other contaminants. However, the burden of biofuel production on growth rate would limit this particularly for model strains not optimised for growth.

With the need for faster growing more productive cyanobacterial hosts, *Synechococcus* UTEX 2973 has been described (Yu et al. 2015; Ungerer et al. 2018) with photoautotrophic growth reported comparable to industrial yeast strains. This strain, a relative of PCC7942, but which grows some two times faster and can be genetically manipulated has been shown to possess a number of single-nucleotide polymorphisms compared to PCC7942 which may cast light on factors that increase its biomass productivity and growth rate (Ungerer et al. 2018).

The use of ultrasonic intensification (periodic ultrasonic treatment during the fermentation process) can also result in a more effective homogenisation of biomass and faster energy and mass transfer to biomass over short time periods, which can result in enhanced microbial growth during fermentation processes (Naveena et al. 2015). Such short ultrasonic pulses have been proposed to increase ethanol yields during the production phase in engineered *Synechocystis sp* PCC6803 (Naveena et al. 2016) and may offer a strategy to manipulate the growth conditions for product release with ethanol and other biofuel products.

The presence of produced biofuel during the production cycle may itself lower production levels perhaps due to feedback loops or more likely due to biofuel stress. During isobutanol production, biomass productivity and isobutanol increased some 1.2 and 2.5 folds, respectively, by removal of the produced isobutanol using a solvent trap (Varman et al. 2013a). The addition of such removal systems may have implications for the production of other biofuels also including ethanol, particularly as contaminants in an open process could lead to product loss. The presence of contaminants could also affect the growth of the biofuel producer by metabolising micronutrients faster, shading the light source or as mentioned even utilising the produced biofuel, which could be a major consideration in the case of ethanol. Another production issue would be the availability of light. Industrial-scale culture would need adequate light supply meaning that only certain global locations may be optimal for natural light supply over a long day period. However, sunlight is just one element, an adequate supply of water to allow culture is also a prerequisite that may impact site location of a production scale plant.

Many of the proof of concept laboratory studies have been carried out under optimal photobioreactor conditions with 24-hour light regimes. Under industrial conditions, only diurnal lighting would exist unless supplemented during the night (by LEDs, e.g. which would add energy cost), hence data from idealised photobioreactors needs to be seen as only an indicator of potential productivity. Temperature maintenance is also an issue, should the temperature spike this could affect growth and limit production or indeed lead to an evaporative loss. Again, to limit CapEx, cooling systems would need to be passive in such situations. Maintenance of such systems can also impact operational expenditure (OpEx). Inevitably, reactor systems will need to be cleaned of biofilm formers and light blockers and reactor leakages maintained. Cleaning will be essential to ensure that future cultures can be maintained as much as possible as a monoculture during the production phase. Contamination from previous culturing will obviously have an effect on productivity and yield. However, adding any complexity or cost to such a production system when the value of the product is $0.40 a litre, in the case of ethanol, will increase OpEx. Thus, the type of biofuel and the profit margin associated with it may have a major impact on production strategies, particularly where the price of the product is low.

Another key issue that is often overlooked at the production stage is being able to rapidly monitor biofuel production levels to monitor operational parameters. In the case of ethanol, or indeed other biofuel candidates, monitoring can be somewhat cumbersome requiring HPLC or enzymatic analysis of product streams. Recently, online systems have been reported (Memon et al. 2017), which may aid the drive towards industrial production in the case of ethanol but which will be needed to monitor the production of other biofuel candidates.

Once optimal production levels are reached, there will then be the need to recover product which will be the case for any biologically produced cyanobacterial biofuel.

In the case of ethanol and other volatile biofuel candidates, the most obvious technique would be evaporation, however, calculation of capital input and energy costs make this unsuitable (Lopes et al. 2019).

Other technological scenarios include controlled natural evaporation and incorporation of collection systems into the reactor design or indeed membrane separation techniques such as pervaporation (Wee et al. 2008). This latter technique involves permeation through specially designed membranes and evaporation to the vapour phase. Such techniques allow the concentration of the product from production broths to commercial levels in an energy-efficient manner. Pervaporation however relies on sufficient initial concentrations of biofuel products in the production stream. Here, improvements in biological manipulations will help. The higher the biofuel concentration the more efficient and cost-effective will be the purification strategy (Kour et al. 2019). Thus, there are a number of challenges to the production cycle that need to be considered. Strategies utilised for the production of ethanol in commercial situations such as in the US companies Joule Unlimited and Algenol are generally not in the public domain. However, they have undoubtedly encountered many of the issues discussed here. Some may have been overcome, while others may still be proving a challenge. Only when commercial production occurs might one get a glimpse of the potential solutions.

3.11 Techno-Economic Evaluation of Biofuel Production from Cyanobacteria

A techno-economic assessment for the direct production of ethanol using metabolically engineered *Synechocystis* sp. PCC6803 has recently been published (Lopes et al. 2019). A number of scenarios and variations on the process were analysed for a 1000 L day^{-1} ethanol plant. This study highlighted issues with overall CapEx, OpEx and capital return on investment as discussed earlier. Because of the current cost structure for ethanol and other low-cost biofuels, a number of biorefinery strategies have also been proposed for the co-production of biofuels and other bioproducts that might make a biofuel process economic. However, a drawback has been the need for further processing of the products such as esterification or fermentation (Trivedi et al. 2015; Moncada et al. 2015; Chew et al. 2017; Moreno-Garcia et al. 2017) or follow on separation of bioproducts which adds to CapEx and OpEx. The current consensus based on process modelling appears to be that ethanol production is currently only economically feasible as a co-product in a biorefinery-based scenario at current cyanobacterial production rates (Lopes et al. 2019) and this may indeed also be the situation for many other biofuel candidates at present.

3.12 Future Perspectives on Cyanobacteria as Engineered Biofuel Producers

The adage *"much done much more to do"* currently applies to biofuel production in cyanobacteria. There are many challenges both biological and technical that need to be addressed and overcome. Strategies utilised with model organisms and engineered model biofuels and lessons learned from these proofs of concept strategies will aid future developments not only in the case of ethanol but also in the case of other biofuels that will pose their own particular challenges. Future biofuels may be gaseous, such as hydrogen, rather than miscible in an aqueous solution such as ethanol. Others may be immiscible and prove easier to purify.

Many of the challenges are not necessarily biological in nature but may be aided by tailoring the biology of the producer to the eventual process. Issues such as developing competitive producers perhaps using thermophilic cyanobacteria could be used to aid monoculture, while faster-growing cyanobacterial species with better partition characteristics could aid flux to biofuel products. More tolerant strains may be less affected by the product. Strains that have less need for co-factor synthesis may generate higher yields. More stable, more productive cassettes in better production backgrounds may generate higher yields and indeed all of these options need to be explored further.

Finally, techno-economic studies (Lopes et al. 2019) suggest that there may be an economic prerogative to co-produce the biofuel product with another high value, non-biofuel product to make the overall production cycle economic. Thus far, such options have been little explored as only model cyanobacteria have been utilised thus far and those do not currently produce other high-value co-products. Modelling such systems and estimating flux patterns and mutation strategies may be useful in developing optimal candidate strains and indeed a co-production strategy may be the key to seeing progress in this exciting though challenging area.

References

Anfelt J, Hallström B, Nielsen J, Uhlén M, Hudson EP (2013) Using transcriptomics to improve butanol tolerance of *Synechocystis* sp. Strain PCC 6803. Appl Environ Microbiol 79:7419–7427. https://doi.org/10.1128/AEM.02694-13

Angermayr SA, Paszota M, Hellingwerf KJ (2012) Engineering a cyanobacterial cell factory for production of lactic acid. Appl Environ Microbiol 78:7098–7106. https://doi.org/10.1128/AEM.01587-12

Angermayr SA, Van Der Woude AD, Correddu D, Vreugdenhil A, Verrone V, Hellingwerf KJ (2014) Exploring metabolic engineering design principles for the photosynthetic production of lactic acid by *Synechocystis* sp. PCC6803. Biotechnol Biofuels 7:99 https://doi.org/10.1186/1754-6834-7-99

Armshaw P, Carey D, Sheahan C, Pembroke JT (2015) Utilising the native plasmid, pCA2.4, from the cyanobacterium *Synechocystis* sp. strain PCC6803 as a cloning site for enhanced product production. Biotechnol Biofuels 8: 201 https://doi.org/10.1186/s13068-015-0385-x

Armshaw P, Ryan MP, Sheahan C, Pembroke JT (2018) Decoupling growth from ethanol production in the metabolically engineered photoautotroph *Synechocystis* PCC6803. In: ECO-BIO 2018

Asayama M Kato, Kato H, Shibato J, Shirai MOT (2002) The curved DNA structure in the 5'-upstream region of the light-responsive genes: its universality, binding factor and function for cyanobacterial *psbA* transcription. Nucleic Acids Res 30:4658–4666. https://doi.org/10.1093/nar/gkf605

Atsumi S, Cann AF, Connor MR, Shen CR, Smith KM, Brynildsen MP, Chou KJY, Hanai T, Liao JC (2008) Metabolic engineering of *Escherichia coli* for 1-butanol production. Metab Eng 10:305–311. https://doi.org/10.1016/j.ymben.2007.08.003

Atsumi S, Higashide W, Liao JC (2009) Direct photosynthetic recycling of carbon dioxide to isobutyraldehyde. Nat Biotechnol 27:1177–1180. https://doi.org/10.1038/nbt.1586

Badger MR, Hanson D, Price GD (2002) Evolution and diversity of CO_2 concentrating mechanisms in cyanobacteria. Funct Plant Biol 29:161–173. https://doi.org/10.1093/jxb/erg076

Baroukh C, Muñoz-Tamayo R, Steyer JP, Bernard O (2015) A state of the art of metabolic networks of unicellular microalgae and cyanobacteria for biofuel production. Metab Eng 30:49–60. https://doi.org/10.1016/j.ymben.2015.03.019

Battchikova N, Vainonen JP, Vorontsova N, Kerãnen M, Carmel D, Aro EM (2010) Dynamic changes in the proteome of *Synechocystis* 6803 in response to CO_2 limitation revealed by quantitative proteomics. J Proteome Res 9:5896–5912. https://doi.org/10.1021/pr100651w

Bentley FK, Zurbriggen A, Melis A (2014) Heterologous expression of the mevalonic acid pathway in cyanobacteria enhances endogenous carbon partitioning to isoprene. Mol Plant 7:71–86. https://doi.org/10.1093/mp/sst134

Borirak O, de Koning LJ, van der Woude AD, Hoefsloot HCJ, Dekker HL, Roseboom W, de Koster CG, Hellingwerf KJ (2015) Quantitative proteomics analysis of an ethanol- and a lactate-producing mutant strain of *Synechocystis* sp. PCC6803. Biotechnol Biofuels 8:111 https://doi.org/10.1186/s13068-015-0294-z

Chen Y, Taton A, Go M, London RE, Pieper LM, Golden SS, Golden JW (2016) Self-replicating shuttle vectors based on pANS, a small endogenous plasmid of the unicellular cyanobacterium *Synechococcus elongatus* PCC 7942. Microbiol (United Kingdom) 162:2029–2041. https://doi.org/10.1099/mic.0.000377

Chew KW, Yap JY, Show PL, Suan NH, Juan JC, Ling TC, Lee DJ, Chang JS (2017) Microalgae biorefinery: high value products perspectives. Bioresour Technol 229:53–62. https://doi.org/10.1016/j.biortech.2017.01.006

Daley SME, Kappell AD, Carrick MJ, Burnap RL (2012) Regulation of the cyanobacterial CO_2-concentrating mechanism involves internal sensing of NADP+ and α-ketogutarate levels by transcription factor CcmR. PLoS ONE 7:e41286. https://doi.org/10.1371/journal.pone.0041286

Dehring U, Kramer D, Ziegler K (2012). Selection of ADH in genetically modified Cyanobacteria for the production of ethanol. US Patent 8,163,516

De Deng M, Coleman JR (1999) Ethanol synthesis by genetic engineering in cyanobacteria. Appl Environ Microbiol 65:523–528

Dexter J, Armshaw P, Sheahan C, Pembroke JT (2015) The state of autotrophic ethanol production in Cyanobacteria. J Appl Microbiol 119:11–24. https://doi.org/10.1111/jam.12821

Dexter J, Fu P (2009) Metabolic engineering of cyanobacteria for ethanol production. Energy Environ Sci 2:857. https://doi.org/10.1039/b811937f

Dienst D, Georg J, Abts T, Jakorew L, Kuchmina E, Börner T, Wilde A, Dühring U, Enke H, Hess WR (2014) Transcriptomic response to prolonged ethanol production in the cyanobacterium *Synechocystis* sp. PCC6803. Biotechnol Biofuels 7:21. https://doi.org/10.1186/1754-6834-7-21

Ducat DC, Avelar-Rivas JA, Way JC, Silvera PA (2012) Rerouting carbon flux to enhance photosynthetic productivity. Appl Environ Microbiol 78:2660–2668. https://doi.org/10.1128/AEM.07901-11

Englund E, Liang F, Lindberg P (2016) Evaluation of promoters and ribosome binding sites for biotechnological applications in the unicellular cyanobacterium *Synechocystis* sp. PCC 6803. Sci Rep 6: 36640.https://doi.org/10.1038/srep36640

Englund E, Pattanaik B, Ubhayasekera SJK, Stensjö K, Bergquist J, Lindberg P (2014) Production of squalene in *Synechocystis* sp. PCC 6803. PLoS One 9: e90270. https://doi.org/10.1371/journal.pone.0090270

Ferreira EA, Pacheco CC, Pinto F, Pereira J, Lamosa P, Oliveira P, Kirov B, Jaramillo A, Tamagnini P (2018) Expanding the toolbox for *Synechocystis* sp. PCC 6803: validation of replicative vectors and characterization of a novel set of promoters. Synth Biol 3: ysy014. https://doi.org/10.1093/synbio/ysy014

Formighieri C, Melis A (2014) Regulation of β-phellandrene synthase gene expression, recombinant protein accumulation, and monoterpene hydrocarbons production in *Synechocystis* transformants. Planta 240:309–324. https://doi.org/10.1007/s00425-014-2080-8

Formighieri C, Melis A (2018) Cyanobacterial production of plant essential oils. Planta 248:933–946. https://doi.org/10.1007/s00425-018-2948-0

Gao Q, Wang W, Zhao H, Lu X (2012a) Effects of fatty acid activation on photosynthetic production of fatty acid-based biofuels in *Synechocystis* sp. PCC6803. Biotechnol Biofuels 5:17.https://doi.org/10.1186/1754-6834-5-17

Gao Z, Zhao H, Li Z, Tan X, Lu X (2012b) Photosynthetic production of ethanol from carbon dioxide in genetically engineered cyanobacteria. Energy Environ Sci 5:9857–9865. https://doi.org/10.1039/C2EE22675H

Griese M, Lange C, Soppa J (2011) Ploidy in cyanobacteria. FEMS Microbiol Lett 323:124–131. https://doi.org/10.1111/j.1574-6968.2011.02368.x

Guerrero F, Carbonell VV, Cossu M, Correddu D, Jones PR (2012) Ethylene Synthesis and Regulated Expression of Recombinant Protein in *Synechocystis* sp. PCC 6803. PLoS One 7:e50470. https://doi.org/10.1371/journal.pone.0050470

Halfmann C, Gu L, Gibbons W, Zhou R (2014) Genetically engineering cyanobacteria to convert CO_2, water, and light into the long-chain hydrocarbon farnesene. Appl Microbiol Biotechnol 98:9869–9877. https://doi.org/10.1007/s00253-014-6118-4

Hendry JI, Prasannan C, Ma F, Möllers KB, Jaiswal D, Digmurti M, Allen DK, Frigaard NU, Dasgupta S, Wangikar PP (2017) Rerouting of carbon flux in a glycogen mutant of cyanobacteria assessed via isotopically non-stationary ^{13}C metabolic flux analysis. Biotechnol Bioeng 114:2298–2308. https://doi.org/10.1002/bit.26350

Hendry JI, Prasannan CB, Joshi A, Dasgupta S, Wangikar PP (2016) Metabolic model of *Synechococcus* sp. PCC 7002: Prediction of flux distribution and network modification for enhanced biofuel production. Bioresour Technol 213:190–197. https://doi.org/10.1016/j.biortech.2016.02.128

Heyer H, Krumbein WE (1991) Excretion of fermentation products in dark and anaerobically incubated cyanobacteria. Arch Microbiol 155:284–287. https://doi.org/10.1007/BF00252213

Holland SC, Artier J, Miller NT, Cano M, Yu J, Ghirardi ML, Burnap RL (2016) Impacts of genetically engineered alterations in carbon sink pathways on photosynthetic performance. Algal Res 20:87–99. https://doi.org/10.1016/j.algal.2016.09.021

Hoppner TC, Doelle HW (1983) Purification and kinetic characteristics of pyruvate decarboxylase and ethanol dehydrogenase from *Zymomonas mobilis* in relation to ethanol production. Eur J Appl Microbiol Biotechnol 17:152–157. https://doi.org/10.1007/BF00505880

Huang HH, Camsund D, Lindblad P, Heidorn T (2010) Design and characterization of molecular tools for a synthetic biology approach towards developing cyanobacterial biotechnology. Nucleic Acids Res 38:2577–2593. https://doi.org/10.1093/nar/gkq164

Jacobsen JH, Frigaard NU (2014) Engineering of photosynthetic mannitol biosynthesis from CO_2 in a cyanobacterium. Metab Eng 21:60–70. https://doi.org/10.1016/j.ymben.2013.11.004

Jones PR (2014) Genetic instability in cyanobacteria: an elephant in the room? Front Bioeng Biotechnol 2:12. https://doi.org/10.3389/fbioe.2014.00012

Joseph A, Aikawa S, Sasaki K, Tsuge Y, Matsuda F, Tanaka T, Kondo A (2013) Utilization of lactic acid bacterial genes in *Synechocystis* sp. PCC 6803 in the production of lactic acid. Biosci Biotechnol Biochem 77:966–970. https://doi.org/10.1271/bbb.120921

Kaiser BK, Carleton M, Hickman JW, Miller C, Lawson D, Budde M, Warrener P, Paredes A, Mullapudi S, Navarro P, Cross F, Roberts JM (2013) Fatty aldehydes in cyanobacteria are a metabolically flexible precursor for a diversity of biofuel products. PLoS ONE 8:e58307. https://doi.org/10.1371/journal.pone.0058307

Kämäräinen J, Nylund M, Aro EM, Kallio P (2018) Comparison of ethanol tolerance between potential cyanobacterial production hosts. J Biotechnol 283:140–145. https://doi.org/10.1016/j.jbiotec.2018.07.034

Kamennaya NA, Ahn SE, Park H, Bartal R, Sasaki KA, Holman HY, Jansson C (2015) Installing extra bicarbonate transporters in the cyanobacterium *Synechocystis* sp. PCC6803 enhances biomass production. Metab Eng 29:76–85. https://doi.org/10.1016/j.ymben.2015.03.002

Kaneko T, Nakamura Y, Sasamoto S, Watanabe A, Kohara M, Matsumoto M, Shimpo S, Yamada M, Tabata S (2003) Structural analysis of four large plasmids harboring in a unicellular Cyanobacterium, *Synechocystis* sp. PCC 6803. DNA Res 10:221–228. https://doi.org/10.1093/dnares/10.5.221

Kaneko T, Sato S, Kotani H, Tanaka A, Asamizu E, Nakamura Y, Miyajima N, Hirosawa M, Sugiura M, Sasamoto S, Kimura T, Hosouchi T, Matsuno A, Muraki A, Nakazaki N, Naruo K, Okumura S, Shimpo S, Takeuchi C, Wada T, Watanabe A, Yamada M, Yasuda M, Tabata S (1996) Sequence analysis of the genome of the unicellular cyanobacterium *Synechocystis* sp. strain PCC6803. II. Sequence determination of the entire genome and assignment of potential protein-coding regions. DNA Res 3:109–136. https://doi.org/10.1093/dnares/3.3.109

Kaplan A, Reinhold L (2002) CO_2 concentrating mechanisms in photosynthetic microorganisms. Annu Rev Plant Physiol Plant Mol Biol 50:539–570. https://doi.org/10.1146/annurev.arplant.50.1.539

Khetkorn W, Rastogi RP, Incharoensakdi A, Lindblad P, Madamwar D, Pandey A, Larroche C (2017) Microalgal hydrogen production—A review. Bioresour Technol 243:1194–1206. https://doi.org/10.1016/j.biortech.2017.07.085

Kiyota H, Okuda Y, Ito M, Hirai MY, Ikeuchi M (2014) Engineering of cyanobacteria for the photosynthetic production of limonene from CO_2. J Biotechnol 185:1–7. https://doi.org/10.1016/j.jbiotec.2014.05.025

Kour D, Rana KL, Yadav N, Yadav AN, Rastegari AA, Singh C et al. (2019) Technologies for biofuel production: current development, challenges, and future prospects. In: Rastegari AA, Yadav AN, Gupta A (eds) Prospects of renewable bioprocessing in future energy systems. Springer International Publishing, Cham, pp 1–50. https://doi.org/10.1007/978-3-030-14463-0_1

Kumar S, Sharma S, Thakur S, Mishra T, Negi P, Mishra S et al (2019) Bioprospecting of microbes for biohydrogen production: current status and future challenges. In: Molina G, Gupta VK, Singh BN, Gathergood N (eds) Bioprocessing for biomolecules production. Wiley, USA, pp 443–471

Kusakabe T, Tatsuke T, Tsuruno K, Hirokawa Y, Atsumi S, Liao JC, Hanai T (2013) Engineering a synthetic pathway in cyanobacteria for isopropanol production directly from carbon dioxide and light. Metab Eng 20:101–108. https://doi.org/10.1016/j.ymben.2013.09.007

Lagarde D, Beuf L, Vermaas W (2000) Increased production of zeaxanthin and other pigments by application of genetic engineering techniques to *Synechocystis* sp. strain PCC 6803. Appl Environ Microbiol 66:64–72. https://doi.org/10.1128/AEM.66.1.64-72.2000

Lan EI, Chuang DS, Shen CR, Lee AM, Ro SY, Liao JC (2015) Metabolic engineering of cyanobacteria for photosynthetic 3-hydroxypropionic acid production from CO_2 using *Synechococcus elongatus* PCC 7942. Metab Eng 31:163–170. https://doi.org/10.1016/j.ymben.2015.08.002

Lan EI, Liao JC (2011) Metabolic engineering of cyanobacteria for 1-butanol production from carbon dioxide. Metab Eng 13:353–363. https://doi.org/10.1016/j.ymben.2011.04.004

Lau NS, Foong CP, Kurihara Y, Sudesh K, Matsui M (2014) RNA-Seq Analysis provides insights for understanding photoautotrophic polyhydroxyalkanoate production in recombinant *Synechocystis* sp. PLoS ONE 9:e86368. https://doi.org/10.1371/journal.pone.0086368

Li H, Liao JC (2013) Engineering a cyanobacterium as the catalyst for the photosynthetic conversion of CO_2 to 1, 2-propanediol. Microb Cell Fact 12:4. https://doi.org/10.1186/1475-2859-12-4

Lin PC, Pakrasi HB (2019) Engineering cyanobacteria for production of terpenoids. Planta 249:145–154. https://doi.org/10.1007/s00425-018-3047-y

Lopes da Silva T, Passarinho PC, Galriça R, Zenóglio A, Armshaw P, Pembroke JT, Sheahan C, Reis A, Gírio F (2018) Evaluation of the ethanol tolerance for wild and mutant *Synechocystis* strains by flow cytometry. Biotechnol Reports 17:137–147. https://doi.org/10.1016/j.btre.2018.02.005

Lopes TF, Cabanas C, Silva A, Fonseca D, Santos E, Guerra LT, Sheahan C, Reis A, Gírio F (2019) Process simulation and techno-economic assessment for direct production of advanced bioethanol using a genetically modified *Synechocystis* sp. Bioresour Technol Reports 6:113–122. https://doi.org/10.1016/j.biteb.2019.02.010

Máté Z, Sass L, Szekeres M, Vass I, Nagy F (1998) UV-B-induced differential transcription of *psbA* genes encoding the D1 protein of photosystem II in the Cyanobacterium *Synechocystis* 6803. J Biol Chem 273:17439–17444. https://doi.org/10.1074/jbc.273.28.17439

Matsusako T, Toya Y, Yoshikawa K, Shimizu H (2017) Identification of alcohol stress tolerance genes of *Synechocystis* sp. PCC 6803 using adaptive laboratory evolution. Biotechnol Biofuels 10:307. https://doi.org/10.1186/s13068-017-0996-5

McCormick AJ, Bombelli P, Lea-Smith DJ, Bradley RW, Scott AM, Fisher AC, Smith AG, Howe CJ (2013) Hydrogen production through oxygenic photosynthesis using the cyanobacterium *Synechocystis* sp. PCC 6803 in a bio-photoelectrolysis cell (BPE) system. Energy Environ Sci 6:2682–2690. https://doi.org/10.1039/c3ee40491a

Memon SF, Ali MM, Pembroke JT, Chowdhry BS, Lewis E (2017) Measurement of ultralow level bioethanol concentration for production using evanescent wave based optical fiber sensor. IEEE Trans Instrum Meas 67:780–788. https://doi.org/10.1109/TIM.2017.2761618

Miao R, Liu X, Englund E, Lindberg P, Lindblad P (2017) Isobutanol production in *Synechocystis* PCC 6803 using heterologous and endogenous alcohol dehydrogenases. Metab Eng Commun 5:45–53. https://doi.org/10.1016/j.meteno.2017.07.003

Moncada J, Cardona CA, Rincón LE (2015) Design and analysis of a second and third generation biorefinery: the case of castorbean and microalgae. Bioresour Technol 198:836–843. https://doi.org/10.1016/j.biortech.2015.09.077

Monshupanee T, Chairattanawat C, Incharoensakdi A (2019) Disruption of cyanobacterial γ-aminobutyric acid shunt pathway reduces metabolites levels in tricarboxylic acid cycle, but enhances pyruvate and poly(3-hydroxybutyrate) accumulation. Sci Rep 9:8184. https://doi.org/10.1038/s41598-019-44729-8

Montenecourt B (1985) *Zymomonas*, a unique genus of bacteria. In: Demain AL, Solomon NA (eds) Biology of industrial microorganisms. Benjamin-Cummings Publishing Co., Inc., Menlo Park, CA, pp 261–289

Moreno-Garcia L, Adjallé K, Barnabé S, Raghavan GSV (2017) Microalgae biomass production for a biorefinery system: Recent advances and the way towards sustainability. Renew Sustain Energy Rev 76:493–506

Mueller TJ, Berla BM, Pakrasi HB, Maranas CD (2013) Rapid construction of metabolic models for a family of Cyanobacteria using a multiple source annotation workflow. BMC Syst Biol 7:142. https://doi.org/10.1186/1752-0509-7-142

Nakahira Y, Ogawa A, Asano H, Oyama T, Tozawa Y (2013) Theophylline-dependent riboswitch as a novel genetic tool for strict regulation of protein expression in cyanobacterium *Synechococcus elongatus* PCC 7942. Plant Cell Physiol 54:1724–1735. https://doi.org/10.1093/pcp/pct115

Naveena B, Armshaw P, Pembroke JT (2015) Ultrasonic intensification as a tool for enhanced microbial biofuel yields. Biotechnol Biofuels 8:140. https://doi.org/10.1186/s13068-015-0321-0

Naveena B, Armshaw P, Pembroke JT, Gopinath KP (2016) Kinetic and optimization studies on ultrasonic intensified photo-autotrophic ethanol production from *Synechocystis sp.* Renew Energy 95:522–530. https://doi.org/10.1016/j.renene.2016.04.061

Neale ADD, Scopes RKK, Kelly JMM, Wettenhall REHE (1986) The two alcohol dehydrogenases of *Zymomonas mobilis*. Eur J Biochem 154:119–124

Ng AH, Berla BM, Pakrasi HB (2015) Fine-tuning of photoautotrophic protein production by combining promoters and neutral sites in the cyanobacterium *Synechocystis* sp. strain PCC 6803. Appl Environ Microbiol 81:6857–6863. https://doi.org/10.1128/AEM.01349-15

Nicolaou SA, Gaida SM, Papoutsakis ET (2010) A comparative view of metabolite and substrate stress and tolerance in microbial bioprocessing: From biofuels and chemicals, to biocatalysis and bioremediation. Metab Eng 12:307–331. https://doi.org/10.1016/j.ymben.2010.03.004

Niederholtmeyer H, Wolfstädter BT, Savage DF, Silver PA, Way JC (2010) Engineering cyanobacteria to synthesize and export hydrophilic products. Appl Environ Microbiol 76:3462–3466. https://doi.org/10.1128/AEM.00202-10

Nobles DR, Brown RM (2008) Transgenic expression of *Gluconacetobacter xylinus* strain ATCC 53582 cellulose synthase genes in the cyanobacterium *Synechococcus leopoliensis* strain UTCC 100. Cellulose 15:691–701. https://doi.org/10.1007/s10570-008-9217-5

Ogata H, Goto S, Sato K, Fujibuchi W, Bono H, Kanehisa M (1999) KEGG: Kyoto encyclopedia of genes and genomes. Nucleic Acids Res 27:29–34. https://doi.org/10.1093/nar/27.1.29

Oliver JWK, Machado IMP, Yoneda H, Atsumi S (2013) Cyanobacterial conversion of carbon dioxide to 2,3-butanediol. Proc Natl Acad Sci 110:1249–1254. https://doi.org/10.1073/pnas.121 3024110

Pembroke JT, Armshaw P, Ryan MP (2019) Metabolic Engineering of the Model Photoautotrophic Cyanobacterium *Synechocystis* for Ethanol Production: Optimization Strategies and Challenges. In: Fuel Ethanol Production from Sugarcane. Ed Basso, T.P.; Basso, L.C.; 2019.p 199–219

Pembroke JT, Lorraine Q, O'Riordan H, Sheahan C, Armshaw P (2017) Ethanol Production in Cyanobacteria: impact of omics of the model organism *Synechocystis* on yield enhancement. In: Los DA (ed) Cyanobacteria: Omics and Manipulation. Caister Academic Press, pp 199–218

Pinto F, Pacheco CC, Oliveira P, Montagud A, Landels A, Couto N, Wright PC, Urchueguía JF, Tamagnini P (2015) Improving a *Synechocystis* -based photoautotrophic chassis through systematic genome mapping and validation of neutral sites. DNA Res 22:425–437. https://doi.org/10.1093/dnares/dsv024

Qiao J, Wang J, Chen L, Tian X, Huang S, Ren X, Zhang W (2012) Quantitative iTRAQ LC-MS/MS proteomics reveals metabolic responses to biofuel ethanol in cyanobacterial *Synechocystis* sp. PCC 6803. J Proteome Res 11:5286–5300. https://doi.org/10.1021/pr300504w

Quinn L, Armshaw P, Soulimane T, Sheehan C, Ryan MP, Pembroke JT (2019) *Zymobacter palmae* pyruvate decarboxylase is less effective than that of *Zymomonas mobilis* for ethanol production in metabolically engineered *Synechocystis* sp PCC6803. Microorganisms 7(11):494. https://doi.org/10.3390/microorganisms7110494

Rastegari AA, Yadav AN, Gupta A (2019) Prospects of Renewable Bioprocessing in Future Energy Systems. Springer International Publishing, Cham

Reinsvold RE, Jinkerson RE, Radakovits R, Posewitz MC, Basu C (2011) The production of the sesquiterpene β-caryophyllene in a transgenic strain of the cyanobacterium *Synechocystis*. J Plant Physiol 168:848–852. https://doi.org/10.1016/j.jplph.2010.11.006

Sarnaik A, Abernathy MH, Han X, Ouyang Y, Xia K, Chen Y, Cress B, Zhang F, Lali A, Pandit R, Linhardt RJ, Tang YJ, Koffas MAG (2019) Metabolic engineering of cyanobacteria for photoautotrophic production of heparosan, a pharmaceutical precursor of heparin. Algal Res 37:57–63. https://doi.org/10.1016/j.algal.2018.11.010

Savakis PE, Angermayr SA, Hellingwerf KJ (2013) Synthesis of 2, 3-butanediol by *Synechocystis* sp. PCC6803 via heterologous expression of a catabolic pathway from lactic acid- and enterobacteria. Metab Eng. https://doi.org/10.1016/j.ymben.2013.09.008

Schirmer A, Rude MA, Li X, Popova E, Del Cardayre SB (2010) Microbial biosynthesis of alkanes. Science (80-) 329:559–562. https://doi.org/10.1126/science.1187936

Segre D, Vitkup D, Church GM (2002) Analysis of optimality in natural and perturbed metabolic networks. Proc Natl Acad Sci 99:15112–15117. https://doi.org/10.1073/pnas.232349399

Shibata M, Katoh H, Sonoda M, Ohkawa H, Shimoyama M, Fukuzawa H, Kaplan A, Ogawa T (2002) Genes essential to sodium-dependent bicarbonate transport in cyanobacteria: Function and phylogenetic analysis. J Biol Chem 277:18658–18664. https://doi.org/10.1074/jbc.M112468200

Shibata M, Ohkawa H, Kaneko T, Fukuzawa H, Tabata S, Kaplan A, Ogawa T (2001) Distinct constitutive and low-CO_2-induced CO_2 uptake systems in cyanobacteria: Genes involved and their phylogenetic relationship with homologous genes in other organisms. Proc Natl Acad Sci 98:11789–11794. https://doi.org/10.1073/pnas.191258298

Shibato J, Agrawal G, Kato H, Asayama M, Shirai M (2002) The 5′-upstream cis-acting sequences of a cyanobacterial *psbA* gene: Analysis of their roles in basal, light-dependent and circadian transcription. Mol Genet Genomics 267:684–694. https://doi.org/10.1007/s00438-002-0704-3

Sleight SC, Sauro HM (2013) Visualization of evolutionary stability dynamics and competitive fitness of *Escherichia coli* engineered with randomized multigene circuits. ACS Synth Biol 2:519–528. https://doi.org/10.1021/sb400055h

Stal LJ, Moezelaar R (1997) Fermentation in cyanobacteria. FEMS Microbiol Rev 21:179–211. https://doi.org/10.1016/S0168-6445(97)00056-9

Takahama K, Matsuoka M, Nagahama K, Ogawa T (2003) Construction and analysis of a recombinant cyanobacterium expressing a chromosomally inserted gene for an ethylene-forming enzyme at the *psbAI* locus. J Biosci Bioeng 95:302–305

Tan X, Yao L, Gao Q, Wang W, Qi F, Lu X (2011) Photosynthesis driven conversion of carbon dioxide to fatty alcohols and hydrocarbons in cyanobacteria. Metab Eng 13:169–176. https://doi.org/10.1016/j.ymben.2011.01.001

Taton A, Unglaub F, Wright NE, Zeng WY, Paz-Yepes J, Brahamsha B, Palenik B, Peterson TC, Haerizadeh F, Golden SS, Golden JW (2014) Broad-host-range vector system for synthetic biology and biotechnology in cyanobacteria. Nucleic Acids Res 42:e136. https://doi.org/10.1093/nar/gku673

Tian X, Chen L, Wang J, Qiao J, Zhang W (2013) Quantitative proteomics reveals dynamic responses of *Synechocystis* sp. PCC 6803 to next-generation biofuel butanol. J Proteomics 78:326–345. https://doi.org/10.1016/j.jprot.2012.10.002

Trivedi J, Aila M, Bangwal DP, Kaul S, Garg MO (2015) Algae based biorefinery—How to make sense? Renew Sustain Energy Rev 47:295–307. https://doi.org/10.1016/j.rser.2015.03.052

Ungerer J, Lin PC, Chen HY, Pakrasi HB (2018) Adjustments to photosystem stoichiometry and electron transfer proteins are key to the remarkably fast growth of the Cyanobacterium *Synechococcus elongatus* UTEX 2973. MBio 9:e02327–17. https://doi.org/10.1128/mBio.02327-17

Varman AM, Xiao Y, Pakrasi HB, Tang YJ (2013a) Metabolic engineering of *Synechocystis* sp. Strain PCC 6803 for isobutanol production. Appl Environ Microbiol 79:908–914. https://doi.org/10.1128/AEM.02827-12

Varman AM, Yu Y, You L, Tang YJ (2013b) Photoautotrophic production of D-lactic acid in an engineered cyanobacterium. Microb Cell Fact 12:117. https://doi.org/10.1186/1475-2859-12-117

Vasudevan R, Gale GAR, Schiavon AA, Puzorjov A, Malin J, Gillespie MD, Vavitsas K, Zulkower V, Wang B, Howe CJ, Lea-Smith DJ, McCormick AJ (2019) CyanoGate: a modular cloning suite for engineering cyanobacteria based on the plant MoClo syntax. Plant Physiol 180:39–55. https://doi.org/10.1104/pp.18.01401

Velmurugan R, Incharoensakdi A (2020) Co-cultivation of two engineered strains of *Synechocystis sp.* PCC 6803 results in improved bioethanol production. Renew Energy 146:1124–1133. https://doi.org/10.1016/j.renene.2019.07.025

Wang J, Chen L, Huang S, Liu J, Ren X, Tian X, Qiao J, Zhang W (2012) RNA-seq based identification and mutant validation of gene targets related to ethanol resistance in cyanobacterial *Synechocystis* sp. PCC 6803. Biotechnol Biofuels 5:89. https://doi.org/10.1186/1754-6834-5-89

Wang W, Liu X, Lu X (2013) Engineering cyanobacteria to improve photosynthetic production of alka(e)nes. Biotechnol Biofuels 6:69. https://doi.org/10.1186/1754-6834-6-69

Wee SL, Tye CT, Bhatia S (2008) Membrane separation process-Pervaporation through zeolite membrane. Sep Purif Technol 63:500–516.https://doi.org/10.1016/j.seppur.2008.07.010

Woods RP, Coleman JR, De Deng M. (2004). Genetically modified cyanobacteria for the production of ethanol, the constructs and method thereof. US Patent 6699696B2

Wu GF, Shen ZY, Wu QY (2002) Modification of carbon partitioning to enhance PHB production in *Synechocystis* sp. PCC6803. Enzyme Microb Technol 30:710–715. https://doi.org/10.1016/S0141-0229(02)00044-3

Yao L, Qi F, Tan X, Lu X (2014) Improved production of fatty alcohols in cyanobacteria by metabolic engineering. Biotechnol Biofuels 7:94. https://doi.org/10.1186/1754-6834-7-94

Yoshino T, Liang Y, Arai D, Maeda Y, Honda T, Muto M, Kakunaka N, Tanaka T (2015) Alkane production by the marine cyanobacterium *Synechococcus* sp. NKBG15041c possessing the α-olefin biosynthesis pathway. Appl Microbiol Biotechnol 99:1521–1529. https://doi.org/10.1007/s00253-014-6286-2

Yu J, Liberton M, Cliften PF, Head RD, Jacobs JM, Smith RD, Koppenaal DW, Brand JJ, Pakrasi HB (2015) Synechococcus elongatus, a fast growing cyanobacterial chassis for biosynthesis using light and CO_2. Sci Rep 5:8132. https://doi.org/10.1038/srep08132

Zhou J, Zhang H, Meng H, Zhu Y, Bao G, Zhang Y, Li Y, Ma Y (2014) Discovery of a super-strong promoter enables efficient production of heterologous proteins in cyanobacteria. Sci Rep 4:4500. https://doi.org/10.1038/srep04500

Zhou J, Zhang H, Zhang Y, Li Y, Ma Y (2012) Designing and creating a modularized synthetic pathway in cyanobacterium *Synechocystis* enables production of acetone from carbon dioxide. Metab Eng 14:394–400. https://doi.org/10.1016/j.ymben.2012.03.005

Chapter 4
Energy and Carbon Balance of Microalgae Production: Environmental Impacts and Constraints

Yachana Jha

Abstract The continuous development of human civilization needs continuous supply of energy, but resources for energy production are limited as it is non-renewable. So, to sustain the pace of development, there is time to opt a sustainable energy production system, i.e., the biofuel. Several options have been assessed for the production of biofuel, but most of them have many limitations. Among them, microalgae are most suitable ones due to their many advantages. Microalgae have already been cultivated for a long time to produce food, feed, and other substances and can also be used to produce biodiesel, bioethanol, biomethane, and biohydrogen as alternatives for fossil resources. Microalgae grow quickly with concentrated CO_2 or reuse CO_2 from other resources to produce bioenergy sources.

4.1 Introduction

Access to the energy is the key requirement for the well-being of the human civilization, and the sun is the ultimate source of energy for the living world, which has been captured in biomass by the photosynthesis by the green plant. Energy requirement initially and even today is mostly fulfilled by fossil fuels, but it produces large amount of carbon dioxide and other greenhouse gases, responsible for global environmental change, having adverse impact on environment. So, a balance needs to be established between development and its environmental impact. For sustainable development to attain high standard of living, access to sufficient energy is required. With increase in world population, there is increased demand of energy, which is responsible for energy crisis globally not only due to growing population but also due to heavy industrialization. The natural resources like coal, natural gas, diesel, and petrol like basic sources for energy are near to exhaustion. Such extensive use of natural resources also has vast impact on our environment due to libration of large

Y. Jha (✉)
N. V. Patel College of Pure and Applied Sciences, S. P. University,
V V Nagar, Anand, Gujarat, India
e-mail: yachanajha@gmail.com

A. N. Yadav et al. (eds.), *Biofuels Production – Sustainability and Advances in Microbial Bioresources*, Biofuel and Biorefinery Technologies 11,
https://doi.org/10.1007/978-3-030-53933-7_4

amount of harmful gases. The level of greenhouse gas in the environment has reached an alarming point in the post-industrialization era.

Natural causes and intensive human activities are the major causes of raised temperature which is responsible for global warming (Rastegari et al. 2019a). To reduce the emission of greenhouse gas related with the energy production, we have to shift energy production from reduced carbon or carbon–neutral resources, like solar, wind, ocean, geothermal, hydroelectric, and biofuel as sustainable alternative resource, to fulfill the energy requirement (Schiermeier et al. 2008). The renewable energy sources need to be derived from microalgae biomass in option of nonrenewable energy systems for effective use of current infrastructure for petroleum-based energy sources, which is an energy dense liquid form as well as efficient in reducing the emission of greenhouse gases and may act as an environmentally sustainable energy source. So, the production and use of biofuels as an alternative source of energy is gaining importance in the world. The commercial production and use of biofuels are already initiated in many developed countries.

There are several resources like agro-wastes, cereal crops, fruits wastes, wastes of timber wood, household wastes, or macro/microalgae which are the best alternative sources for the production of biofuels like bioethanol and biodiesel. The sustainable energy solutions with reduced global carbon emission are the biofuels. Biomass produced by microalgae has been fermented to generate ethanol, but it has only half energy density compared to fossil fuel. The mass production of biomass for biofuel production requires energy input as various stages as soil tillage, irrigation, increased soil respiration, use of pesticides, fertilizers, herbicides, and transportation of the feedstock require additional input of energy, and two-thirds of the carbon in the biomass has been emitted as CO_2, when fermented for production of ethanol (Hill et al. 2006). Also, the production of biofuel has limitations of available crop land because land is limited for fulfillment of the demands for food, fiber, and other important things necessary for the humankind. So, the alarming concern is selection of alternative source of energy having high potential for enhanced energy as well as low CO_2 emission to reduce environmental pollution and such alternative is algae. Algae are wide groups of marine photosynthetic plant with high ability for oil production and in extenuating CO_2 emissions. It can efficiently grow in oceans, ponds, lakes, rivers as well as in wastewater and has high tolerance against salinities, high light intensities, extreme temperatures, and pH. Any reservoirs can be used for the growth of algae to be used for the production of biofuel, where it either grows independently or in association with other organisms.

Algae are classified on the basis of size into microalgae or macroalgae. As name indicates, microalgae are single cellular microscopic in size, while macroalgae are multicellular large in size and can be visualized by naked eyes. Photosynthetic microalgae having ability for carbon fixation are rich source of carbon and must be a suitable source for production of biofuels and mitigation of atmospheric CO_2. The potential of microalgae enhances interest as a sustainable and renewable feedstock for production of biofuel. Such microalgae at the same time having ability to grow efficiently in wastewater can be used for wastewater treatment. Microalgae are fast growing organisms found in marine or freshwater which have important positions in aquatic ecosystems and establishes the basis of aquatic food chains. The fast growing microalgae have source of large enough biomass for biofuel production and rapidly influence global climate. Microalgae are fast growing organism and highly adaptable to the surrounding environment, due to its unicellular it is useful to capture nutrients as well as concentrate useful chemicals in an economical way.

4.2 Carbon and Nitrogen Metabolism of Algae

Microalgae are microscopic plant having same metabolic activities as photosynthetic plant cell. The metabolic activity of microalgae depends on the surrounding it grows, as it decides the types of nutrient for its growth and development. Nitrogen and carbon are the two most important elements in metabolic pathway of photosynthetic organism, which directly influence the mass of cell, carbon assimilation in biomass, protein, chlorophyll, nucleotides, and other important biomolecules (Jha 2019a, b, c). Like photosynthetic plant, in algae also carbon gets incorporated in the glucose during photosynthesis and converted into glucose-6-phosphate for growth, respiration, and storage in sunlight. No carbon assimilation takes place in absence of light as well as no fermentation of stored glucose takes place in algae in absence of enzyme lactate dehydrogenase.

In algae, predominantly the only pathway that takes place for carbon assimilation is Embden–Meyerhof and Pentose phosphate pathway (PPP), and conversion of available glucose into oligosaccharides and polysaccharides takes place. At the same time, nitrogen is one of the most abundant nutrients in the environment and is important contributor to the dry weight of algae. There is interconnection between C and N metabolism in all photosynthetic organism including algae. Nitrogen in the form of ammonia is the contributor for amino acids required carbon skeleton (Jha, 2019a, b, c). So, incorporation of C and N requires coordination for the synthesis of important metabolites for the survival of the algae. Most important feature of microalgae is their ability to trap carbon in form of CO_2 from the atmosphere and surrounding water source in the form of bicarbonate. Photosynthetic algae have bicarbonate transporters on its plasma membrane and on chloroplast envelope for efficient use of bicarbonate from water, which has been converted into CO_2 inside chloroplast for photosynthetic dark reaction (Enamala et al. 2018).

4.3 Bioenergy and Microalgae

There is continuous increase in the demand of fuel energy with increase in population and rigorous use of fossil fuels. It will be at the merge of finish due to its non-renewable nature and non-sustainability. So, alternative renewable and sustainable option for fossil fuels, that is, biofuels are now drawing attention (Kour et al. 2019a;

Rastegari et al. 2020). Biofuels such as bioethanol and biodiesel are exceptional substitute of fossil fuels, which have been generated from variable resources of biomass, such as agriculture wastes, fruits, food, crops, woods, and algae (Rana et al. 2019). Burning of fossil fuels generate about 29 gigatons CO_2 per year and other greenhouse gases responsible for global warming. The biofuel has about 10–45% of oxygen and very less sulfur content in comparison to fossil fuel. So, advantage of biofuel is its sustainability, renewability, accessibility, non-polluting, and locally available. Microalgae can fix 1.83 kg of CO_2 per kg of biomass and are non-toxic and eco-friendly.

Microalgae species having potential for high lipid content (50–70%) accumulation in biomass can accumulate about 60,000 L oil per hectare biomass, and is capable of producing biofuel of about 121,104 L per hectare biodiesels (Gardner Dale et al. 2017).The bioethanol is an ecologically clean fuel which has several benefits over fossil fuels as (i) burning of bioethanol produces minimum amount of greenhouse gases due to the presence of high oxygen contents in it, (ii) bioethanol can be directly used without any further modification, in present energy infrastructure/automobile industry due to similar feature, (iii) bioethanol can be directly mixed with classical fuel due to similar nature, (iv) bioethanol can reduce the wear and tear of the engines due to having high octane content, which prevents knocking of oil cylinders. Production of bioethanol has been dynamically enhanced worldwide up to 100 billion soon from 1 to 39 billion within a few years (Basso et al. 2011). Carbohydrates like agar, starch, cellulose, and glycogen which can be easily transformed to fermentable sugars for production of bioethanol are present in high amount in microalgae.

4.4 The Growth of Algae for Biomass Production

The growth of algae is influence by several factors like carbon source, nutrient source, light and optimum temperature has major impact. Major nutrients like phosphate, nitrate, and carbohydrate and trace nutrients like zinc, cobalt, molybdenum, and manganese are necessary for desired growth. For cultivation, some additional parameters which also play important role are proper mixing in the photoreactor, optimal pH, uptake of CO_2, and removal of O_2 in equal amount. These parameters need be controlled and coordinated properly to achieve desired algal biomass production. Among all these, the temperature is the most significant and sensitive factor for large-scale production, and optimal growth temperatures for the growth are in range of 20–30 °C (Dragone et al. 2010). Increase in temperature results in decrease in algal cell volume, and frequent variation in temperature significantly decreases the algal lipid production. The major advantages of microalga for biomass and energy production are its very short doubling time; it easily grows in any aquariums; for its growth, cheap media can be used (including wastewater); its ability to utilize CO_2 as it grows; helps in cleaning of the environment; it can be grown in a non-arable land on a large scale, and on small scale can be grown in our own houses; it has great potential of competency for food vs. fuel; and is considered as tough competitors for biofuel production.

4.5 Environmental Impacts on Biofuel Production

For the production of biofuel, microalga is an option, but it requires large-scale production of microalgae, having large multiple impact on environment with consumption of energy. Such limitation could limit system design and operation of microalgae biomass production (Fig. 4.1).

4.5.1 Impact of Light

As algae is a photosynthetic organism and requires proper intensity of light for its growth and development, at proper light intensity, only biosynthesis of appropriate biomass takes place. Mostly, algae require only 90–100 μmol of light intensity for photosynthesis to run dark reaction properly. So, cultivation of microalgae in open pond system requires constant monitoring of light intensity for proper biomass production (Carvalho 2010). Naturally, algae form only a few mm layers and do not shade each other remarkably. But during mass cultivation of shading, it is definitely a problem. So, mixing of cultivated microalgae is required to bring each cell on the surface of the pool to get illuminated regularly. Algae can harvest best under flashing light effect in which cell is illuminated with very high light intensities for short time duration, which has been efficiently used completely in darkness (Panjiar et al. 2017).

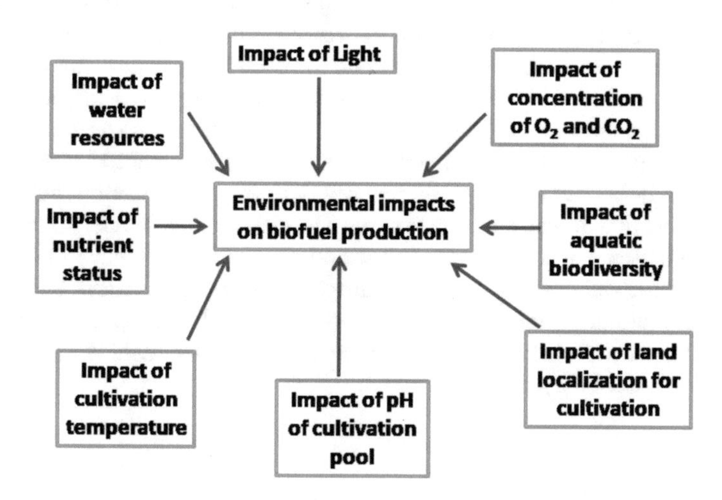

Fig. 4.1 Various environmental impacts on biofuel production by mass cultivation of microalgae

4.5.2 Impact of Concentration of O_2 and CO_2

As algae are photosynthetic microorganisms, they constantly require CO_2 for the production of biomass. At the same time, photosynthesis results in production of O_2 which gradually accumulates in the cultivating pond and inhibits the enzyme RUBISCO. RUBISCO is the main enzyme for the dark reaction of the photosynthesis and has more affinity for O_2. So, to achieve fast and efficient growth of algae, it requires external supply of CO_2 to the culture, which has been coupled with removal of O_2. As the O_2 produced during light reaction has inhibitory effect on photosynthesis, proper gaseous exchange is achieved by aeration of the cultivation pond, which also helps in proper mixing of culture component due to turbulence of air bubbles. This turbulence helps microalgae to assess required amount of CO_2 and nutrient due to proper mixing of substance, thus enhances mass transfer rates.

4.5.3 Impact of Water Resources

A consistent, continuous low cost water supply is necessary for the cultivation of microalgae for biofuel production. Although there is large volume of water present on the earth and microalgae are able to grow in marine as well as freshwater, but for the cultivation of microalgae is require well define boundary of the water body used for the growth. The water level and temperature of the cultivating pool need to be maintained, as open environment results in water loss due to evaporation (Teter et al. 2018). So, freshwater needs to be added at uniform time difference to maintain water at constant condition, but it is costly and energy demanding. For efficient growth of microalgae, continuous addition of oxygen in the water is also necessary; this is achieved by pumping, and significant amount of energy produced by the algae is consumed in it. So, the location with reduced pumping due to natural tidal flows is the choice location to feed cultivation pond. The distance to the water source is also an important factor in locating the cultivation site. Microalgae have the potential to grow in wastewater as well, but it required pretreatment to remove contaminant and growth-inhibiting components, such as metabolites of dead algae, inorganic and organic chemicals, etc. Pretreatment and recycling of water could raise the energy demand and cost of the process.

4.5.4 Impact of Cultivation Temperature

Temperature is one of the important factors for the growth of microalgae and very sensitive parameter for large-scale production, especially as open pool cultivation. Continuous variation in environmental temperature is normal phenomenon, but for the cultivation of microalgae such variation is not desirable. Increase in temperature significantly results in decrease in algal cell volume; all the enzymes of the photosynthetic dark reaction perform well within narrow range of temperature variation. Such variation in cultivation temperature remarkably reduced the lipid synthesis efficiency of the algae, which has direct effect on production of energy (Kassem and Çamur 2017). The optimum growth temperature for microalgae is in the range of 20–30 °C. But, during summer, when the light intensity increased, it caused increase of cultivation pool temperature. At the same time, during winter or evening, low light intensity results in reduced cultivation pool temperature, and both conditions remarkably reduced the microalgae biomass production. Even increase in temperature by few degrees can lead to the mass death of the microalgae.

4.5.5 Impact of pH of Cultivation Pool

The pH is another important physical parameter, which has direct effect on microalgae. Alkaline condition is desirable for the growth of microalgae, while acidic condition has highly deleterious effect on the growth of microalgae. Under alkaline conditions, microalgae photosynthetic rate gets increased many folds and yields additional biomass, as it enhanced the ability of microalgae to capture the CO_2 from the atmosphere. With increase in the rate of photosynthesis, gradual accumulation of OH^- ions takes place, which gradually changes the pH of the cultivation pool from basic to acidic. The change in pH from basic to acidic also altered the permeability of the algal cell and effects the transportation of important minerals and ions, necessary for the growth of algal biomass (Agasteswar et al. 2017). Acidic pH also results in hydronium forms of the inorganic salt and amalgamation of the inorganic salts, while cultivation of specific algae strains having ability to grow at extreme pH has the advantage to overcome the contamination.

4.5.6 Impact of Nutrient Status

Large-scale production of algae requires sufficient amount of nutrient for fast biomass production, which is directly proportional to the energy production. There are large number of elements like N, O_2, C, H_2, K, Ca, Mg, Fe, P, and S which are the main mineral nutrients for the growth of microalgae, and trace minerals are also required. Among these, the mineral nutrients like O_2, H_2, and carbon are directly obtained from atmosphere, and mineral nutrients like K, P, and N are essential for the growth of microalgae. Nitrogen and phosphorus are more important as they participate in lipid production and are necessary to maintain high growth rate. Nitrogen and phosphorus are most essential elements for the microalgae cultivation as they are necessary for formation of amino acid, DNA, RNA, etc., for growth, cell division, and other biochemical functions (Jha 2017). Potassium is also an essential element as it maintains membrane permeability of the microalgae for the efficient growth. So, to get good yield of biomass and lipid accumulation, all these nutrients need to be supplied in proper proportion (Elsayed et al. 2017). Mass cultivation of microalgae requires nutrients supplement, primarily nitrogen, phosphorus, and potassium, which can be supplied in the form of dry algae powder (Hein and Leemans 2012). Supply of recycled nutrients from the wastewater has the potential to reduce the cost, but it will enhance the rate of contamination.

4.5.7 Impact of Land Localization for Cultivation

Cultivation of microalgae requires large land area generally has competition for land for food production. So, use of marginal land for the cultivation is good option, but topology and soil porosity/permeability are important factors and affect the growth of microalgae. Cultivation land should be neither shaded nor fully open as it directly affects the temperature of cultivation pond. At the same time, cultivation pond needs to be established in pollution-free environment, as matter and poisonous gas have direct effect on growth of algae (Gasparatos et al. 2018). The most suitable locations are warm countries close to the equator or in low-latitude regions, where least variation of temperature takes place.

4.5.8 Impact of Aquatic Biodiversity

For energy production, mass cultivation of microalgae is required known as "regulated eutrophication," which required custom harvesting and sufficient supply of air. However, regulated eutrophication also has remarkable risk to the biodiversity. Mass cultivation of microalgae regularly causes decomposition of dead algal biomass, which consumes dissolved oxygen causing asphyxiation for its own growth and for other aerobic aquatic organisms. Absence of oxygen is cause for death of aquatic organisms, which results in water turbidity and toxicity due to degradation of dead organism. In absence of oxygen during anaerobic condition, there are production of methane and other greenhouse gases, responsible for global warming. Due to emission of such gases, there are bad odorous in the surrounding environment. Accidental release of water from cultivation pond into the surrounding area or specifically near large water body can lead to large-scale eutrophication and ultimately loss of large number of aquatic organisms. This impact is directly proportional to the amount of leaking and quality of the receiving water body. But it has positive impact also if the algae production is integrated with the treatment of water bodies already suffering from excess nutrient.

4.6 Constrain in Algal Production Technology

The objective for production of microalgal biofuel is inspired by the motivation to replace the existing conventional nonrenewable source of energy with renewable one, as a cheap and eco-friendly approach. But practically there are several limitations in mass scale production of biofuel by microalgae (Fig. 4.2).

4.6.1 Strain Selection

The first and the main issue in microalgal production technology is selection of specific correct strain of microalgae as per the requirement. So, screening and selection of such microalgae strain is quite tedious and requires lot of screening in hope to find specific strain that can work efficiently and give desirable product. There are many species of microalgae with high potential for production technology, but till date very less numbers of families has been evaluated and there is lack of phenotypic information. So, it is need of hour to characterize genetic diversity of microalgae for its domestication and wild species as well as identification of potential strain, protection of improved strains, and conservation of precious germplasm (Allen et al. 2018). Also, conservation of genetic diversity is necessary for genetic basis for various breeding programs for the development of new strain having high potential in different production systems or climates. Selection of precise germplasm for future use requires genetic and phenotypic characterization. Lack of phenotypic information is most hindering factor for genetic improvement of algal strain (Venteris et al. 2014). There are numerous species of microalgae with high potential for genetical engineering, but are in early stages of development. The working conditions of

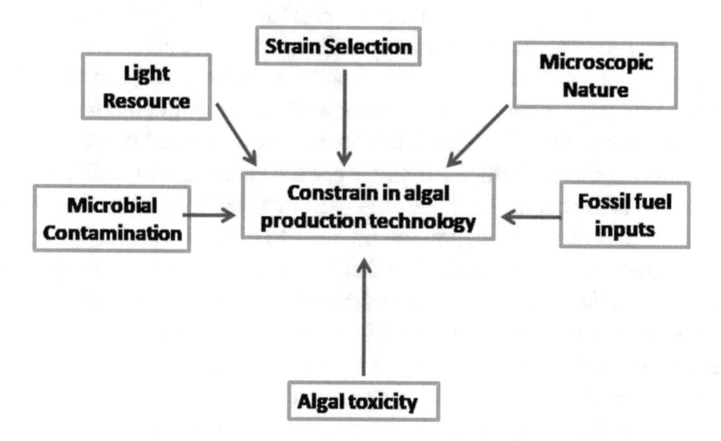

Fig. 4.2 Various constrains on biofuel production by mass cultivation of microalgae

specific system constraints must be considered for each strain to narrow the list down to a group that can be used on a mass production.

4.6.2 Light Resource

Microalgae production system driven by photosynthesis based on cell factory directly depends on light/solar source. Sunlight intensity varies across the globe, so maximum solar to chemical energy conversion potential also varies. Therefore, the low solar light region has to use artificial light source for the efficient microalgal production system, to conquer the restrictions posed by the low competence of photosynthesis. For microalgal production system, first and cheap light option is natural sunlight only, as sunlight is critical to autotrophic growth of algae. But there is regular fluctuation in quality and intensity in natural sunlight at daily, regionally, and seasonally. Like any photosynthetic organism, the chlorophylls of microalgae also show best light absorption at around 440 and 680 nm wavelengths (Schuurmans et al. 2015). The bottom of the water column received reduced light intensity due to inhibitive property of light by the surface layer, so there is insufficient light intensity for photosynthesis (Barry et al. 2015). For 24 h functioning of microalgae production system, an artificial lighting system is necessary. So, light is one of the main constraining factors of productivity and growth of microalgae even in presence of sufficient nutrition and suitable temperature.

4.6.3 Microscopic Nature

Microalgae production system uses microalgal strains typically 3–20 μm and grows in low concentrations, so harvesting in typically less than 2 g algae/L in conventional way is very difficult. Microalgae have negatively charged surfaces to form stable suspensions, and separations from suspension add more difficulty in its harvesting. Not only that, many microalgae cell walls are very sensitive and get damaged during separation process like centrifugation which can result in leaching of the cell contents. Several method of harvesting has been used as centrifugation, flocculation, foam fractionation, ultrasonic separation, and membrane filtration, which finally increase the cost of algal biomass. Most common and efficient method of harvesting is filtration with the help of cellulose membrane, but the membrane tends to become clogged and needs application of filter to draw liquid through it. Although filtration is simple, it is highly time consuming (Khan et al. 2018). With regard to time, centrifugation appears more suitable, but it is highly energy intensive, and for large-scale production, it is not very suitable. Other option is flotation using gas bubble the algae suspension, creating a froth of algae that can be skimmed off.

4.6.4 Microbial Contamination

The isolation and identification of microalgae in lab of unialgal and pure species is possible. But during mass cultivation there is always chance of contamination of algae with lower metazoan and other aquatic organisms (Lian et al. 2018). One of the main sources of contamination is bacteria, which inhibit the growth of microalgae due to secretion of toxic substance, which interferes with algal metabolism. But bacteria from plant growth promoting group have positive effect on the growth of microalgae (Jha, 2019a, b, c). Such growth promoting bacteria have the ability to produce growth promoting beneficial biomolecules, but till date these beneficial biomolecules are unknown (Kumar et al. 2019; Rastegari et al. 2019b, c; Yadav et al. 2019).

4.6.5 Fossil Fuel Inputs

Every step of mass cultivation of microalgae requires energy as for mixing, harvesting, aeration, etc. Microalgae are temperature-sensitive organism, and for efficient production, it required controlled temperature to maintain high productivity. Maintenance of temperature demands both cooling and heating, which require input of energy either in the form of electricity or fossil fuel. Microalgae production system optimization is a means to minimize energy demand (Cotton et al. 2015). Even production system efficiency has been enhanced by integrating options as using waste heat from power generation/direct heat for the process like to dry the algae.

4.6.6 Algal Toxicity

The most important perspective of biofuels production is production of co-products, which have been used by the human and are safe. Many species of microalgae at certain stage of its lifecycle may produce toxins. These toxins have been produced by the microalgae to protect itself and compete with its competitor or to reduce competition. Such toxin may be simple as some gas or complex as physiologically active biomolecules (Marc 2012). Toxin production is strain and species specific and also depends on cultivation/environmental conditions. The prediction of presence or absence of toxin production ability of particular microalgae is quite difficult.

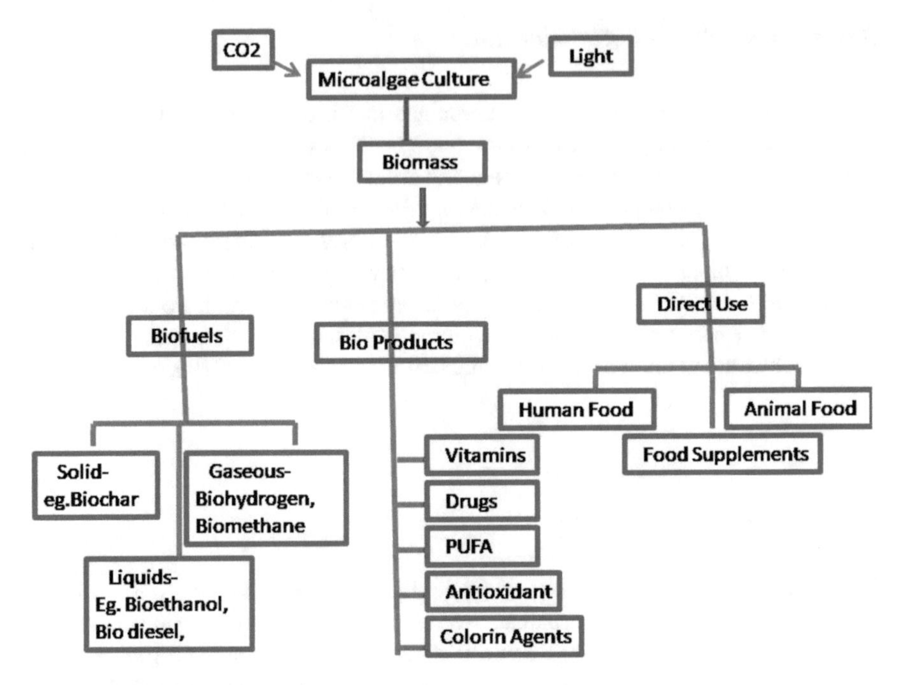

Fig. 4.3 Various applications of biofuel produced by mass cultivation of microalgae

4.7 Applications of Microalgae

The microalgae have been used by the humans from many decades ago, and nowadays there are varieties of commercial as well as industrial applications of the algae. With biofuel, microalgae also produce many important products like polyunsaturated fatty acids, pigments, natural colorants for cosmetics, antioxidants, pharmaceuticals, proteins and carbohydrates, and too many other products (Fig. 4.3). Microalgae also produce variety of animal feed, biohydrogen, biofertilizer, stabilizers, bioelectricity, and essential food supplement (Kour et al. 2019b; Yadav et al. 2017, 2020). Microalgae efficiently contribute in wastewater treatment pollution control and reduce greenhouse gases (Pienkos and Darzins 2009). The genetic modification of the microalgal genes can be a pathway to get new potential products.

4.8 Conclusion and Future Perspectives

With industrialization, the consumption of fossil fuel increased at the highest level, which resulted in increase in atmospheric CO_2 at an alarmingly situation as well as cause atmospheric pollution and depletion of ozone layer. And now fossil fuel reserves are at the wedge of depletion, and for continuous function of industries, there is requirement of alternative source of fuel. This is possible by shifting on renewable energy source as production of biofuel from microalgae, where continuous carbon fixation and long-term biomass production for production of biofuel. For efficient production of biofuel, selection of specific species with good fatty acid profile is important. The future challenge in this field is improvement through the genetic engineering techniques in the lipid profiles of important microalgal strains, having high lipid productivity.

For the mass scale production of biofuel, microalgae are used which definitely have several environmental benefits when compared to other energy sources. Wastewater treatment during the microalgae cultivation and production of various food and pharmaceutical compounds are certain important benefits of microalgae. Microalgae can provide a lower cost alternative to wastewater treatment, which reduces the demand of chemical use as well as reduce the energy input (Stephens et al. 2010). For biofuel production, large-scale production of microalgae is required, which utilized large amount of atmospheric CO_2 and burning of biofuel at the same time which resulted in more impartial level of CO_2 emissions when compared to fossil fuel. This will result in reduced contribution of CO_2 to environment responsible for global climate change.

References

Agasteswar V, Sridhar V, Brahmaiah P, Sasidhar V (2017) Cultivation of microalgae at extreme alkaline pH conditions: a novel approach for biofuel production. ACS Sustain Chem Eng 5(8):7284–7294

Allen J, Unlu S, Demirel Y (2018) Integration of biology, ecology and engineering for sustainable algal-based biofuel and bioproduct biorefinery. Bioresour Bioprocess 5:47

Barry AN, Starkenburg SR, Sayre RT (2015) Strategies for optimizing algal biology for enhanced biomass production. Front Energy Res 3:1

Basso LC, Basso TO, Rocha SN (2011) Recent developments and prospects in biofuel production. In: Bernardes MA (ed), pp 85–100

Carvalho AP (2010) Light requirements in microalgal photobioreactors. Springer, Berlin

Cotton CAR, Douglass JS, De Causmaecker S, Brinkert K, Cardona T, Fantuzzi A et al (2015) Photosynthetic constraints on fuel from microbes. Front Bioeng Biotechnol 3:36

Dragone G, Fernandes B, Vicente A, Teixeira JA (2010) Third generation biofuels from microalgae, current research, technology and education. Appl Microbiol Biotechnol 2:1355–1366

Elsayed KNM, Kolesnikova TA, NokeA, Klöck G(2017) Imaging the accumulated intracellular microalgal lipids as a response to temperature stress. 3 Biotech 7:41

Enamal MK, Enamala S, Chavali M, Donepudi J, Yadavalli R, Kolapalli B et al (2018) Production of biofuels from microalgae - A review on cultivation, harvesting, lipid extraction, and numerous applications of microalgae. Renew Sust Energ Rev 94:49–68

Gardner-Dale DA, Bradley IM, Guest JS (2017) Infuence of solids residence time and carbon storage on nitrogen and phosphorus recovery by microalgae across diel cycles. Water Res 121:231–239

Gasparatos A, von Maltitz G, Johnson F (2018) Survey of local impacts of biofuel crop production and adoption of ethanol stoves in southern Africa. Sci Data 5:180186

Hein L, Leemans R (2012) The impact of first-generation biofuels on the depletion of the global phosphorus reserve. Ambio 41:341–349

Hill J, Nelson E, Tilman D, Polasky S, Tiffany D (2006) Environmental, economic, and energetic costs and benefits of biodiesel and ethanol biofuels. Proc Natl Acad Sci (USA) 103:11206–11210
Jha Y (2017) Potassium mobilizing bacteria: enhance potassium intake in paddy to regulate membrane permeability and accumulate carbohydrates under salinity stress. Brazil J Biol Sci 4:333–344
Jha Y (2019) Endophytic bacteria as a modern tool for sustainable crop management under stress. In: Giri B, Prasad R, Wu QS, Varma A (eds) Biofertilizers for sustainable agriculture and environment. Soil biology, vol 55: Springer, Cham, pp 203–223
Jha Y (2019) Mineral mobilizing bacteria mediated regulation of secondary metabolites for proper photosynthesis in maize under stress. In: Ahmad P, Abass MA, Alyemeni MN, Alam P (eds) Photosynthesis, productivity and environmental stress, John Wiley & Sons Ltd, pp 197-293
Jha Y (2019) Regulation of water status, chlorophyll content, sugar, and photosynthesis in maize under salinity by mineral mobilizing bacteria. In: Ahmad P, Abass MA, Alyemeni MN, Alam P (eds) Photosynthesis, productivity and environmental stress, John Wiley & Sons Ltd, pp 75–95
Kassem Y, Çamur H (2017) A laboratory study of the effects of wide range temperature on the properties of biodiesel produced from various waste vegetable oils. Waste Biomass Valorization 8:1995–2007
Khan MI, Shin JH, Kim JD (2018) The promising future of microalgae: current status, challenges, and optimization of a sustainable and renewable industry for biofuels, feed, and other products. Microb Cell Fact 17:36
Kour D, Rana KL, Yadav N, Yadav AN, Rastegari AA, Singh C et al (2019a) Technologies for biofuel production: current development, challenges, and future prospects. In: Rastegari AA, Yadav AN, Gupta A (eds) Prospects of renewable bioprocessing in future energy systems. Springer International Publishing, Cham, pp 1–50. https://doi.org/10.1007/978-3-030-14463-0_1
Kour D, Rana KL, Yadav N, Yadav AN, Singh J, Rastegari AA et al. (2019b) Agriculturally and industrially important fungi: current developments and potential biotechnological applications. In: Yadav AN, Singh S, Mishra S, Gupta A (eds) Recent advancement in white biotechnology through fungi, Volume 2: Perspective for value-added products and environments. Springer International Publishing, Cham, pp 1–64. https://doi.org/10.1007/978-3-030-14846-1_1
Kumar S, Sharma S, Thakur S, Mishra T, Negi P, Mishra S et al (2019) Bioprospecting of microbes for biohydrogen production: Current status and future challenges. In: Molina G, Gupta VK, Singh BN, Gathergood N (eds) Bioprocessing for biomolecules production. Wiley, USA, pp 443–471
Lian J, Wijffels RH, Smidt H, Sipkema D (2018) The effect of the algal microbiome on industrial production of microalgae. Microb Biotechnol 11:806–818
Marc YM (2012) An overview of algae biofuel production and potential environmental impact. Environ Sci Technol 46:7073–7085
Panjiar N, Mishra S, Yadav AN, Verma P (2017) Functional foods from cyanobacteria: an emerging source for functional food products of pharmaceutical importance. In: Gupta VK, Treichel H, Shapaval VO, Oliveira LAd, Tuohy MG (eds) Microbial functional foods and nutraceuticals. John Wiley & Sons, USA, pp 21–37. https://doi.org/10.1002/9781119048961.ch2
Pienkos PT, Darzins A (2009) The promise and challenges of microalgal derived biofuels. Biofuels Bioprod Bioref 3:431–440
Rana KL, Kour D, Sheikh I, Yadav N, Yadav AN, Kumar V et al (2019) Biodiversity of endophytic fungi from diverse niches and their biotechnological applications. In: Singh BP (ed) Advances in endophytic fungal research: present status and future challenges. Springer International Publishing, Cham, pp 105–144. https://doi.org/10.1007/978-3-030-03589-1_6
Rastegari AA, Yadav AN, Gupta A (2019a) Prospects of renewable bioprocessing in future energy systems. Springer International Publishing, Cham
Rastegari AA, Yadav AN, Yadav N (2019b) Genetic Manipulation of secondary metabolites producers. In: Gupta VK, Pandey A (eds) New and future developments in microbial biotechnology and bioengineering. Elsevier, Amsterdam, pp 13–29. https://doi.org/10.1016/B978-0-444-63504-4.00002-5

Rastegari AA, Yadav AN, Yadav N, Tataei Sarshari N (2019c) Bioengineering of secondary metabolites. In: Gupta VK, Pandey A (eds) New and future developments in microbial biotechnology and bioengineering. Elsevier, Amsterdam, pp 55–68. https://doi.org/10.1016/B978-0-444-63504-4.00004-9

Rastegari AA, Yadav AN, Yadav N (2020) New and future developments in microbial biotechnology and bioengineering: trends of microbial biotechnology for sustainable agriculture and biomedicine systems: diversity and functional perspectives. Elsevier, Amsterdam

Schiermeier Q, Tollefson J, Scully T, Witze A, Morton O (2008) Electricity without carbon. Nature 454:816–823

Schuurmans RM, van Alphen P, Schuurmans JM, Matthijs HCP, Hellingwerf KJ (2015) Comparison of the photosynthetic yield of cyanobacteria and green algae: different methods give different answers. PLoS ONE 10(9):e0139061

Stephens E, Ross IL, King Z, Mussgnug JH, Kruse O, Posten C et al (2010) An economic and technical evaluation of microalgal biofuels. Nat Biotechnol 28:126–128

Teter J, Yeh S, Khanna M, Berndes G(2018) Water impacts of U.S.biofuels: Insights from an assessment combining economic and biophysical models. PLoS ONE 13: e0204298

Venteris ER, Wigmosta MS, Coleman AM, Skaggs RL (2014) Strain selection, biomass to biofuel conversion, and resource colocation have strong impacts on the economic performance of algae cultivation sites. Front Energy Res 2:37

Yadav AN, Kumar R, Kumar S, Kumar V, Sugitha T, Singh B et al (2017) Beneficial microbiomes: biodiversity and potential biotechnological applications for sustainable agriculture and human health. J Appl Biol Biotechnol 5:45–57

Yadav AN, Rastegari AA, Yadav N (2020) Microbiomes of extreme environments: biodiversity and biotechnological applications. CRC Press, Taylor & Francis, Boca Raton, USA

Yadav AN, Singh S, Mishra S, Gupta A (2019) Recent advancement in white biotechnology through fungi. Volume 2: Perspective for value-added products and environments. Springer International Publishing, Cham

Chapter 5
Impact of Climate Change on Sustainable Biofuel Production

Shiv Prasad, Ajar Nath Yadav, and Anoop Singh

Abstract Global energy crisis and climate change have forced to find alternative energy sources to serve in the transition from fossil fuel based economy to a sustainable bio-based economy. In this context, biofuels are a key opportunity for governments, researchers, and industry. They can work together to achieve the goal of the global energy crisis and climate change through large-scale production and use of advanced biofuels. The basic concept of defining biomass as a renewable energy resource includes the capturing of solar energy and carbon from ambient CO_2 in increasing biomass. Production of biofuels from biomass has the potential to boost sustainable development and mitigate climate change issues with socio-economic benefits.

5.1 Introduction

Currently, around 90% of the world's energy requirement is satisfied by the application of non-renewable fossil fuels, such as petroleum, natural gas, and coal (Rodrigues et al. 2017). At present, especially oil and natural gas are the most valuable input for the production of commodities and various types of petrochemicals (Liu et al. 2010). However, unlimited fossil fuel use is well-thought-out unsustainable because of its determinate supply and inequitable distribution reserves and high GHG (Greenhouse Gase) emissions (Tan et al. 2013). Besides, these are also non-renewable that leads

S. Prasad (✉)
Centre for Environment Science & Climate Resilient Agriculture, ICAR-Indian Agricultural Research Institute, New Delhi, India
e-mail: shiv_drprasad@yahoo.co.in

A. N. Yadav
Department of Genetics, Plant Breeding and Biotechnology, Dr. Khem Singh Gill Akal College of Agriculture, Eternal University Baru Sahib, Sirmour, Himachal Pradesh, India

A. Singh
Department of Scientific and Industrial Research (DSIR), Ministry of Science and Technology, Government of India, Technology Bhawan, New Mehrauli Road, New Delhi, India

A. N. Yadav et al. (eds.), *Biofuels Production – Sustainability and Advances in Microbial Bioresources*, Biofuel and Biorefinery Technologies 11,
https://doi.org/10.1007/978-3-030-53933-7_5

to adverse effects like global warming which poses a severe warning to humans and makes it questionable for further usage (Prasad et al. 2014). These all aspects of using fossil fuels are the driving force toward a transition from fossil fuel based economy to a sustainable bio-based economy (Sadik et al. 2010; Prasad et al. 2020b).

Today, every nation is pledging to decrease its carbon footprint, GHG emissions, and trying to lower the rate of rising global warming (Prasad et al. 2014). These all problems can be solved by using biofuel which is the most fabulous alternative to non-renewable source of energy like fossil fuels. Biofuel can be prepared by using biomass (field crops and other lignocellulosic materials), which is existing in plenty on Earth in various forms (Rastegari et al. 2019) and can be transformed into solid, liquid, or gaseous form fuels (Sheetal et al. 2019). The biomass valorization is now considered as the most potent biorefinery component where waste generation is zero or nearly zero.

5.2 CO_2 Emissions by Country

CO_2 is the primary GHG, emitted from the burning of fossil fuels. Along with carbon dioxide emission, smaller amounts of methane (CH_4) and nitrous oxide (N_2O) are also emitted. Usually, GHG absorbs and emits thermal radiation and creates a greenhouse effect. CO_2 is essential to keep the Earth in a habitable temperature. However, the excessive CO_2 emissions from fossil fuels used are disrupting Earth's carbon cycle and accelerating global warming. In the nineteenth century, before industrial era, global mean CO_2 level was nearly 280 ppm. An exceptional rise in the global average atmospheric CO_2 concentration has been witnessed in approaching a record level of 407.4 ppm in 2018 (Lindsey 2019). Carbon dioxide (CO_2) is the most potent GHG in terms of its emitted volume. The list of top greenhouse gases [kilotonnes CO_2 equivalent] emitters in the world in 2015 is shown in Table 5.1. Among the top ten greenhouse gases emitters, China was the highest CO_2 emitter. The United States was the second highest CO_2 emitter (Table 5.1). Figure 5.1 below exhibits a global level of CO_2 emissions from the burning of fossil fuels, which is indicated by China-emissions in red shading color, India-emissions in yellow color, the U.S.-emissions in bright blue color, E.U.-emissions in dark blue color, and the rest of the world-emissions in gray color. It is expected that emissions will cross to a new high of 37.15 billion (bn) tons of CO_2 ($GtCO_2$), with these two largest emitters, mainly China and the U.S. (Hausfather 2018). The world is looking at those nations to lead the initiatives for lowering CO_2 emissions.

Subsequently, a fast rise in worldwide emissions of CO_2 is nearly 3% annually from 2000 to 2013; emissions just rose by 0.4% annually from 2013 to 2016. That was overturned in 2017 with emission increment by 1.6% and expected to rise by 2.7% in 2018 (within 1.8–3.7%, uncertainty range). Developing nations, including India, observed emission increments in 2018 because of economic growth, but it is not yet decoupling from GHG emissions. It is estimated that India's emissions are

Table 5.1 Top greenhouse gases [kilotonnes CO_2 equivalent] emitters in the world in 2015

S. No.		GHGs [kilotonnes CO_2 equivalent]
1.	China	13,067,691
2.	United States	6,444,396
3.	European union	4,499,851
4.	India	3,346,954
5.	Russia	2,233,876
6.	Japan	1,359,553
7.	Brazil	1,229,246
8.	Indonesia	897,152
9.	Iran	815,652
10.	Canada	779,870

Source JRC report on fossil CO_2 and GHG emissions of all world countries (2019): European Parliament 2019. https://www.europarl.europa.eu/news/en/headlines/society/20180301STO98928/greenhouse-gas-emissions-by-country

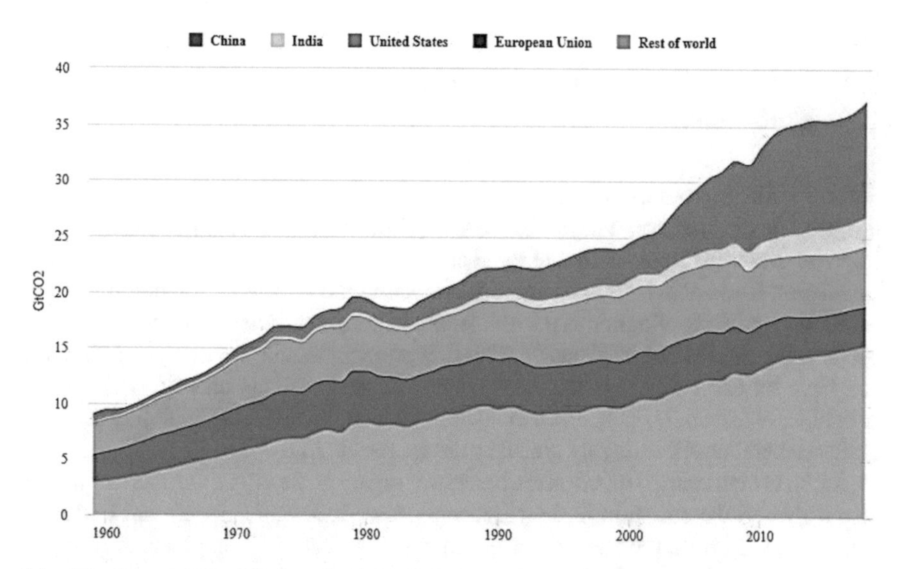

Fig. 5.1 Annual CO_2 emissions from fossil fuels by chief nations from 1959 to 2018. Annual emissions of CO_2 from the burning of fossil fuels and industrial activities by allied nations and the remainder of the world from 1959 to 2018 in bn tons of CO_2 year^{-1} ($GtCO_2$) *Source* Global Carbon Project by C-Brief using High charts

growing with a range of 4.3–8.3%, while the world's emission is supposed to rise with a range of 0.5–3.0% only (Hausfather 2018).

5.3 Global Warming and CO_2

Today, the signs of global warming are throughout the world, and it is the most prominent issue. It is created by the enhanced concentration of GHG in the atmosphere. Joos et al. (2013) showed a directly proportional relationship to global warming and CO_2. According to IPCC (2018) (SR15) special report, global warming has led to increase in the Earth's temperature by 1.5 °C above preindustrial levels, with a likely range of 0.8–1.2 °C. Even considering the complete implementation and contributions submitted by nations in the Paris Agreement, net emissions would rise compared to 2010, leading to a warming of around 3 °C by 2100.

In contrast, restricting warming below or close to 1.5 °C would need to reduce net emissions by nearly 45% by 2030 and approach net zero by 2050. Even just for restricting global warming to below 2 °C, CO_2 emissions must reduce by 25% by 2030 and by 100% by 2075. Global warming is accelerating to various regional and global changes such as high temperature, heavy rainfall, floods, droughts, soil moisture, and rising sea levels (IPCC 2018).

5.4 Worldwide Initiative to Reduce GHG and CO_2 Emissions

Several nations promote the advanced liquid and gaseous biofuels obtained from lignocellulosic biomass. The main reason behind that is their co-benefits, which improve a nation's long-term energy security and lessen dependency on imported petroleum. It can also help to promote socio-economic growth by providing income for rural people's livelihoods as a whole, nonetheless, if sufficient environmental and social protection is not in place. The biofuel generation and use in a given domain can have unreasonable consequences such as adverse effects on soil, water, food supply, or biodiversity. Hence, when deciding whether to establish a biofuel project, policymakers must thoroughly consider the trade-offs (STAP 2015).

At the G7 summit 2016, climate change was again at the top of the agenda, which also triggered the attention in the extensive use of renewable energy and biofuels. Currently, CO_2 from the auto sector is contributing about 25% in global emissions. The biofuel was being a viable alternative to fossil fuels and witnessed as an essential for shifting to low carbon fuel economy. It would not only help in bringing sustainable transport systems but also in rapidly phasing out the dependency on coal, oil, and gas from the global economy. IPCC has shifted its view on biofuels for the first time. It is affirming that they may have some adverse impacts that may take away their advantage in decreasing GHG emissions. The PCC's report, 2014: impacts, adaptation, and vulnerability offer a subtle but vital caveat to the IPCC's viewpoint of past view of biofuels as one of the "key mitigation technologies" for decreasing fossil fuel usage and GHG emissions, as articulated in this 2007 IPCC report. While there are concerns over sustainability of biodiesel made from, for example, palm oil or used cooking oil in the EU; however, most biofuel is sustainably produced from crops grown according to prevailing farming standards in the US, and Europe. Recent IPCC special report on 1.5 °C warming states that biofuel use in transportation will likely require to expand by a factor of 7 if catastrophic climate change is to be avoided. The report shows that, in 2050, biofuels will still be as crucial as electro mobility in the displacement of carbon-emitting fossil fuels. IPCC report addressing climate change, land management, and food security states that almost all of the climate actions assessed for limiting global warming to below 1.5 °C require large-scale bioenergy programs to succeed (UN-AR5 climate science report 2019).

Effective GHG mitigation requires a range of behavioral alterations and the application of alternative technologies. Among mitigation alternatives, renewable energy is observed to decarbonize energy sources and stabilize the climate at a safe level of atmospheric GHG concentrations. EU Energy and Climate Change Package (CCP) 2009 details the guidelines for biofuel usage in automotive transport (Ruiz et al. 2016). The CCP directed that, by 2020, 20% of the entire energy demand would be satisfied by bioenergy sources. Freshly, a new EU energy strategy (Jonsson et al. 2015) has asked for a reflective Europe's energy system transformation, based on energy security, more sustainability, and having a low carbon economy, with the promise to accomplish 40% GHG emission lessening relative to emissions level in 1990 and further to achieve at least 27% share of renewable by 2030 in EU's energy consumption (Giuntoli et al. 2016). European Union Renewable Energy Directive (RED) 2009/28/EC, which mandates levels of renewable energy use, also specified that biofuels must be sustainable. However, quite a lot of factors must be well thought out, not only limited to GHG decrease but also include environmental apprehensions, land use, and many other socio-economic aspects (USDA Foreign Agricultural Service 2016).

The European Commission, in 2015, mandated as per the 2009/28/EC directive that by 2020 first-generation biofuel use in the transport sector would be restricted to a maximum of 7% of the entire EU energy consumption (USDA Foreign Agricultural Service 2019). However, RED mandated 10% biofuel blending (7% from first generation) by 2017 for altogether to the member states and offered a voluntary 5% blending target by advanced biofuels at the national level. Recently, the Indian government has enacted National Policy on Biofuels 2018, which aims to reduce import dependency on fossil fuel and to move toward a green energy economy to mitigate global climate change. The Indian government has targeted the bioenergy contribution around 10 gigawatts (GW) by the year 2022. Further, policy pursues to attain 20% ethanol blending in petrol and 5% biodiesel blending in the diesel by the year 2030.

In India, both the national policy on biofuel 2018 and earlier approved biofuels policy 2009 have a clear-cut mandate on food, energy, and environment to fight biofuel trilemma and policy focused on waste utilization and cultivation of non-edible oilseeds on only degraded forest and non-forest lands to produce feedstock for biodiesel generation. Ethanol is produced in India by using molasses. The policy also focused on ensuring the second-generation biofuel from non-food feedstocks so that fuel versus food security can be tracked efficiently in India (MNRE 2009; PIB 2018).

The new national policy on biofuel 2018 not only economically helping farmers to sell out the surplus stock but also strategically strengthening ongoing initiatives like skill development and make in India. These initiative scans also help in employment generation, waste-to-wealth creation, doubling of farmer income, and reduction in crude oil import bill (PIB 2018).

The government of India is committed under the Paris Agreement to cut GHG intensity by 20–25% by the year 2020 and 33–35% by the year 2030 over 2005 levels. Moving toward a low carbon, a dedicated transport sector will likely support GHG reduction goals. In India, Indian Oil Company (IOC) plans to invest almost $3.5 billion in green energy projects across a wide-range portfolio of renewable fuels, including second- and third-generation biofuels as part of its roadmap to moderate its impact on climate change (Biofuels Digest newsletter 2019).

5.5 Biofuels from Biomass

Biomass is an organic material that includes crop residues and other lignocellulosic waste. On Earth, biomass is available in every form and in plenty amount. They can be transformed into solid, liquid, or gaseous form of fuels (Prasad et al. 2007). Worldwide several nations are blessed with suitable fertile soil, rainfall, sunshine, and water, including India. India has more than half of its land productive, on a universal average of 11%, which represens India as an agrarian economy (IEA 2014). Many of them hold an essential place in the cultivation of commodities and farm produce at the global level. That makes India to have a massive potential to produce renewable energy resources. Alternative first-generation feedstock resources such as cane juice, sweet sorghum are a promising source for ethanol production (Prasad et al. 2006, 2012, 2013).

Additionally, second-generation feedstock resources such as straw of rice, wheat, and many other available lignocellulosic biomasses need to be used for ethanol production and would be promoted for achieving the blending target (Prasad et al. 2018, 2020a). The residue is burned after harvesting of crops, which leads to loss of organic matter from the soil. Biomass burning in the agricultural field generates many environmental problems (Wyman 1996). Even today, biomass is a vital energy resource, contributing approximately 10% of the world's entire main energy supply. Additionally, it may deliver a unique effect in meeting the growing power need sustainably (Prasad et al. 2012; IPCC 2014; IEA 2014; Szarka et al. 2017).

In 2008, the Roundtable for sustainable biofuels production released its standards for sustainable biofuels which mainly focused on (i) Biofuel projects intend to be designed and operated in a participatory mode that involves entire relevant stakeholders during planning and monitoring. (ii) Biofuel making intends to follow national laws and international treaties such as air quality, water resources, farming practices, and labor conditions. (iii) Biofuel means to decrease GHGs as compared to fossil fuels significantly that endeavors to set a standard methodology for examining Greenhouse Gases' (GHGs) benefits. (iv) Air pollution means to be declined along with the supply chain. (v) Biofuels production shall not violate human or labor rights and shall ensure decent work and the well-being of workers. (vi) Biofuels production shall not impair food security. (vii) Biofuels production means to avoid adverse impacts on biodiversity, ecosystems, and areas of high conservation value. (viii) Biofuels production shall promote practices that improve soil health and minimize degradation. (ix) Surface and groundwater use would be optimized, and contamination or depletion of water resources would be reduced. (x) Biofuels production shall contribute to the social and economic development of local, rural, and indigenous peoples and communities (xi) Biofuels intend to be cost effective, with a commitment to improving production efficiency, the socio-environmental enactment in all stages of the biofuel making value chain. (xii) Biofuel production shall not violate land rights (Schill 2008).

5.6 Biofuel Production Pathways

Biofuels from biomass resources are obtained in forms of liquid, gas, and solid. Among these, liquid biofuels are the most important in the current scenario of the world transportation sector. Biofuels can reduce/replace non-renewable petroleum fuels. The primary ways to make cellulosic biofuels are presented in Fig. 5.2.

5.6.1 Process of Thermochemical Conversion of Biomass to Biofuel

The biomass conversion to biofuel via the thermochemical process under controlled heating or oxidation is a promising alternative form of modern bioenergy. That covers direct combustion to produce heat, as well as gasification and pyrolysis, to produce gaseous, liquid fuel and precursors to upgrading advanced liquid fuels and electricity (Balat and Kırtay 2010). Utilizing biomass resources for bioenergy through the thermochemical process is considered as modern bioenergy and a valuable part of its future energy mix (Kour et al. 2019; Rastegari et al. 2020; Yadav et al. 2020). Direct biomass combustion is an old practice since ancient times. Combustion involved the generation of heat as a result of the oxidation of C- and H-rich biomass into CO_2 and H_2O (Balat 2009). In developing nations, underlying combustion technology is the open, three-stone biomass fire that is used for cooking food or heating water. Animal waste is also burned underneath a cooking pot supported by the stones. The efficiency of these methods is destitute at roughly 15%, and its users are straight exposed to smoky gaseous pollutants containing CO, SPM, NO_2, and CH_4.

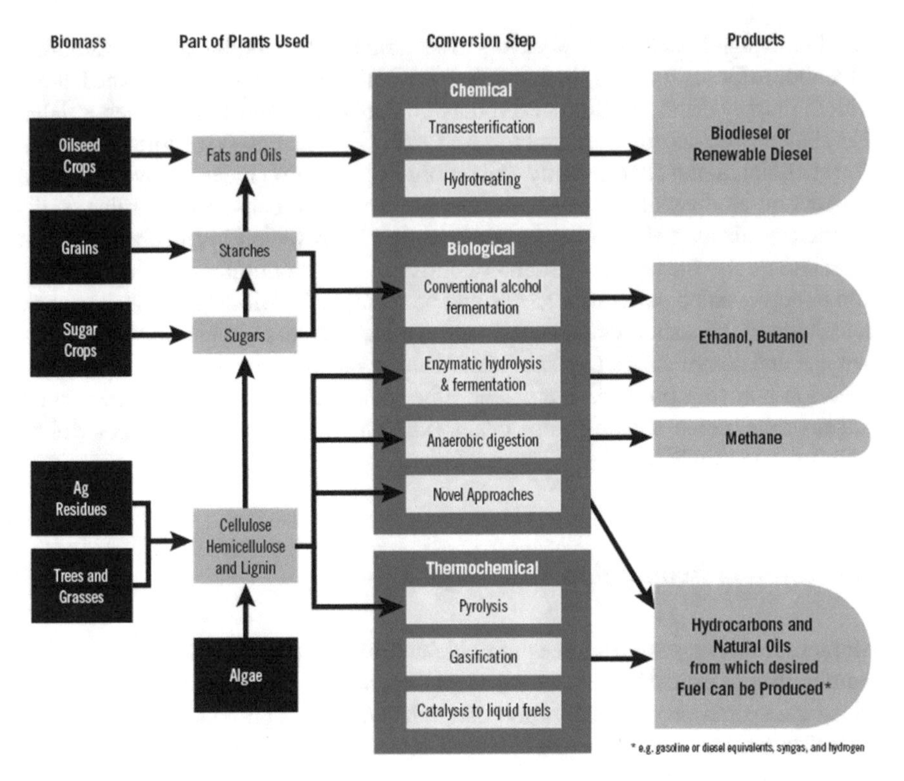

Fig. 5.2 First- and second-generation biofuel pathways. *Source* Pena and Sheehan (2007). Center for Climate and Energy Solutions

Pyrolysis is primarily the thermal decomposition of biomass to bioenergy under inert airy conditions or a limited supply of air. The thermal breakdown of organic ingredients in biomass begins at 350–550 °C and reaches up to 700–800 °C in the absence of O_2. During the process, the long chains of C, H, and O chain in biomass turned into small molecules in very heterogeneous gaseous, liquid, and solid intermediates form. The pyrolytic bio-oil is a heterogeneous mixture of high O_2 content and resembles a very viscous tar, which can be upgraded to fuels or chemicals. Solid product (char) is produced in reactors during pyrolysis processes, which is used as biofuel or applied as a soil amendment (Sánchez et al. 2009).

Gasification is an effective means of producing green power (Pavlas et al. 2010). This process is an exothermic partial oxidation of biomass with optimized conditions for higher yields of syngas or producer gas which contains CO, H_2, CH_4, and CO_2 (Hickman and Schmidt 1993). These gaseous products are used to run a diesel engine in dual fuel mode with minimal changes in the air inlet. The major challenge of gasification is the managing of higher molecular weight volatiles that condense into tars. The tars are considered as potential air pollutants. The recent advancements have made it possible to operate a spark-ignited engine using gas alone. An alternator is attached to the engine for electric generation that facilitates it to local consumption or for grid synchronization. Biomass gasification based power generation offers a solution for producing off-green grid power to provide electricity at smaller scales, especially in remote areas and hilly terrains of India (EAI 2012).

5.6.2 *Process of Biochemical of Biomass to Biofuel*

The biochemical process of biomass to biofuel offers the most secure and eco-friendly way to sustainability. Raw matters comprising sugars that can be transformed into sugars are used for ethanol fermentation. The raw materials containing sugars are classified as (i) sugary juices and molasses (ii) starchy material grains (iii) polysaccharides-lignocellulosic biomass residues (Prasad et al. 2007, 2020a). Sugary juices and molasses from sugarcane, sugar beet, and sweet sorghum can be directly fermented to produce ethanol (Prasad et al. 2009). Starchy materials need the least costly pretreatment, whereas lignocellulosic materials need expensive pretreatment to convert it into ethanol (Prasad et al. 2012, 2020b). The main aim of Indian agriculture, however, would remain to satisfy the food demands of the ever-increasing population. We have to explore options to simultaneously meet the ethanol requirements for the transport sector (Prasad et al. 2014, 2020a).

Lignocellulosic biomass, as a renewable feedstock, has been extensively examined for ethanol production. It is estimated that around 73.9 Tg dry waste crop residues produced worldwide could generate 49.1 GL year^{-1} of ethanol, nearly 16 times higher than the current world ethanol production. This amount of ethanol has the potential to replace 353 GL of gasoline, which is 32% of the global gasoline consumption. Making use of one-third of the 189 Mt of surplus biomass will yield ethanol nearly 19 billion liters of petrol equivalent, which is the equivalent to India's entire annual petrol consumption.

Chemically, biodiesel is a monoalkyl ester of long-chain fatty acids obtained from renewable lipid or oil by transesterification process. Biodiesel produced from edible and non-edible oil or fat is a suitable substitute for diesel. Worldwide many projects have been launched to produce biodiesel through non-edible oil transesterification. In India, Aatmiya Biofuels Pvt. Ltd., Gujarat has a biodiesel creation capacity of 1000 L/day from Jatropha curcas. Southern Online Biotechnologies (P) Ltd., Andhra Pradesh industrial biodiesel plant is in progress, which is planned with an initial 30 tons capacity and expandable to 100b tons of biodiesel per day.

Hydrogen (H_2) is sustainable energy produced from biomass sources and can be used as a substitute for fossil fuels (Kumar et al. 2019). It is produced via various processes such as water electrolysis, biomass gasification, and photo-bio fermentation. Presently, H_2 is formed solely via steam reformation of methane and water electrolysis. The thermocatalytic process is also used to produce H_2 via steam reforming, supercritical water partial oxidation, and biomass gasification. In the last decade, the defining economics issues of H_2 have improved histrionically. However, refineries presently become exclusive H_2 consumers to reduce pollutants and encounter Indian environmental compliant and regulations. The roadmap envisages taking up of H_2

Table 5.2 World fuel ethanol production by Country/Region (Million Gallons) (2014–2017)

World rank (2017)	Country/Region	2014	2015	2016	2017
1	United States	14,313	14,807	15,329	15,800
2	Brazil	6,190	7,093	7,295	7,060
3	European Union	1,445	1,387	1,377	1,415
4	China	635	813	845	875
5	Canada	510	436	436	450
6	Thailand	310	334	322	395
7	Argentina	160	211	264	310
8	India	155	211	225	280
*	Rest of World	865	391	490	465

Source RFA analysis of public and private data sources (2018)

energy technologies research and development activities in various sectors and forecasted goal line of one million H_2-fuelled vehicles and 1000 MW H_2-based energy generation capacity to be established in the country by 2020 (Nouni 2012).

Biogas technology is a sustainable and efficient process to convert organic wastes into clean bioenergy, which provides excellent opportunities to reduce GHG emissions. Biogas can be generated from biodegradable resources such as cattle dung wastes, biomass from farms, gardens, including kitchen wastes via anaerobic digestion. Biogas is a mixture of about 60% CH_4 and 40% CO_2 gas (Prasad et al. 2017). CH_4 is a combustible ingredient of biogas. It is combusted directly as a source of heat for cooking or used for internal combustion engines for various applications. The biogas technology is the most suitable option for families having cattle and other wastage feed material. This technology is the opportunity to become self-dependent on cooking gas and highly enriched organic fertilizer. It also offers the solution to protect the families from indoor air pollution problems and saving on the refilling cost of LPG cylinders.

5.6.3 World Liquid Biofuel Production by Country or Region

Liquid biofuels used in automobiles are ethanol and biodiesel as a substitute for petrol and diesel fuel, respectively. The world's top ethanol producers in 2017 were the US with 115,800 Million Gallons and Brazil with 7,060 Million Gallons, accounting together for 84% of world production of 27,050 Million Gallons (Table 5.2). Significant incentives, coupled with various industry progress initiatives, are giving rise to fledgling ethanol enterprises in countries such as China, Canada, Thailand, Argentina, and India (RFA 2018). The countries' shared global ethanol production in 2017 is presented in Fig. 5.3.

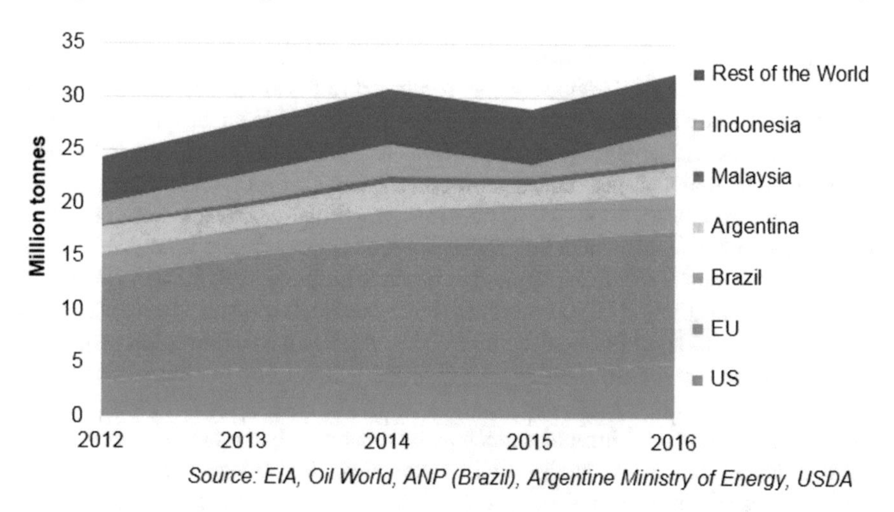

Fig. 5.3 Biodiesel production in leading countries of world

Biodiesel includes biofuel (methyl ester generated from vegetable or animal fat), bio-dimethyl ether (dimethyl ether generated from biomass), Fischer Tropsch (bio-oil generated from biomass). They can be blended with or used straight (unblended) as automotive diesel. Figure 5.3 shows that global biodiesel production reached record levels in 2016, following a dip in 2015. The EU and the US are responsible for a sizeable chunk (roughly 56%) of global biodiesel production. As a result, policy decisions in these regions will have a bearing on both sectors as a whole and overall vegetable oil demand.

5.7 Greenhouse Gas Management by Biofuels

There are several ways to manage GHG and CO_2 emission reduction. Among many of the options, a clean energy alternative such as biofuel is an excellent way to limit the greenhouse effect. Many efforts have been taken worldwide in this direction. According to Kartha (2006), to reduce carbon emissions through biofuels, two approaches are considered. Firstly, across the life cycle, biofuels produced from plants absorb as well as liberate Carbon (C) from the atmosphere pool without adding any Carbon (C) in contrast to fossil fuels. Secondly, they replace fossil fuels use through their blending. However, biofuels generation does, in utmost cases, involve, to some extent, the burning of non-renewable fuels.

Sugarcane- or molasses-based ethanol and oil-based biodiesel may realize about 70–100% decline in GHG emissions as compared to fossil fuel. In contrast, grain-based ethanol showed a moderate decline or flat increases in the situation of inefficient initial practices with ethanol production from corn (Dufey 2006). However, the high usage of fertilizers and mechanical farm operation during production is linked with more emissions (Peters 2006). GHG balances vary broadly between crops and sites, depending on feedstock production approaches, conversion know-hows, and its use. Resources such as N fertilizers applied to produce biofuels from biomass may have a varied concentration of GHG emissions. Furthermore, it may vary from one region to another (Hanaki and Portugal-Pereira 2018). Studies have shown that producing first- and second-generation biofuels can reduce 20–60% GHG relative to fossil fuels (de Jong et al. 2017).

A net emission GHG from biofuels may be theoretically equal to zero because C emitted during its burning is absorbed by plants' photosynthesis. In many studies, first-generation biofuels have been confirmed to decrease net emission by 20–60% of CO_2eq as compared to fossil-based fuels. Commercial second-generation biofuels may decrease 70–90% of CO_2eq as compared to fossil fuels. The lignocellulosic biomass utilization for biofuel is anticipated to succeed in higher C saving and sequestration as compared to cereal grain based starch and sugar-based biofuel generation (Tilman et al. 2006).

Many investigations have shown that biodiesel production from Jatropha and its use can decrease nearly 8–88% GHG emissions as compared to diesel. An investigation by Francis et al. (2005) has predicted a CO_2 sequestration potential in Jatropha curcas biomass of 4.6 and 22.9 Mt yr^{-1} if 2 and 10 Mha of wasteland in India is used for Jatropha cultivation. They have also assessed a mean annual C-sequestration rate of 2.25 CO_2 tons ha^{-1} $year^{-1}$ from wastelands cultivated with Jatropha curcas. In many investigations, the potential of microalgae for carbon sequestration is also assessed using nutrients from industrial effluents and wastewater streams. Sahoo et al. (2012) summarized that the use of macroalgae could add an average of 0.26×10^6 tons C into the harvested microalgal biomass annually. Therefore, these biomass sources can be utilized for biofuel generation and climate change abatements.

Larson (2006) have estimated GHG emissions on a well-to-wheel basis from various fossil fuel, including first-generation biofuels (ethanol from sugar and starch-based feedstocks, biodiesel from oilseeds) and selected second-generation biofuels obtained from cellulosic biomass (ethanol and diesel or bio-oil from the process of Fischer–Tropsch). Biofuels were observed to have the high potential to decrease entire life-cycle GHG emissions associated with the whole fuel supply chain. Second-generation biofuels (with life-cycle GHG emissions between –10 and 38 g CO_2eq/MJ) were recorded, which present immense mitigation potential over first-generation biofuels (with entire life-cycle GHG emissions between –19 and 77 g CO_2eq/MJ) as compared to 85–109 g CO_2eq/MJ for fossil fuels (Larson 2006). Figure 5.4 displays a range of GHG emissions which decreases in per km from the vehicle (v-km).

The third-generation biofuel feedstock, mainly microalgae, is well known to produce a renewable and green fuel source that can help to mitigate climate change

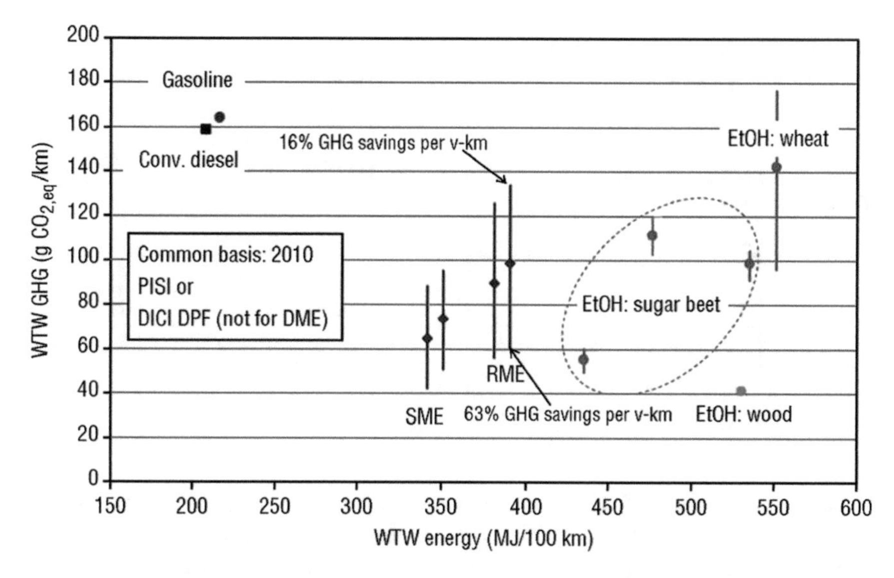

Fig. 5.4 A range in decreases of GHG emissions by biofuel versus gasoline and diesel. *Note* EtOH = Ethyl Alcohol (ethanol); SME = Soy Methyl Ester; RME = Rape Methyl Ester; PISI = Port Injection Spark Ignition; DICI DPF = Direct Injection Compression Ignition with Diesel Particulate Filter

impact (Patil et al. 2008). Microalgae can be grown everyplace as they do not need arable land for cultivation and can be harvested any time throughout the year (Williams et al. 2007). It provides non-toxic and extremely biodegradable biofuels. Various types of research programs are going on to increase the biofuel rate by improving the performance of algal species through molecular engineering. In contrast to other biofuel crops, the microalgae-originated biofuel is recognized as more eco-friendly due to its high fuel transformation rate.

5.8 Biofuels—Carbon Cycle, Net Energy Balances

The carbon cycle is directly connected with terrestrial biomass production because it is absorbed by plant biomass through photosynthesis for their growth and development. Biofuel is produced from biomass; its real benefits depend on its net energy contents and C-balances, which can be realized through corresponding fossil fuel GHG emissions and C savings by its use as an alternative fuel. A study is directed by the CII (Indian Industry), for particular biofuel groups to estimate net energy contents and C-balance. The results are summarized in Table 5.3. The results data showed that the Jatropha curcas based biodiesel has the highest net energy and carbon balance annually. During biodiesel production, the co-products (seed coat, de-oiled cakes, and glycerol after transesterification) have significantly contributed

Table 5.3 Net energy balance and carbon balance for selected categories of biofuels

Biofuel Type	Feedstock	Net energy ratio	Net energy balance (GJ/kl)	Net carbon balance (tCO2e/kl)	% carbon emission reduction (%)
Ethanol	Molasses	4.57	19.11	−1.1	75
	Sweet Sorghum	7.06	21.57	−1.4	86
	Cellulosic (Bagasse)	4.39	25.41	−1.7	70
	Cellulosic (Rice straw)	3.32	22.79	−1.6	68
Biodiesel	Jatropha—Transesterification	53.41	63.76	−4.0	30
	Jatropha—SVO	4.38	66.73	−4.5	50

Source CII (2010)

to biodiesel, nearly half of the entire biodiesel produced throughout the end-use stage. Juice-based ethanol production from sweet sorghum stood to have the highest transformation efficacy regarding output energy to input energy (CII 2010).

5.9 A Recent Case Study on Advanced Biofuels

California Air Resources Board and the US-EPA categorize biodiesel as an advanced biofuel, competent in reducing GHG emissions by at least 50% equivalents to fossil fuel. Blending of 20% biodiesel and 80% Low-Sulfur Diesel fuel (ULSD) can help to reduce GHGs which is also extensively supported by auto-engine makers and manufacturers. The use of biodiesel with ULSD diesel can decrease GHG emissions by 50–85% without any investments. According to the latest data obtained from the Low Carbon Fuel Standard (LCFS) program authorized by the California Air Resources Board (CARB), biodiesel delivers the state's most notable decrease in transport-related GHG emissions (Sacramento 2019). In the year 2018, the application of biodiesel in California removed 4.3 million tons (Mt) of CO_2, higher than the decreases brought by ethanol. Subsequently, in 2011, the LCFS program was launched; till now, biodiesel has removed more than 18 Mt of CO_2 (Fig. 5.5).

Domestically produced biodiesel provides a cost-effective fuel to customers allowing them to fleets without any modification in existing vehicles. It can help to reduce GHG emissions extensively. According to recent reports, bio-based diesel decreases around 20 million metric tons CO_2 in California annually. That signifies a win–win situation for the community to protect their environment and dependency on energy security (Sacramento 2019).

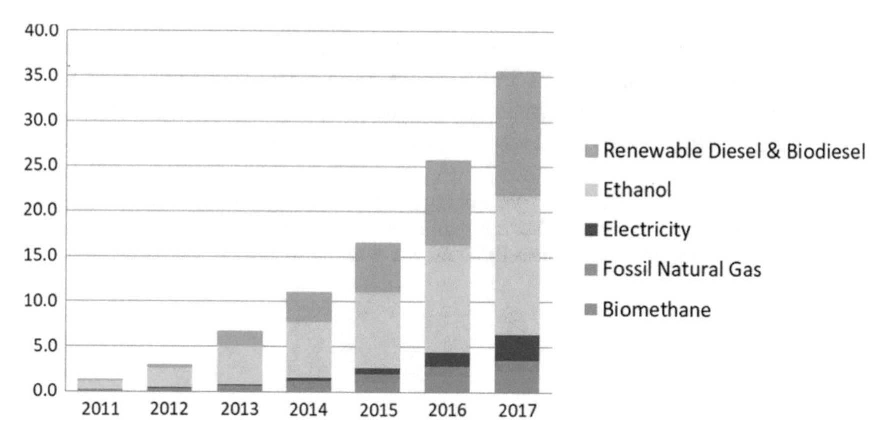

Fig. 5.5 Cumulative CO_2 reductions in million tons. *Source* California Energy Commission, Low Carbon Fuel Standard Dashboard Diesel *Technology Forum*

5.10 Potential of Biofuels to Mitigate Climate Change

Biofuels have gained massive attention for several reasons, one of which is their potential to reduce GHG emissions from the transportation sector (Prasad et al. 2007). The fossil energy balance of a biofuel, feedstock characteristics, conversion process production location, and agricultural practices are vital in terms of their contribution to reduce GHG emissions. It is considered that the replacement of fossil fuels with biofuel would have significant and positive climate change impacts by lowering GHGs in the atmosphere. Biofuel crops can decrease or offset GHG emissions by directly extracting CO_2 from the air as they grow, and store it in biomass and soil. Direct or indirect land use changes can also emit GHGs. For instance, while maize is used to produce ethanol, it can make GHG savings of nearly 1.8 tonnes of CO_2 ha^{-1} $year^{-1}$. Switchgrass, a second-generation energy crop, can make GHG savings of 8.6 t ha^{-1} $year^{-1}$; the change of grassland to produce those crops releases 300 t ha^{-1}, and transformation of forestland can release 600–1000 t ha^{-1} (Fargione et al. 2008; Searchinger et al. 2008).

5.11 Conclusion and Future Prospects

Energy is an essential part of the overall growth and progress of any society and country as a whole. However, fossil fuel use has triggered the enhancement of potential greenhouse gases, which is responsible for global warming and many other environmental issues. Today, the world is facing challenges to mitigate climate change and crude oil price fluctuations due to instability in geopolitics. These circumstances have forced to find clean and alternative sources of energy. Biofuel has been identified as potential fuel to restrict the worsening of human-induced climate. Now it has been accepted as a substitute for fossil fuels, which can help to reduce GHG emissions from transport sectors. Furthermore, biofuel can fulfill the objectives of the Kyoto Protocol and other climate change initiatives. Biofuel can also act to serve in the transition from fossil fuel based economy to a sustainable bio-based economy.

Acknowledgments The authors are grateful to ICAR-Indian Agricultural Research Institute (IARI), New Delhi, and the Indian Council of Agricultural Research for providing facilities and financial support to undertake these investigations. There are no conflicts of interest.

References

Balat M (2009) Gasification of biomass to produce gaseous products. Energy Sour Part A 31:516–526

Balat H, Kırtay E (2010) Hydrogen from biomass–present scenario and future prospects. Int J Hydrogen Energy 35:7416–7426

Biofuels Digest newsletter (2019) Indian oil company to invest nearly $3.5 billion in green projects including 2G and 3G ethanol. https://www.biofuelsdigest.com/bdigest/tag/green-projects/

CII (2010) Estimation of energy and carbon balance of biofuels in India. Confederation of Indian Industry. http://www.cii.in/webcms/Upload/Energy_Balance.pdf

de Jong S, Hoefnagels R, Wetterlund E, Pettersson K, Faaij A, Junginger M (2017) Cost optimization of biofuel production–the impact of scale, integration, transport, and supply chain configurations. Appl Energy 195:1055–1070

Dufey A (2006) Biofuels production, trade, and sustainable development: emerging issues. In: Dufey A (ed) Environmental economics program/sustainable markets group discussion paper 2. International Institute for Environment and Development, London, UK, pp 1–17

EAI (2012) India Biomass Gasification Power Production, Entrepreneurs Association of India. http://www.eai.in/ref/reports/biomass_gasification.html

Fargione J, Hill J, Tilman D, Polasky S, Hawthorne P, Jason (2008) Land clearing and the biofuel carbon debt. American Association for the Advancement of Science New York, Washington DC 319:1235

Francis G, Edinger R, Becker K (2005) A concept for simultaneous wasteland reclamation, fuel production, and socio-economic development in degraded areas in India: Need, potential, and perspectives of *Jatropha* plantations. Nat Res Forum 29:12–24

Giuntoli J, Agostini A, Caserini S, Lugato E, Baxter D, Marelli L (2016) Biomass and bioenergy climate change impacts of power generation from residual biomass. Biomass Bioenerg 89:146–158

Hanaki K, Portugal-Pereira J (2018) The effect of biofuel production on greenhouse gas emission reductions. In: Takeuchi K, Shiroyama H, Saito O, Matsuura M (eds) Biofuels and sustainability: holistic perspectives for policy-making. Springer Japan, Tokyo, pp 53–71

Hausfather Z (2018) Carbon brief. https://www.carbonbrief.org/analysis-fossil-fuel-emissions-in-2018-increasing-at-fastest-rate-for-seven-years

Hickman D, Schmidt L (1993) Production of syngas by direct catalytic oxidation of methane. Science 259:343–346

IEA (2014) Energy technology perspectives, Paris: International Energy Agency, organization for economic co-operation and development. http://www.iea.org/etp/

IPCC (2014) Climate change: synthesis report, contribution of working groups I, II, and III to the fifth assessment report of the Intergovernmental Panel on Climate Change [Core Writing Team, RK Pachauri, LA Meyer (eds.)]. IPCC, Geneva, Switzerland, 151

IPCC (2018) Summary for Policymakers of IPCC Special Report on Global Warming of 1.5 °C approved by governments. https://www.ipcc.ch/2018/10/08/summary-for-policymakers-of-ipcc-special-report-on-global-warming-of-1-5c-approved-by-governments/
Jonsson DK, Johansson B, Månsson A, Nilsson LJ, Nilsson M, Sonnsjö H (2015) Energy security matters in the EU energy roadmap. Energy Strategy Rev 6:48–56
Joos F, Roth R, Fuglestvedt J, Peters G, Enting I andBloh W et al (2013) Carbon dioxide and climate impulse response functions for the computation of greenhouse gas metrics: a multi-model analysis. Atm Chem Phys 13:2793–2825
Kartha S (2006) Environmental effects of bioenergy, Brief 4, Bioenergy, and agriculture: promises and challenges. IFPRI 2020 vision, Focus 14, Washington, International Food Policy Research Institute, DC
Kour D, Rana KL, Yadav N, Yadav AN, Rastegari AA, Singh C et al (2019) Technologies for biofuel production: current development, challenges, and future prospects. In: Rastegari AA, Yadav AN, Gupta A (eds) Prospects of renewable bioprocessing in future energy systems. Springer International Publishing, Cham, pp 1–50. https://doi.org/10.1007/978-3-030-14463-0_1
Kumar S, Sharma S, Thakur S, Mishra T, Negi P, Mishra S et al (2019) Bioprospecting of microbes for biohydrogen production: current status and future challenges. In: Molina G, Gupta VK, Singh BN, Gathergood N (eds) Bioprocessing for biomolecules production. Wiley, USA, pp 443–471
Larson ED (2006) A review of life-cycle analysis studies on liquid biofuel systems for the transport sector. Energy Sustain Develop 10:109–126
Lindsey R (2019) Climate Change: Atmospheric CO_2. https://www.climate.gov/news-features/understanding-climate/climate-change-atmospheric-carbon-dioxide
Liu C, Li F, Ma LP, Cheng HM (2010) Advanced materials for energy storage. Adv Mater 22:28–62
MNRE (2009) *GOI national policy on biofuels.* http://mnre.gov.in/file-manager/UserFiles/biofuel_policy.pdf%3e
Nouni MR (2012) Hydrogen energy and fuel cell technologies. Ministry New Renew Energy, GOI AkshayUrja 5(5):10–15
Patil V, Tran K-Q, Giselrød HR (2008) Towards sustainable production of biofuels from microalgae. Int J Mol Sci 9:1188–1195
Pavlas M, Stehlík P, Oral J, Klemeš J, Kim JK, Firth B (2010) Heat integrated heat pumping for biomass gasification processing. Appl Thermal Eng 30:30–35
Pena N, Sheehan J (2007) Biofuels for transportation. In: CDM investment newsletter. https://www.c2es.org/site/assets/uploads/2007/11/cdm-investment-newsletterbiofuelstransportation.pdf
Peters U (2006) Biofuels for transportation: global potential and implications for agriculture and sustainable energy in the 21st century. In: InBiofuels Conference (2006), pp 16–17
PIB (2018) Press information bureau of India, government cabinet, cabinet approves national policy on biofuels (2018). http://pib.nic.in/newsite/PrintRelease.aspx?relid=179313
Prasad S, Dhanya MS, Gupta N, Kumar A (2012) Biofuels from biomass: a sustainable alternative to energy and environment. Biochem Cell Arch 12:255–260
Prasad S, Joshi HC, Jain N, Kaushik R (2006) Screening and identification of forage sorghum (*Sorghum bicolor*) cultivars for ethanol production from stalk juice. Indian J Agric Sci 76:557–560
Prasad S, Kumar A, Muralikrishna KS (2013) Assessment of ethanol yields associated character in sorghum biomass. Maydica 58:299–303
Prasad S, Kumar A, Muralikrishna KS (2014) Biofuels production: a sustainable solution to combat climate change. Indian J Agric Sci 84:1443–1452
Prasad S, Kumar S, Yadav KK, Choudhry J, Kamyab H, Bach QV et al (2020a) Screening and evaluation of cellulytic fungal strains for saccharification and bioethanol production from rice residue. Energy 25:116422
Prasad S, Lata Joshi HC, Pathak H (2009) Selection of efficient *Saccharomyces cerevisiae* strain for ethanol production from sorghum stalk juice. Curr Adv Agric Sci 1:70–72
Prasad S, Malav MK, Kumar S, Singh A, Pant D, Radhakrishnan S (2018) Enhancement of bio-ethanol production potential of wheat straw by reducing furfural and 5-hydroxymethylfurfural (HMF). Bioresour Technol Rep 4:50–56

Prasad S, Rathore D, Singh A (2017) Recent advances in biogas production. Chem Engin Process Tech 3:1038

Prasad S, Singh A, Joshi HC (2007) Ethanol as an alternative fuel from agricultural, industrial, and urban residues. Resources. Conserv Recycl 50:1–39

Prasad S, Singh A, Korres NE, Rathore D, Sevda S, Pant D (2020b) Sustainable utilization of crop residues for energy generation: A Life Cycle Assessment (LCA) perspective. Bioresour Technol 303:122964

Rastegari AA, Yadav AN, Yadav N (2020) New and Future Developments in Microbial Biotechnology and Bioengineering: Trends of Microbial Biotechnology for Sustainable Agriculture and Biomedicine Systems: Diversity and Functional Perspectives. Elsevier, Amsterdam

Rastegari AA, Yadav AN, Gupta A (2019) Prospects of renewable bioprocessing in future energy systems. Springer International Publishing, Cham

RFA (2018) Renewable Fuels Association (RFA) analysis of public and private data sources (2018) https://ethanolrfa.org/resources/industry/statistics/

Rodrigues A, Bordado JC, Santos RG (2017) Upgrading the glycerol from biodiesel production as a source of energy carriers and chemicals a technological review for three chemical pathways. Energies 10:1–36

Ruiz V, Boon-Brett L, Steen M, van den Berghe L (2016) Putting science into standards - driving towards decarbonization of transport: safety, performance, second life, and recycling of automotive batteries for e-Vehicles. https://ec.europa.eu/jrc/sites/jrcsh/files/jrc104285_jrc104285_final_report_psis_2016_pubsy_revision.pdf/

Sacramento C (2019) Bio-based diesel fuels deliver the biggest reductions in transportation-related GHGS in California–ever. https://finance.yahoo.com/news/bio-based-diesel-fuels-deliver-100500080.html

Sadik MW, El Shaer HM, Yakot HM (2010) Recycling of agriculture and animal farm wastes into compost using compost activator in Saudi Arabia. J Int Environ Appl Sci 5:397–403

Sahoo D, Elangbam G, Devi SS (2012) Using algae for carbon dioxide capture and biofuel production to combat climate change. Phykos 42:32–38

Sánchez M, Lindao E, Margaleff D, Martínez O, Morán A (2009) Pyrolysis of agricultural residues from rape and sunflowers: production and characterization of bio-fuels and biochar soil management. J Anal Appl Pyrol 85:142–144

Schill SR (2008) Roundtable for Sustainable Biofuels releases proposed standards for review. Biomass Magazine

Searchinger T, Heimlich R, Houghton RA, Dong F, Elobeid A, Fabiosa J et al (2008) Use of US croplands for biofuels increases greenhouse gases through emissions from Land-Use change. Science 319:1238–1240

Sheetal KR, Prasad S, Renjith PS (2019) Effect of cultivar variation and *Pichia stipitis* NCIM 3498 on cellulosic ethanol production from rice straw. Biomass Bioeng 127:105253

STAP (2015) Optimizing the global environmental benefits of transport biofuels. Scientific and technical advisory panel of the global environment facility, Washington, D.C. Authored and edited by Bierbaum R, Cowie A, Gorsevski V, Sims R, Rack M, Strapasson A, Woods J (Imperial College, London) and Ravindranath N (Indian Institute of Science, Delhi)

Szarka N, Eichhorn M, Ronny K, Alberto B, Daniela T (2017) Interpreting long-term energy scenarios and the role of bioenergy in Germany. Renew Sust Energ Rev 68:1222–1233

Tan HW, Aziz AR, Aroura MK (2013) Glycerol production and its applicationsas a raw material: a review. Renew Sustain Energy Rev 27:118–127

Tilman D, Hill J, Lehman C (2006) Carbon-negative biofuels from low-input high-diversity grassland biomass. Science 314:1598

UN AR5 climate science report (2019) UN science reports show biofuels are essential to climate change, https://www.climatechangenews.com/2019/08/08/un-science-reports-show-biofuels-essential-climate-action/

USDA (2016)Foreign Agricultural Service, Malaysia: Biofuels Annual. USDA Foreign Agricultural Service. https://www.fas.usda.gov/data/eu-28-biofuels-annual-0

USDA (2019) Foreign Agricultural Service, EU-28: Biofuel Mandates in the EU by Member State. USDA Foreign Agricultural Service. https://www.fas.usda.gov/data/eu-28-biofuel-mandates-eu-member-state-1
Williams C, Black I, Biswas T, Heading S (2007) Pathways to prosperity: second-generation biomass crops for biofuels using saline lands and wastewater. Agri Sci 21:28–34
Wyman CE (1996) In: Wyman CE (ed) Handbook on bioethanol: production and utilization. CRC press
Yadav AN, Rastegari AA, Yadav N (2020) Microbiomes of extreme environments: biodiversity and biotechnological applications. CRC Press, Taylor and Francis, Boca Raton, USA

Chapter 6
Photosynthetic Production of Ethanol Using Genetically Engineered Cyanobacteria

F. P. De Andrade, M. L. F. De Sá Filho, R. R. L. Araújo, T. R. M. Ribeiro, A. E. Silva, and C. E. De Farias Silva

Abstract The increasing global energy demand and the advance of new technologies to produce biofuel from CO_2 led to the expansion in research using genetically modified cyanobacteria as biocatalyst to produce ethanol. The expression of the enzymes pyruvate descarboxylase (PDC) and alcohol dehydrogenase (ADH) from *Zymomonas mobilis* in cyanobacteria is the main strategy used to redirect the carbon fixed by photosynthesis into ethanol. This chapter emphasizes the genetic modification used in metabolic engineering and their effects on ethanol production, as well as, the bottlenecks of the technology.

6.1 Introduction

The development of new Technologies to produce biofuels has intensified due to the issue of oil depletion and environmental concerns resulting from the emission of greenhouse gases (GHGs) by burning fossil fuels. The biological conversion of CO_2 to biofuels using photosynthetic microorganisms, such as cyanobacteria, has several advantages when compared to the conventional production of biofuels using vegetable biomass due to high growth rates, as well as the requirement for simple nutrients (namely water, solar light and CO_2), adaptable genetics and independence from fertile land for cultivation (Machado and Atsumi 2012; Silva and Bertucco 2016). Among the biofuels available, bioethanol stands out due to its low toxicity, as well as for being easily biodegradable, for the low emission of pollutants and for being

F. P. De Andrade · M. L. F. De Sá Filho · A. E. Silva · C. E. De Farias Silva (✉)
Chemical Engineering, Federal University of Alagoas, Maceió, Brazil
e-mail: eduardo.farias.ufal@gmail.com

R. R. L. Araújo
Federal Institute of Pernambuco, Campus Caruaru, Recife, Brazil

T. R. M. Ribeiro
Chemical Engineering, Federal University of Sergipe, São Cristóvão, Brazil

A. N. Yadav et al. (eds.), *Biofuels Production – Sustainability and Advances in Microbial Bioresources*, Biofuel and Biorefinery Technologies 11,
https://doi.org/10.1007/978-3-030-53933-7_6

more sustainable when compared to fossil fuels. In addition, there are different generations of bioethanol, depending on the raw material used and technology available (Kour et al. 2019b; Rastegari et al. 2020, 2019a).

The conventional process for the production of ethanol is based on the fermentation of sugar present in traditional crops, such as sugarcane (Brazil), corn (USA) and beet (Europe), denominated first generation bioethanol (saccharide and starchy raw materials). In turn, second generation biofuels are obtained from the hemicellulosic and cellulosic fractions of agro-industrial waste (lignocellulosic biomass, such as sugarcane bagasse, corn straw or wood). On the other hand, third generation bioethanol uses macroalgal and microalgal/cyanobacterial biomass as substrate for ethanolic fermentation. Finally, fourth generation bioethanol refers to the use of genetically modified cyanobacteria for bioethanol production.

Therefore, several recent studies have focused on biotechnology with cyanobacteria for the controllable and adjustable production of biofuels (such as biodiesel and bioethanol), in a low-cost form by using metabolic engineering and synthetic biology techniques (Singh et al. 2016). With this in mind, this Chapter aims at focusing on how genetically engineered cyanobacteria can be used in the production of ethanol (biochemical aspects), pointing out to the main works carried out and the main species used, as well as the main technological bottlenecks.

6.2 The Market and the Use of Bioethanol Fuel

The United States is the greatest worldwide producer of ethanol, with a production of approximately 16 billion gallons in 2017, followed by Brazil, which produced 7 billion gallons in the same year, and Europe (1.4 billion gallons), while the rest of the world produced approximately 3 billion gallons (AFDC 2019).

Since it is an alcohol, ethanol (CH_3CH_2OH) shows a polar fraction in its molecular fraction due to the hydroxyl radical and a non-polar fraction in its carbon chain, which explains why ethanol can be dissolved in both gasoline (non-polar) and water (polar) (Costa and Sodré 2010). Therefore, ethanol for fuel is mixed in gasoline in various proportions to be used in vehicles worldwide. For instance, in the USA, E10 consists of 10% ethanol and 90% gasoline, while E15 contains between 10.55% and 15% ethanol, and E85 (or flexible fuel) is a mixture of ethanol-gasoline, containing from 51% to 83% ethanol (AFDC 2019). In Brazil, 25–27% of anhydrous ethanol is mandatorily added to gasoline, with hydrous ethanol (E100) being used alone in engines especially developed for this purpose (ANP 2019). In turn, in Europe, 5.75% of ethanol is mandatorily added to gasoline for use in spark ignition vehicles, with E4.5 being used in the United Kingdom, while E5 is used in India, with New South Wales and Queensland, in Australia, using E6 and E3, respectively (Subramanian 2017).

6.3 Microalgae and Cyanobacteria

Cyanobacteria are prokaryote and photosynthesising microorganisms, part of the monophyletic taxon, with great morphological, genomic and metabolic diversity (Braakman 2019; Vijay et al. 2019). They were previously known as blue algae due to their similarity with microalgae, in terms of cell pigments and photosynthetic capacity (Hitzfeld et al. 2000). However, microalgae are eukaryotes, containing karyotheca and mitochondria, for instance (Molina et al. 2003). Microalgae and cyanobacteria are promising for chemical and biological applications, being capable of retaining 40% more carbon during photosynthesis when compared to higher plants (Pierobon et al. 2018). These microorganisms carry out a synthesis of various materials with commercial value from only one structure, depending on the condition factors, namely carbon dioxide concentration, nitrogen, phosphorus, pH, temperature and salinity. Furthermore, they are a sustainable alternative for the production of biofuels, nutrients and medications (Lau et al. 2015). Moreover, they can be applied in wastewater treatment (Silva et al. 2019). Cyanobacteria are able to grow in various ecosystems due to their robust physiological features, tolerating water at different salinity levels and various temperatures (Vijay et al. 2019), though more favourable growth conditions are possible in neutral-alkaline freshwater environments, with pH between 6 and 9, and temperatures between 15 and 30 °C (Paerl and Paul 2012).

These organisms are considered the first primary producers of organic matter to release elemental oxygen (Wang et al. 2016; Abramson et al. 2018). They use solar light to convert carbon dioxide (CO_2), nitrogen and (mainly) phosphorus into several products of interest for the chemical industry (Dvořák et al. 2015). Moreover, cyanobacteria contain chlorophyll *a* and various other accessory pigments, predominantly phycocyanin and allophycocyanin, with cyanophycean and lipids included as their reserve products. Cyanobacteria reproduce asexually by cell division or spore formation. Besides, cyanobacteria have two different photosystems (PS) in series, type I (PS I) and type II (PS II), which, in the presence of oxygen, has water as the usual proton-donor (Cohen et al. 1986; Stal 1995).

With cyanobacteria, photosynthesis is characterised by the light capture complex (chlorophyll and others), which absorbs solar light photons that are used by photosystem II for catalytic oxidation of water, while NADPH and ATP are the result of the PSs. NADPH and ATP are substrates of the cycle of Calvin-Benson-Bassham, or simply Calvin-Benson Cycle, in which CO_2 fixation takes place in the form of assimilated molecules which form sugars, lipids and other biomolecules required for cellular growth (Silva and Bertucco 2016). Furthermore, several cynobacterial lineages exhibit a heterotrophic metabolism in the absence of light, consuming organic molecules, such as sugar and organic acids (Mata et al. 2010).

They also have a diverse morphology, ranging from unicellular to colonial forms with irregular, radial or regular planes, filamentous or pseudoparenchymatous morphology, also varying from microscopic to macroscopic dimensions (Wanterbury 2006). Photosynthetic efficiency is higher when compared to terrestrial plants and algae, with rapid growth which enables the production of biomass (Wang et al.

2016; Abramson et al. 2018), besides a high CO_2 assimilation rate and the possibility of genetic treatment, which allows to obtain high-added value products. Therefore, cyanobacteria can grow on non-arable land, requiring less water, enabling their cultivation with saline water and wastewater in harsh conditions (Milano et al. 2016). The different cultivation environments lead to changes in the intracellular media, with the maintenance of operational conditions promoting the biosynthesis of compounds and influencing cell growth, as well as biomass composition and products of interest (Mata et al. 2010).

Two strategies are commonly used to favour desirable metabolic pathways, with the first being genetic mutation for the control of enzyme activity, which is specific of certain pathways and strains, while the second includes the use of nutrient-restricted media that lead to changes in the formation of the main metabolites (Rana et al. 2019; Rastegari et al. 2019b, c). Metabolism inhibition due to the restriction of nutrients can lead to issues in the process at an industrial scale (Abramson et al. 2018). Another important microbial feature to enable the use of cyanobacteria at an industrial scale includes growth rate which, when increased, can decrease bioprocessing costs and the risk of contamination (Vijay et al. 2019).

These microorganisms form a group of highly diversified organisms with an unexplored genetic potential, as various genes have not yet been identified. Furthermore, they are functionally diverse, despite their similar morphological features. Thus, these organisms have been explored in the development of cyanobacterial lineages for industrial applications with excellent yields (Silva and Bertucco 2016).

Over the past years, genetically modified cyanobacteria have been studied as an alternative to the production of glucose or sucrose, with some strains producing higher volumes than sugarcane (laboratory results), also enabling the recovery of a great fraction of protein (Smachetti et al. 2019). Their prokaryotic characteristics, with the absence of the nuclear envelope and greater diversity when compared to microalgae, facilitate genetic manipulations (Cohen et al. 1986; Chisti 2007). In addition, the *Synechocystis* sp. PCC 6803, *Synechococcus* sp. PCC 7002, *Synechococcus elongatus* and *Nostoc* sp. PCC 7002 lineages are the most used in genetic manipulation studies (Lea-Smith et al. 2017).

6.4 Methods for Obtaining Ethanol from Microalgae and Cyanobacteria

There are three methods for obtaining ethanol from microalgae/cyanobacteria, as follows: biomass hydrolysis and fermentation; dark fermentation and photofermentation. The first method involves the cultivation of biomass with high carbohydrate content (Silva and Sforza 2016; Silva et al. 2017, 2018a) (starch in the case of some microalgae and glycogen for cyanobacteria), followed by (chemical and/or enzymatic) hydrolysis for the conversion of monosaccharides and subsequent fermentation by microorganisms (Silva et al. 2018b, c). In turn, in dark fermentation there is

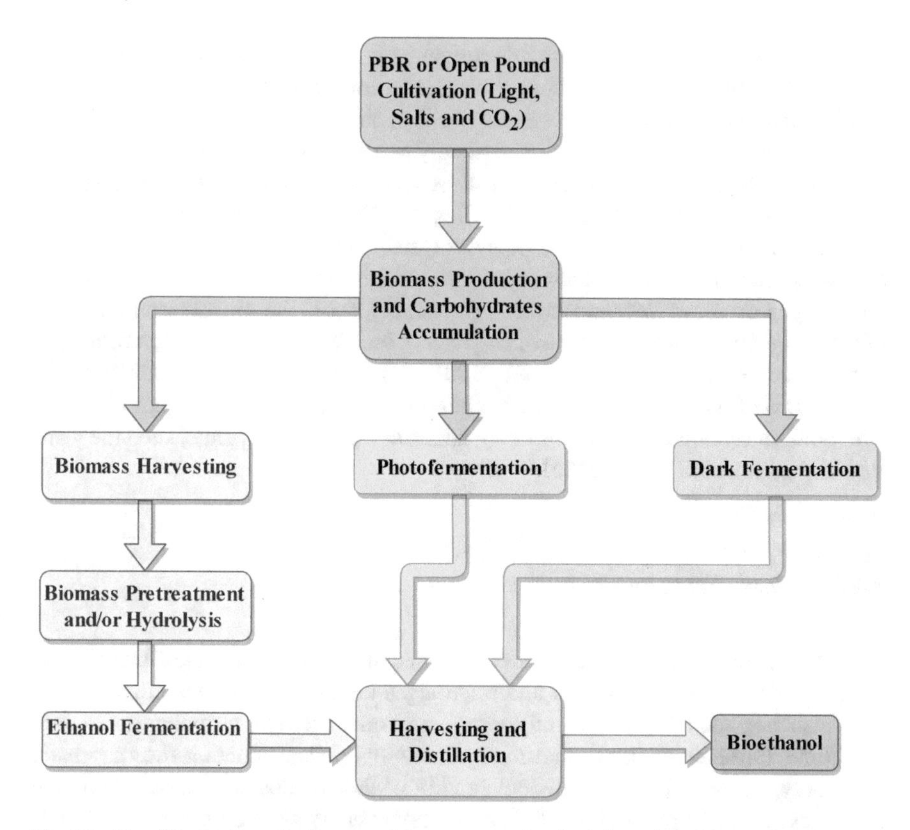

Fig. 6.1 Simplified schematic diagram of the processes for obtaining ethanol from microalgae and cyanobacteria *Source* Modified from Silva and Bertucco (2016)

an initial accumulation of sugars with the carbohydrate being then fermented, in the absence of light, producing acids and alcohols, including ethanol (Abo-Hashesh et al. 2011). As for photofermentation, cyanobacteria are genetically modified in order to carry out photosynthesis and, even in the presence of light, simultaneously produce ethanol and excrete it from the cell (Angermayr et al. 2009). The basic difference between these processes is illustrated in Fig. 6.1.

6.4.1 Hydrolysis and Fermentation

The main challenge faced with this method includes carrying out the conversion of complex sugars, present in biomass, in fermentable sugars. The hydrolysis of microalgae/cyanobacteria biomass produces a liquor of carbohydrates rich in simple sugars, fermentable monosaccharides, which is originated from the rupture of cell walls and energy reserves of these microorganisms (starch in the case of microalgae and glycogen in the case of cyanobacteria as aforementioned).

Chemical and enzymatic hydrolysis are the most common methods employed. The first one uses chemical products, mainly acids such as chloridric and sulfuric acids. The enzymatic method uses mostly cellulases, pectinases and amylases for the degradation of the polysaccharides (Silva and Bertucco 2017; Silva et al. 2018b, c). The simple sugars released in the hydrolysis phase can be easily converted into ethanol through fermenting microorganisms, which can be either bacteria, filamentous fungus or yeast. *Saccharomyces cerevisiae* yeast is the most commonly used microorganism for the conversion of sugars into bioethanol, given its high fermentation rates and ethanol yields (Jambo et al. 2016). However, pentoses (usually between 10 and 20% of the carbohydrates in biomass) constitute part of the liquor obtained, with other microorganisms strains being necessary to increase yields, such the yeast *Pichia stipitis* (Silva et al. 2018b).

6.4.2 Dark Fermentation

Dark fermentation consists of the conversion of organic substrates into biohydrogen. However, cyanobacteria and microalgae are able to produce ethanol from simple sugars, such as glucose and sucrose, through anaerobic fermentation in dark conditions (absence of light), that is, these microorganisms contain the enzymatic machinery, although in a less efficient condition when compared to other metabolic processes. Thus, the production of ethanol is favoured by the accumulation of carbohydrates in microalgae cells through photosynthesis, with microalgae/cyanobacteria being then forced to synthesize ethanol through a fermentative metabolism, when autotrophic/mixotrophic condition is changed to a dark condition (absence of light) (Silva and Bertucco 2016). Although possible, this process is not often used, as it is not highly efficient for bioethanol production (Cardoso et al. 2014).

6.4.3 Photofermentation

Photofermentation, or photanol, when applied to the production of ethanol, is the natural process of capturing solar light for the conversion of this energy into final fermentation products, through highly efficient metabolic pathways. Photanol is not only limited to the production of ethanol, but it is also used in a great amount of natural products resulting from glucose-based fermentation. Each stage has main factors which determine the efficiency of the process and the metabolic needs of cyanobacteria, as shown in Fig. 6.2. Nevertheless, this route requires the use of genetically modified microorganisms (Silva and Bertucco 2016; Rai and Singh 2016).

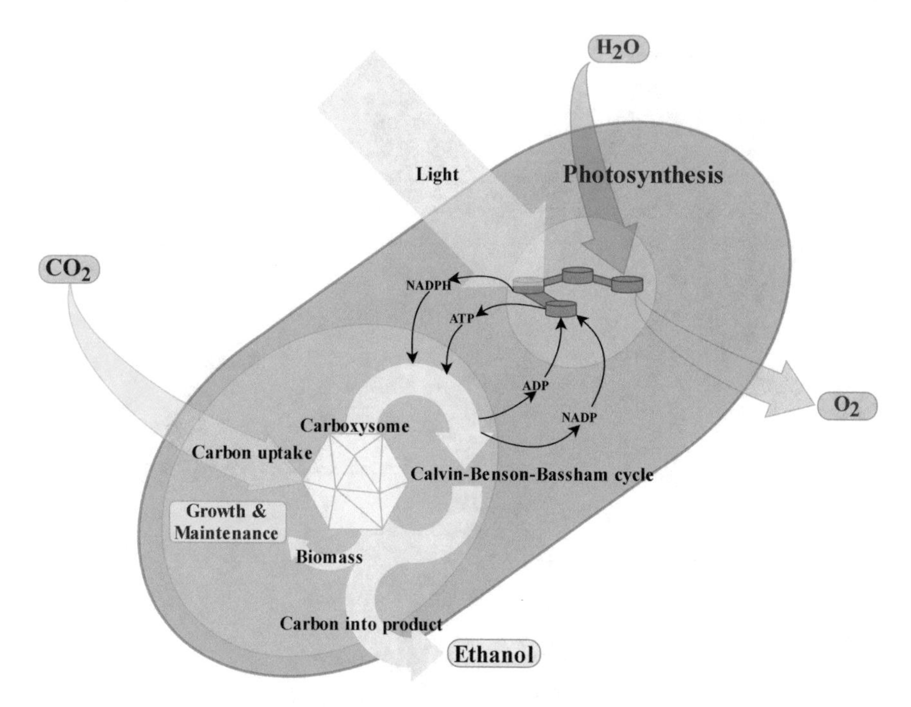

Fig. 6.2 Schematic diagram showing the process for obtaining ethanol from genetically modified cyanobacteria. *Source* Modified from Savakis and Hellingwerf (2015)

6.4.3.1 Photofermentation Biochemistry

The Calvin-Benson-Bassham (CCB) cycle is the main CO_2 pathway through photosynthesis. Inorganic carbon fermentation in ethanol with cyanobacteria consists in the uptake of CO_2 from the CCB through the enzyme RuBisCO (ribulose-1,5-bisphosphate carboxylase/oxygenase), present in carboxysome to carboxylate ribulose-1,5-bisphosphate (RuBP), forming two molecules of 3-phosphoglycerate, commonly known as glyceraldehyde-3-phosphate. Glyceraldehyde-3-phosphate, as a glucose intermediate, can be converted into pyruvate and, from the action of the enzymes PDC (pyruvate decarboxylase) and ADH (adenine dehydrogenase), produce ethanol. Alternatively, glyceraldehyde-3-phosphate can integrate the pentose phosphate pathway in the form of xylulose, ribose and erythrose, which can be converted into ribulose-5-phosphate—a CCB precursor. Finally, glucose can be converted into fructose-6-phosphate, then into dihydroxyacetone-phosphate which, in turn, can be sequentially converted into glyceraldehyde-3-phosphate and consequently into pyruvate to form ethanol through glycolytic pathways (Liang et al. 2018). Figure 6.3 provides an overview of the heterologous pathway added to cyanobacteria to produce ethanol.

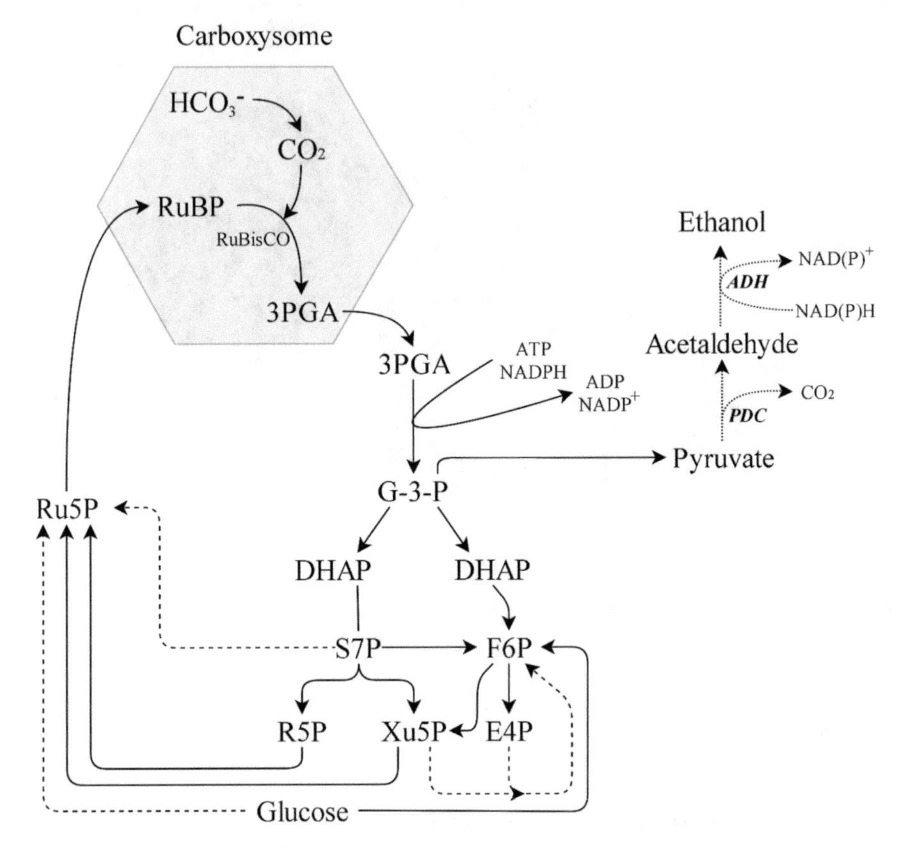

Fig. 6.3 Abbreviations of intermediates: RuBP, ribulose-1,5-bisphosphate; 3PGA, 3-phosphoglycerate; G-3-P, glyceraldehyde-3-phosphate; DHAP, dihydroxyacetone phosphate; F6P, fructose-6-phosphate; S7P, sedoheptulose-7-phosphate; E4P, erythrose-4-phosphate; Xu5P, xylulose-5-phosphate; R5P, ribose-5-phosphate; Ru5P, ribulose-5-phosphate. Abbreviations of enzymes: RuBisCO, Ribulose-1,5-bisphosphate carboxylase/oxygenase; PDC, pyruvate decarboxylase; ADH, alcohol dehydrogenase. *Source* Modified from Liang et al. (2018)

6.5 Production Process: Challenges and Opportunities

In several cases, the combined optimisation of abiotic (such as light intensity, organic carbon source, pH and growth, CO_2 concentration) and biotic factors (such as genetic engineering and synthetic biology) lead to greater efficiency and yield of fermentation processes (Liang et al. 2018). Cyanobacteria do not have a complete/efficient biosynthetic pathway for ethanol production (Kumar et al. 2019; Yadav et al. 2017). Thus, it is necessary to apply synthetic biology techniques in order to produce this biofuel, using cyanobacteria at an industrial scale and competitive prices (Singh et al. 2016). Photosynthetic cyanobacteria can be modified for an efficient ethanol production using the following approaches: the combination of gene transformation, strain/process development and metabolic modelling/profiling analysis (Pamar et al.

2011). Therefore, cyanobacterial metabolic pathways have been used to increase the yields of sugars, alcohols and other substances produced and excreted by these microorganisms (Frigaard 2018).

Photofermentation is carried out by photosynthetic bacteria that use solar light and biomass for production processes, namely cyanobacteria that can be genetically modified to metabolically convert metabolites of organic carbon into ethanol. As aforementioned, this process consists of two steps: photosynthesis and fermentation. In the presence of light, CO_2 is fixed through the Calvin-Benson cycle, forming phosphoglycerate. This sugar is then converted into pyruvate and, through the action of two enzymes (pyruvate decarboxylase (PDC) and adenine dehydrogenase (ADH)), thus ethanol is produced (Silva and Bertucco 2016). The first cyanobacterium to have a whole genome sequenced was *Synechocystis* sp. PCC 6803, being considered a model organism, enabling to better understand its genetics and molecular mechanisms (Singh et al. 2016). Most cyanobacteria engineering studies for synthesizing carbon-based products use three model species: *Synechocystis sp.* PCC 6803 (*Synechocystis* PCC 6803), *Synechococcus elongatus* sp. PCC 7942 and *Synechococcus* sp. PCC 7002 (Vijay et al. 2019).

The enzymes PDC and ADH are the main enzymes that catalyse ethanol synthesis, in which PDC catalyses the non-oxidative decarboxylation of pyruvate, which, in turn, produces acetaldehyde and CO_2, with acetaldehyde then converted into ethanol by ADH. The fermentation bacteria *Zymonomas mobilis* is one of the few prokaryotes that are able to generate ethanol as a product predominantly resulting from fermentation, with PDC and ADH being greatly present in its structure (Deng and Coleman 1999).

In 1999, Deng and Coleman carried out the first study in which oxygenic photoautotrophic microorganisms were genetically modified to produce ethanol, having added the enzymes PDC and ADH from *Zymonomas mobilis* in *Synechocystis* sp. PCC 6803. Ten years after this study was carried out, Dexter and Fu cloned the same group of genes in *Synechococcus* sp. PCC 7942. Since then, many strategies and methods have been adopted for efficiently rerouting the carbon fixed in the Calvin cycle for ethanol production, including deletion or weakening of competitive pathways, improvement of photosynthesis activities, strengthening of precursor supplies and engineering of ethanol-tolerance (Luan et al. 2015).

For Frigaard (2018), the main biological challenges for genetically modified bacteria include genetic stability, low production yield per cell and per volume (as seen in Table 6.1), while the main technological challenges are related to the increase in the cultivation scale and product recovery. In addition, Savakis and Hellingwerf (2015) highlight the importance of the separation between cell replication phases and the ethanol production phase, so that mutations which take place during the multiplication of cyanobacteria, leading to decreased productivity, will not be positively selected for, being preferable to maintain the production pathways under control.

Joule Unlimited estimated an ethanol production from the cultivation of cyanobacteria greater than 230.000 L ha^{-1} $year^{-1}$. In turn, Algenol estimated 60.000 L ha^{-1} $year^{-1}$for ethanol production using cyanobacteria, while traditional crops led to a production lower than 9,500 L ha^{-1} $year^{-1}$. Therefore, ethanol produced

Table 6.1 Examples of fermentation processes carried out with genetically modified cyanobacteria

Species	Strategy	Ethanol concentration (g/L)	Reference
Synechococcus sp. PCC 7942	*Zymomonas mobilis* PDC/ADH II, 21-days cultivation	0.23	Deng and Coleman (1999)
Synechocystis sp. PCC 6803	*Z. mobilis* PDC/ADH II, 6-days cultivation	0.46	Dexter and Fu (2009)
Synechocystis sp. PCC 6803	*Z. mobilis* PDC/*slr1192*, 37-days cultivation	3.6	Algenol Biofuel Inc. (2012) (US8163516B2)
Synechococcus sp. PCC 7002	Isolated JCC1581_B,13. 7-days cultivation	5.62	Joule Unlimited (2012) (US20120164705A1)
Synechocystis sp. PCC 6803	*Z. mobilis* PDC/*slr1192*, 26-days cultivation	5.5	Gao et al. (2012)
Synechococcus sp. PCC 7002	*Z. mobilis* PDC/*slr1192 corT* promoter, 20-days cultivation	4.7	Algenol Biofuel Inc. (2013) (WO2013098267A1)
Synechocystis sp. PCC 6803	*Z. mobilis* PDC/*slr1192* ziaA promoter construct #1318, 30-days cultivation	7.1	Algenol Biofuel Inc. (2013) (WO2013098267A1)
Synechocystis sp. PCC 6803	*Z. mobilis* PDC/*slr1192*, 18-days cultivation	4.7	Dienst et al. (2014)
Synechocystis sp. PCC 6803	Various constructs	Various	Algenol Biofuel Inc. (2014a) (20140154762)
ABICyano1	Plasmid TK504 Copper inducible promoter (3 μmol L^{-1} Cu^{2+}), 1-day cultivation	0.552	Algenol Biofuel Inc. (2014b) (20140178958)
Synechococcus elongatus PCC 7942	*ictB* overexpression, *acaA* and *acsB* and co-fermentation with *Z. mobilis*	7.2	Chow et al. (2015)
Synechocystis sp. PCC 6803	*Z. mobilis* PDC/*slr1192*, 9-days cultivation	2.3	Luan et al. (2015)
Synechocystis sp. PCC 6803	*Z. mobilis* PDC and ADH from slr1192 from *Synechocystis,* proteins from CBB cycle, RuBisCO, FBA, FBP/SBPase, TK, 7-days cultivation	0.4 – 0.7	Liang et al. (2018)
Synechococcus sp. PCC 7002	PDC overexpression of *Z. mobilis* and ADH, 6-days cultivation	0.6	Wang et al. (2019)

with cyanobacteria has a more interesting production ratio per m^2 when compared to ethanol produced from traditional crops, such as sugarcane and corn, because cyanobacteria/microalgae are able to fixe carbon faster than higher plants (Silva and Bertucco 2016). Nevertheless, it is important to point out that these values were not accurately verified by the scientific community.

The technology currently available for the production of ethanol through photofermentation still leads to much greater costs when compared to the production costs of fossil fuels, being also higher than other ethanol sources, such as corn and sugarcane. Therefore, aiming at developing a more accessible technology based on direct conversion, it may be wise to initially concentrate on the production of higher value-added compounds than ethanol (Savakis and Hellingwerf 2015).

6.5.1 The Case of Algenol

Algenol is a North American biotechnology company specializing in various patent-holding cyanobacterial biofuels. Since its founding in 2006, the company has received about $ 35–$ 50 million in government investment and tens of millions from the private sector with the promise that cyanobacteria could economically transform CO_2 into ethanol, a process that has been described as "holy grail" of bioenergy production. However, it was unable to present a product to the market. Algenol's system involved closed vertical photobioreactors filled with seawater and genetically modified cyanobacteria to secrete significantly more ethanol using CO_2 as the raw material. This ethanol mixed with seawater evaporates to the top of the photobioreactor, then condensed and drained.

Productive (such as poor cyanobacterial performance, contamination by ethanol-consuming bacteria and genetic stability) and economic barriers (for example, FBR scale-up expenses) led Algenol to decide to close its ethanol research in 2015. In addition, modified cyanobacteria pose a high environmental risk due to Algenol's inability to maintain genetic stability, raising additional concerns about unforeseen genetic alterations or gene alterations if released in natural ecosystems (Biofuelwatch 2017).

6.6 Other Purposes of Genetically Modified Cyanobacteria

Some genetic modifications carried out in cyanobacteria contribute to the increase in the capacity of synthesizing and accumulating sucrose, glycogen and other carbohydrates, for instance (Kour et al. 2019a; Yadav et al. 2020). In turn, these compounds are potential sources of fermentable sugar for the production of biofuels and its accumulation on cyanobacteria is considered a result of salt stress, in the case of sucrose (Xu et al. 2013). Regarding the potential production of sucrose using genetically modified cyanobacteria, it is estimated that this production is higher than that observed in other sources of sucrose, such as sugarcane (Du et al. 2013).

Du et al. (2013) analysed the production of sucroseusing the species *Synechocystis* sp. PCC 6803, *Synechococcus elongatus* PCC 7942 and *Anabaena* sp. PCC 7120, obtaining different growth curves and accumulation rates of sucrose for the three species under salt stress conditions. The co-overexpression of *sps* (*slr0045*), *spp* (*slr0953*) and *ugp* (*slr0207*) in *Synechocystis* sp. resulted in a twofold increase in the accumulation of sucrose, while the knockout of *ggpS* (*sll1566*) led to a 1.5-fold increase in the production of this sugar. Sanz Smachetti et al. (2019) used *Anabaena* sp. PCC 7120 for over-expressing the *spsB* gene, resulting in the accumulation of sucrose up to 10% (w/w). In addition, glycogen synthase null mutants (*glgA-I glgA-II*) were constructed in the cyanobacterium *Synechococcus* sp. PCC 7002 (Xu et al. 2013), accumulating 1.8 times more soluble sugar in hypersaline conditions, with these cyanobacteria being able to spontaneously excrete soluble sugars in the medium at high levels without the need for additional transporters. Finally, Chow et al. (2015) carried out the co-expression of *ictB*, *ecaA* and *acsAB* in *S. elongatus* PCC 7942, with a 4.9-fold increase in glucose production and a four-fold increase in the production of total carbohydrates when compared to the wild species.

References

Abo-Hashesh M, Wang R, Hallenbeck PC (2011) Metabolic engineering in dark fermentative hydrogen production; theory and practice. Biores Technol 102:8414–8422

Abramson BW, Lensmire J, Lin Y, Jennings E, Ducat DC (2018) Redirecting carbon to bioproduction via a growth arrest switch in a sucrose-secreting cyanobacterium. Algal Res 33:248–255

AFDC–Alternative Fuels Data Center (2019). Available in: https://afdc.energy.gov/data/10331, https://afdc.energy.gov/fuels/ethanol_production.html, https://afdc.energy.gov/fuels/ethanol_fuel_basics.html

Algenol Biofuels, Inc (2013) Genetically enhanced Cyanobacteria for the production of a first chemical compound harbouring Zn^{2+}, Co^{2+} or Ni^{2+} -inducible promoters, Patent Publication Number: WO2013098267A1

Algenol Biofuels, Inc (2014a) Cyanobacterium sp. for Production of Compounds, Patent application number: US20140178958A1

Algenol Biofuels, Inc (2014b) Genetically Enhanced Cyanobacteria Lacking Functional Genes Conferring Biocide Resistance for the Production of Chemical Compounds, Patent application number: US20140154762A1

Algenol Biofuels, Inc. (2012) Selection of ADH in genetically modified Cyanobacteria for the production of ethanol, Official Gazette of the United States Patent and Trademark Office Patents, Patent Publication Number: US8163516B2

Angermayr SA, Hellingwerf KJ, Lindblad P, Mattos MJT (2009) Energy biotechnology with cyanobcteria. Curr Opin Biotechnol 20:257–263

ANP–Agência Nacional do Petróleo, Gás natural e Biocombustíveis (2019). Available in: http://www.anp.gov.br/,in

Biofuelwatch (2017) Algenol: Case Study of an Unsuccessful Algae Biofuels Venture (2019). Available in: https://www.biofuelwatch.org.uk/2017/algenol-report/

Braakman R (2019) Evolution of cellular metabolism and the rise of a globally productive biosphere. Free Radic Biol Med 140:172–187

Cardoso V, Romão BB, Silva FTM, Santos JG, Batista FRX, Ferreira JS (2014) Hydrogen production by dark fermentation. Chem Eng Trans 38:481–486

Chisti Y (2007) Biodiesel from microalgae. Biotechnol Adv 25:294–306

Chow T, Su H, Tsai T, Chou H, Lee T, Chang J (2015) Using recombinant cyanobacterium (*Synechococcus elongatus*) with increased carbohydrate productivity as feedstock for bioethanol production via separate hydrolysis and fermentation process. Biores Technol 184:33–41

Cohen Y, Jörgensen BB, Revsbeck NP, Poplawskil R (1986) Adaptation to hydrogen sulfide of oxygenic and anoxygenic photosynthesis among cyanobacteria. Appl Environ Microbiol 51:398–407

Costa RC, Sodré JR (2010) Hydrous ethanol versus Gasoline-ethanol blend: Engine performance and emissions. Fuel 89:287–293

Deng M, Coleman JR (1999) Ethanol synthesis by genetic engineering in cyanobacteria. Appl Environ Microbiol 65:523–528

Dexter J, Fu P (2009) Metabolic engineering of Cyanobacteria for ethanol production. Energy Environ Sci 2:857–864

Dienst D, Georg J, Abts T, Jakorew L, Kuchmina E, Borner T, et al (2014) Transcriptomic response to prolonged ethanol production in the cyanobacterium *Synechocystis* sp. PCC6803. Biotechnol Biofuels 7:21

Du W, Liang F, Duan Y, Tan X, Lu X (2013) Exploring the photosynthetic production capacity of sucrose by cyanobacteria. Metab Eng 19:17–25

Dvořák P, Poulíčková A, Hašler P, Belli M, Casamatta DA, Papini A (2015) Species concepts and speciation factors in cyanobacteria, with connection to the problems of diversity and classification. Biodivers Conserv 24:739–757

Frigaard N (2018) Sugar and sugar alcohol production in genetically modified cyanobacteria. In: Holban AM, Grumezescu AM (eds) Genetically engineered foods. Academic Press, Handbook of Food Bioengineering 6:31–47

Gao Z, Zhao H, Li Z, Tan X, Lu X (2012) Photosynthetic production of ethanol from carbon dioxide in genetically engineered Cyanobacteria. Energy Environ Sci 5:9857–9865

Hitzfeld BC, Höger SJ, Dietrich DR (2000) Cyanobacterial toxins: removal during drinking water treatment and human risk assessment. Environ Heath Perspect 108:113–122

Jambo SA, Abdulla R, Azhar SHM, Marbawi H, Gansau JA, Ravindra P (2016) A review on third generation bioethanol feedstock. Renew Sustain Energy Rev 65:756–769

Joule Unlimited Technologies Inc (2012) Metabolic Switch, Patent Publication Number: US20120164705A1

Kour D, Rana KL, Yadav N, Yadav AN, Rastegari AA, Singh C et al (2019a) Technologies for biofuel production: current development, challenges, and future prospects. In: Rastegari AA, Yadav AN, Gupta A (eds) Prospects of renewable bioprocessing in future energy systems. springer international publishing, Cham, pp 1–50. https://doi.org/10.1007/978-3-030-14463-0_1

Kour D, Rana KL, Yadav N, Yadav AN, Singh J, Rastegari AA et al (2019b) Agriculturally and industrially important fungi: current developments and potential biotechnological applications. In: Yadav AN, Singh S, Mishra S, Gupta A (eds) Recent advancement in white biotechnology through fungi, Volume 2: perspective for value-added products and environments. Springer International Publishing, Cham, pp 1–64. https://doi.org/10.1007/978-3-030-14846-1_1

Kumar S, Sharma S, Thakur S, Mishra T, Negi P, Mishra S et al (2019) Bioprospecting of microbes for biohydrogen production: current status and future challenges. In: Molina G, Gupta VK, Singh BN, Gathergood N (eds) Bioprocessing for biomolecules production. Wiley, USA, pp 443–471

Lau N, Matsui M, Abdulah A (2015) Cyanobacteria: photoautotrophic microbial factories for the sustainable synthesis of industrial products. Biomed Res Int 2015:1–9

Lea-Smith DL, Howe DJ, Love J, Bryant JA (2017) The use of cyanobacteria for biofuel production. In: Love J, Bryant JA (eds) Biofuels and bioenergy. Wiley, New York, pp 143–155

Liang F, Englund E, Lindberg P, Lindblad P (2018) Engineered cyanobacteria with enhanced growth show increased ethanol production and higher biofuel to biomass ratio. Metab Eng 46:51–59
Luan G, Qi Y, Wang M, Li Z, Duan Y, Tan X et al (2015) Combinatory strategy for characterizing and understanding the ethanol synthesis pathway in cyanobacteria cell factories. Biotechnol Biofuels 8:184
Machado IMP, Atsumi S (2012) Cyanobacterial biofuel production. J Biotechnol 162:50–56
Mata TM, Martins AA, Caetano NS (2010) Microalgae for biodiesel production and other applications: a review. Renew Sust Energ Rev 14:217–232
Milano J, Ong HC, Masjuki HH, Chong WT, Lam MK, Loh PK et al (2016) Microalgae biofuels as an alternative to fossil fuel for Power generation. Renew Sust Energ Rev 58:180–197
Molina GE, Belarbi EH, Fernández FGA, Medina AR, Chisty Y (2003) Recovery of microalgal biomass and metabolites: process options and economics. Biotechnol Adv 20:491–515
Paerl HW, Paul VJ (2012) Climate change: links to global expansion of harmful cyanobacteria. Water Res 46:1349–1363
Pamar A, Singh NK, Pandey A, Gnansounou E, Madamwar D (2011) Cyanobacteria and microalgae: a positive prospect for biofuels. Biores Technol 102:10163–10172
Pierobon SC, Cheng X, Graham PJ, Nguyen B, Karalolis EG, Sinton D (2018) Emerging microalgae technology: a review. Sust Energ Fuels 2:13–38
Rai PK, Singh SP (2016) Integrated dark- and photo- fermentation: recent advances and provisions for improvement. Int J Hydrogen Energ 41:19957–19971
Rana KL, Kour D, Sheikh I, Yadav N, Yadav AN, Kumar V et al (2019) Biodiversity of endophytic fungi from diverse niches and their biotechnological applications. In: Singh BP (ed) Advances in endophytic fungal research: present status and future challenges. Springer International Publishing, Cham, pp 105–144. https://doi.org/10.1007/978-3-030-03589-1_6
Rastegari AA, Yadav AN, Yadav N (2020) New and Future Developments in Microbial Biotechnology and Bioengineering: Trends of Microbial Biotechnology for Sustainable Agriculture and Biomedicine Systems: Diversity and Functional Perspectives. Elsevier, Amsterdam
Rastegari AA, Yadav AN, Gupta A (2019a) Prospects of renewable bioprocessing in future energy systems. Springer International Publishing, Cham
Rastegari AA, Yadav AN, Yadav N (2019b) Genetic Manipulation of Secondary metabolites producers. In: Gupta VK, Pandey A (eds) New and future developments in microbial biotechnology and bioengineering. Elsevier, Amsterdam, pp 13–29. https://doi.org/10.1016/B978-0-444-63504-4.00002-5
Rastegari AA, Yadav AN, Yadav N, Tataei Sarshari N (2019c) Bioengineering of secondary metabolites. In: Gupta VK, Pandey A (eds) New and future developments in microbial biotechnology and bioengineering. Elsevier, Amsterdam, pp 55–68. https://doi.org/10.1016/B978-0-444-63504-4.00004-9
Savakis P, Hellingwerf KL (2015) Engineering cyanobacteria for direct biofuel production from CO_2. Curr Opin Biotechnol 33:8–14
Silva CEF, Bertucco A (2016) Bioethanol from microalgae and cyanobacteria: a review and technological outlook. Process Biochem 51:1833–1842
Silva CEF, Bertucco A (2017) Dilute acid hydrolysis of microalgal biomass for bioethanol production: an accurate kinetic model of biomass solubilization, sugars hydrolysis and nitrogen/ash balance. React Kinet, Mech Catal 122:1095–1114
Silva CEF, Cerqueira RBO, Monteiro CC, Oliveira CF, Tonholo J (2019) Microalgae and wastewaters: from ecotoxicological interactions to produce a carbohydrate-rich biomass towards biofuel application. In: Gupta S, Bux F (eds) Application of microalgae in wastewater treatment. Springer, Cham, pp 495–529
Silva CEF, Meneghello D, Abud AKS, Bertucco A (2018a) Pretreatment of microalgal biomass to improve the enzymatic hydrolysis of carbohydrates by ultrasonication: Yield versus energy consumption. J King Saud Univ Sci 32:606–613
Silva CEF, Meneghello D, Bertucco A (2018b) A systematic study regarding hydrolysis and ethanol fermentation from microalgal biomass. Biocatal Agric Biotechnol 14:172–182

Silva CEF, Sforza E (2016) Carbohydrate productivity in continuous reactor under nitrogen limitation: effect of light and residence time on nutrient uptake in *Chlorella vulgaris*. Process Biochem 51:2112–2118

Silva CEF, Sforza E, Bertucco A (2017) Effects of pH and Carbon Source on *Synechococcus* PCC 7002 cultivation: biomass and carbohydrate production with different strategies for pH control. Appl Biochem Biotechnol 181:682–698

Silva CEF, Sforza E, Bertucco A (2018c) Stability of carbohydrate production in continuous microalgal cultivation under nitrogen limitation: effect of irradiation regime and intensity on *Tetradesmus obliquus*. J Appl Phycol 30:261–270

Singh V, Chaudhary DK, Mani I, Dhar PK (2016) Recent advances and challenges of use of cyanobacteria towards the production of biofuels. Renew Sust Energ Rev 60:1–10

Smachetti MES, Cenci MP, Salerno GL, Curatti L (2019) Ethanol and protein production from minimally processed biomass of a genetically-modified cyanobacterium over-accumulating sucrose. Biores Technol Rep 5:230–237

Stal LJ (1995) Physiological ecology of cyanobacteria in microbial mats and other communities. New Phytol 131:1–31

Subramanian KA (2017) Biofueled Reciprocating Internal Combustion Engines.CRC Press: Taylor & Francis Group.ISBN 9781315116785

Vijay D, Akhtar MK, Hes WR (2019) Genetic and metabolic advances in the engineering of cyanobacteria. Curr Opin Biotechnol 59:150–156

Wang Y, Ho SH, Cheng CL, Guo WQ, Nagarajan D, Lee DJ, Chang JS (2016) Perspectives on the feasibility of using microalgae for industrial wastewater treatment. Biores Technol 222:485–497

Wang M, Luan G, Lu X (2019) Systematic identification of a neutral site on chromosome of *Synechococcus* sp. PCC7002, a promising photosynthetic chassis strain. J Biotechnol 295:37–40

Waterbury JB (2006) The cyanobacteria: isolation, purification and identification. In: Dworkin M, Falkow S, Rosenberg E, Schleifer KH, Stackebrandt E (eds) The prokaryotes. Springer, New York, NY

Xu Y, Guerra LT, Li Z, Ludwig M, Dismukes GC, Bryant DA (2013) Altered carbohydrate metabolism in glycogen synthase mutants of *Synechococcus* sp. strain PCC 7002: Cell factories for soluble sugars. Metab Eng 16:56–67

Yadav AN, Kumar R, Kumar S, Kumar V, Sugitha T, Singh B et al (2017) Beneficial microbiomes: biodiversity and potential biotechnological applications for sustainable agriculture and human health. J Appl Biol Biotechnol 5:45–57

Yadav AN, Rastegari AA, Yadav N (2020) Microbiomes of extreme environments: biodiversity and biotechnological applications. CRC Press, Taylor and Francis, Boca Raton, USA

Chapter 7
Biofuel Synthesis by Extremophilic Microorganisms

Salma Mukhtar and Mehwish Aslam

Abstract Microbial biofuel production has gained great interest over the last 3 decades due to an increase in global energy demand. Fossil fuels are not considered good as they release large volumes of greenhouse gas into the environment and ultimately cause global warming. Microorganisms from extreme environments are especially important because they have enzymes and proteins that can work properly in extreme environmental conditions, such as, extreme temperatures, pH, salinity, drought, and pressure. These microorganisms can be used in different biotechnological applications, providing great momentum for biofuel production. Extremophilic microorganisms including thermophiles, psychrophiles, halophiles, alkaliphiles, and acidophiles have the ability to produce biofuels, such as bioethanol, biobutanol, biodiesel, and biogas or methane, by using various starting materials, such as sugars, starch crops, plant seeds, lignocellulosic agricultural waste, and animal waste, under extreme environments. With progress being made with bioinformatics and gene-editing tools, microorganisms such as *Saccharomyces cerevisiae, Escherichia coli, Clostridium thermocellum, Pyrobaculum calidifontis,* and *Thermococcus kodakarensis* have been genetically engineered to upscale biofuel production. This chapter provides an overview of the various types of biofuels produced by extremophiles, their commercial scale production, and research conducted to improve current technologies. Biofuel production by thermophiles, psychrophiles, halophiles, alkaliphiles, and acidophiles is explained thoroughly. Finally, we discuss the metabolic engineering of extremophiles for upscaling biofuel production.

7.1 Introduction

The global population explosion caused an increase in industry and transport that ultimately led to an increased demand for fossil fuels. This led to their depletion, making them unsecure and expensive (Agrawal 2007; Uzoejinwa et al. 2018). Burning most

S. Mukhtar (✉) · M. Aslam
School of Biological Sciences, University of the Punjab, Lahore, Pakistan
e-mail: salmamukhtar85@gmail.com

A. N. Yadav et al. (eds.), *Biofuels Production – Sustainability and Advances in Microbial Bioresources*, Biofuel and Biorefinery Technologies 11,
https://doi.org/10.1007/978-3-030-53933-7_7

fossil fuels causes an increase in greenhouse gas emissions and contributes to global pollution and climate change (Escobar et al. 2009; Singh et al. 2010). Research on microbial biofuel production, by the degradation of cellulose and other organic compounds, has been undertaken since the mid-20th century. Currently, biofuel production using microorganisms has become an area of interest for scientists around the world due to the increased demand for petroleum-based fuels relative to their availability.

Biofuel production by the conversion of plant-based and algal-based biomass, such as corn, wheat, beets, sugar cane, and other lignocellulosic agricultural waste, has been reported in several studies over the last few years (Decker 2009; Linger et al. 2014). Microbial biofuel production has received great interest over the last decade. Extremophilic microorganisms have great biotechnological potential because they have special physiological and genetic characteristics that allow them to survive in extreme environments (Demain 2009; Gerday and Glansdorff 2007). These organisms can thrive under various extreme environments, including conditions of high salinity, acidity, aridity, and pressure, as well as high and low temperatures. Extremophiles have novel enzymes that can efficiently work under extreme conditions of temperature, salinity, pressure, radiation, etc. (Kour et al. 2019a; Yadav et al. 2016). These enzymes are eco-friendly and efficient, offering a good alternative to current industrial biocatalysts. They can be used in different biotechnological and industrial applications like biofuel production (Egorova and Antranikian 2005; Gurung et al. 2013).

Among the different extremophilic microorganisms, thermophiles are the most commonly used, providing a number of industrial applications. These organisms are able to work at high temperatures and pH levels. Thermophiles have the ability to degrade complex biomass, like carbohydrates, and ferment pentose or hexose sugars to produce biofuels (Gerday and Glansdorff 2007; Jiang et al. 2017; Zaldivar et al. 2001). Moderate thermophiles including *Clostridium, Geobacillus*, and *Sulfobacillus*, and hyperthermophiles including *Thermococcus, Pyrobaculum, Pyrococcus*, and *Pyrolobus* play an important role in the production of biofuels—especially ethanol, butanol, and methane (Barnard et al. 2010; Wagner et al. 2008). Enzymes from halophiles (*Halobacillus* spp. and Haloarchaea) have contributed to the production of bioethanol and biobutanol by the degradation of lignocellulosic compounds (Miriam et al. 2017). Acidophilic microorganisms including *Acidithiobacillus, Pseudomonas*, and *Pyrococcus furiosus* have been used for the degradation of agricultural waste and the production of biodiesel and biogas (Hu et al. 2014; Kernan et al. 2016; Sonntag et al. 2014). Psychrophilic bacteria including *Bacillus, Pseudomonas, Methanosarcina,* and *Methylobacterium* are capable of producing bioethanol and biodiesel by the degradation of lignocellulosic agricultural waste (Lidstrom 1992; Mukhtar et al. 2019b; Sonntag et al. 2014).

Research on microbial biofuel production has been reported extensively. However, only a few studies have focussed on the production of biofuels by extremophiles (Gurung et al. 2013; Jiang et al. 2017; Kernan et al. 2016; Miriam et al. 2017). This chapter provides an overview of the different types of biofuels produced by extremophilic microorganisms. The role of different extremophilic enzymes in the production of biofuels, such as biogas, ethanol, butanol, hydrogen, and biodiesel, is discussed. The chapter explains developments in this area during the last decade and considers the current applications and future implications of using extremophilic microorganisms and their enzymes for the production of biofuels.

7.2 Types of Biofuel Produced by Extremophiles

Biofuels can be divided into two different generations according to their starting materials. First-generation biofuels can be defined as those that utilize readily available crops, such as sugarcane, corn, wheat, and soybean, ultimately being subjected to bioethanol, biobutanol, and biodiesel production using conventional technologies (Luque et al. 2008; Taylor et al. 2009; Kour et al. 2019b, c; Kumar et al. 2019). Second-generation biofuels can be produced using raw materials such as natural/perennial growing plants and agricultural waste that contains lignocellulosic material (Carere et al. 2008; Dutta et al. 2014). Marine or freshwater microalgal biofuels are often considered as third-generation (Dragone et al. 2010). Genetically modified algae is considered a fourth-generation biofuel that may require evaluation of its effects in terms of hazards to the environment and human health. Bioethanol and biodiesel are the main biofuels produced on a large scale, comprising more than 90% of total global biofuel (Fig. 7.1).

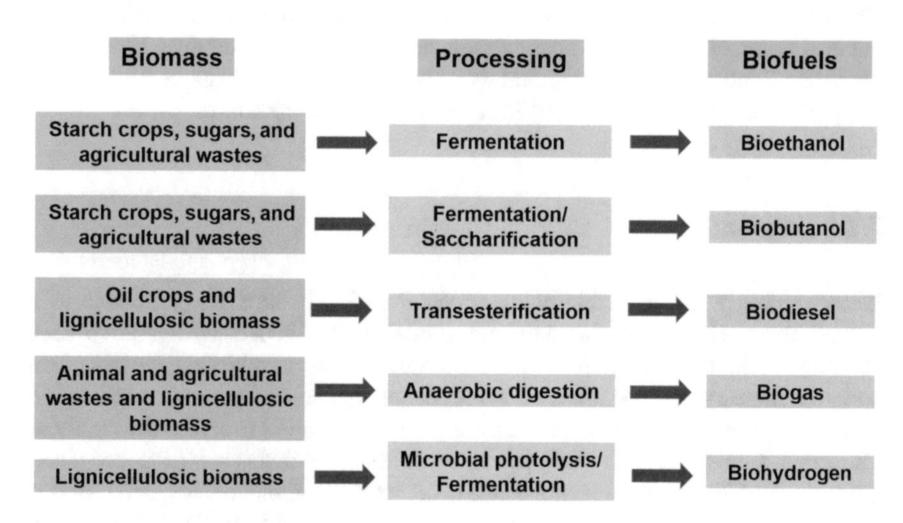

Fig. 7.1 Types of biofuel produced by extremophiles

7.2.1 Bioethanol

From the mid-20th century, many studies have considered the microbial production of ethanol. Many facultative anaerobic bacteria including *Lactobacillus, Clostridium, Alloiococcus, Pediococcus, Aerococcus, Carnobacterium, Streptococcus,* and *Weissella* have been reportedly used for ethanol production using various waste materials, such as corncob, paper, pine cones, and rice straw (Rogers et al. 1982; Sommer et al. 2004; Sun et al. 2003; Tan et al. 2010; Wagner et al. 2008). Some genetically modified strains of *Zymomonas mobilis* and *S. cerevisiae* have been used on an industrial scale for the production of bioethanol from starch crops such as corn, sugar cane, and wheat (Fig. 7.1). *Zymomonas mobilis* produce about 20% more ethanol compared with *S. cerevisiae.* This usually involves the processes of fermentation and saccharification being undertaken independently while the addition of lignocellulosic-degrading microorganisms allows simultaneous fermentation and saccharification (Glazer and Nikaido 1995; Ho et al. 1998; Lynd et al. 2002; Sanchez and Cardona 2006).

Ethanol production by extremophilic microorganisms using lignocellulosic agricultural waste material is more economic compared with the traditional production of ethanol using starch crops (Rastegari et al. 2019a). Xylose-degrading, genetically modified strains of *Erwinia, Geobacillus* and *Klebsiella* have the ability to produce ethanol more efficiently using pure substrates as well as sugars obtained from waste plant materials (Gulati et al. 1996; Hartley and Shama 1987; Kuyper et al. 2005; Sedlak et al. 2004; Wouter et al. 2009). Several extremophilic archaeal, bacterial, or fungal strains can survive under different abiotic stress conditions and produce ethanol efficiently under extreme conditions of temperature, pH, and salt concentration (Yadav et al. 2019a). These strains have the ability to produce biofuels by degrading lignocellulosic agricultural waste, such as sugarcane bagasse, corn stover, and pine cones (Fig. 7.1) (Lau and Dale 2009; Luli et al. 2008).

7.2.2 Biobutanol

Water solubility and available energy content makes butanol less attractive as a biofuel. Butanol has been industrially produced since the 1960s as an organic solvent, however, in the last few decades has it been used more as a biofuel for the transportation industry because it has a 25% higher energy content than bioethanol (Lee et al. 2008; Zheng et al. 2009). Recently, a group of scientists from the University of California, Los Angeles (UCLA) produced different alcohols such as isopropanol, n-butanol, and 2-methyl-1-butanol by the genetic modification of *E. coli* and *C acetobutylicum* (Atsumi et al. 2008; Hanai et al. 2007; Shen and Liao 2008). Biobutanol production from lignocellulosic agricultural waste, using non-fermentable pathways, was a major discovery and attracted a number of multinational companies wishing to fund research on an industrial scale (Fig. 7.1). Some studies have reported on the production of biobutanol from syngas using thermophilic and halophilic bacteria such as *C carboxidivorans, Bacillus*, and *Synechococcus* (Bengelsdorf et al. 2013; Durre 2005, 2016).

7.2.3 *Biodiesel*

Biodiesel can be defined as a non-petroleum-based diesel fuel that mainly contains alkyl esters including methyl, ethyl, and propyl groups. Most importantly, biodiesel does not emit carbon monoxide or carbon dioxide, or cause environmental pollution (Gerpen 2005; Singh and Singh 2010). Biodiesel is biodegradable, sulfur-free, and non-toxic in comparison to petroleum diesel (Demain 2009). It also extends engine life as it contains desirable aromatic compounds with appropriate lubricity (Luque et al. 2008). Different extremophiles can produce biodiesel using animal, plant, and algal biomass (Fig. 7.1). This process involves the esterification of triglycerides and alcohols (Chisti 2007; Fukuda et al. 2001). Recently, biodiesel production by microalgae from different extreme environments, especially marine algae, have attracted a great deal of interest and have been called third-generation biofuels (Tollefson 2008). Biodiesel production using microalgae offers several advantages such as rapid growth compared with other algae and plants and very rich lipid content (80% of dry weight). Some companies in the United States use carbon dioxide–emitting coal for the growth of different acidophilic microalgae (Metting 1996; Spolaore et al. 2006; Tollefson 2008). A number of bacterial (*P. fluorescens, B. cepacian,* and *Rhizopusoryzae*) and yeast strains (*Lipomyces starkeyi, Yarrowia lipolytica, Rhodotorula glutinis*, and *Cryptococcus albidus*) have the ability to produce biodiesel from animal and plant sources (Fig. 7.1) (Al-Zuhair 2007; Du et al. 2004; Meng et al. 2009).

7.2.4 *Biogas*

Biogas or methane can be produced from anaerobic degradation or the methanogenic decomposition of organic waste (Barnard et al. 2010; Schink 1997; Youssef et al. 2007). On a large scale, biogas is usually produced using a defined culture of a syntroph, an acetoclastic or acetate-degrading microorganism, and hydrogenotrophic methanogens. A lot of biogas-producing extremophilic bacteria, including *Lactobacilli, Clostridia, Bifidobacteria,* and *Bacteriocides,* have been isolated from different waste materials including activated sludge, cow dung, slaughter waste, and household organic waste (Chandra et al. 2011; Gao et al. 2018; Narihiro and Sekiguchi 2007; Singh et al. 2000). These bacteria have the ability to degrade complex organic waste material into soluble small organic molecules, such as glucose, maltose, amino acids, and fatty acids, from which acetogenic and hydrogenotrophic bacteria produce acetate and carbon dioxide (Fig. 7.1). Finally, archaeal methanogenic strains, including *Metanonococcus mazei, Methanosarcina thermophile, M lacustri,*

M. barkerican Methanothermococcus okinawensis, Methanosaet aconcilii, and *Methanolobus psychrophilus*, and *Ma. barker*ican, produce methane and carbon dioxide by the process of methanogensis (Franzmann et al. 1997; Nozhevnikova et al. 2003; Ronnow and Gunnarsson 1981; Takai et al. 2002; Zhang et al. 2008). For industrial applications, thermophilic or psychrophilic methanogens can be used, depending upon the anaerobic digestion process and temperature of the fermenter. Recently, several studies have reported the use of mixed bacterial and archaeal methanogenic communities to maximise biogas production (Holm-Nielsen et al. 2009; McKeown et al. 2009).

7.2.5 Biohydrogen

Biohydrogen is a better alternative to petroleum-based fuels as it is the cleanest, non-toxic, cost-effective biofuel producing no emissions of carbon monoxide or carbon dioxide gas (Figs. 7.1 and 7.2). Biohydrogen also has the ability to convert chemical energy into electrical energy in fuel cells (Das and Veziroglu 2001; Malhotra 2007). Hydrogen is produced in many naturally occurring chemical reactions as a final product or a side product, like during the process of photosynthesis (Esper et al. 2006; Vignais and Billoud 2007). The idea of utilization of unused biomass to produce biohydrogen has gained the attention of many scientists (Figs. 7.1 and 7.2). Many bacteria, archaea, and fungi have a variety of hydrogenases that are involved in hydrogen production (Rastegari et al. 2020; Yadav et al. 2017, 2019b). Different approaches have been used for microbial production of hydrogen, for example,

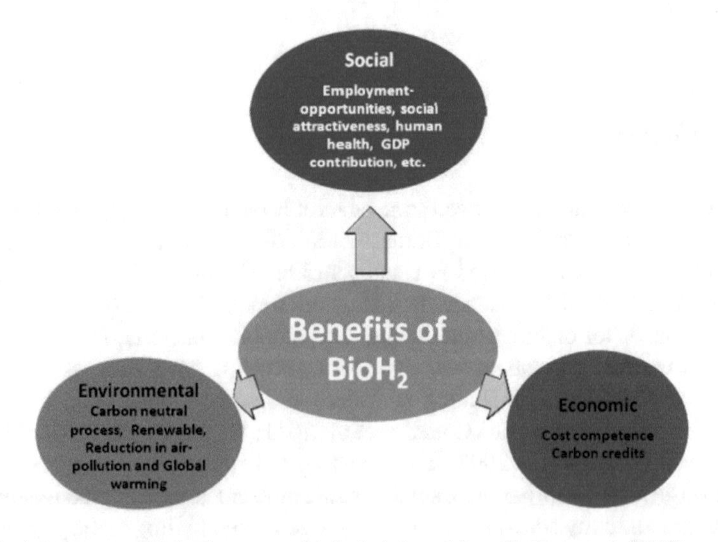

Fig. 7.2 Advantages of biohydrogen as a biofuel Adapted from Rathore et al. (2019)

hydrogen is produced as a side product during cyanobacteria and algal photosynthesis processes as well as during the anaerobic fermentation of organic substances by using anaerobic bacteria and archaea (*Enterobacter, Megasphaera, Lactobacillus,* and *Prevotella*) (Cheng and Zhu 2013; Claassen et al. 2004; Lopez-Hidalgo et al. 2018).

Thermophilic microorganisms including *C thermocellum, Thermotogoelfii, P furiosus, Caldicellulos iruptorsaccharolyticus, T kodakarensis,* and *Aeropyrum camini* contain different hydrogenases and can be used in the production of biohydrogen (Baker et al. 2009; Cheng et al. 2014; Claassen et al. 2004; de Vrije et al. 2002; Dien et al. 2003). Microbial hydrogenases can generate hydrogen from glucose, maltose, starch, or some animal carbohydrate sources (Sommer et al. 2004; Zaldivar et al. 2001). Hydrogenases are mostly metal-dependent (nickel and iron) enzymes that can catalyze reactions in reversible conditions, for example, they produce protons from hydrogen gas by using direct sunlight or organic molecules (Barnard et al. 2010; Rogers et al. 1982; Yun et al. 2018). Recently, many multinational companies in United States have funded the production of biohydrogen on a commercial scale.

7.3 Biofuel Production by Thermophiles

Several thermophilic bacterial and archaeal species including *Clostridium, Thermoanaerobacter, Thermococcus,* and *Pyrococcus* are well known for their role in biofuel production (Table 7.1). Alcohol dehydrogenase enzymes, involved in ethanol production, are widely present in hyperthermophilic arachea strains, including *T.s kodakarensis* (Wu et al. 2013), *P. furiosus* (Van-der Oost et al. 2001; Machielsen et al. 2006), *T. litoralis* (Ma et al. 1994), *T. sibiricus,* and *Thermococcus* strain ES1 (Stekhanova et al. 2010). Primarily, the end products of carbohydrate metabolism in *P. furiosus* are hydrogen, carbon dioxide, and acetate (Kengen et al. 1996). Recently, a report on the conversion of acetate into ethanol in *P. furiosus* (Basen et al. 2014; Nguyen et al. 2015) showed the potential of this organism to produce bioethanol. The AAA pathway in *P. furiosus,* involving aldehyde oxidoreductase (AOR), acetyl-CoA synthetase (ACS), and alcohol dehydrogenase (AdhA), also showed ethanol production via the formation of acetyl-CoA from other metabolic pathways (Keller et al. 2017). When *adhA* (bacterial alcohol dehydrogenase) and CODH (carbon monoxide dehydrogenase) were introduced to *P. furiosus* the engineered strain was able to convert glucose, various organic acids, C2–C6 aldehydes, and phenyl acetaldehyde into various alcoholic products. An engineered strain of *P. furiosus* was able to produce ethanol up to 70 °C (Basen et al. 2014). *T. kodakarensis* enzymes can be useful to degrade chitin and cellulose from raw shrimp shell and rice straw waste to produce ethanol (Chen et al. 2019). This makes cellulose and chitin waste an attractive and potentially valuable future bioethanol source. Some archaeal strains have also been reported to produce butanol from glucose. In the case of *P. furiosus,* when butyrate/isobutyrate was supplied to the growth media (Basen et al. 2014) a large amount of butanol was produced compared with ethanol. An engineered *P. furiosus*

Table 7.1 Biofuel production using different extremophilic bacterial and archaeal strains

Abiotic stress	Extremophiles	Biofuel production	Biomass	Reference
Heat	*Thermococcus kodakarensis*	Ethanol and biohydrogen	Chitin, sugars, starch	Kanai et al. (2005), Aslam et al. (2017)
	Pyrococcus furiosus	Biohydrogen	Sugars, starch crops	Basen et al. (2014)
	Sulfolobus solfataricus	Ethanol	Wood, straw, grass, lignocellulose	Quehenberger et al. (2017)
	Sulfolobus acidocaldarius	Ethanol	Lignocellulose	Keasling et al. (2008), Quehenberger et al. (2017)
	Thermotoga maritima	Biohydrogen	Starch and xylan polymers	Auria et al. (2016)
	Thermoanaerobacterium saccharolyticum	Ethanol	Xylan polymers, hemicellulose	Liu et al. (1996)
	Clostridium thermohydrosulfuricum	Ethanol, hydrogen	Starch, xylose	Wagner et al. (2008)
	Clostridium thermocellum	Ethanol	Lignocellulosic waste	Lynd et al. (2002), Wagner et al. (2008)
	Geobacillus stearothermophilus	Ethanol	Xylan polymers	Hartley and Shama (1987)
Cold	*Rhodobacter ovatus*	Ethanol and biohydrogen	Starch crops and sugars	Srinivas et al. (2008)
	Bacillus pumilus	Ethanol and butanol	Starch crops	Siddiqui and Cavicchioli (2006)
	Pseudomonas fluorescens	Biodiesel	Lignocellulosic agricultural waste and seeds	Luo et al. (2010)
	Sejongia marina	Biohydrogen	Starch crops and sugars	Zhang et al. (2008)
	Brevumdimonas sp.	Biohydrogen	Lignocellulosic agricultural waste	Bao et al. (2012)
	Trichococcus collinsii	Biohydrogen	Starch crops and sugars	Bottos et al. (2014)
	Methanosarcina barkeri	Biogas/methane	Animal and agricultural waste	Nozhevnikova et al. (2003)

(continued)

Table 7.1 (continued)

Abiotic stress	Extremophiles	Biofuel production	Biomass	Reference
	Methanosaeta concilii	Biogas/methane	Lignocellulosic agricultural waste	Zhang et al. (2008)
Salinity	*Nesterenkonia* sp.	Ethanol and butanol	Starch crops and sugars	Amiri et al. (2016)
	Aquisalibacillus elongatus	Ethanol	Starch crops and sugars	Rezaei et al. (2017)
	Kocuria varians	Biohydrogen	Starch crops and sugars	Taroepratjeka et al. (2019)
	Enterobacter aerogenes	Biohydrogen	Starch crops and sugars	Ike et al. (1999)
	Vibrio furnissii	Butanol	Starch crops and sugars	Park et al. (2007)
	Flammeovirga pacifica	Biohydrogen	Lignocellulosic agricultural waste	Cai et al. (2018)
	Bacillus atrophaeus	Biodiesel	Lignocellulosic agricultural waste and seeds	Amiri et al. (2016)
	Dunaliella salina	Biodiesel	Lignocellulosic agricultural waste and seeds	Rasoul-Amini et al. (2014)
	Salinivibrio sp.	Biodiesel	Lignocellulosic agricultural waste and seeds	Amoozegar et al. (2008)
	Arthrospira maxima	Biogas/methane	Animal and agricultural waste	Varel et al. (1988)
	Clostridium carboxidivorans	Butanol	Lignocellulosic agricultural waste	Liou et al. (2005)
	Halolamina pelagica	Biohydrogen	Lignocellulosic agricultural waste	Gaba et al. (2017)
	Methanosaeta concilii	Biogas/methane	Animal and agricultural waste	Barber et al. (2011)
Alkalinity	*Bacillus alcalophilus*	Ethanol and butanol	Starch crops and sugars	Meng et al. (2009)
	Clostridium cellulovorans	Ethanol and butanol	Starch crops and sugars	Wen et al. (2014)
	Butyribacterium methylotrophicum	Bioethanol	Lignocellulosic agricultural waste	Kumari and Singh (2018)
	Carboxydibrachium pacificus	Biohydrogen	Starch crops and sugars, lignocellulosic agricultural waste	Sokolova et al. (2001)

(continued)

Table 7.1 (continued)

Abiotic stress	Extremophiles	Biofuel production	Biomass	Reference
	Pseudomonas nitroreducens	Biodiesel	Lignocellulosic agricultural waste and seeds	Watanabe et al. (1977)
	Halanaerobium hydrogeniformans	Biohydrogen	Lignocellulosic agricultural waste	Begemann et al. (2012)
	Methanosalsus zhilinaeae	Biogas/methane	Animal and agricultural waste	Kevbrin et al. (1997)
Acidity	*Alicyclobacillus acidoterrestris*	Bioethanol	Starch crops and sugars, lignocellulosic agricultural waste	Wisotzky et al. (1992)
	Thiobacillus acidophilus	Bioethanol	Starch crops and sugars, lignocellulosic agricultural waste	Guay and Silver (1975)
	Acidiphilium angustum	Biohydrogen	Lignocellulosic agricultural waste	Wichlacz et al. (1986)
	Acidobacterium capsulatum	Biohydrogen	Lignocellulosic agricultural waste	Kishimoto et al. (1991)
	Sulfolobus solfataricus	Biohydrogen	Lignocellulosic agricultural waste	Schelert et al. (2006)
	Methylacidiphilum infernorum	Biogas/methane	Animal and agricultural waste	Hou et al. (2008)
	Methylococcus capsulatus	Biogas/methane	Animal and agricultural waste	Islam et al. (2015)
	Methylocaldum szegedienseare	Biogas/methane	Animal and agricultural waste	Takeuchi et al. (2014)

strain has been reported to produce 1-butanol and 2-butanol with high yields at 60 °C (Keller et al. 2015). Several bacterial and archaeal strains, as well as isolated/purified enzymes from thermophilic environments, have been investigated in the last decade. Several archaeal strains have been reported to evolve hydrogen from surplus/unused biomass, including *T. kodakarensis* (Kanai et al. 2005; Aslam et al. 2017), *P. furiosus* (Schicho et al. 1993), and *T. onnurineus* NA1 (Kim et al. 2010).

The utilization of hyperthermophilic archaea and their enzymes at high temperatures make them highly attractive for biohydrogen production. Some archaeal strains can utilize the crude glycerol phase (CGP), which can easily be obtained from biodiesel production and is an inexpensive surplus product. It can be converted into polyhydroxyalkanoate (PHA) co- and ter-polyesters (Hermann-Krauss et al. 2013).

7.4 Biofuel Production by Psychrophiles

Psychrophilic microorganisms have been isolated and characterized from different cold environments around the world, especially from Antarctic and Arctic regions (Bottos et al. 2014; Margesin and Miteva 2011). Psychrophilic enzymes have been used for several biotechnological applications due to their ability to function properly at very low temperatures (Feller et al. 2003; Margesin and Feller 2010). Cold-adapted cellulases, lipases, and esterases can produce biofuels using cellulosic plant materials from cold environments. For example, yeast cellulases have the potential to produce ethanol directly from cellulosic materials in cold environments or at low temperatures (Tutino et al. 2009; Ueda et al. 2010). Psychrophilic bacterial strains, including *Arthrobacter, Bacillus, Sejongia, Polaromonas,* and *Pseudomonas* isolated from cold environments, have the ability to produce ethanol and butanol using starch crops, sugars, and lignocellulosic agricultural waste, as shown in Table 7.1 (Cavicchioli et al. 2010; Garcıa-Echauri et al. 2011; Singh et al. 2016; Yadav and Saxena 2018; Yadav et al. 2019c).

Most of the anaerobic fermenters for biohydrogen production operate at room temperature (mesophilic) or high temperatures (thermophilic). However, psychrophilic microorganisms produce biohydrogen at low temperatures and therefore save energy heating the digesters (Weng et al. 2008; Zazil et al. 2015). A large number of bacterial genera including *Klebsiella, Clostridium, Brevumdimonas, Carnobacterium, Trichococcus, Polaromonas, Rhodobacter,* and *Pseudomonas* have the potential to produce biohydrogen at low temperatures (Rathore et al. 2019; Yadav and Saxena 2018; Zazil et al. 2015). Psychrophilic members of the Firmicutes, such as *Bacillus, Carnobacterium, Clostridium,* and *Trichococcus*, can produce a high volume of hydrogen at low temperatures (Margesin and Miteva 2011; Zazil et al. 2015). Gram-negative bacteria including members of *Rhodobacter, Klebsiella, Brevumdimonas,* and *Pseudomonas* produce hydrogen under aerobic conditions in the dark using lignocellulosic waste material. These bacteria can also work in anaerobic conditions in the presence of sunlight (Table 7.1) (Bao et al. 2012; Srinivas et al. 2008).

Several studies have described cold-adapted lipases and esterases for the production of biodiesel at low temperatures (Luo et al. 2010; Tutino et al. 2009). Psychrophilic microbial biodiesel production has been reported in different environments, e.g., Arctic and Antarctic sediments, mountainous rocks and soil from cold environments, deep-sea sediments, and mangrove soils (Couto et al. 2010; Heath et al. 2009; Jeon et al. 2009a; Park et al. 2007; Wei et al. 2009). Methanogens, such as *Methanosarcina, Methanosaeta,* and *Methanolobus,* isolated and characterized from cold environments, play an important role in the production of biogas at low temperatures (Table 7.1) (Franzmann et al. 1997; Nozhevnikova et al. 2003; Ronnow and Gunnarsson 1981; Zhang et al. 2008).

7.5 Biofuel Production by Halophiles

Halophilic bacteria and archaea are widely distributed in hypersaline environments such as salt lakes, saline soils, salt marshes, and marine water and sediments (Irshad et al. 2014; Mukhtar et al. 2018, 2019a, b), and have the ability grow in high salt concentrations. They are classified as slight halophiles, with salt requirements of 0.21–0.85 M NaCl; moderate halophiles, with salt requirements of 0.85–3.4 M NaCl; and extreme halophiles, with salt requirements of 3.4–5.1 M NaCl. Halophilic microorganisms have developed special physiological and genetic modifications to live under hypersaline environments (Irshad et al. 2014; Mukhtar et al. 2019a, c).

Several halophiles have the ability to synthesize biofuels, such as bioethanol, butanol, biodiesel, biohydrogen, and biogas, using plant and animal biomass under extreme conditions of salinity (Amoozegar et al. 2019). Bioethanol is the most promising biofuel produced by halophilic microorganisms. Halophilic bacterial genera including *Nesterenkonia, Aquisalibacillus,* and *Clostridium* can produce bioethanol from the decomposition of plant and agriculture biomass (Table 7.1) (Amiri et al. 2016; Marriott et al. 2016; Rezaei et al. 2017). Some bacterial genera, such as *Vibrio furnissii* and *C carboxidivorans,* can produce butanol using lignocellulosic or hemicellulosic agricultural waste (Liou et al. 2005; Park et al. 2007). The production of ethanol or butanol includes four major steps: (1) pretreatment of plant biomass; (2) enzymatic hydrolysis of biomass; (3) fermentation; and (4) distillation and purification of biofuels (Indira et al. 2018; Khambhaty et al. 2013).

Some halophilic microalgae such as *Dunaliella salina* are considered a safe source of fuel production, such as biodiesel (Table 7.1). They provide the largest biomass for energy production and decrease environmental pollution and global warming (Rasoul-Amini et al. 2014; Tandon and Jin 2017). Halophilic bacterial strains including *Salinivibrio* sp. and *B. atrophaeus* can also produce biodiesel using lignocellulosic and hemicellulosic agricultural waste and seeds in hypersaline environments (Amiri et al. 2016; Amoozegar et al. 2008).

Halophilic bacterial strains including *K varians, E aerogenes, Flammeovirga pacifica,* and archaeal strain *Halolaminapelagica* are capable of producing hydrogen from starch crops and lignocellulosic or hemicellulosic agricultural waste under conditions of high salinity (Table 7.1) (Cai et al. 2018; Gaba et al. 2017; Ike et al. 1999; Taroepratjeka et al. 2019). Some halophilic methanogenic bacterial and archaeal strains including *Arthrospira maxima* and *Methanosaeta concilii* produce biogas or methane from animal and lignocellulosic agricultural waste (Barber et al. 2011; Varel et al. 1988). Some halophilic methanogenic archaeal strains can produce methane using brown algae biomass in marine environments (Miura et al. 2015).

7.6 Biofuel Production by Alkaliphiles

It is mostly mesophilic microorganisms that can produce ethanol and butanol at pH levels between 4.0 and 7.2. However, alkaliphiles can produce biofuels at pH levels between 8.0 and 9.0. A number of bacteria and archaea, including *B alcalophilus, C cellulovorans, Alkalibaculumbacchi,* and *Butyribacterium methylotrophicum,* have cellulases and glucanases that break down lignocellulosic agricultural waste into ethanol and butanol (Table 7.1) (Allen et al. 2010; Kumari and Singh 2018; Meng et al. 2009; Wen et al. 2014). *Carboxydibrachium pacificus* and *Halanaerobium hydrogeniformans* are novel alkaliphilic and thermophilic bacteria that can produce hydrogen using starch crops and lignocellulosic agricultural waste (Liu et al. 2012; Sokolova et al. 2001; Rana et al. 2019).

Biodiesel is well known as a first-generation biofuel that can be produced by transesterification processes of vegetable oils and lignocellulosic agricultural waste. *P. nitroreducens* and *B. alcalophilus* are alkaliphilic bacteria that produce biodiesel using bio-transesterification processes under alkaline conditions (Table 7.1). These bacteria are also involved in the biodegradation of xylan and lignin under alkaline conditions (Meng et al. 2009; Watanabe et al. 1977). Methanogens, such as *Arthrospira maxima* and *M. zhilinaeae,* isolated and characterized from alkaline environments, play an important role in the production of biogas at high pH levels (Begemann et al. 2012; Kevbrin et al. 1997; Varel et al. 1988).

7.7 Biofuel Production by Acidophiles

Acidophilic bacteria and archaea are widely distributed in acidic water found in mines and the acidic springs around the world. They can grow in environments with pH levels between 2.5 and 6.3, but their optimum pH is 4 (Schelert et al. 2006; Sharma et al. 2012). Acidophiles produce biofuels such as bioethanol, biobutanol, biohydrogen, and biogas/methane and greatly reduce carbon emissions to the environment (Yadav et al. 2020). Many acidophiles have been reported for biofuel production. Acidophilic bacterial and archaeal genera, including *Alicyclobacillus, Acidianus, Sulfolobus, Thermotoga, Desulphurolobus,* and *Pyrococcus,* can produce cellulases, amylases, xylanases, and esterases (Table 7.1). Bacterial strains, such as *Alicyclobacillus, Thiobacillus, Sulfolobus,* and *Picrophilus*, can produce ethanol or butanol using starch crops and lignocellulosic agricultural waste under acidic environments (Bertoldo et al. 2004).

Sulfolobussolfataricus is a well-known acidophilic bacterium used for the production of butanol and hydrogen at a pH of 4.1 (Table 7.1). *Acidiphilium angustum* and *Acidobacterium capsulatum* can produce hydrogen as a biofuel from lignocellulosic plant biomass at low pH levels between 4.0 and 6.0 (Kishimoto et al. 1991; Limauro et al. 2001; Wichlacz et al. 1986). *Methylacidiphilum infernorum, Methylococcus capsulatus,* and *Methylocaldum szegediensis* are biogas and methane producers (Table 7.1). They have the ability to produce methane under acidic conditions using different carbon sources, such as animal and plant biomass (Hou et al. 2008; Islam et al. 2015; Takeuchi et al. 2014). Acidophilic bacteria and thermostable enzymes are a better combination for biofuel production on an industrial scale than acidophilic bacteria and mesophilic enzymes (Galbe and Zacchi 2007).

7.8 Metabolic Engineering of Extremophiles to Upscale Biofuel Production

Several extremophiles have been engineered for different types of catalytic enzymes used for biofuel production. Genetic and adaptive engineering approaches have provided new insights into the manipulation of cellulose and chitin metabolic pathways to produce biohydrogen using surplus chitinous biomass (Aslam et al. 2017; Chen et al. 2019; Rastegari et al. 2019b; Rastegari et al. 2019c). Such modifications provide an example of how to manipulate metabolic pathways across many archaea as well as bacteria. Another example of genetic manipulation includes that ethanol and butanol produced by *P. furious* by genetic engineering techniques made it possible to enhance their yields from trace levels to 35% (Basen et al. 2014; Keller et al. 2017).

Yeast (*S. cerevisiae*) and *E. coli* are the most used microorganisms for the commercial production of biofuels through genetic engineering (Fig. 7.3). *S. cerevisiae* can produce ethanol directly from the decarboxylation of pyruvate (Liao et al.

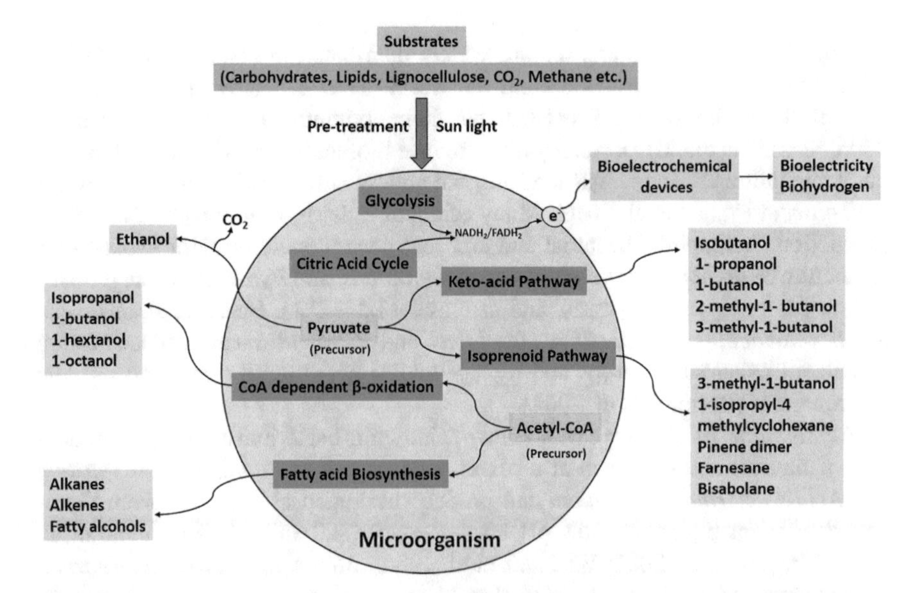

Fig. 7.3 An overview of the microbial metabolic pathways for biofuel production

2016). Other microorganisms have been genetically engineered using this metabolic pathway to produce ethanol.

The overexpression of certain genes involved in biofuel production increases the catalytic activity of both enzyme and substrate and helps to produce more biofuel (Fig. 7.3). Recently, artificial metabolic pathways or mRNAs have been used for the efficient production of biofuels. For example, microbial electrolysis cells (MECs) are used for biohydrogen and bioelectricity production (Dai et al. 2016; Kracke et al. 2015). Use of MECs provides a platform for biofilm formation and develops microbe–metal interactions which transfer electrons from bacterial cell walls/membranes to an electrode (Kracke et al. 2015; Kumar and Kumar 2017). Certain proteins and enzymes produced by exoelectrogens are used to enhance this process. However, the MEC technique is not capable of producing biofuels on a commercial scale.

Despite the great potential archaeal enzymes have for biofuel production they require harsh conditions for optimum growth and enzyme functionality. This has made them unsuitable for industrial fermentation and downstream processing. However, recent developments involving several genetic engineering/manipulation techniques, i.e., pop-in/pop-out, development of archaea–*E. Coli* shuttle vectors, and site-directed mutagenesis (Rashid and Aslam 2019), have provided breakthroughs in utilizing their hyper-thermostable enzymes in thermophilic/mesophilic organisms and environments. CRISPR–CAS approaches can also be used to improve specific biofuel production and downstream processing in both archaea and bacteria.

7.9 Conclusions and Future Prospects

Microbial biofuel production is still particularly challenging since it is difficult to produce a large amount of fuel more economically and efficiently from raw biomass than conventional fossil fuels. With progress being made in the strategies used for biofuel production, such as biomass based on lignocellulosic agricultural waste, the process has become relatively economic compared to production based on the biomass of sugars or starch crops. Bioethanol, biobutanol, biodiesel, and biogas are important biofuels produced by extremophilic microorganisms. Different sequencing approaches have been used to understand the complexity of microbial communities in various extreme environments. The advances in sequencing technology make it possible to study microbial enzymes and proteins using genomics, transcriptomics, and proteomics. Enzymes from extremophilic microorganisms are especially important because they can work properly in extreme environmental conditions, such as extremes of temperature, pH, salinity, drought, and pressure. Continued research on genetic manipulation of various extremophilic bacterial and archaeal strains will create innovations to produce economically available biofuels. In the near future a wide range of extremophilic enzymes, with the ability to degrade or utilize lignocellulosic waste materials, will be successfully used for biofuel production on a commercial scale.

References

Agrawal AK (2007) Biofuels (alcohols and biodiesel) applications as fuels for internal combustion engines. Prog Energy Combust Sci 33:233e71

Al-Zuhair S (2007) Production of biodiesel: possibilities and challenges. Biofuels Bioprod Biorefin 1:57–66

Allen TD, Caldwell ME, Lawson PA, Huhnke RL, Tanner RS (2010) *Alkalibaculum bacchi* gen. Nov., sp. nov., a CO-oxidizing, ethanol-producing acetogen isolated from livestock-impacted soil. Int J Syst Evol Microbiol 60:2483–2489

Amiri H, Azarbaijani R, Yeganeh LP, Fazeli AS, Tabatabaei M, et al. (2016) *Nesterenkonia* sp. strain F, a halophilic bacterium producing acetone, butanol, and ethanol under aerobic conditions. Sci Rep 6:18408–18418

Amoozegar MA, Salehghamari E, Khajeh K, Kabiri M, Naddaf S (2008) Production of an extracellular thermohalophilic lipase from a moderately halophilic bacterium, *Salinivibrio* sp. strain SA-2. J Microbiol Methods 48:160–167

Amoozegar MA, Safarpour A, Noghabi KA, Bakhtiary T, Ventosa A (2019) Halophiles and their vast potential in biofuel production. Front Microbiol 10:1895

Aslam M, Horiuchi A, Simons JR, Jha S, Yamada M, Odani T, Atomi H (2017) Engineering of the hyperthermophilic archaeon thermococcus kodakarensis for chitin-dependent hydrogen production. Appl Environ Microbiol 83:e00280

Atsumi S, Hanai T, Liao JC (2008) Non-fermentative pathways for synthesis of branched-chain higher alcohols as biofuels. Nature 451:86–89

Auria R, Boileau C, Davidson S, Casalot L, Christen P, Liebgott PP, Combet-Blanc Y (2016) Hydrogen production by the hyperthermophilic bacterium *Thermotoga maritima* Part II: modeling and experimental approaches for hydrogen production. Biotechnol Biofuels 9:268

Baker SE, Hopkins RC, Blanchette CD, Walsworth R, Sumbad NO, et al. (2009) Hydrogen production by a hyperthermophilic membrane-bound hydrogenase in water-soluble nanolipoprotein particles. J Am Chem Society 131:7508–7509

Bao M, Su H, Tan T (2012) Biohydrogen production by dark fermentation of starch using mixed bacterial cultures of *Bacillus* sp. and *Brevumdimonas* sp. Energy Fuels 26:5872e8

Barber RD, Zhang L, Harnack M, Olson MV, Kaul R, Ingram-Smith C, Smith KS (2011) Complete genome sequence of *Methanosaeta concilii*, a specialist in aceticlastic methanogenesis. J Bacteriol 193:3668–3669

Barnard D, Casanueva A, Tuffin M, Cowan D (2010) Extremophiles in biofuel synthesis. Environ Technol 31:871–888

Basen M, Schut GJ, Nguyen DM, Lipscomb GL, Benn RA, et al (2014) Single gene insertion drives bioalcohol production by a thermophilic archaeon. Proc Natl Acad Sci USA 111:17618–17623

Begemann MB, Mormile MR, Sitton OC, Wall JD and Elias DA (2012) A streamlined strategy for biohydrogen production with *Halanaerobium hydrogeniformans*, an alkaliphilic bacterium. Front Microbio 3:93

Bengelsdorf F, Straub M, Durre P (2013) Bacterial synthesis gas (syngas) fermentation. Environ Technol 34:1639–1651

Bertoldo C, Dock C, Antranikian G (2004) Thermoacidophilic microorganisms and their novel biocatalysts. Eng Life Sci 4:521–531

Bottos EM, Scarrow JW, Archer SDJ, McDonald IR, Cary SC (2014) Bacterial community structures of antarctic soils. In: Cowan DA (ed) Antarctic terrestrial microbiology: physical and biological properties of antarctic soils. Springer, Heidelberg, p 9e34

Cai ZW, Ge HH, Yi ZW, Zeng RY, Zhang GY (2018) Characterization of a novel psychrophilic and halophilic b-1, 3-xylanase from deep-sea bacterium, *Flammeovirga pacifica* strain WPAGA1. Int J Biol Macromol 118:2176–2184

Carere CR, Sparling R, Cicek N, Levin DB (2008) Third generation biofuels via direct cellulose fermentation. Int J Mol Sci 9:1342–1360

Cavicchioli R, Charlton T, Ertan H, Omar SM, Siddiqui KS Williams TJ (2010) Biotechnological uses of enzymes from psychrophiles. Microb Biotechnol 4:449–460

Chandra R, Vijay V, Subbarao P, Khura T (2011) Performance evaluation of a constant speed IC engine on CNG, methane enriched biogas and biogas. Appl Energy 88:3969–3977

Chen L, Wei Y, Shi M, Li Z and Zhang SH (2019) An archaeal chitinase with a secondary capacity for catalyzing cellulose and its biotechnological applications in shell and straw degradation. Front Microbiol 10:1253

Cheng J, Zhu M (2013) A novel anaerobic co-culture system for bio-hydrogen production from sugarcane bagasse. Bioresour Technol 144:623–631

Cheng J, Yu Y, Zhu M (2014) Enhanced biodegradation of sugarcane bagasse by *Clostridium thermocellum* with surfactant addition. Green Chem 16:2689–2695

Chisti Y (2007) Biodiesel from microalgae. Biotechnol Adv 25:294–306

Claassen PAM, de Vrije T, Budde MAW (2004) Biological hydrogen production from sweet sorghum by thermophilic bacteria. In: 2nd World conference on biomass for energy, ETA-Florence and WIP-Munich, Rome

Couto GH, Glogauer A, Faoro H, Chubatsu LS, Souza et al (2010) Isolation of a novel lipase from a metagenomic library derived from mangrove sediment from the south Brazilian coast. Genet Mol Res 9:514–523

Dai H, Yang H, Liu X, Jian X, Liang Z (2016) Electrochemical evaluation of nano-Mg(OH)2/graphene as a catalyst for hydrogen evolution in microbialelectrolysis cell. Fuel 174:251–256

Das D, Veziroglu TN (2001) Hydrogen production by biological processes: a survey of literature. Int J Hydrogen Energy 26:13–28

de Vrije T, de Haas GG, Tan GB, Keijsers ERP, Claassen PAM, (2002) Pretreatment of Miscanthus for hydrogen production by *Thermotoga elfii*. Int J Hydrogen Energy 27:1381–1390

Decker J (2009) Going against the grain: Ethanol from lignocellulosics. Renew Energy World Mag 11

Demain A (2009) Biosolutions to the energy problem. J Ind Microbiol Biotechnol 36:319–332

Dien BS, Cotta MA, Jeffries TW (2003) Bacteria engineered for fuel ethanol production: current status. Appt Microbiol Biotechnol 63:258–266

Dragone G, Fernandes BD, Vicente AA, Teixeira JA (2010) Third generation biofuels from microalgae. Curr Res Technol Edu Top Appl Microbiol Microb Biotechnol 2:1355–1366

Du W, Xu Y, Liu D, Zeng J (2004) Comparative study on lipase-catalyzed transformation of soybean oil for biodiesel production with different acyl acceptors. J Mol Catal Enzym 30:125–129

Durre P (2005) Formation of solvents in clostridia. In: Durre P (ed) Handbook on *Clostridia*. CRC Press-Taylor and Francis Group, Boca Raton, USA, pp 671–93

Durre P (2016) Butanol formation from gaseous substrates. FEMS Microbiol Lett 363:pii:fnw040

Dutta K, Daverey A, Lin J-G (2014) Evolution retrospective for alternative fuels: first to fourth generation. Renew Energy 69:114–122

Egorova K, Antranikian G (2005) Industrial relevance of thermophilic archaea. Curr Opin Microbiol 8:649–655

Escobar JC, Lora ES, Venturini OJ, Yanez EE, Castillo EF, Almazan O (2009) Biofuels: environment, technology and food security. Renew Sustain Energy Rev 13:1275e87

Esper B, Badura A, Rögner M (2006) Photosynthesis as a power supply for (bio-) hydrogen production. Trends Plant Sci 11:543–549

Feller G, Gerday C (2003) Psychrophilic enzymes: hot topics in cold adaptation. Nat Rev Microbiol 1:200e8

Franzmann PD, Liu Y, Balkwill DL, Aldrich HC, Conway De MHC, Boone DR (1997) *Methanogenium frigidum* sp. nov., a psychrophilic, H2-using methanogen from Ace Lake, Antarctica. Int J Syst Bacteriol 47:1068–1072

Fukuda H, Kondo A, Noda H (2001) Biodiesel fuel production by transesterification of oils. J Biosci Bioeng 92:405–416

Gaba S, Singh RN, Abrol S, Yadav AN, Saxena AK, Kaushik R (2017) Draft genome sequence of *Halolamina pelagica* CDK2 isolated from natural salterns from Rann of Kutch, Gujarat, India. Genome Announc 5:1–2

Galbe M, Zacchi G (2007) Pretreatment of lignocellulosic materials for efficient bioethanol production. Adv Biochem Eng Biotechnol 108:41–65

Gao Y, Jiang J, Meng Y, Yan F, Aihemaiti A (2018) A review of recent developments in hydrogen production via biogas dry reforming. Energy Convers Manag 171:133–155

Garcıa-Echauri SA, Gidekel M, Gutierrez-Moraga A, Santos L, De Leon-Rodriguez A (2011) Isolation and phylogenetic classification of culturable psychrophilic prokaryotes from the Collins glacier in the Antarctica. Folia Microbiol 56:209e14

Gerday C, Glansdorff N (2007) Physiology and biochemistry of extremophiles. ASM Press, Washington, DC

Gerpen JV (2005) Biodiesel processing and production. Fuel Process Technol 86:1097–1107

Glazer AN Nikaido H (1995) Microbial biotechnology: fundamentals of applied microbiology W.H. Freeman, New York

Guay R, Silver M (1975) *Thiobacillus acidophilus* sp. nov., isolation and some physiological characteristics. Can J Microbiol 21:281–288

Gulati M, Kohlmann K, Ladisch MR, Hespell R, Bothast RJ (1996) Assessment of ethanol production options for corn products. Bioresour Technol 58:253–64

Gurung N, Ray S, Bose S, Rai V (2013) A broader view: microbial enzymes and their relevance in industries, medicine, and beyond. Biomed Res Int 2013:329121

Hanai T, Atsumi S, Liao JC (2007) Engineered synthetic pathway for isopropanol production in Escherichia coli. Appl Environ Microbiol 73:7814–7818

Hartley BS, Shama G (1987) Novel ethanol fermentations from sugar cane and straw. Philso Trans Roy Soc Lond A 321:555–568

Heath C, Hu XP, Cary SC, Cowan D (2009) Identification of a novel alkaliphilic esterase active at low temperatures by screening a metagenomic library from Antarctic desert soil. Appl Environ Microbiol 75:4657–4659

Hermann-Krauss C, Koller M, Muhr A, Fasl H, Stelzer F, Braunegg G (2013) Archaeal production of polyhydroxyalkanoate (pha) co- and terpolyesters from biodiesel industry-derived by-products. Archaea 2013:129268

Ho NWY, Chen Z, Brainard AP (1998) Genetically engineered *Saccharomyces* yeast capable of effective cofermentation of glucose and xylose. Appl Environ Microbiol 64:1852–1859

Holm-Nielsen JB, Al Seadi T, Oleskowicz- Popiel P (2009) The future of anaerobic digestion and biogas utilization. Bioresour Technol 100:5478–5484

Hou S, Makarova KS, Saw JHW, Senin P, Ly BV, Zhou Z, et al. (2008) Complete genome sequence of the extremely acidophilic methanotroph isolate V4, *Methylacidiphilum infernorum*, a representative of the bacterial phylum Verrucomicrobia. Biol Direct 3:26

Hu B, Lidstrom ME (2014) Metabolic engineering of *Methylobacterium extorquens* AM1 for 1-butanol production. Biotechnol Biofuels 7:156

Ike A, Murakawa T, Kawaguchi H, Hirata K, Miyamoto K (1999) Photoproduction of hydrogen from raw starch using a halophilic bacterial community. J Biosci Bioeng 88:72–77

Indira D, Das B, Balasubramanian P, Jayabalan R (2018) Sea water as a reaction medium for bioethanol production. Science 2:171–192

Irshad A, Ahmad I, Kim SB (2014) Cultureable diversity of halophilic bacteria in foreshore soils. Braz J Microbiol 45:563–571

Islam T, Larsen Ø, Torsvik V, Øvreås L, Panosyan H, Murrell JC (2015) Novel methanotrophs of the family Methylococcaceae from different geographical regions and habitats. Micro Organisms J 3:484–499

Jeon J, Kim JT, Kang S, Lee JH, Kim SJ (2009a) Characterization and its potential application of two esterases derived from the Arctic sediment metagenome. Mar Biotechnol 11:307–316

Jiang Y, Xin F, Lu J, Dong W, Zhang W, Zhang M, Wu H, Ma J, Jiang M (2017) State of the art review of biofuels production from lignocellulose by thermophilic bacteria. Bioresour Technol 245:1498–1506

Kanai T, Imanaka H, Nakajima A, Uwamori K, Omori Y, Fukui T, Atomi H, Imanaka T (2005) Continuous hydrogen production by the hyperthermophilic archaeon, *Thermococcus kodakaraensis* KOD1. J Biotechnol 116:271

Keasling JD, Chou H (2008) Metabolic engineering delivers nextgeneration biofuels. Nat Biotech 26:298–299

Keller MW, Lipscomb GL, Loder AJ, Schut GJ, Kelly RM, Adams MW (2015) A hybrid synthetic pathway for butanol production by a hyperthermophilic microbe. Metabolic Eng 27:101–6

Keller MW, Lipscomb GL, Nguyen DM, Crowley AT, Schut GJ, Scott I, Adams M (2017) Ethanol production by the hyperthermophilic archaeon *Pyrococcus furiosus* by expression of bacterial bifunctional alcohol dehydrogenases. Microbial Biotechnol 10:1535–1545. https://doi.org/10.1111/1751-7915.12486

Kengen SM, Stams AJ, Vos WD (1996) Sugar metabolism of hyperthermophiles. FEMS Microbiol Rev 18:119–138

Kernan T, Majumdar S, Li X, Guan J, West AC, Banta S (2016) Engineering the iron-oxidizing chemolithoautotroph *Acidithiobacillus ferrooxidans* for biochemical production. Biotechnol Bioeng 113:189–97

Kevbrin VV, Lysenko AM, Zhilina TN (1997) Physiology of the alkaliphilic methanogen Z-7936, a new strain of *Methanosalsus zhilinaeae* isolated from Lake Magadi. Microbiology 66:261–266

Khambhaty Y, Upadhyay D, Kriplani Y, Joshi N, Mody K, Gandhi MR (2013) Bioethanol from macroalgal biomass: utilization of marine yeast for production of the same. Bioenergy Res 6:188–195

Kim YJ, Lee HS, Kim ES, Bae SS, Lim JK, Matsumi R, Lebedinsky AV, et al (2010) Formate-driven growth coupled with H(2) production. Nature 467:352

Kishimoto N, Inagaki K, Sugio T, Tano T (1991) Purification and properties of an acidic b-glucosidase from *Acidobacterium capsulatum*. J Fermen Bioeng 71:318–321

Kour D, Rana KL, Kaur T, Singh B, Chauhan VS, Kumar A et al (2019a) Extremophiles for hydrolytic enzymes productions: biodiversity and potential biotechnological applications. In: Molina G, Gupta VK, Singh B, Gathergood N (eds) Bioprocessing for biomolecules production, pp 321–372. https://doi.org/10.1002/9781119434436.ch16

Kour D, Rana KL, Yadav N, Yadav AN, Rastegari AA, Singh C et al (2019b) Technologies for biofuel production: current development, challenges, and future prospects. In: Rastegari AA, Yadav AN, Gupta A (eds) Prospects of renewable bioprocessing in future energy systems. Springer International Publishing, Cham, pp 1–50. https://doi.org/10.1007/978-3-030-14463-0_1

Kour D, Rana KL, Yadav N, Yadav AN, Singh J, Rastegari AA et al (2019c) Agriculturally and industrially important fungi: current developments and potential biotechnological applications. In: Yadav AN, Singh S, Mishra S, Gupta A (eds) Recent advancement in white biotechnology through fungi, Volume 2: perspective for value-added products and environments. Springer International Publishing, Cham, pp 1–64. https://doi.org/10.1007/978-3-030-14846-1_1

Kracke F, Vassilev I, Krömer JO (2015) Microbial electron transport and energy conservation-the foundation for optimizing bioelectrochemical systems. Front Microbiol 6:575

Kumar R, Kumar P (2017) Future microbial applications for bioenergy production: a perspective. Front Microbiol 8:450

Kumar S, Sharma S, Thakur S, Mishra T, Negi P, Mishra S et al (2019) Bioprospecting of microbes for biohydrogen production: current status and future challenges. In: Molina G, Gupta VK, Singh BN, Gathergood N (eds) Bioprocessing for biomolecules production. Wiley, USA, pp 443–471

Kumari D, Singh R (2018) Pretreatment of lignocellulosic wastes for biofuel production: a critical review. Renew Sust Energ Rev 90:877–891

Kuyper M, Toirkens MJ, Diderich JA, Winkler AA, Van Dijken JP, Pronk JT (2005) Evolutionary engineering of mixed-sugar utilization by a xylose-fermenting Saccharomyces cerevisiae strain. FEMS Yeast Res 5:925–934

Lau MW, Dale BE (2009) Cellulosic ethanol production from AFEX-treated corn stover using Saccharomyces cerevisiae 424A (LNH-ST). Proc Nat Acad Sci USA 106:1368–1373

Lee SY, Park JH, Jang SH, Nielson LK, Kim J, Jung KS (2008) Fermentative butanol production by *Clostridia*. Biotechnol Bioeng 101:209–228

Liao JC, Mi L, Pontrelli S, Luo S (2016) Fuelling the future: microbial engineering for the production of sustainable biofuels. Nat Rev Microbiol 14:288–304

Lidstrom ME (1992) The genetics and molecular biology of methanol-utilizing bacteria. In: Murrell JC, Dalton H (eds) Methane and methanol utilizers. Springer, Boston, MA, pp 183–206

Limauro DR, Cannio G, Fiorentino MR, Bartolucci S (2001) Identification and molecular characterization of an endoglucanase gene, CelS, from the extremely thermophilic archaeon *Sulfolobus solfataricus*. Extremophiles 5:213–219

Linger JG, Vardon DR, Guarnieri MT, Karp EM, Hunsinger GB, Franden MA, et al (2014) Lignin valorization through integrated biological funneling and chemical catalysis. Proc Natl Acad Sci USA 111:2013–12018

Liou JSC, Balkwill DL, Drake GR, Tanner RS (2005) *Clostridium carboxidivorans* sp. nov., a solvent-producing *clostridium* isolated from an agricultural settling lagoon, and reclassification of the acetogen *Clostridium scatologenes* strain SL1 as *Clostridium drakei* sp. nov. Int J Syst Evol Microbiol 55:2085–2091

Liu K, Atiyeh HK, Tanner RS, Wilkins MR, Huhnke RL (2012) Fermentative production of ethanol from syngas using novel moderately alkaliphilic strains of *Alkalibaculum bacchi*. Bioresour Technol 104:336–341

Liu S-Y, RAINEY FA, Morgan HW, MAYER F, Wiegel J (1996) *Thermoanaerobacterium aotearoense* sp. nov., a slightly acidophilic, anaerobic thermophile isolated from various hot springs in New Zealand, and emendation of the genus *Thermoanaerobacterium*. Int J Syst Bacteriol 46(2):388–396

Lopez-Hidalgo AM, Alvarado-Cuevas ZD, De Leon-Rodriguez A (2018) Biohydrogen production from mixtures of agro-industrial wastes: chemometric analysis, optimization and scaling up. Energy 159:32–41. https://doi.org/10.1016/j.energy.2018.06.124

Luli GW, Jarboe L, Ingram LO (2008) The development of ethanologenic bacteria for fuel production. In: Wall JD, Harwood CS, Demain AL (eds) Bioenergy. ASM Press, Washington, DC

Luo G, Xie L, Zou Z, Wang W, Zhou Q (2010) Evaluation of pretreatment methods on mixed inoculum forboth batch and continuous thermophilic biohydrogen and biodiesel production from cassava stillage. Bioresour Technol 101:959–964

Luque R, Herrero-Davila L, Campelo JM, Clark JH, Hidalgo JM, Luna D, Marinas JM, AA Romero (2008) Biofuels: a technological perspective. Energy Environ Sci 1:542–564

Lynd LR, Weimer PJ, van Zyl WH, Pretorius IS (2002) Microbial cellulose utilization: Fundamentals and biotechnology. Microbiol Mol Biol Rev 66:506–577

Ma K, Robb FT, Adams MWW (1994) Purification and characterization of NADP-specific alcohol dehydrogenase and glutamate dehydrogenase from the hyperthermophilic archaeon *Thermococcus litoralis*. Appl Environ Microbiol 60:562–568

Machielsen R, Uria AR, Kengen SW, Van-der Oost J (2006) Production and characterization of a thermostable alcohol dehydrogenase that belongs to the aldo-keto reductase superfamily. Appl Environ Microbiol 72:233–238

Malhotra R (2007) Road to emerging alternatives-biofuels and hydrogen. J Petrotech Soc 4:34–40

Margesin R, Feller G (2010) Biotechnological applications of psychrophiles. Environ Technol 31:835e44

Margesin R, Miteva V (2011) Diversity and ecology of psychrophilic microorganisms. Res Microbiol 162:346e61

Marriott PE, Góme LD, McQueen-Mason SJ (2016) Unlocking the potential of lignocellulosic biomass through plant science. New Phytol 209:1366–1381

McKeown RM, Scully C, Enright AM, Chinalia FA, Lee C, Mahony T, Collins G, O'Flaherty V (2009) Psychrophilic methanogenic community development during long-term cultivation of anaerobic granular biofilms. ISME J 3:1231–1242

Meng X, Yang J, Xu X, Zhang L, Nie Q, Xian M (2009) Biodiesel production from oleaginous microorganisms. Renew Energy 34:1–5

Metting FB (1996) Biodiversity and application of microalgae. J Ind Microbiol Biotechnol 17:477–489

Miriam LRM, Raj RE, Kings AJ, Visvanathan MA (2017) Identification and characterization of a novel biodiesel producing halophilic *Aphanothece halophytica* and its growth and lipid optimization in various media. Energy Convers Manag 141:93–100

Miura T, Kita A, Okamura Y, Aki T, Matsumura Y, Tajima T, et al (2015). Improved methane production from brown algae under high salinity by fedbatch acclimation. Bioresour Technol 187:275–281

Mukhtar S, Mehnaz, S, Malik KA (2019b) Microbial diversity in the rhizosphere of plants growing under extreme environments and its impact on crops improvement. Environ Sustain. https://doi.org/10.1007/s42398-019-00061-5

Mukhtar S, Mirza BS, Mehnaz S, Mirza MS, Mclean J, Kauser AM (2018) Impact of soil salinity on the structure and composition of rhizosphere microbiome. World J Microbiol Biotechnol 34:136

Mukhtar S, Mirza MS, Mehnaz S, Malik KA (2019a) Isolation and characterization of halophilic bacteria from the rhizosphere of halophytes and non-rhizospheric soil samples. Braz J Microbiol 50:85–97

Mukhtar S, Ahmad S, Bashir A, Mirza MS, Mehnaz S, Malik KA (2019c) Identification of plasmid encoded osmoregulatory genes from halophilic bacteria isolated from the rhizosphere of halophytes. Microbiol Res 228:126307

Narihiro T, Sekiguchi Y (2007) Microbial communities in anaerobic digestion processes for waste and wastewater treatment: A microbiological update. Curr Opin Biotechnol 18:273–278

Nguyen DM, Lipscomb GL, Schut GJ, Vaccaro BJ, Basen M, Kelly RM, Adams MW (2015) Temperature-dependent acetoin production by *Pyrococcus furiosus* is catalyzed by a biosynthetic acetolactate synthase and its deletion improves ethanol production. Metab Eng 34:71–79

Nozhevnikova AN, Zepp K, Vazquez F, Zehnder AJB, Holliger C (2003) Evidence for the existence of psychrophilic methanogenic communities in anoxic sediments of deep lakes. Appl Environ Microbiol 69:1832–1835

Park HJ, Jeon JH, Kang SG, Lee JH, Lee SA, Kim HK (2007) Functional expression and refolding of new alkaline esterase, EM2L8 from deep-sea sediment metagenome. Protein Expression Purif 52:340–347

Quehenberger J, Shen L, Albers SV, Siebers B, Spadiut O (2017) *Sulfolobus*—A potential key organism in future biotechnology. Front Microbiol 8:2474

Rana KL, Kour D, Sheikh I, Yadav N, Yadav AN, Kumar V et al (2019) Biodiversity of endophytic fungi from diverse niches and their biotechnological applications. In: Singh BP (ed) Advances in endophytic fungal research: present status and future challenges. Springer International Publishing, Cham, pp 105–144. https://doi.org/10.1007/978-3-030-03589-1_6

Rashid N, Aslam M (2019) An overview of 25 years of research on *Thermococcus kodakarensis*, a genetically versatile model organism for archaeal research. Folia Microbiol. https://doi.org/10.1007/s12223-019-00730-2

Rasoul-Amini S, Mousavi P, Montazeri-Najafabady N, Mobasher MA, Mousavi SB, et al (2014) Biodiesel properties of native strain of *Dunaliella salina*. Int J Renew Energy Res 4:39–41

Rastegari AA, Yadav AN, Gupta A (2019a) Prospects of renewable bioprocessing in future energy systems. Springer International Publishing, Cham

Rastegari AA, Yadav AN, Yadav N (2019b) Genetic manipulation of secondary metabolites producers. In: Gupta VK, Pandey A (eds) New and future developments in microbial biotechnology and bioengineering. Elsevier, Amsterdam, pp 13–29

Rastegari AA, Yadav AN, Yadav N, Tataei Sarshari N (2019c) Bioengineering of secondary metabolites. In: Gupta VK, Pandey A (eds) New and future developments in microbial biotechnology and bioengineering. Elsevier, Amsterdam, pp 55–68

Rastegari AA, Yadav AN, Yadav N (2020) New and future developments in microbial biotechnology and bioengineering: Trends of microbial biotechnology for sustainable agriculture and biomedicine systems: diversity and functional perspectives. Elsevier, Amsterdam

Rathore D, Singh A, Dahiya D, Nigam PS (2019) Sustainability of biohydrogen as fuel: Present scenario and future perspective. AIMS Energy 7:1–19

Rezaei S, Shahverdi AR, Faramarzi MA (2017) Isolation, one-step affinity purification, and characterization of a polyextremo-tolerant laccase from the halophilic bacterium *Aquisalibacillus elongatus* and its application in the delignification of sugar beet pulp. Bioresour Technol 230:67–75

Rogers PLK, Lee J, Skotnicki ML, Tribe DE (1982) Ethanol production by *Zymomonas mobilis*. Adv Biochem Eng 23:37–84

Ronnow PH, Gunnarsson LAH (1981) Sulfide dependent methane production and growth of a thermophilic methanogenic bacterium. Appl Environ Microbiol 42:580–584

Sanchez OJ, Cardona CA (2006) Trends in biotechnological production of fuel ethanol from different feedstocks. Bioresour Technol 99:5270–5295

Schelert J, Drozda M, Dixit V, Dillman A, Blum P (2006) Regulation of mercury resistance in the crenarchaeote *Sulfolobus solfataricus*. J Bacteriol 188:7141–7150

Schicho RN, Ma K, Adams MW, Kelly RM (1993) Bioenergetics of sulfur reduction in the hyperthermophilic archaeon *Pyrococcus furiosus*. J Bacteriol 175:1823

Schink B (1997) Energetics of syntrophic cooperation in methanogenic degradation. Microbiol MolBiol Rev 61:262–280

Sedlak M, Ho NWY (2004) Production of ethanol from cellulosic biomass hydrolysates using genetically engineered *Saccharomyces* yeast capable of co-fermenting glucose and xylose. Appl Biochem Biotechnol 114:403–416

Sharma A, Kawarabayasi Y, Satyanarayana T (2012) Acidophilic bacteria and archaea: acid stable biocatalysts and their potential applications. Extremophiles 16:1–19

Shen CR, Liao JC (2008) Metabolic engineering of Escherichia coli for 1-butanol and 1-propanol production via the keto-acid pathways. Metab Eng 10:312–320

Siddiqui KS, Cavicchioli R (2006) Cold-adapted enzymes. Ann Rev Biochem 75:403–433

Singh SP, Singh D (2010) Biodiesel production through the use of different sources and characterization of oilsand their esters as the substitute of diesel: a review. Renew Sustain Energy Rev 14:200–216

Singh BP, Panigrahi MR, Ray HS (2000) Review of biomass as a source of energy for India. Energy Sour 22:649–658

Singh A, Pant D, Korres NE, Nizami AS, Prasad S, Murphy JD (2010) Key issues in life cycle assessment of ethanol production from lignocellulosic biomass: challenges and perspectives. Bioresour Technol 101:5003e12

Singh RN, Gaba S, Yadav AN, Gaur P, Gulati S, Kaushik R, Saxena AK (2016) First, High quality draft genome sequence of a plant growth promoting and Cold Active Enzymes producing psychrotrophic *Arthrobacter agilis* strain L77. Stand Genomic Sci 11:54. https://doi.org/10.1186/s40793-016-0176-4

Sokolova T, Gonzalez J, Kostrikina N, Chernyh N, Tourova T, Kato C, Bonch-Osmolovskaya E, Robb F (2001) *Carboxydobrachium pacificum* gen. nov., sp. nov., a new anaerobic, thermophilic, CO-utilizing marine bacterium from Okinawa Trough. Int J Syst Evol Microbiol 51:141–149

Sommer P, Georgieva T, Ahring BK (2004) Potential for using thermophilic anaerobic bacteria for bioethanol production from hemicellulose. Biochem Soc Trans 32:283–289

Sonntag F, Buchhaupt M, Schrader J (2014) Thioesterases for ethylmalonylCoA pathway derived dicarboxylic acid production in *Methylobacterium extorquens* AM1. Appl Microbiol Biotechnol 98:4533–44

Spolaore P, Cassan CJ, Duran E, Isambert A (2006) Commercial applications of microalgae. J Biosci Bioeng 101:87–96

Srinivas TNR, Kumar PA, SasikalaCh, Sproer C, Ramana CV (2008) *Rhodobacterovatus* sp. nov., a phototrophic alphaproteobacterium isolated from a polluted pond. Int J Syst Evol Microbiol 58:1379e83

Stekhanova TN, Mardanov AV, Bezsudnova EY, Gumerov VM, Ravin NV, Skryabin KG, Popov VO (2010) Expression, purification and crystallization of a thermostable short-chain alcohol dehydrogenase from the archaeon *Thermococcus sibiricus*. Appl Environ Microbiol 76:4096–4108

Sun X, Robert RG (2003) Synthesis of higher alcohols in a slurry reactor with cesium-promoted zinc chromite catalyst in decahydronaphthalene. Appl Catal A Gen 247:133–142

Takai K, Inoue A, Horikoshi K (2002) *Methanothermococcus okinawensis* sp. nov., a thermophilic, methane-producing archaeon isolated from a Western Pacific deep-sea hydrothermal vent system. Int J Syst Evol Microbiol 52:1089–1095

Takeuchi M, Kamagata Y, Oshima K, Hanada S, Tamaki H, Marumo K, et al (2014) *Methylocaldum marinum* sp. nov., a thermotolerant, methane-oxidizing bacterium isolated from marine sediments, and emended description of the genus Methylocaldum. Int J Syst Evol Microbiol 64 3240–3246

Tan HT, Lee KT, Mohamed AR (2010) Second-generation bio-ethanol (SGB) from Malaysian palm empty fruit bunch: Energy and exergy analyses. Bioresour Technol 101:5719–5727

Tandon P, Jin Q (2017) Microalgae culture enhancement through key microbial approaches. Renew Sustain Energy Rev 80:1089–1099

Taroepratjeka DAH, Imai T, Chairattanamanokorn P, Reungsang A (2019) Investigation of hydrogen-producing ability of extremely halotolerant bacteria from a salt pan and salt-damaged soil in Thailand. Int J Hydrogen Energy 44:3407–3413

Taylor MP, Eley KL, Martin S, Tuffin MI, Burton SG, Cowan DA (2009) Thermophilic ethanologenesis: future prospects for second-generation bioethanol production. Trends Biotechnol 27:398–405

Tollefson J (2008) Energy: not your father's biofuels. Nature 451:880–883

Tutino ML, di Prisco G, Marino G, de Pascale D (2009) Cold-adapted esterases and lipases: from fundamentals to application. Protein Pept Lett 16:1172–1180

Ueda M, Goto T, Nakazawa M, Miyatake K, Sakaguchi M, Inouye K (2010) A novel cold-adapted cellulase complex from *Eiseniafoetida*: Characterization of a multienzyme complex with carboxymethylcellulase, betaglucosidase, beta-1,3 glucanase, and beta-xylosidase. Comp Biochem Physiol B Biochem Mol Biol 157:26–32

Uzoejinwa BB, He X, Wang S, Abomohra AE-F, Hu Y, Wang Q (2018) Co-pyrolysis of biomass and waste plastics as a thermochemical conversion technology for high-grade biofuel production: Recent progress and future directions elsewhere worldwide. Energy Conv Manag 163:468–492

Van-der Oost J, Voorhorst WG, Kengen SW, Geerling AC, Wittenhorst V, Gueguen Y, de Vos WM (2001) Genetic and biochemical characterization of a short-chain alcohol dehydrogenase from the hyperthermophilic archaeon *Pyrococcus furiosus*. Eur J Biochem 268:3062–8

Varel V, Chen T, Hashimoto A (1988) Thermophilic and mesophilic methane production from anaerobic degradation of the cyanobacterium *Spirulina maxima*. Resour Con Recy 1:19–26

Vignais PM, Billoud B (2007) Occurrence, classification and biological function of hydrogenases: an overview. Chem Rev 107:4206–4272

Wagner ID, Wiegel J (2008). Diversity of thermophilic anaerobes. In: Incredible anaerobes: From physiology to genom ics fuels. Annals of the New York Academy of Sciences, vol 1125, pp 1–43

Watanabe N, Ota Y, Minoda Y, Yamada K (1977) Isolation and identification of alkaline lipase-producing microorganisms, culture conditions and some properties of crude enzymes. Agric Biol Chem 41:1353–1358

Wei P, Bai L, Song W, Hao G (2009) Characterization of two soil metagenome-derived lipases with high specificity for p-nitrophenyl palmitate. Arch Microbiol 191:233–240

Wen Z, Wu M, Lin Y, Yang L, Lin J, Cen P (2014) Artificial symbiosis for acetone-butanol-ethanol (ABE) fermentation fro m alkali extracted deshelled corn cobs by co-culture of *Clostridium beijerinckii* and *Clostridium cellulovorans*. Microb Cell Fact 13:92

Weng JK, Li X, Bonawitz ND, Chapple C (2008) Emerging strategies of lignin engineering and degradation for cellulosic biofuel production. Curr Opin Biotechnol 19:166–172

Wichlacz PL, Unz RF, Langworthy TA (1986) *Acidiphiliumangustum* sp. nov., *Acidiphiliumfacilis* sp. nov., *Acidiphiliumrubrum* sp. nov. Acidophilic heterotrophic bacteria isolated from acidic coal mine drainage. Int J Syst Bacteriol 36:197–201

Wisotzkey JD, Jurtshuk P, Fox GE, Deinhard G, Poralla K (1992) Comparative sequence analyses on the 16S rRNA (rDNA) of *Bacillus acidocaldarius*, *Bacillus acidoterrestris*, and *Bacillus cycloheptanicus* and proposal for creation of a New Genus, *Alicyclobacillus* gen. nov. Int J Syst Bacteriol 42(2):263–269

Wouter WH, Toirkens MJ, Wu Q, Pronk JT, Van Maris AJA (2009) Novel evolutionary engineering approach for accelerated utilization of glucose, xylose and arabinose mixtures by engineered *Saccharomyces cerevisiae* strains. Appl Environ Microbiol 75:907–914

Wu X, Zhang C, Orita I, Imanaka C, Fukui T (2013) Thermostable alcohol dehydrogenase from *Thermococcus kodakarensis* KOD1 for enantioselective bioconversion of aromatic secondary alcohols. Appl Environ Microbiol 79:2209–2217

Yadav AN, Saxena AK (2018) Biodiversity and biotechnological applications of halophilic microbes for sustainable agriculture. J Appl Biol Biotechnol 6:1–8

Yadav AN, Sachan SG, Verma P, Kaushik R, Saxena AK (2016) Cold active hydrolytic enzymes production by psychrotrophic *Bacilli* isolated from three sub-glacial lakes of NW Indian Himalayas. J Basic Microbiol 56:294–307

Yadav AN, Kumar R, Kumar S, Kumar V, Sugitha T, Singh B et al (2017) Beneficial microbiomes: biodiversity and potential biotechnological applications for sustainable agriculture and human health. J Appl Biol Biotechnol 5:45–57

Yadav AN, Gulati S, Sharma D, Singh RN, Rajawat MVS, Kumar R et al (2019a) Seasonal variations in culturable archaea and their plant growth promoting attributes to predict their role in establishment of vegetation in Rann of Kutch. Biologia 74:1031–1043 https://doi.org/10.2478/s11756-019-00259-2

Yadav AN, Singh S, Mishra S, Gupta A (2019b) Recent advancement in white biotechnology through fungi. Volume 2: Perspective for value-added products and environments. Springer International Publishing, Cham

Yadav AN, Yadav N, Sachan SG, Saxena AK (2019c) Biodiversity of psychrotrophic microbes and their biotechnological applications. J Appl Biol Biotechnol 7:99–108

Yadav AN, Rastegari AA, Yadav N (2020) Microbiomes of extreme environments: biodiversity and biotechnological applications. CRC Press, Taylor and Francis, Boca Raton, USA

Youssef N, Simpson DR, Duncan KE, McInerney MJ, Folmsbee M, et al (2007) In situ biosurfactant production by Bacillus strains injected into a limestone petroleum reservoir. Appl Environ Microbiol 73:1239–1247

Yun Y-M, Lee M-K, Im S-W, Marone A, Trably E, Shin S-R, Kim M-G, Cho S-K, Kim D-H (2018) Biohydrogen production from food waste: current status, limitations, and future perspectives. Bioresour Technol 248:79–87

Zaldivar J, Nielsen J, Olsson L (2001) Fuel ethanol production from lignocellulose: a challenge for metabolic engineering and process integration. Appl Microbiol Biotechnol 56:17–34

Zazil DAC, Angel MLH, Leandro GO, Edén OC, José TOS, et al (2015) Biohydrogen production using psychrophilic bacteria isolated from Antarctica. Int J Hydrog Energy 40:7586–7592

Zhang G, Jiang N, Liu X, Dong X (2008) Methanogenesis from methanol at low temperatures by a novel psychrophilic methanogen, *Methanolobuspsychrophilus* sp. nov., prevalent in Zoige wetland of Tibetan plateau. Appl Environ Microbiol 74:6114–6120

Zheng YN, Li LZ, Xian M, Ma YJ, Yang JM, Xu X, He DZ (2009) Problems with the microbial production of butanol. J Ind Microbiol Biotechnol 36:1127–1138

Chapter 8
Microbial Biofuel and Their Impact on Environment and Agriculture

Archita Sharma and Shailendra Kumar Arya

Abstract It is a paramount concern to make certain the proper and impervious disposal of organic matter due to piling of organic wastes, changes in the climate, energy security which result in the pollution prevailing in the environment that leads to emerging issues such as epidemics, diseases, obnoxious odors, ammonia release, etc. The emergence of such issues has grabbed the attention of researchers to perform investigations regarding the applications of organic wastes with respect to the concepts of the biotechnology. Researches in these particular domains propose a wide range of advantages both economically and ecologically, which include restrained utilization of fossil fuel, reductions in the emissions o greenhouse gases (GHGs), generation of economical raw materials, substrate development required for numerous microbes. Energy obtained from the biofuels (end-product) is an appealing elucidation for legitimate dumping of feedstocks. This chapter puts spotlights on the production of biofuel from numerous microorganisms and their impact on the agriculture and environment. This also gives insights on the certain examples, which describe the biofuel generation from different microbes or agricultural residues via different mechanisms and concepts. This chapter also describes the commercialization of the biofuels and associated concepts.

8.1 Introduction

The global deadlock of energy has put the universe into erratic and agitated situations because of the increasing demands and rapid reduction of resources on daily basis and it has been observed that soon these available resources will vanish out completely. In situations like these, much attention is required for the concept of the exploitation of renewable resources for the production of energy. With global consumption of fossil fuels, the demerit of fossil fuels is that they are not unsustainable in nature

A. Sharma · S. K. Arya (✉)
Department of Biotechnology, University Institute of Engineering and Technology (UIET), Panjab University (PU), Chandigarh, Punjab, India
e-mail: skarya_kr@yahoo.co.in

A. N. Yadav et al. (eds.), *Biofuels Production – Sustainability and Advances in Microbial Bioresources*, Biofuel and Biorefinery Technologies 11,
https://doi.org/10.1007/978-3-030-53933-7_8

since they elevate the levels of carbon dioxide and hence results in the accumulation of greenhouse gases (GHGs) and thus leading to a noxious climate. In order to keep certain concepts on point like a clean and pure environment, sustainable development, etc., it is a necessity to develop and produce eco- friendly fuels, namely biofuels (Schenk et al. 2008). These eco-friendly biofuels are produced from numerous types of biomass obtained from distinctive products of agriculture and forest domain and biodegradable wastes of various industries (Dufey 2006; Yadav et al. 2017, 2019). Examples of biofuels like biodiesel (Shay 1993), butanol (Dürre 2007), *Jatropha curcas* (Becker and Makkar 2008) and algae (Sheehan et al. 1998). It has been estimated that the production of biofuels will be 35 billion liters, approximately. Brazil, the United States (US), and the European nations are considered as the world's biggest producers of biofuels (precisely biodiesel) (Khan et al. 2017).

An alternative source of energy (biofuels) is being considered at a large extent since (a) there is no requirement to modify the engine, (b) helps in the reduction of emissions of greenhouse gases (GHGs), (c) endows security of energy, (d) assure sustainable environment, (e) push the development of rural areas due to switching on the power obtained from the agricultural industries rather than petroleum industries. The exploitation of biofuels is comparatively much more compliant and alluring with respect to the present energy (Hassan and Kalam 2013). The goal of the investigations and studies associated with the development and production of the biofuels is to generate products from numerous sources of biological origin which produces energy like alcohols (ethanol, propanol, butanol, propanediol, and butanediol), biodiesel, biohydrogen, biogas, etc. (Elshahed 2010).

From the burning of the biofuels, there has been a reduction in the emission of greenhouse gases such as carbon dioxide, methane, carbon monoxide, etc., as compared to the fossil fuels and thus they are perceived as an eco-friendly approach to produce energy rather than from the fossil fuels. Additionally, the initial raw materials required for the production the biofuels are present in ample amount in the United States (US) and other nations with developed industries and economies. Hence, from the political, sustainable development, and environment point of view, it is pretty much appropriate to raise the concerns in each and every society regarding the research on the development and production of the biofuels and thus eliminating the reliance on foreign nations for oil (Elshahed 2010; Rastegari et al. 2020).

In accordance with the renewable energy policy network (REN 21), in the year of 2011, it has been estimated that around 78% of energy is consumed from fossil fuels all over the globe, 3% of the energy was consumed from nuclear energy and the rest 19% of the consumption was from renewable sources such as wind energy, solar energy, geothermal energy, hydrothermal energy, and biomass from agricultural sector or industries. It has been acknowledged that approximately 13% of the renewable source of energy is exploited from materials that were rich in carbon and was feasible on earth either by direct burning of the biomass or by converting biomass into heat and power via a thermochemical process (Balan 2014; Mohr and Raman 2013).

Presently, approximately 10% of the demand for energy all over the globe has been transformed by biomass. With the increase in the prices of crude oil, exhaustion of the resources, environmental instability, and biomass has the ability to meet the demands of energy prevailing all over the globe. Biomass obtained from plants is present in ample amount and is the renewable energy source which is rich in carbohydrates and thus effective for conversion into biofuels via microbes. Till date bioethanol is one such commercialized product in the industries but not yet exploited for transportation purposes. This chapter gives insights into biofuels produced from numerous types of microbes and their impact on the agriculture and environment, respectively. Additionally, this chapter deals with certain examples of microbial biofuels to gain a more and clear knowledge of different aspects of microbial biofuels (Antoni et al. 2007).

8.2 Microbial Setup for Biofuels

By consuming organic substrates with the aid of microorganisms and also their exploitation in the processes dealing with the metabolic help in the generation of the favorable products, which will be the source of energy production (Kumar and Kumar 2017). Figure 8.1 gives an insight into the generation of microbial biofuel from numerous microorganisms and pathways.

There are certain factors such as (a) choice of microorganisms, (b) substrates to be utilized, (c) process required to produce biofuels are essential with respect to the synthesis of biofuels. For example, the production of ethanol from corn (substrate) requires more intake of energy from fossil fuels in comparison to the production of ethanol when sugarcane is being used as a substrate (Goldemberg et al. 2008). Furthermore, it is required to have a positive balance of energy (on an average) with a viewpoint of commercialization. The other important concern is the selection of an efficient substrate for microbes (Chang et al. 2013). Lignocellulosic biomass can be exploited and transformed into biofuels by dismantling sugars. This approach usually initiates first with the step of pretreatment of the lignocelluloses following hydrolysis by employing enzymes or by centralized bioprocessing approaches (Kumar et al. 2009; Mosier et al. 2005). The biomass will further be hydrolyzed either by simple cocktails of enzyme cellulose or by a cellulolytic microbe (Lynd et al. 2002).

When compared to carbon dioxide, there is less emission of methane, a component of natural gas, but it is more persuasive in nature (Yvon-Durocher et al. 2014). The production of methane is from landfills or from anaerobic digestion of numerous wastes with organic content. There has been a dramatic flow of methane in the recent past and thus grabs the attention of the researchers to look for an effective source of carbon. By employing methanotrophs which will feed on methane (from landfills and natural gas) directly and produce or another approach is to transform methane into to methanol via methanotrophs and hence the biofuel production (Liao et al. 2016). The oxidation of methane via methanotrophs was done by first reducing the oxygen atoms of hydrogen peroxide followed by the conversion of methane to

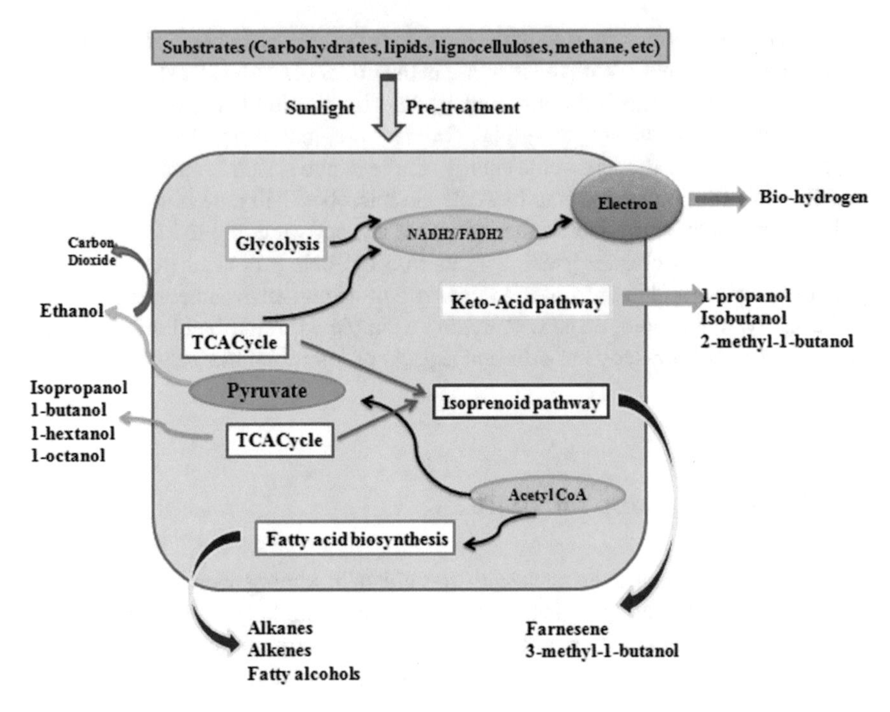

Fig. 8.1 Synopsis of biofuels from different pathways (Kumar and Kumar 2017)

methanol by utilizing methane monooxygenases (Fuerst 2013). There are two types of methane monooxygenases: (a) soluble methane monooxygenases and (b) particulate methane monooxygenases. It has been noticed that the cells which consist of particulate methane monooxygenases have high abilities to grow along with the high affinity for methane than cells which consist of soluble methane monooxygenases (Kumar and Kumar 2017).

8.3 Biofuels from Microbes

The two basic categories of biofuels are primary biofuels and secondary biofuels. The primary biofuels are basically the unprocessed form of fuel used chiefly during heating, cooking, generation of electricity like fuelwood, wood chips, etc., whereas production of secondary biofuels are from the processing of biomass, for example, ethanol, biodiesel, etc., which can later be exploited in the transportation sector and for numerous processes of industries (Rana et al. 2019; Rastegari et al. 2019a). The secondary biofuels are also categorized, on the grounds of raw material and the technology to employ to produce biofuels, into first-generation biofuels, second-generation biofuels, and third-generation biofuels, respectively. There has been

Table 8.1 Various microorganisms and their significance in the production of biofuels

Microorganism's name	Importance in terms of biofuel production
Anaebaena variabilis	It is a cyanobacteria which produces biohydrogen
Clostridium acetobutylicum	It is considered as one of the significant microorganism which produces butanol
Micrococcus luteus	This source generates alkenes of long chains
Zymomonas mobilis	It has a role in the fermentation process of ethanol with a high level of ethanol tolerance
Saccharomyces cerevisiae	Significant source for the production of ethanol
Rhodospeudomonas palustris	This microorganism which belongs to the phototroph family and produces producing hydrogen gas
Saccharophagus degradans	This microorganism helps in the degradation of the various biopolymers

Source Wackett (2008)

extensive research done in the recent past regarding the production of biofuels from numerous microorganisms (Table 8.1) (Nigam and Singh 2011).

Present-day technologies which include biotechnology are doing huge exercises to convert biomass into substances with a potential to utilize them as a fuel at a particular cost, which can go up against in the contest of the high prices of the crude oil. There will be a release of carbon dioxide into the atmosphere when these substances of the biomass are burned that were fixed by the process of photosynthesis and thus resolving the issue of global warming (Fig. 8.2). Additionally, the costs and yield associated with the production of biofuels from biomass is a matter of concern. The favorable outcome of biofuels depends on the economic process of the conversion of biomass into biofuel having physicochemical features which function as a substitute for fuels generated from fossil fuels (Nigam and Singh 2011).

While burning fossil fuels, there is a release of carbon dioxide into the atmosphere resulting in an increase in the concentration of carbon dioxide and thus contribution to global warming. Plant or certain photosynthetic microorganisms help in fixing some portion of the carbon dioxide prevailing in the atmosphere. There are certain microorganisms that either employ carbon dioxide (photosynthetic microorganisms) or biomass as a source of carbon or produce various carbon substances which are later utilized as fuels (Fig. 8.2). The biggest obstacle for the biotechnology is to generate such substances in an economical, sustainable, and appropriate way. One can derive biomass by cultivating dedicated crops that generate energy either by reaping residues of forests and plants or from the wastes of biomass (Gullison et al. 2007). There are numerous feedstocks of biomass such as crops of sucrose and starch (sugarcane and corn), lignocellulosic materials (rice straw and switchgrass), which can be utilized for the production of biofuels but the cost associated with the hydrolysis of the lignocellulosic materials is a matter which one must consider (Sharma and Arya 2017).

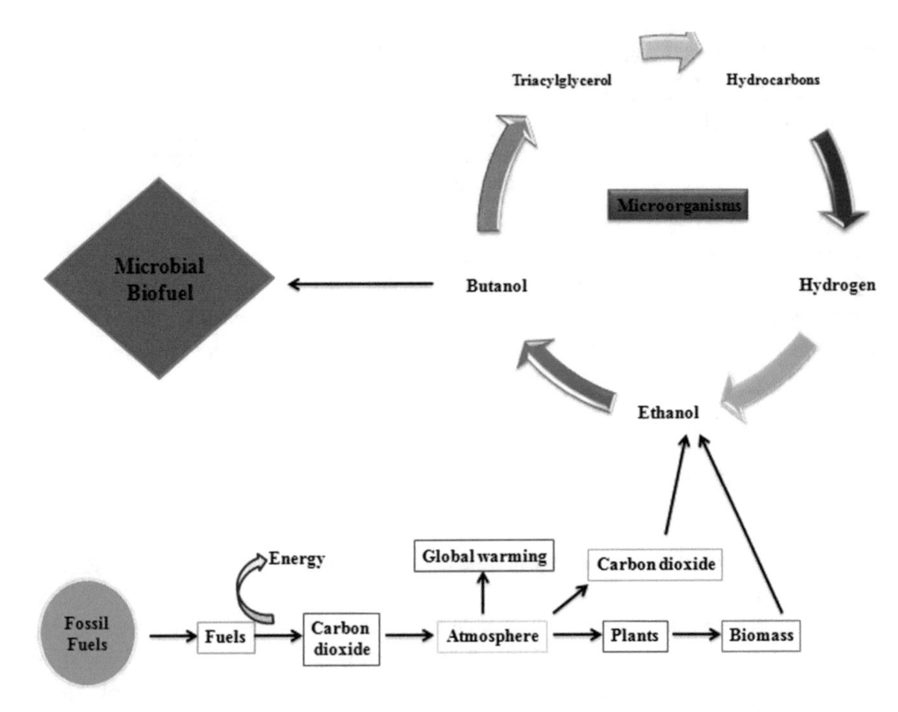

Fig. 8.2 Generation of biofuels from numerous microorganisms (Rojo 2008)

8.3.1 Generation of Biofuels from Microbes

In the recent past, the advancements have depicted that there are certain species of microbes like yeast, fungi, microalgae that have the potential to utilize them for the production of biofuels like biodiesel, biohydrogen, bioethanol, etc., since it is possible to synthesize biologically along with the storage of huge content of fatty acids in the biomass (Kour et al. 2019a; Yadav et al. 2020; Xiong et al. 2008). It has been reported by the team of researchers back in the year 2009 (Huang et al. 2009) about the production of microbial oil from the wastes of rice straw. The microbial oil can also be generated from hydrolysate of the rice straw, which was treated with sulfuric acid (H_2SO_4) by cultivating *Trichosporon fermentans*. It has been observed that the fermentation of rice straw which was treated with sulfuric acid (H_2SO_4) when detoxification was not done results in low yields of lipid, that is approximately 0.17% w/v (1.7 g l^{-1}). Group of researchers (Huang et al. 2009) exercised to enhance the yield of the process. The pretreatment stages (detoxification) consist of (a) over-liming, (b) concentration, and (c) adsorption via Amberlite XAD-4 have enhanced the fermentation capability of rice straw, which was treated with sulfuric acid (H_2SO_4) in a significant manner.

The concept of pretreatment has assisted in augmenting the yield of lipids via removal of the inhibitors present in the rice straw treated with sulfuric acid. It has been recognized that the biomass of the microbe in total was 28.6 g l^{-1} after a fermentation process for 8 days. The lipid content of the rice straw which was treated with sulphuric acid (H_2SO_4) observed was 40.1% which corresponds to the yield of lipid to around 11.5 g l^{-1} after cultivating with *Trichosporon fermentans*. Furthermore, apart from rice straw which was treated with sulphuric acid (H_2SO_4), *Trichosporon fermentans* has the ability to metabolize more sugars like mannose, galactose, cellobiose, etc., which are present in the hydrolysates of additional lignocellulosic materials of natural origin and can be utilized as a source of carbon. *Trichosporon fermentans* has the ability to evolve and employ hydrolysate of the rice straw in order to increase the content of lipid within the biomass of the cell and thus resulted in increased yields (10.4 g l^{-1}). Hence, this particular microorganism can be utilized as a potential candidate for the production of microbial biofuels (Nigam and Singh 2011).

Another group of researchers have performed experiments (Zhu et al. 2008) and produced biofuel microbial from the wastes of molasses. It has been published that the lipids generated in the biomass of the microbe can later be utilized for producing biodiesel. There has been optimization of the constituents of the growth medium in order to cultivate the culture of interest and after that studies have been done to check the consequence of conditions which are required by the culture on microbial biomass and generation of the lipid via *Trichosporon fermentans*. The favorable source of nitrogen and carbon and the molar ratio of carbon to nitrogen (C: N) with respect to the yield of the lipids was peptone, glucose, respectively. Also, the favorable pH and temperature required for the growth medium for cultivation was 6.5 and 25 °C, respectively. Within such a favorable environment, there has been a cultivation of the culture for seven days with an outcome of the 28.1 g l^{-1} biomass yield from the microbial strain which consists of an amount of lipids to approximately 62.4% which was found to be more when compared to the original data, that is 19.4 g l^{-1} and 50.8%, respectively (Xiong et al. 2008; Zhu et al. 2008). It is also possible to cultivate the strain *Trichosporon fermentans* in a medium which consist of wastes of molasses collected from the sugar industries. From the already published reports, the yield of the lipids was approximately 12.8 g l^{-1} and the total concentration of the sugar (in terms of w/v) was 15% when converted biologically from the wastes of the at a pH value of 6.0 (Zhu et al. 2008).

It is possible to improve the assembly of lipids inside the cells of the microbes by adding numerous sugars into the molasses (pretreated) (Chen et al. 1992; Fakas et al. 2007). It has also been noticed that the amount of the lipid was augmented to 50% of the mass of the cells. The lipid present in the microbes major includes palmitic acid, stearic acid, oleic acid, and linoleic acid along with the unsaturated fatty acids approximately 64% of the total content of the fatty acids (Zhu et al. 2008). The reports have suggested that yeast has the capability to evolve and cultivate pretty well over the lignocellulosic biomass (which was already pretreated). This evolution has resulted in an increase in the accumulation of the lipids in an effective way and thus a potential candidate to produce microbial biofuel from residues of the agriculture sector in an economic and eco-friendly way (Nigam and Singh 2011).

8.3.2 *Generation of Biofuels from Algae*

Algae being the oldest form of life are existent in all the ecosystems surrounding the earth, and thus represent a huge diversity of species active in a broad range of conditions of the environmental (Mata et al. 2010). They are basically called the primitive plants named thallophyte that is no roots, no stems, and no leaves and even does not have a sterile cell covering surrounding the reproductive cells. The primary pigment of photosynthesis of algae is chlorophyll a (Farrell et al. 1998). When growth conditions are innate, the sunlight was absorbed by phototrophic algae and thus the assimilation of carbon dioxide (CO_2) from the air and obtains nutrients from the aquatic biosphere. It has also been observed that there can be a production of lipids, protein carbohydrates in bulk amounts from microalgal species in a very brief time which can later be exploited for the production of the biofuels and other worthy related products (Brennan and Owende 2010). But the production of biofuels from lipids, proteins is a limited affair because of the restrained availability of the sunlight and hence results in the commercialization of such biofuels only in those regions where sunlight is available in bulk amount (Pulz and Scheibenbogen 2007). The capability of microalgal species for carbon dioxide fixation has been considered as an alternative to diminish the emissions of greenhouse gases (GHGs). Furthermore, algal species are highly rich in the oil and thus can be utilized for the production of the biodiesel (Gislerød et al. 2008).

Researchers have observed that (Widjaja et al. 2009) when a particular or appropriate nitrogen source was absent there has been the production of oil, in the majority, while when the sunlight was present there has been the production of sugars, proteins, etc., from carbon dioxide (CO_2). One of the species of microalgae, namely, *Chlorella protothecoides* accumulates lipids when grown in the presence of autotrophic and heterotrophic environment later which can be utilized for the production of biodiesel. In order to enhance the accumulation of the lipids via microalgal species, the best approach is to limit the source of nitrogen. Apart from the accumulated lipids, there has been a progressive change in the arrangement of the lipids from free fatty acids into triacylglycerols (TAGs) which are comparatively better for the production of biofuels (Meng et al. 2009; Tsukahara and Sawayama 2005).

There are primarily two approaches to convert microalgal species for their utilization and they are (a) thermochemical conversion approach and (b) biochemical conversion approach. The former approach of conversion deals with the decomposition of the organic constituents into the fuel thermally like direct combustion, gasification, thermochemical liquefaction, pyrolysis, etc. (Energy 2002). The later one deals with the conversion of energy obtained from the biomass into certain other fuels via processes like anaerobic digestion, fermentation by employing alcohols, production of hydrogen from photobiological approach, etc. (Grant 2009).

Presently, there are only certain multinational companies (MNCs) owned privately and certain research groups (funded publicly) which are engaged onto the cultivation of algal species and on the condition to lower down the cost of production of oil from the microalgal species at a modest range. A Colorado-based company, Solix

Biofuels, have manufactured a closed-tank bioreactor which employs the generated carbon dioxide (waste) and produces beer. A company of New Zealand named Aqua flow Bionomics is engaged in the biofuels generation through harvesting wild algal species obtained from the foul waterways (Nikolić et al. 2009). Researchers have reported that only the algal species are not capable enough to produce an acceptable amount of fuels so later they shifted their focus on the heterotrophic species of algae and exploited the substances which employ carbon instead of carbon dioxide fixation into the environment. All the algal strains have the ability to use up everything from the glycerol wastes and the wastes of the sugar cane and convert it into the pulps of sugar beets and molasses, respectively (Nikolić et al. 2009).

8.4 Upscaling the Production of Biofuels via Metabolic Engineering

For producing biofuels from microbes, a particular metabolic pathway and various groups of catalytic enzymes are required. For instance, it has been observed that there has been a direct decarboxylation of pyruvate in *Saccharomyces cerevisiae* (baker's yeast), which results in the generation of the ethanol whereas, in *Escherichia coli*, there has been an activation of the acyl group by coenzyme A (CoA) in the course of pyruvate decarboxylation and ultimately leads to ethanol. Hence, the concept of metabolic engineering of aforementioned pathways can be considered as a worthy opportunity to enhance the production of biofuels and be exploited in numerous ways in order to improve the production of biofuels by utilizing microbes. One approach is to produce ethanol from two numerous pathways (mentioned above) in *Saccharomyces cerevisiae* (baker's yeast) and in *Escherichia coli* (*E. coli*), respectively. It has been reported that the best and effective way to produce ethanol is when coenzyme A (CoA) is absent (Liao et al. 2016). For this reason, it is possible to express such pathways in certain microbes via approaches which make use of genetic engineering for producing ethanol. Likewise, the microbes in which the metabolic routes are absent for producing specific biofuel, an injection of imperative genes, enzymes (extracted from particular microbes with an efficiency to produce biofuel) can be given which will help in converting the microbes which do not produce biofuel into microbes which will produce biofuels (Rastegari et al. 2019c). This very particular approach is advantageous to utilize numerous substrates for producing biofuels in the near future.

The second approach can be thought of the competing pathways which will drain either the products, that are biofuels or the precursors like pyruvate, acetyl-Coenzyme A. Additionally, certain enzymes create hindrances in the synthesis process of the biofuel that can be knocked out by exploiting metabolic engineering approaches. For instance, in *Escherichia coli*, an acyl carrier protein (ACP) impedes the route of synthesis of fatty acids (Davis and Cronan 2001). In order to overcome this inhibition, the over-expression of thioesterase enzymes proves beneficial and thus allows the synthesis of fatty acids in free form, which eventually leads to the production of a precursor, that is acyl-Coenzyme A (for the synthesis of fatty alcohol).

Furthermore, to improve the catalytic activity of the enzymes that are specific to a particular substrate and to improve the turnover number, maneuvering the genetic material of an enzyme by utilizing progressive design tools and certain experimental methods. Also, proteins manipulated via computational tools can also be exploited to support amino acids of unnatural origin in order to fabricate and imitate enzymes with all the desired features and properties which later can be exploited for producing biofuels. Although, performing such manipulation is a challenging task which requires a high level of tools with an efficiency to control the proteins at a particular stage of mRNA levels in order to perform very well in an artificial route or environment (Kumar and Kumar 2017).

8.5 Examples of Microbial Biofuels

8.5.1 Producing 1,3-Propanediol from Microbe Klebsiella Pneumoniae *via Utilization of Crude Glycerol*

Biodiesel a derivative of triacylglycerols (TAGs) through a transesterification reaction which utilizes alcohols (short-chains) have gained a lot of attention in the recent past due to certain properties such as renewability, biodegradability, non-toxicity, etc. (Andrade and Vasconcelos 2003; Xu et al. 2003). There has been a generation of byproducts called glycerol during the biodiesel synthesis, which is regarded as a 10% (w/w) of an ester which can be utilized further to enhance the desirability of the whole process (Mu et al. 2006).

The process of conversion of glycerol into 1,3-Propanediol via microbe is an alluring approach and grabbing huge attention of the researchers since it is a comparatively easy process with no generation of lethal byproducts. There are various applications of 1,3-Propanediol such as in polymers, cosmetics section, foods, lubricants, medicines, etc. Production of 1,3-propanediol industrially is considered as an important approach for the synthesis of an advanced class of polyester, namely, polytrimethylene terephthalate (PTT) (Zeng and Biebl 2002). One of the significant constraints for producing 1,3-propanediol via microbe at an industrial scale is the increased costs of the raw materials. It is advised to employ crude glycerol with no purification beforehand with respect to the economic point of view. More research has been going on about the study of the bacterial growth on glycerol of low grade apart from the study of using them as a substrate (Papanikolaou et al. 2004).

In this very particular example, the work has been done on performing shake-flask fermentations and fed-batch fermentations by *Klebsiella pneumonia* for the production of 1,3-Propanediol either by utilizing the pure form of glycerol or crude form of glycerol acquired during methanolysis of soybean oil. There are no reports till date on the production of 1,3-propanediol by *Klebsiella pneumonia* from crude glycerol. The concentration of 1,3-propanediol from crude glycerol obtained during the methanolysis of soybean oil using alkali catalyst was 51.3 g/l^{-1} whereas the concentration of 1,3 propanediol was from crude oil when lipase catalyst was utilized 53 g/l^{-1}. The yield when crude oil was employed was 1.7 $g\ l^{-1}\ h^{-1}$ whereas when pure glycerol was employed the yield was 2 $g\ l^{-1}\ h^{-1}$. Thus, in conclusion, crude oil has the ability to directly transform into 1,3-Propanediol with no requirement of purification beforehand. Also, this work also suggests that the fermentation cost is less which is one of the important factors while employing byproducts and transforms it into certain significant substances (Mu et al. 2006).

8.5.2 Microbial Biofuels from Fatty Acids and Chemicals Obtained from Biomass of Plants

With the increase in the costs of the crude oil and associated concerns of environment, much effort has been put by the researchers for the production of fuels that obey the concept of sustainability (Fortman et al. 2008). The significant efforts have been made on the production of biofuels from various microbes from an economic viewpoint (Lynd et al. 2005). Fatty acids are derivatives of long-chain alkyl and are considered as a primary metabolite being utilized by cells for certain functions. They are the energy-rich compounds being separated from the oils obtained from plants and animals and are considered as a broad group with a range of fuels to oleochemicals. For this particular class of compounds, another pathway to produce an economical biofuel is from the utilization of microbes and which will help in the transformation of feedstocks (renewable) into fuels (Steen et al. 2010).

This example demonstrates the engineering of *Escherichia coli* for the production of artificial fatty esters (called biodiesel), fatty alcohols, waxes, etc., from simple sugars by a direct route. Additionally, this example also provides information regarding the engineering of the cells that generate biodiesel in order to assert hemicelluloses—a significant constituent of biomass which is obtained from plants. Oils obtained from plants and animals are considered as raw materials for the production of biofuels like biodiesel, surfactants, solvents, lubricants, etc. (Hill et al. 2006). A substitute for sustainable development is the production of such oils with a direct route to produce these products directly from sufficient and economical sources which are renewable in nature via the fermentation process. *Escherichia coli* is one such microbe of an industry which constitutes about 9.7% lipid and generates metabolites of fatty acids with productivity at a commercial level of 0.2 $g\ l^{-1}\ h^{-1}$ per gram of mass of the cell in order to grow and accomplish the yield of mass to around 30–35% (Rude and Schirmer 2009). Another merit is that it is possible to manipulate the genetic make-up of the microbe in a flexible way. This particular line of work provides strength to such products enabling them to excel at commercial levels. Also, extensive research has been going on regarding the enhancements of strain and development of advanced process via scale-up bioprocesses keeping in mind the point of view of commercialization (Tsuruta et al. 2009).

8.5.3 Ionic Liquid for the Production of Microbial Biofuel

It has been well reported that biomass of lignocelluloses is present in ample amount and can be readily used for producing biofuels in a sustainable with high commercial values. In recent days, microbial engineering (Liu and Khosla 2010; Wen et al. 2013) is grabbing a lot of attention as it has a potential to produce biofuel along with the utilization of a broad range of feedstocks such as biomass from woods, residues of native grass, agricultural products like corn stover, etc. (Bokinsky et al. 2011; de Jong et al. 2012). As it is a prerequisite that the pretreatment approach is required in order to take care of the recalcitrant biomass and leads to free polysaccharides free from lignin either through enzymatic hydrolysis or through the use of chemicals for fermenting sugars. There are some ionic liquids of hydrophilic nature, which can be exploited for solubilizing lignocellulosic biomass. These ionic liquids are very efficient and eco-friendly candidate to utilize in the pretreatment process and result in the generation of inhibitors obtained from biomass in very less amount in comparison to low numerous traditional methods of pretreatment (Liu et al. 2012; Mora-Pale et al. 2011). A common ionic liquid named imidazolium has certain demerits like the generation of which leads to impairment in the growth of hosts (*Escherichia coli* and *Saccharomyces cerevisiae*), which will produce biofuel inherent toxicity of microbe thus inhibition in the efficiency of the production of the biofuel (Ouellet et al. 2011). Furthermore, another issue is the severe reduction of the product yield at the end of the process of the production of biofuel (Park et al. 2012).

In this example, a mechanism has been developed that will help in resisting the ionic liquids. This mechanism includes two adjoining genes from the strain *Enterobacter lignolyticus* (soil bacteria), which can tolerate ionic liquids having imidazolium. Such genes have the ability to hold their complete functional property during their transform into *Escherichia coli* which will ultimately produce biofuel with resistant ionic liquid which have been established by a transporter present in the inner membrane which is further regulated by an ionic liquid inducible repressor. The transporter is adjusted in such a way so that the expression will be directly through ionic liquids, which will enable the growth and production of biofuel at a particular stage of ionic liquids which is lethal for indigenous strains. Such original autoregulatory mechanisms (by EilR repressor) are efficient for converting lignocellulosic biomass into biofuels. The researchers have chosen the targeted functional screening method for identification of important genetic elements which are subjected to Cl tolerance and to uncover that such genes have efflux pump and regulator. Researchers have transferred such genes into an *Escherichia coli* (engineered host) and demonstrated the improved production of biofuel based on terpenes (secondary metabolites) in the presence of ionic liquids (Ruegg et al. 2014).

This mechanism of efflux has enhanced the production of biofuel when ionic liquid is present in a low amount. We anticipate that engineering IL-tolerant biofuel pathway enzymes and production strains with tolerance to inhibitors originating from biomass breakdown (Klinke et al. 2004) are needed to further increase yields. Furthermore, the tenacity of the strain *Escherichia coli* has been strengthened via an autoregulatory mechanism of efflux of ionic liquids via repressor EilR. Apart from the strain improvement, the production of biofuels has also been efficiently improved for fermentation processes at the industry level in which the levels of ionic liquid fumbles within the batches of the biomass. Additionally, such mechanism forestalls the requirement of expensive molecules for an induction process. It renders the fermentation process cost-effective with the aseptic environment by inhibiting the growth of the contaminants of the microbes (Ruegg et al. 2014).

8.6 Influence of Microbial Biofuel on the Agricultural Sector

With an increase in the requirement of energy and fuel, the global economy is increasing at a fast pace in order to scrutinize the strength of advanced stage biofuels (Kour et al. 2019b; Yang et al. 2009). These forms of bioenergy will help nations to curb the import of petroleum reserves from foreign countries rendering to provide an elucidation regarding dual obstacles which are the issue of security of energy and changes in the climate. However, the major matter is the escalation of emissions of greenhouse gases (GHGs) coming from biofuels produced from the crops with expanded lands (Gibbs et al. 2008). The production of biofuel all over the world is increasing and has stretched to the remarkable levels in the recent past (Gerber et al. 2008). According to the reports, from the year 2001 to the year 2007, the global production of ethanol gets world ethanol production has been intensified, that is, from 20 to 50 billion liters and that of the biodiesel increased from 0.8 to 4 billion liters (Banse and Meijl 2008).

It has been a tough task to make an estimate of numerous calculations of the generation of biofuels on the prices of food commodities and agricultural commodities, respectively (Gerber et al. 2008). Furthermore, compared to the countries with well-developed industries, the developing nations will be more competitive for producing biofuels because of the reduced production and the opportunity of availing the reasonable agricultural land for cultivating feedstocks for the generation of biofuels. With necessary global trades and investments for developing nations, there has been a prediction of numerous challenges. Also, increased prices already exist in the markets of feedstocks (sugar, rapeseed, soybeans, jatropha) for the production of biofuel (Ottinger 2007).

In opposition, poor belonging to urban and rural areas of the nation which imports food to other countries have to pay increased prices of simple and important food with less availability of grains to feed the humanity (Cassman 2007). It has been recorded that trading of certain food products like wheat was high in terms of pricing in the year 2006–2008. There are numerous nations which have adopted certain strict policies regarding the promotion of biofuels as an alternative of gasoline in the transport section. For example, around 10% of the use of gasoline for the United States is generated from corn ethanol which will grow to 30% in the year 2022 (Charles 2012). Such hikes will create disturbances in the weak and poor sections of society by spending crooked shares of their monthly salary at a high rate on food commodities in order to meet the cliché requirements of nutrition (Charles 2012). It has been investigated globally that the production of biofuels from first-generation feedstocks along with the use of agricultural land for the generation of foods will have an adverse and serious implication on the supplies of agriculture and food (Ottinger 2007). Universally, the land is the scarce source and thus more pressure on the effective allocation of land with innovative agricultural practices (Lambin and Meyfroidt 2011).

The significant difference in food crops and energy crops is the relationship between yield and the input. In case of food crops, the major concern is the yield and to clinch the same, there should be a willingness to people have been willing to boost the supply of inputs like water, fertilizer, labor, machinery, etc. (Sang 2011). There are certain crops that produce biofuels which are in need of bulk amount of water required to cultivate the crops and this a major concern specifically to the areas where water supplies are scarce. Additionally, reduction of supply of water and contamination of resources of water has an acute effect on the health of humans and animals (Ottinger 2007).

From the study of World Bank, it has been found out that in the year 2006–2008 there has been an increase in the prices of the food commodities for the production of biofuel by 70–75% and thus more worldwide focus regarding the relation between biofuels and prices of the commodities of the food. Since the already utilized crops are not sufficient enough to meet the aims and requirements of the production of energy in a sustainable way, so the extensive research is going on the contemporary crops that will provide energy. In order to overcome the aforementioned issue, it is required to cultivate and grow such contemporary crops on marginal lands which are not appropriate for the production of food. These marginal lands must have increase yield of biomass which requires a very little requirement of irrigation facilities. There should be a minimization of certain inputs which requires energy such as tilling, planting, harvesting, storage, transportation, etc. Such crops are known as second-generation energy crops (Heaton et al. 2008; Karp and Shield 2008; Oliver et al. 2009). The most promising advantage of biofuels is their positive effect on the employment rates in agricultural lands and practices and enhancement in the livelihoods of poor farmers in rural sectors. Apart from this potential merit, the serious implications of biofuels are on the global market of agriculture. There is strength in the production of biofuels to generate employment opportunities in rural sectors but the majority of shares of employment are for those agricultural workers that are not extremely skilled or knowledgeable and such workers are precisely more in jeopardy (Kumar et al. 2019; Rastegari et al. 2019b).

There are numerous risks associated with the health of the farmers while working in the agricultural fields or lands majorly because of exploiting the agrochemicals by improper means like no proper and full information regarding their use and no means of safety equipment to them while working. A better environment for sound work should be considered while mentioning the constituents of the standard protocol for the production of biofuels along with their trading (Rosegrant 2008). The production of biofuels from second-generation feedstocks such as residues of animals, crops, timber, food, etc., provide an edge to overcome the existing competition with respect to the food for human consumption but such residues are a vital source of nutrients with respect to the growth and development of plants. With the burning of such residues of crops results in the decrease in the amount of organic matter from the soils and increased utilization of fertilizers such as ammonia. All these practices are exploited under high-energy usage (Bisth et al. 2015).

Considering all the aspects, merits, demerits of biofuel production from second-generation feedstocks, currently biofuel production from third-generation feedstocks are grabbing a lot of attention of researchers. There has been the production of biofuels such as biodiesel from microorganisms like cyanobacteria, microalgae, etc. Microorganisms are pretty much better alluring feedstock for the production of biofuel as compared to the traditional feedstocks such as oil from crops since they have high efficiency of photosynthesis and a high amount of lipids. Photosynthetic microorganisms such as algae and cyanobacteria are the major and significant producer in aquatic animals and cover approximately 71% earth space (Andersen 2005).

These photosynthetic microorganisms have the ability to transform carbon dioxide into various hydrocarbons like lipids. It has also been published that these algal lipids can be considered as an alternative of fossil fuels in the near future because of the building up of them in the cells present at the end stage of growth (Abdeshahian et al. 2010; Kenthorai et al. 2011). Another advantage of these microorganisms are that (a) they have the capability to grow in the nutrient medium with minimum supplements like water, photon, carbon dioxide, (b) the simple mass cultivation, (c) simple process of extraction and purification and thus appropriate for the production of biofuel in bulk (Hu et al. 2008). Since the impact of agriculture is both on the climate and production of food. Thus, the adverse impacts will be more (a) on the nations which are highly vulnerable to the changes in the climate (b) on the farmers with low-income source and thus high chances of poverty. The developing nations have a positive impact of biofuels, that is poor farmers of the rural sector may get employment for producing biofuels from microbes and thus helps in improving the development of the area and their livelihoods. Thus, on the whole, biofuels, when compared to the traditional fuels, provide many advantages such as security of energy in the near future, less serious implications on the environment, savings during the foreign exchange, and prevention of certain socioeconomic problems (Mohammady 2007).

8.7 Influence of the Microbial Biofuel on the Environment

From the period of an automobile long ago, oil was considered as an exclusive source of energy of the transport sector. In the year 2007, 95% of the energy for transportation worldwide was from petroleum resources (Hill et al. 2009). Extensive research regarding the search of alternative source of energy is going at full pace. The forces behind such extensive research and urgency are the elusive prices of oil, the global increase in the demand of the fuel and so the energy, more dependence on the imports of fuels from unsettled regions, awareness regarding harmful implications of greenhouse gases and pollution in the air, etc. (Maclean and Lave 2003). Numerous nations are trying to enhance the security of energy along with the economic aspects to diminish the emissions of the harmful greenhouse gases to lessen the effect on the changes in the climate worldwide. Research has been done to develop biofuels from biomass for transportation. Such biofuels will help in replacing the fuels generated from the petroleum reserves, which consist of atmospheric carbon instead of fossil carbon, hence addresses the concerns related to the emissions of greenhouse gases (Frank et al. 2012).

Recent policies associated with energy labeling the problems of the environment such as the development of eco-friendly technologies for increasing the supplies of energy and to support clean and effective utilization of energy which considers certain issues such as air pollution, global warming, and changes in the climate (Demirbas 2009). Traditional fuels are answerable toward certain problems associated with the presence of the pollutants and greenhouse gases in the atmosphere which are contributing to global warming day by day. The ramification of greenhouse gases is a natural event where a portion of the infrared radiations are kept by the atmosphere of the earth because the greenhouse gases get reflected back. Back in the 1990s, The International Panel on Climate Change (IPCC) classified three significant preferences in order to mitigate the concentration of carbon dioxide in the atmosphere via agricultural domain and they are (a) reducing the emissions associated with the agriculture, (b) fabrication of carbon sinks in the soil, (c) generation of biofuels in order to take the place of fossil fuels (Bisth et al. 2015).

There has been a firm conviction that biofuels can be considered as an alternative to mollify the current changes in the climate. This faith has forced the government to think of biofuel by promoting the ethanol production, biodiesel production via certain policies that will ensure the market and grants incentives to producers and consumers, respectively (Sexton and Zilberman 2008). One of the significant reasons that forced to shift the gears on the production of biofuels as a substitute of energy is with respect to the perks being provided to the climate like sequestration of harmful gases and thus called as greenhouse gas neutral. This particular thought of clean and green energy has led one to include biofuel in numerous nations, where industrialization is significant like Unites States of America (USA), European Union (EU), etc. (German and Schoneveld 2011). A significant notion regarding biofuel is the limit up to which biofuel will help in reducing the emissions of carbon dioxide in case of deforestation. Other questions like damage to local worthy goods and services obtained from the forests. While using cultivated land, there are chances that there will be a loss of production of food along with the reduction of the security of food. There are certain potential risks while producing biofuel that might alter the conventional patterns of land, social alliances, favorable circumstances of livelihood, precisely in case of production at a large scale instead of small scale (Lima and Skutsch 2011). All over the globe, the problem of soil erosion is prevailing at a big platform. Biofuel is helpful in accelerating the aspect of geology if the productivity of the cultivated lands is increased (Bisth et al. 2015).

Keeping in mind such issues, much effort has been made on the production of biofuels and cut down the exploitation and consumption of petroleum reserves, increasing the exploitation of the land use, reduction in the emissions of the greenhouse gases (GHGs). With respect to this microalgal species are gaining interest since they have the potential to produce large yields when compared to the grass, grains, trees, etc. Also, they can also generate oils that can be transformed into products like diesel, gasoline, etc. It is mandatory to quantify the products obtained from algal biofuel ad fuels from reserves of petroleum as it requires enough energy to produce fuel (Frank et al. 2012). There is a requirement of cost-effective feedstocks for the production of biofuels but resources such as land, water, nutrients required to produce biofuel from the crops should be readily available. The production of biofuel from woods has serious implications in the form of deforestation or the impact of the emissions of greenhouse gases. It has been reported that productivity of biofuel from the photosynthetic microorganism such as microalgae, cyanobacteria, etc., was large when the comparison was made with terrestrial plants which consist of cellulose such as grass, grains, trees, etc. Such quality for agricultural practices suggests a high yield of biomass per acre. Furthermore, energy can be stored in the lipid of algae which later can be transformed into products like diesel and gasoline with the help of advanced technologies which will play a significant role to recycle carbon (Frank et al. 2012).

8.8 Conclusion and Future Prospects

Presently, the universe is dealing with three crucial issues such as increased prices of fuels, changes in the climate and environmental pollution. Recently, issues like the high demand of energy, rapid depletion of nonrenewable sources of energy, elevated levels of pollution in the environment, etc. have to be resolved as soon as possible. To this, biofuels come to the rescue which provides the security of energy worldwide, make environment amiable, and sustainable (reduction in the emissions of the greenhouse gases), huge savings during the foreign exchange, development of rural areas, etc. For many decades, biofuels are the part of many debates of various nations but in the past years, there has been a shift in the debate and considers the increase in the prices of crude oil. Not because of the prices, there are several reasons (mentioned above) why the government has a keen interest in the development of biofuels even though the subsidies are required for their commercialization. It is difficult to calculate the impact of the biofuels on agricultural commodities with respect to the midterm projections. Biofuels generated from the microbes have enough potential in the domain dealing with the research on the development and generation of the biofuels as an alternative energy source. Much research is required in the biology domain in order to enhance the production of biofuels via breeding energy plants, hydrolysis by using enzymes, strains to treat the wastes and to exploit during the fermentation process.

Acknowledgments The authors thankfully acknowledged Professor Sanjeev Puri for believing in them and giving them this opportunity to explore and gain knowledge and excel.

References

Abdeshahian P, Dashti M, Kalil MS, Yusoff WMW (2010) Production of biofuel using biomass as a sustainable biological resource. Biotechnology 9:274–282

Andersen RA (2005) Algal culturing techniques. Elsevier

Andrade JC, Vasconcelos I (2003) Continuous cultures of *Clostridium acetobutylicum*: culture stability and low-grade glycerol utilisation. Biotechnol Lett 25:121–125

Antoni D, Zverlov VV, Schwarz WH (2007) Biofuels from microbes. Appl Microbiol Biotechnol 77:23–35

Balan V (2014). Current challenges in commercially producing biofuels from lignocellulosic biomass. In ISRN Biotechnology (Vol. 2014). https://doi.org/10.1155/2014/463074

Banse M, van Meijl H, Woltjer G (2008) The impact of first and second generation biofuels on global agricultural production, Trade and Land Use. In: 11th Annual GTAP conference, Helsinki, Finland, June 1–14. Finland, pp 1–14

Becker K, Makkar HPS (2008) *Jatropha curcas*: a potential source for tomorrow's oil and biodiesel. Lipid Technol 20:104–107. https://doi.org/10.1002/lite.200800023

Bisth TS, Panwar A, Pandey M, Pande V (2015) Algal biofuel and their impact on agriculture and environment. Int J Curr Microbiol Appl Sci 4:586–604

Bokinsky G, Peralta-Yahya PP, George A, Holmes BM, Steen EJ, Dietrich J, Keasling JD (2011) Synthesis of three advanced biofuels from ionic liquid-pretreated switchgrass using engineered *Escherichia coli*. Proc Natl Acad Sci (PNAS) 108:19949–19954. https://doi.org/10.1073/pnas.1106958108

Brennan L, Owende P (2010) Biofuels from microalgae: a review of technologies for production, processing, and extractions of biofuels and coproducts. Renew Sustain Energy Rev 14:557–577. https://doi.org/10.1016/j.rser.2009.10.009

Cassman K (2007) Climate change, biofuels, and global food security. Environ Res Lett 2. https://doi.org/10.1088/1748-9326/2/1/011002

CC (2012) Should we be concerned about competition between food and fuel? https://www.iisd.org/library/should-we-be-concerned-about-competition-between-food-and-fuel-analysis-biofuel-consumption

Chang JJ, Ho FJ, Ho CY, Wu YC, Hou YH, Huang CC, Li WH (2013) Assembling a cellulase cocktail and a cellodextrin transporter into a yeast host for CBP ethanol production. Biotechnol Biofuels 6. https://doi.org/10.1186/1754-6834-6-19

Chen J, Ishii T, Shimura S, Kirimura K, Usami S (1992) Lipase production by *Trichosporon fermentans* WU-C12, a newly isolated yeast. J Fermen Bioeng 73:412–414. https://doi.org/10.1016/0922-338X(92)90290-B

Davis MS, Cronan J (2001) Inhibition of *Escherichia coli* acetyl coenzyme a carboxylase by acyl-acyl carrier protein. J Bacteriol 183:1499–1503. https://doi.org/10.1128/JB.183.4.1499-1503.2001

de Jong B, Siewers V, Nielsen J (2012) Systems biology of yeast: enabling technology for development of cell factories for production of advanced biofuels. Curr Opin Biotechnol 23:624–630. https://doi.org/10.1016/j.copbio.2011.11.021

Demirbas A (2009) Political, economic and environmental impacts of biofuels: a review. Appl Energ 86:108–117. https://doi.org/10.1016/j.apenergy.2009.04.036

Dufey A (2006) Biofuels production, trade and sustainable development: emerging issues. International Institute for Environment and Development, London

Dürre P (2007) Biobutanol: an attractive biofuel. Biotechnol J 2:1525–1534. https://doi.org/10.1002/biot.200700168

Elshahed MS (2010) Microbiological aspects of biofuel production: current status and future directions. J Adv Res 1:103–111. https://doi.org/10.1016/j.jare.2010.03.001

Energy, USDE (2002) A national vision of Americas transition in a hydrogen economy- To 2030 and beyond

Fakas S, Galiotou-Panayotou M, Papanikolaou S, Komaitis M, Aggelis G (2007) Compositional shifts in lipid fractions during lipid turnover in *Cunninghamella echinulata*. Enzyme Microb Technol 40:1321–1327. https://doi.org/10.1016/j.enzmictec.2006.10.005

Farrell E, Bustard M, Gough S, McMullan G, Singh P, Singh D, McHale A (1998) Ethanol production at 45 C by *Kluyveromyces marxianus* IMB3 during growth on molasses pre-treated with Amberlite® and non-living biomass. Bioprocess Eng 19:217–219

Fortman JL, Chhabra S, Mukhopadhyay A, Chou H, Lee TS, Steen E, Keasling JD (2008) Biofuel alternatives to ethanol: pumping the microbial well. Trends Biotechnol 26:375–381. https://doi.org/10.1016/j.tibtech.2008.03.008

Frank ED, Han J, Palou-Rivera I, Elgowainy A, Wang MQ (2012) Methane and nitrous oxide emissions affect the life-cycle analysis of algal biofuels. Environ Res Lett 7. https://doi.org/10.1088/1748-9326/7/1/014030

Fuerst J (2013) Planctomycetes: cell structure, origins and biology. (J. A. Fuerst, ed.). https://doi.org/10.1007/978-1-62703-502-6

Gerbe N, Van Eckert M, Breuer T (2008) The impacts of biofuel production on food prices: a review. In: University of Bonn, Center for Development Research (ZEF), Discussion Papers. https://doi.org/10.2139/ssrn.1402643

German L, Schoneveld GC, Gumbo D (2011) The local social and environmental impacts of smallholder-based biofuel investments in Zambia. Ecol Soc 4

Gibbs H, Johnston M, Foley JA, Holloway T, Monfreda C, Ramankutty N, Zaks D (2008) Carbon payback times for crop-based biofuel expansion in the tropics: The effects of changing yield and technology. Environ Res Lett 3. https://doi.org/10.1088/1748-9326/3/3/034001

Goldemberg J, Coelho ST, Guardabassi P (2008) The sustainability of ethanol production from sugarcane. Energy Policy 36:2086–2097. https://doi.org/10.1016/j.enpol.2008.02.028

Grant B (2009) Biofuels made from algae are the next big thing on the alternative energy horizon. But can they free us from our addiction to petroleum? (Vol. 23)

Gullison RE, Frumhoff PC, Canadell JG, Field CB, Nepstad DC, Hayhoe K, Nobre C (2007) Tropical forests and climate policy. Science 316(5827), 985 LP–986. https://doi.org/10.1126/science.1136163

Hassan MH, Kalam MA (2013) An overview of biofuel as a renewable energy source: development and challenges. Proc Eng 56:39–53. https://doi.org/10.1016/j.proeng.2013.03.087

Heaton EA, Flavell RB, Mascia PN, Thomas SR, Dohleman FG, Long SP (2008) Herbaceous energy crop development: Recent progress and future prospects. Curr Opin Biotechnol 19:202–209. https://doi.org/10.1016/j.copbio.2008.05.001

Hill J, Nelson E, Tilman D, Polasky S,Tiffany D (2006) Environmental, economic, and energetic costs, and benefits of biodiesel and ethanol biofuels. In: Proceedings of the National Academy of Sciences of the United States of America, vol 103. https://doi.org/10.1073/pnas.0604600103

Hill J, Polasky S, Nelson E, Tilman D, Huo H, Ludwig L, Bonta D (2009) Climate change and health costs of air emissions from biofuels and gasoline. In: Proceedings of the National Academy of Sciences of the United States of America, vol 106. https://doi.org/10.1073/pnas.0812835106

Hu Q, Sommerfeld M, Jarvis E, Ghirardi M, Posewitz M, Seibert M, Darzins A (2008) Microalgal triacylglycerols as feedstocks for biofuel production: perspectives and advances. Plant J 54:621–639. https://doi.org/10.1111/j.1365-313X.2008.03492.x

Huang C, Zong MH, Wu H, Liu Q (2009) Microbial oil production from rice straw hydrolysate by *Trichosporon fermentans*. Bioresour Technol 100:4535–4538. https://doi.org/10.1016/j.biortech.2009.04.022

Karp A, Shield I (2008) Bioenergy from plants and the sustainable yield challenge. New Phytol 179. https://doi.org/10.1111/j.1469-8137.2008.02432.x

Kenthorai Raman J, Chan E, Ravindra P (2011) Biotechnology in biofuels-A cleaner technology. J Appl Sci 11:2421–2425. https://doi.org/10.3923/jas.2011.2421.2425

Khan S, Siddique R, Sajjad W, Nabi G, Hayat KM, Duan P, Yao L (2017) Biodiesel production from algae to overcome the energy crisis. HAYATI J Biosci 24:163–167. https://doi.org/10.1016/j.hjb.2017.10.003

Klinke H, Bjerre A, Ahring B (2004) Inhibition of ethanol-producing yeast and bacteria by degradation products produced during pre-treatment of biomass. Appl Microbiol Biotechnol 66:10–26. https://doi.org/10.1007/s00253-004-1642-2

Kour D, Rana KL, Yadav N, Yadav AN, Rastegari AA, Singh C et al (2019a) Technologies for biofuel production: current development, challenges, and future prospects. In: Rastegari AA, Yadav AN, Gupta A (eds) Prospects of renewable bioprocessing in future energy systems. Springer International Publishing, Cham, pp 1–50. https://doi.org/10.1007/978-3-030-14463-0_1

Kour D, Rana KL, Yadav N, Yadav AN, Singh J, Rastegari AA et al (2019b) Agriculturally and industrially important fungi: current developments and potential biotechnological applications. In: Yadav AN, Singh S, Mishra S, Gupta A (eds) Recent advancement in white biotechnology through Fungi, Volume 2: Perspective for value-added products and environments. Springer International Publishing, Cham, pp 1–64. https://doi.org/10.1007/978-3-030-14846-1_1

Kumar P, Barrett DM, DelwicheMJ Stroeve P (2009) Methods for pretreatment of lignocellulosic biomass for efficient hydrolysis and biofuel production. Ind Eng Chem Res 48:3713–3729. https://doi.org/10.1021/ie801542g

Kumar R, Kumar P (2017) Future microbial applications for bioenergy production: a perspective. Front Microbiol 8:1–4. https://doi.org/10.3389/fmicb.2017.00450

Kumar S, Sharma S, Thakur S, Mishra T, Negi P, Mishra S et al (2019) Bioprospecting of microbes for biohydrogen production: current status and future challenges. In: Molina G, Gupta VK, Singh BN, Gathergood N (eds) Bioprocessing for biomolecules production. Wiley, USA, pp 443–471

Lambin EF, Meyfroidt P (2011) Global land use change, economic globalization, and the looming land scarcity. Proc Natl Acad Sci 108:3465–3472. https://doi.org/10.1073/pnas.1100480108

Liao JC, Mi L, Pontrelli S, Luo S (2016) Fuelling the future: microbial engineering for the production of sustainable biofuels. Nat Rev Microbiol 14:288–304. https://doi.org/10.1038/nrmicro.2016.32

Lima M, Skutsch MCG (2011) Deforestation and the social impacts of soy for biodiesel: perspectives of farmers in the South Brazilian Amazon. Ecol Soc 4(16)

Liu CZ, Wang F, Stiles AR, Guo C (2012) Ionic liquids for biofuel production: opportunities and challenges. Appl Energ 92:406–414. https://doi.org/10.1016/j.apenergy.2011.11.031

Liu T, Khosla C (2010) Genetic engineering of *Escherichia coli* for biofuel production. Annu Rev Genet 44:53–69. https://doi.org/10.1146/annurev-genet-102209-163440

Lynd L, van Zyl W, McBride J, Laser M (2005) Consolidated bioprocessing of cellulosic biomass: an update. Curr Opin Biotechnol 16:577–583. https://doi.org/10.1016/j.copbio.2005.08.009

Lynd L, Weimer PJ, van Zyl W, Pretorius I (2002) Microbial cellulose utilization: fundamentals and biotechnology. Microbiol Mol Biol Rev (MMBR) 66. https://doi.org/10.1128/MMBR.66.3.506-577.2002

Maclean H, Lave LB (2003) Evaluating automobile fuel/propulsion system technologies. Prog Energ Combust Sci 29:1–69. https://doi.org/10.1016/S0360-1285(02)00032-1

Mata TM, Martins AA, Caetano NS (2010) Microalgae for biodiesel production and other applications: a review. Renew Sust Energ Rev 14:217–232. https://doi.org/10.1016/j.rser.2009.07.020

Meng X, Yang J, Xu X, Zhang L, Nie Q, Xian M (2009) Biodiesel production from oleaginous microorganisms. Renew Energy 34:1–5. https://doi.org/10.1016/j.renene.2008.04.014

Mohammady N (2007) Different light spectral qualities influence sterol pool in *Porphyridium cruentum* (Rhodophyta). Am J Plant Physiol 2. https://doi.org/10.3923/ajpp.2007.115.121

Mohr A, Raman S (2013) Lessons from first generation biofuels and implications for the sustainability appraisal of second generation biofuels. Energy Policy 63:114–122. https://doi.org/10.1016/j.enpol.2013.08.033

Mora-Pale M, Meli L, Doherty T, Linhardt R, Dordick J (2011) Room temperature ionic liquids as emerging solvents for the pretreatment of lignocellulosic biomass. Biotechnol Bioeng 108). https://doi.org/10.1002/bit.23108

Mosier N, Wyman C, Dale B, Elander R, Lee YY, Holtzapple M, Ladisch M (2005) Features of promising technologies for pretreatment of lignocellulosic biomass. Bioresour Technol 96:673–686

Mu Y, Teng H, Zhang DJ, Wang W, Xiu ZL (2006) Microbial production of 1,3-propanediol by *Klebsiella pneumoniae* using crude glycerol from biodiesel preparations. Biotechnol Lett 28:1755–1759. https://doi.org/10.1007/s10529-006-9154-z

MW, R. (2008). Biofuels grain prices: impacts and policy responses. Washington, USA

Nigam PS, Singh A (2011) Production of liquid biofuels from renewable resources. Prog Energy Combust Sci 37:52–68. https://doi.org/10.1016/j.pecs.2010.01.003

Nikolić S, Mojović L, Rakin M, Pejin D, Nedović V (2009) Effect of different fermentation parameters on bioethanol production from corn meal hydrolyzates by free and immobilized cells of Saccharomyces cerevisiae var. ellipsoideus. J Chem Technol Biotechnol 84:497–503. https://doi.org/10.1002/jctb.2068

Oliver R, Fich JON, Taylor G (2009) Second generation bioenergy crops and climate change: a review of the effects of elevated atmospheric CO_2 and drought on water use and the implications for yield. GCB Bioenergy 1:97–114. https://doi.org/10.1111/j.1757-1707.2009.01011.x

Ottinger RL (2007) Biofuels—potential, problems & solutions. Biofuels conference. Pace University, School of Law, August 1–9, pp 1–9

Ouellet M, Datta S, Dibble DC et al (2011) Impact of ionic liquid pretreated plant biomass on *Saccharomyces cerevisiae* growth and biofuel production. Green Chem 13. https://doi.org/10.1039/C1GC15327G

Papanikolaou S, Fick M, Aggelis G (2004) The effect of raw glycerol concentration on the production of 1,3-Propanediol by *Clostridium butyricum.* J Chem Technol Biotechnol 79:1189–1196. https://doi.org/10.1002/jctb.1103

Park J, Steen EJ, Burd H et al (2012) A thermophilic ionic liquid-tolerant cellulase cocktail for the production of cellulosic biofuels. PloS One 7. https://doi.org/10.1371/journal.pone.0037010

Patil V, Tran KQ, Giselrød H (2008) Towards sustainable production of biofuels from microalgae. Int J Mol Sci 9:1188–1195. https://doi.org/10.3390/ijms9071188

Pulz O, Scheibenbogen K (2007) Photobioreactors: design and performance with respect to light energy input. Adv Biochem Eng/Biotechnol 38:123–152. https://doi.org/10.1007/BFb0102298

Rana KL, Kour D, Sheikh I, Yadav N, Yadav AN, Kumar V et al (2019) Biodiversity of endophytic fungi from diverse niches and their biotechnological applications. In: Singh BP (ed) Advances in endophytic fungal research: present status and future challenges. Springer International Publishing, Cham, pp 105–144. https://doi.org/10.1007/978-3-030-03589-1_6

Rastegari AA, Yadav AN, Yadav N (2020) New and future developments in microbial biotechnology and bioengineering: trends of microbial biotechnology for sustainable agriculture and biomedicine systems: diversity and functional perspectives. Elsevier, Amsterdam

Rastegari AA, Yadav AN, Gupta A (2019a) Prospects of renewable bioprocessing in future energy systems. Springer International Publishing, Cham

Rastegari AA, Yadav AN, Yadav N (2019b) Genetic manipulation of secondary metabolites producers. In: Gupta VK, Pandey A (eds) New and future developments in microbial biotechnology and bioengineering. Elsevier, Amsterdam, pp 13–29. https://doi.org/10.1016/B978-0-444-63504-4.00002-5

Rastegari AA, Yadav AN, Yadav N, Tataei Sarshari N (2019c) Bioengineering of secondary metabolites. In: Gupta VK, Pandey A (eds) New and future developments in microbial biotechnology and bioengineering. Elsevier, Amsterdam, pp 55–68. https://doi.org/10.1016/B978-0-444-63504-4.00004-9

REN 21, 21st Century. (n.d.). REN21-Renewable Energy Policy Network for the 21st century. Retrieved from http://www.ren21.net/REN21Activities/GlobalStatusReport%0A.aspx

Rojo F (2008) Biofuels from microbes: a comprehensive view. Microb Biotechnol 1:208–210. https://doi.org/10.1111/j.1751-7915.2008.00024.x

Rude MA, Schirmer A (2009) New microbial fuels: a biotech perspective. Curr Opin Microbiol 12:274–281. https://doi.org/10.1016/j.mib.2009.04.004

Ruegg TL, Kim EM, Simmons BA, Keasling JD, Singer SW, Soon Lee T, Thelen MP (2014) An auto-inducible mechanism for ionic liquid resistance in microbial biofuel production. Nat Comm 5:1–7. https://doi.org/10.1038/ncomms4490

Sang T (2011) Toward the domestication of lignocellulosic energy crops: learning from food crop domestication free access. J Integ Plant Biol 53:96–104. https://doi.org/10.1111/j.1744-7909.2010.01006.x

Schenk PM, Thomas-Hall SR, Stephens E et al (2008) Second generation biofuels: high-efficiency microalgae for biodiesel production. BioEnergy Res 1:20–43. https://doi.org/10.1007/s12155-008-9008-8

Sexton SE, Zilberman D (2008) Biofuel impacts on climate change. Report to the Renewable Fuels Agency

Sharma A, Arya SK (2017) Hydrogen from algal biomass: a review of production process. Biotechnol Rep 15:63–69. https://doi.org/10.1016/j.btre.2017.06.001

Shay EG (1993) Diesel fuel from vegetable oils: status and opportunities. Biomass Bioenergy 4:227–242. https://doi.org/10.1016/0961-9534(93)90080-N

Sheehan J, Dunahay T, Benemann J, Roessler P (1998) Look back at the U.S. Department of Energy's aquatic species program: biodiesel from Algae; Close-Out Report. https://doi.org/10.2172/15003040

Steen EJ, Kang Y, Bokinsky G et al (2010) Microbial production of fatty-acid-derived fuels and chemicals from plant biomass. Nature 463:559–562. https://doi.org/10.1038/nature08721

Tsukahara K, Sawayama S (2005) Liquid fuel production using microalgae. J Jpn Pet Inst 48:251–259. https://doi.org/10.1627/jpi.48.251

Tsuruta H, Paddon C, Eng D et al (2009). High-level production of amorpha-4,11-diene, a precursor of the antimalarial agent artemisinin, in *Escherichia coli*. PloS One 4. https://doi.org/10.1371/journal.pone.0004489

Wackett LP (2008) Microbial-based motor fuels: science and technology. Microb Biotechnol 1:211–225. https://doi.org/10.1111/j.1751-7915.2007.00020.x

Wen M, Bond-Watts BB, Chang MCY (2013) Production of advanced biofuels in engineered *E. coli*. Curr Opin Chem Biol 17:472–479. https://doi.org/10.1016/j.cbpa.2013.03.034

Widjaja A, Chien CC, Ju YH (2009) Study of increasing lipid production from fresh water microalgae *Chlorella vulgaris*. J Taiwan Inst Chem Eng 40:13–20. https://doi.org/10.1016/j.jtice.2008.07.007

Xiong W, Li X, Xiang J, Wu Q (2008) High-density fermentation of microalga *Chlorella protothecoides* in bioreactor for microbio-diesel production. Appl Microbiol Biotechnol 78:29–36. https://doi.org/10.1007/s00253-007-1285-1

Xu Y, du W, Liu D, Zeng J (2003) A novel enzymatic route for biodiesel production from renewable oils in a solvent-free medium. Biotechnol Lett 25:1239–1241. https://doi.org/10.1023/A:1025065209983

Yadav AN, Kumar R, Kumar S, Kumar V, Sugitha T, Singh B et al (2017) Beneficial microbiomes: biodiversity and potential biotechnological applications for sustainable agriculture and human health. J Appl Biol Biotechnol 5:45–57

Yadav AN, Rastegari AA, Yadav N (2020) Microbiomes of extreme environments: biodiversity and biotechnological applications. CRC Press, Taylor & Francis, Boca Raton, USA

Yadav AN, Singh S, Mishra S, Gupta A (2019) Recent advancement in white biotechnology through fungi. In: Perspective for value-added products and environments, vol 2. Springer International Publishing, Cham

Yang H, Zhou Y, Liu J (2009) Land and water requirements of biofuel and implications for food supply and the environment in China. Energy Policy 37:1876–1885. https://doi.org/10.1016/j.enpol.2009.01.035

Yvon-Durocher G, Allen A, Bastviken D et al (2014) Methane fluxes show consistent temperature dependence across microbial to ecosystem scales. Nature 507:488–491. https://doi.org/10.1038/nature13164

Zeng AP, Biebl H (2002) Bulk chemicals from biotechnology: the case of 1,3-propanediol production and the new trends. Adv Biochem Eng/Biotechnol 74:239–259. https://doi.org/10.1007/3-540-45736-4_11

Zhu LY, Zong MH, Wu H (2008) Efficient lipid production with Trichosporonfermentans and its use for biodiesel preparation. Bioresour Technol 99:7881–7885. https://doi.org/10.1016/j.biortech.2008.02.033

Chapter 9
Biofuels Production from Diverse Bioresources: Global Scenario and Future Challenges

I. Abernaebenezer Selvakumari, J. Jayamuthunagai, K. Senthilkumar, and B. Bharathiraja

Abstract Roadmaps toward bioeconomy strategy included biofuel production from sustainable biomass. This is due to the worldwide increasing environmental concerns, fast fossil fuel depletion, and the need for energy security. Although complete replacement of petroleum-derived fuels is not possible, the marginal substitution of diesel with biofuel could prolong the depletion of oil resources. The biofuel produced as an alternate energy source is currently a top priority in many nations' research and development sectors. Biofuels are produced by the fermentation process using various starch or sugar-containing feedstocks by microorganisms. Lignocellulosic biomass sources like oilseeds, oils, agricultural residues, forest wastes, paper industrial wastes, municipal solid wastes, and microalgae were potential abundant feedstocks widely used for biofuel production at low cost. This chapter mainly focuses on the diverse significant bioresources used for biofuel production, global scenario in biofuel development, biofuel policies, challenges, and future perspectives in biofuel production across the world. The first segment explains the need for biofuels, the next segment presents a detail presentation on different potential substrates used for biofuel generation and the last section deals with the current biofuel policies and concerns of biofuel.

9.1 Introduction

The concern for the world's dwindling petroleum demand and price volatility has been increasing globally. The ultimate aim of the petroleum market is to expand nominally from 50% by the next 10 years to 118 million barrels per day (mbd),

I. Abernaebenezer Selvakumari · B. Bharathiraja (✉)
Department of Chemical Engineering, Vel Tech High Tech Dr. Rangarajan Dr. Sakunthala Engineering College, Chennai, India
e-mail: btrbio@gmail.com

J. Jayamuthunagai
Centre for Biotechnology, Anna University, Chennai, India

K. Senthilkumar
Kongu Engineering College, Perundurai, Erode, India

A. N. Yadav et al. (eds.), *Biofuels Production – Sustainability and Advances in Microbial Bioresources*, Biofuel and Biorefinery Technologies 11,
https://doi.org/10.1007/978-3-030-53933-7_9

with a lead consumption rate of 28, 16, and 15 mbd by United States, Europe, and China, respectively. In India since 1990, fuel production has raised from roughly 650 thousand barrels per day (tbd) to around 1 mbd. In the meantime, utilization has expanded from 1.2 mbd from 1990 to about 3 mbd in 2008 (Pathak et al. 2012). In a global context, various nations are mindful of their energy surveillance because of the substantial depletion of crude oil reserves and the controls of global petroleum reserves are declining continuously. Consequently, the ever-growing populace and rapid industrial growth expand the gap between the production and consumption of petroleum resources in the past few years. In India, the demand for oil is increasing annually and so the import has lifted from US$ 6 to 15 billion during 1990–2000 and to US$77 billion in 2008. Under this circumstance, different nations have stepped in suitable actions principally intended in improving energy security by endowing in renewable resources and developing policies for exploiting alternative energy sources. Thus, these uncertainties in future energy supply, unfeasible patterns of energy consumption, and the price of expanding as certain fossil fuel reserves have for various energy analysts and researchers around the world to explore alternative and renewable sources of energy such as biofuel (Dufey 2006).

Biofuels are obtained from biological components, primarily from microorganisms, plants, animals, and wastes. Every type of biofuels possesses similar basic as well as sustainable origin. Biofuels offer numerous priorities, including renewability, availability, lower CO_2 emissions, provide energy security, regional development with social structure. Biofuels have the potential to manage two major issues. At first sight, they are known to be carbon–neutral (the carbon emitted by biofuels is neutralized by the atmospheric plants by absorbing from the atmosphere while growing), renewable (surplus supplies can be grown), and suitable for being cultivated in different environments (Dutta et al. 2014). The full picture, yet, is further complex as different biofuels impose different economic, social, and environmental impacts. Due to extensive available opportunities for biomass resources, fossil-fuel-based technology could be possibly replaced by bio-based fuel technology.

9.2 Biofuel and Types

Biofuels are attributed to renewable kind of fuels integrated from biomass originated from organic matters that has been processed to play a valuable role in providing a viable energy source. Distinctive biofuels bring about a different variety of fuel types (liquid, gaseous, and solid forms) for generating energy seems as a signifying alternative energy source with related properties to petroleum fuel. The prevalent types of biofuel include biodiesel, bioethanol, and biohydrogen (Kumar et al. 2019; Rastegari et al. 2019).

Biodiesel is obtained as the result of transesterification of vegetable oils or animal fat with alcohol (methanol/ethanol) in the existence of catalytic agents that can be utilized in pure as well as blended forms with vehicle diesel. The second most common one is bioethanol, the product of fermented sugars/starch biomass is utilized in its purest form in the specially designed vehicles as well as blended form mixed with gasoline in a specific ratio and the other fuel requirements are performed corresponding to regulations. Biohydrogen is a third kind of biofuel produced by living microorganisms as the source of energy via fermentation and photolysis process in a specialized container or a bioreactor and known to be an advanced biofuel (Patni et al. 2011).

9.3 Biofuels Feedstocks

The biomass feedstock utilized in the production of biofuel can be sorted into the following groupings. The first-generation biofuels were produced from feedstocks grown for starch, sugar, and oil such as corn, barley, wheat, soybean, cassava, rye, sugar beet, sugarcane, or sweet sorghum using conventional technologies such as fermentation, distillation, and transesterification (Msangi et al. 2007). Bioethanol, an additive to gasoline are primarily produced through fermentation of starch and sugar substrates with by-products of butanol and propanol. The major advantage of bioethanol is that it burns cleaner with zero carbon emission and hence produces negligible greenhouse gases. Another important biofuel called biodiesel, produced from plant oil or animal fat through a process called transesterification in which the oil exposes with an alcohol (methanol/ethanol) in the existence of a catalyst (acid/alkali). This was followed by the distillation process in which the biodiesel separation from other by-products takes place. Biodiesel can be used as an alternative fuel in many diesel engines in a proportionate mixture of petroleum diesel and biodiesel. These first-generation biofuels symbolize a step toward energy independence and promote rural communities and agricultural industries through increased demand for crops. The first-generation biofuel production has also counter effect in contributing global price increase for food and animal feeds and have a possible negative impact on biodiversity and competition for water in several regions (Singh and Singh 2010). Additionally, they provide only a minimal advantage over fossil fuels in regards to greenhouse gases since anyhow they require a large degree of energy for feedstock cultivation, collection, and processing. Prevailing production practices employ fossil fuels for power generation in the production process of first-generation biofuels. Thus, they are a more expensive choice than gasoline, concluding it economically unfeasible.

Researchers are then aimed at promoting second-generation technologies in the production of biofuels from nonedible dedicated energy crops such as agricultural waste, forest residue, organic residue, food waste, and industrial waste (Sims et al. 2010). These feedstocks need to undergo thermochemical or biochemical pretreatment steps to unlock the sugars embedded in the plant fibers. Forest residues such as straw have to encounter thermochemical pretreatment in order to generate syngas (a mixture of carbon monoxide + hydrogen + methane). The hydrogen formed in this manner is the biohydrogen, employed as a biofuel. The biochemical pretreatment route converts the various polymeric sugars (cellulose and hemicellulose) present in the crop feedstock to sugar monomers, fermented by microorganisms to biofuels. No compete between fuels and food crops has been noted in the second-generation biofuels as they were derived from independent biomass. Additionally, they endorse the usage of poor quality land where food crops fail to grow. Recent estimates show that second-generation biofuels production costs are double the times to petroleum fuels on the basis of energy equivalence as they requires more energy and materials.

To cut down the biofuel formulation cost, the third generation feedstocks are based on distinctively engineered crops specifically algae as the energy source. The oil extracted from algal species is converted into biodiesel through transesterification process, or it can be enriched into other fuels as petroleum alternatives. This field is presently under far-reaching research toward enhancing the production as well as the separation of bio-oil from nonfuel elements and to further reduce the manufacturing costs. Algae are highly beneficial in the following manner that they can be cultivated as cheapest, immense-energy, and absolutely renewable source of energy. This can grow in municipal as well as industrial wastewater, saltwater, such as oceans or salt lakes, and can deliver the dual purpose of biofuel production along with phytoremediation. The ability of these microorganisms to develop under both oxygen consuming and anaerobic conditions has made them less demanding to move inside various cultivating modes to start biohydrogen generation (Suali and Sarbatly 2012). In this manner, the development of wastewater microalgae affords the numerous focal points in the treatment of wastewater, production of algal biomass, and greenhouse gas mitigation all the while. However, further research still needs for further extraction process in order to make it economically competitive to petroleum-based fuels. The various feedstocks used for biofuel production are tabulated in Table 9.1.

Table 9.1 Various bioresources for biofuel production

Biofuel	Country	Bioresource	References
Biodiesel Bioethanol	Australia	Sugarcane, Molasses, Wheat, Palm oil, Cotton oil	Araújo et al. (2017)
Biodiesel Bioethanol Biohydrogen	Brazil	Sugarcane, Soybean, Palm oil, Wheat straw, Vinasse wastewater	Bajpai and Tyagi, (2006), Kaparaju et al. (2009)
Bioethanol	Canada	Corn, Wheat	Araújo et al. (2017), Demirbas (2009)
Biodiesel	Malaysia	Palm oil, Waste cooking oil	Dufey (2006), Elbehri et al. (2013)
Bioethanol	Thailand	Cassava, Molasses, Sugarcane	Balat et al. (2008)
Bioethanol	Indonesia	Sugarcane, Cassava	Balat et al. (2008)
Biodiesel Bioethanol	China	Corn, Cassava, Sweet potato, Rice, Jatropha	Bajpai and Tyagi (2006), Demirbas (2009)

(continued)

Table 9.1 (continued)

Biofuel	Country	Bioresource	References
Biodiesel Bioethanol Biohydrogen	EU	Rapeseed, Sunflower, Wheat Sugar beet, Barley, Sewage manure, Food wastes,	Araújo et al. (2017), Bajpai and Tyagi (2006), Elbehri et al. (2013)
Biodiesel Bioethanol	India	Molasses, Sugarcane, Jatropha	Elbehri et al. (2013), Ghosh and Ghose (2003)
Biodiesel Bioethanol	USA	Corn, Switchgrass, Soybean, Sunflower	Demirbas (2009). Dufey (2006)

9.4 Biodiesel

Biodiesel also known as monoalkyl esters of long-chain fatty acids, supposed as a viable equivalent of conventional petroleum diesel could be derived from various renewable feedstocks, such as vegetable oil, animal fats, microbial oils, etc. These fatty acid methyl/ethyl esters are generally attained from triglycerides by the process of transesterification with respective alcohol (methanol/ethanol). In the beginning, diglycerides and alkyl esters were produced from the triglycerides, followed by the production of monoglycerides, and later biodiesel (alkyl esters) and glycerol were formed. Various catalysts were investigated for the transesterification process that includes acids, bases, both in heterogeneous and liquid forms using free as well as immobilized enzymes as catalysts (Haas et al. 2003).

9.4.1 Substrates for Biodiesel

9.4.1.1 Biodiesel from Vegetable Oil

Several plants are highly effective in transforming solar energy toward reduced form of hydrocarbons or oils. Therefore, vegetable oils have come in advance for biodiesel production due to their feasibility. The association between the composition of vegetable oils and petroleum-derived diesel fuel made the vegetable oils as a suitable substrate for biodiesel conversion (Demirbas 2009; Tiwari et al. 2007). They are made up of one glycerol to three fatty acids, so that commonly referred to as triglycerides. The vegetable oils comprise edible oils, nonedible oils, and waste/used edible oil. The selection of vegetable oil for biodiesel production depends on availability and locality.

9.4.1.2 Biodiesel from Tree Born Oils

These are nonedible oil including jatropha (*Jatrophacurcas*), castor (*Ricinus communis* L.), rubber seed (*Hevea brasiliensis*), Paradise Tree (*Simarouba glauca*), sea mango (*Cerbera manghas*), and Indian Beech Tree (*Pongamia pinnata*). One of the major limitations in converting this nonedible oil into biodiesel associates with their abundant free fatty acid (FFA) content. India is one of the leading jatropha cultivators and set aside around 1.72 million hectares of land for cultivation of jatropha and few pilot plants of jatropha biodiesel are being handed over to oil companies belonging to public sector (Bajpai and Tyagi 2006).

9.4.1.3 Biodiesel from Animal Fats

Animal fats acquired from poultry, beef, and pork and are the common substrates used for biodiesel production (Sharma et al. 2008). Researchers have also attempted to produce biodiesel from fish oil like salmon oil and animal fat residue. As it might not be cost-effective to nurture fish and different animals merely for fat, the utilization of by-products of fat residues from cattle, poultry, and hogs increase the profit of the livestock industries (Reyes et al. 2006).

9.4.1.4 Biodiesel from Microbial Oils

Microbial oils of micro- and macroalgae, bacteria, and fungi have been examined for the production of biodiesel by many researchers (Kour et al. 2019; Schenk et al. 2008; Raju et al. 2009). Microalgae are regarded as a promising candidate for biodiesel production as they are highly rich in oils (over 80% of their dry weight) (Chisti 2008; Manzanera 2011). Moreover, microalgal cultures demand minimal maintenance and could even cultivate in non-potable water, waste effluents, and water sources regarded as unfit for agriculture, and also in the seawater (Mata et al. 2010). This microalgal biodiesel production could also be connected with the greenhouse gas removal from power stations or the synthesis of several value-added products (Harun et al. 2010; Banerjee et al. 2002). Various investigations have exhibited hat the oil composition of algae obtained per hectare is 200 times higher than the fertile land crops. Thus, it is a hopeful eminence for new generation biofuels, devoid of perplexing the food supply as microalgae could be grown on nonagricultural lands. Additionally, diverse prokaryotes and eukaryotes can also incorporate an increased amount of lipids in terms of TAGs. The most important prokaryote included *Mycobacterium* sp., *Rhodococcus* sp., *Nocardia* sp., *Dietzia* sp., *Micromonospora* sp., and *Gordonia* sp., accompanying streptomycetes that incorporate TAGs in their cells as well as mycelia. Within eukaryotes, apart from microalgae, yeasts of the genera Candida (Waltermann and Steinbüchel 2010), Saccharomyces (Maity et al. 2014), and Rhodotorula (Benson et al.2014) are also the most significant candidates for the production of biodiesel. Global biodiesel production is depicted in Fig. 9.1.

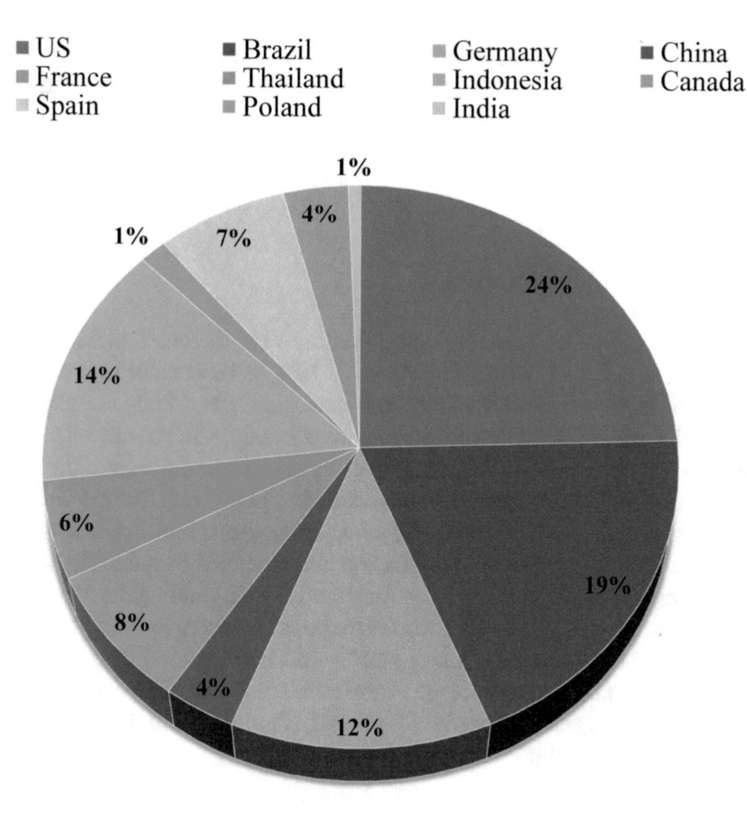

Fig. 9.1 Global status of biodiesel production till 2018

9.5 Bioethanol

Ethanol produced from renewable substrates is known as bioethanol. It is considered as eco friendly and renewable and considered to be one of the excellent substitutes for petroleum-based fossil fuels. The bioethanol producing substrates include sugar-loaded crops such as sugarcane, sugar beet, starch-loaded crops (corn and cassava), lignocellulosic residual biomass, and microbial consortia. The choice of feedstock relies on the countries' agricultural policies. Presently, around 60% of bioethanol is produced from sugar-based crops and the remaining is starch-based.

9.5.1 *Substrates for Bioethanol Production*

The well-known commercial technology for bioethanol production is crop based, making use of molasses, corn starch, sugarcane juice, and beet juice. As the expense of these raw feedstocks accounts for above 40% of the bioethanol production cost (Balat et al. 2008), researchers started focused on employing lignocelluloses substrates since the late 90s. This naturally abundant cheap polymer is endowed as an agricultural residue (wheat and rice straw, sugarcane bagasse, soybean residues, corn stalks), industrial wastes (paper and pulp industry), forestry residues, municipal solid wastes, etc. (Wyman 1999).

9.5.1.1 Bioethanol from Sugars

In general, the sugarcane juice and cane molasses are the chief substrates for the production of bioethanol. Brazil accounts for 79% of bioethanol production from fresh juice of sugarcane and the rest from cane molasses (Seelke and Yacobucci 2007). In India, sugarcane molasses is the major raw source for bioethanol production (Ghosh and Ghose 2003); Molasses as well as sugar beet juices are the alternate sources of fermentable substrates for bioethanol fermentation in Europe. In the large-scale industries, bioethanol is produced by*saccharomyces cerevisiae*, as it hydrolyzes cane sucrose into easily assimilable glucose and fructose. In the midst of various bacteria, the highly significant one is *Zymomonas mobilis*, yielding bioethanol around 97% of theoretical maximum with a narrow range of fermentable sugars like glucose, sucrose, and fructose. The bioethanol yield could be increased by supplementing additional growth factors that include ergosterol, soy flour, oleic acid, chitin, fatty acids, vegetable oils, and skimmed milk powder (Patil and Patil 1989; Wilkie et al. 2000; Shigechi et al. 2004; Pimentel and Patzek 2005).

9.5.1.2 Bioethanol from Starch

Starch is the best yielding feedstock for bioethanol production, but the only limitation is that the yeast *S. cereviciae* is unable to exploit it directly. Prior hydrolysis is necessary to synthesis bioethanol from starch through fermentation. Earlier, starch hydrolysis was done by acids, but later, due to the enzyme specificity at milder reaction conditions skipping the secondary reactions has made use of the amylases as catalysts for the bioethanol fermentation process. Amylase hydrolysis includes two steps, namely, liquefaction and saccharification. In the first step, the starch suspensions are subjected to high temperatures of 90–110 °C for collapsing the starch kernels. At the end of the liquefaction process, the resultant liquid contains dextrines and fewer quantity of glucose. In the next step, the melted starch is subjected to saccharification at moderate temperatures in the range of 60–70 °C through glucoamylase obtained from *Aspergillus niger* or *Rhizopus species* (Pandey et al. 2000; Kaparaju et al. 2009).

9.5.1.3 Bioethanol from Corn

In the US, bioethanol is produced solely from corn substrate. Corn is grounded for starch extraction, and also enzymatically hydrolyzed for collecting glucose syrup that was further fermented to bioethanol. Corn milling can be done by both wet and dry methods in industries. During the process of wet milling, corn grain is detached allowing the starch to convert into bioethanol and other fermented co-products. In the course of dry-milling, grains are not evenly fragmented and their source of nourishments is condensed as a distillation co-product employed as animal feed (Dried Distiller's Grains Soluble (DDGS)) (Gulati et al. 1996).

9.5.1.4 Bioethanol from Wheat

The most common method of bioethanol production in Europe is from beet molasses whereas, in France, wheat is used as a primary substrate. In order to increase the productivity and yield of bioethanol, attempts have been done for improving the fermentation process conditions. Wang et al. (1999) have optimized the temperature as well as specific gravity for the fermentation of the wheat mash and Soni et al. (2003) have determined the optimal process parameters using α- amylase and glucoamylase for starch hydrolysis of wheat bran in solid-state fermentation.

9.5.1.5 Bioethanol from Cassava

Cassava, a substitute source of starch widely preferred for bioethanol and glucose syrup production. Cassava is the tuber that grabs keen attention by various researchers as it is available abundantly in tropical countries and ranked to be one among the top ten significant tropical crops. Bioethanol could be produced using either the whole cassava tuber or the extracted starch. Starch extraction could be attained by the Alfa Laval extraction method (FAO 2004) in the industrial-scale process or through the conventional process in small- and mid-scale industrial plants.

9.5.1.6 Bioethanol from Other Feedstocks

Apart from corn and wheat, bioethanol can also be synthesized from sorghum (Prasad et al. 2007), barley, rye, triticale (Wang et al. 1997) with pretreatments. Abd-Aziz (2002) recommended the employment of sago palm for bioethanol production. Bioethanol from bananas and their peels have been investigated by Hammond et al. (1996) with commercial α- amylase and glucoamylase. The malt processing of starch-containing food wastes has been patented in 2002 (Chung and Nam 2002). Other highly assuring widely used crops for the production of bioethanol are sweet sorghum, which produces seed granules (high starch), shaft (high sucrose), leaves, and bagasse (high lignocellulosic).

9.5.1.7 Bioethanol from Lignocellulosic Biomass

Various lignocellulosic feed stocks have been approved for the synthesis of bioethanol. Generally, lignocellulosic substrates can be classified into six major groups that are involved in bioethanol production: crop residues (wheat straw, barley and rice straw, corn stover, rice hulls, pulps, olive stones, and sweet sorghum bagasse), hardwood (poplar and aspen), softwood (pine and spruce), herbaceous biomass (switchgrass, alfalfa hay, coastal Bermuda grass, reed canary grass, timothy grass), cellulose wastes (newsprint, recycled paper sludge, waste office paper,), and municipal solid wastes (MSW). However, a large extent of complexity implicit in feedstock processing is the only major limiting factor. This is associated with the nature and their distribution of lignocellulosic biomass and so the fermentation process using these substrates is quite complicated and energy-consuming.

9.5.1.8 Bioethanol from Algal Biomass

First studies as algal biofuels are concentrated on biodiesel production. However, the carbohydrates in the structure of algae made to consider as potential substrate utilized for bioethanol production after hydrolysis. Marine algae can exhibit a large amount of carbohydrates every year. Also, it is expected that algal species could meet the future biofuel demand by harvesting at short time and regarded more highly reproducible than other raw materials. Microalgae with a high amount of starch in the cell walls include *Dunaliella, Chlorella, Chlamydomonas, Scenedesmus* are widely used in bioethanol production. Like microalgae, macroalgae could also serve as potential renewable feedstock for bioethanol production. The absence or less amount of complex lignin molecules in their structure, simplifies the hydrolysis treatment process (Araújo et al. 2017). The fermentation of algal polysaccharides such as starch, sugar, and cellulose yields bioethanol and the carbohydrate content (mostly starch) of microalgae can be enhanced up to 70% under specific conditions. Global bioethanol production is depicted in Fig. 9.2.

9.6 Biobutanol

Butanol ($C_4H_{10}O$) contains a higher number of hydrogen as well as carbon related to ethanol (Ramey 2004). In the early twentieth century, Chaim Weizman identified a bacterial strain able to produce promising amounts of acetone as well as butanol and was labeled as *Clostridium acetobutylicum*. Biobutanol could readily intermix with gasoline and other hydrocarbons, and also possess extended heat energy, not as great corrosive as ethanol, and can be conveyed across functioning pipelines and fueling stations. 85% butanol–gasoline-blended mixture could be employed directly in automotive engines and is regarded as minimal evaporative compared to both gasoline and ethanol. Thus butanol is highly safer to utilize and generates mild

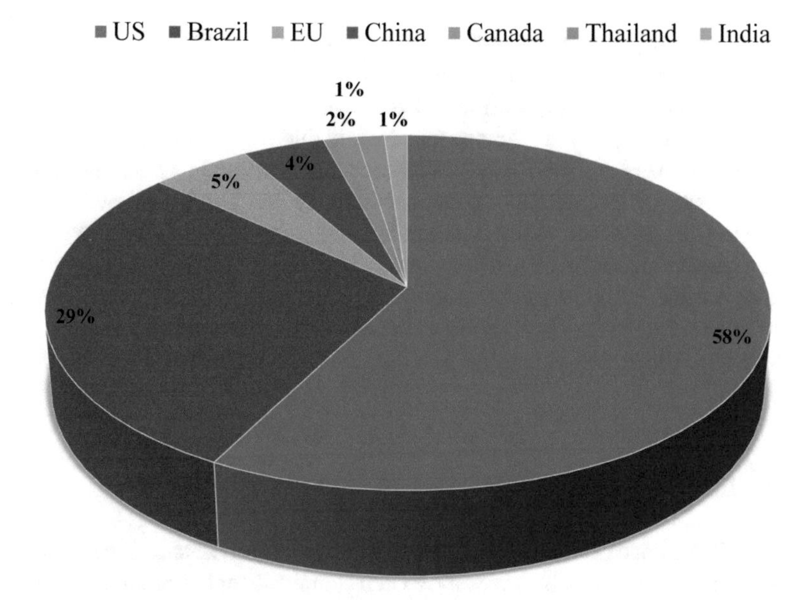

Fig. 9.2 Global status of bioethanol production till 2018

volatile organic compound (VOC) emissions (Qureshi et al. 2005). It is composed of 22% oxygen that originating it as an environmentally friendly fuel that burns cleanly and produces only carbon dioxide.

9.6.1 Substrates for Biobutanol Production

Biobutanol productive microorganisms can make use of an extensive variety of carbohydrates containing glucose, lactose, fructose, arabinose, mannose, sucrose, starch, xylose, dextrin, and inulin obtained from raw fermentable substrates, namely, whey refine, sugar beet, maize, wheat, millet, oats, rye, paper industry residues such as Jerusalem artichoke, and sulfite waste liquor. The ability of microbes to ferment all these kinds of carbohydrates as well as cellulosic substrates (pentose sugars) including xylose and arabinose makes it feasible to utilize variety of agricultural feedstocks such as agricultural residues, woody biomass, and energy crops.

9.6.1.1 Biobutanol from Cane Molasses and Whey Permeate

Just as feedstock, sugarcane molasses posses' higher superiority than maize together with effortless handling and contains simple sucrose molecules that could be easily broken down by solventogenic Acetone–Butanol–Ethanol (ABE) clostridia in accordance with the conversion of sugars into biobutanol. Whey permeate is another valuable feedstock approximately with 45–50 g/l lactose. Butanol synthesizing microbial cultures can easily hydrolyze lactose into assimilable sugar without any additional enzymes (Maddox et al.1993). Soya molasses could also use as a potential substrate for the fermentation of butanol. It contains relatively 745 g carbohydrates per kg of which 58% (434 g/kg) are easily fermentable into galactose, glucose, fructose, and sucrose (Jesse et al. 2002). Sugars like verbascose, raffinose, pinitol, stachyose, and melibiose are incapable of fermentation by *Clostridia,* and therefore require prior enzyme or acid hydrolysis. In addition, fruit industry wastes and contaminated maize also have been illustrated as a useful substrate for biobutanol fermentation (Qureshi et al. 2001).

9.6.1.2 Biobutanol from Starch

In the consideration of *Clostridia,* it could efficiently hydrolyze starch in the range of 45–48 g/l, potatoes and their wastes have been examined for biobutanol production. To investigate the effect of hydrolysis prior to fermentation, each of two hydrolyzed and unhydrolyzed potatoes were subjected to fermentation. The starch content of unhydrolyzed potato yields 12 g/l ABE, much as hydrolyzed potato produced 10–11.4 g/l ABE indicating that hydrolysis is not strictly required for ABE fermentation. In addition, maize starch also could be easily bioconverted to ABE.

9.6.1.3 Biobutanol from Lignocellulose

Besides all the previous traditional fermentable substrates, there exist few promising lignocellulosic feedstocks including switchgrass, maize fiber, maize stover, rice straw, wheat straw, barley straw, corn cobs, hemp waste, DDGE, and sunflower husks usable for ABE generation. Maize fiber, a maize residual obtained during the wet milling process contains 60–70% carbohydrates and produces around 25% of butanol by the fermentation process. Ezeji and Blaschek (2008) employed hydrolyzed DDGS for the biosynthesis of biobutanol using *C. beijerinckii* P260, *C. beijerinckii* BA101, *C. sacchrobutylicum* P262, *C. acetobutylicum* 824, *and C. butylicum* 592. Pretreated wheat straw in dilute (1%, v/v) sulfuric acid followed by enzyme hydrolysis is used as a potential substrate for the bioconversion of ABE using *C. beijerinckii* P260. Switchgrass is another energy crop that can also be applied for the generation of biofuels including ABE. In order to minimize the utilization of food and feed-grade fermentable substrates such as rye flour and molasses, efforts were made by researchers to employ agricultural residues such as corn cobs, and sunflower shells. They are rich in fermentable pentose and reduced hexose sugars and are easily hydrolyzed by dilute sulfuric acid at reaction temperatures extending from 115 to 125 °C. The substantial corn cob hydrolysis and fermentation for the synthesis of biobutanol using *C. acetobutylicum* was demonstrated by Marchal et al. (1992).

The Jerusalem Artichoke juice was also used by several researchers for biobutanol production.

9.6.1.4 Microbial Production of Biobutanol

Biobutanol can be produced by anaerobically by solventogenic Clostridia; the rod-shaped, spore-forming Gram-positive bacteria which can ferment diverse feed-stocks from monosaccharides made up of pentoses and hexoses sugars (Jones and Woods 1986). Besides various solventogenic Clostridia, *Clostridia beijerinckii, Clostridia acetobutylicum, Clostridia saccharobutylicum,* and *Clostridia saccharoperbutyl acetonicum* are the dominant solvent producers. On the other hand, the biobutanol yield is certainly minimal as two moles of CO_2 is derived from per mole of glucose with the formation of various by-products, namely ethanol, acetic acid, acetone, butyric acid, and gaseous hydrogen with solvent toxicity. Considering the yield enhancement and solvent tolerance, genetically engineered recombinant microbial strains would be established (Chen et al. 2013). However, many strains displayed no makeable increase in biobutanol yield except in few hyper-butanol producing strains, such as *Clostridium beijerinckii* P260 and *C.beijerinckii* BA101. (Ezeji and Blaschek 2008). *Clostridia strain* TU-103and *Clostridium cellulolyticum* are capable of direct fermenting cellulose to biobutanol by the anaerobic process without any pretreatment (Qureshi and Blaschke 2005). Some genetically engineered non-Clostridial stains such as *Saccharomyces cerevisiae* BY4742, *Ralstoniaeutropha* H16, *Escherichia coli, Bacillus subtilis* KS438, *Pseudomonas putida* S12 can convert various substrates to biobutanol with superior solvent tolerance (Schenk et al. 2008). For instance, between the non-Clostridial strains, *P. putida* S12 can manage solvent concentration up to 6% (v/v). Microalgal species, such *as Dunaliella, Chlorella, Spirulina, Chlamydomonas,* and *Scenedesmus* are known to possess high amount (>50% of the dry weight) of starch, cellulose, and glycogen that could be used as potential raw material for biobutanol production (Surriya et al. 2015).

9.7 Biohydrogen

Hydrogen is one among the most assuring alternative forms of energy carriers. Similar to electricity, hydrogen is not regarded as a primary form of energy but considered as a secondary source of energy produced from natural as well as bioresources. Hydrogen is witnessed to be a clean fuel with zero toxic emissions as well as been regarded as the future energy source practiced in the fuel cells for the electric current generation. Utilization of hydrogen in the automotive sector either as fuel in combustion engines or fuel cell in electrical energy has earned benign consideration in an energy policy issue (Sorda et al. 2010). Hydrogen utilization is highly eco friendly as it is devoid of noxious gas as well as CO_2 emission whereas the only co-product formed is water vapor. Thus vehicles running by hydrogen energy remarkably decrease the dependency on fossil fuel in the near future.

9.7.1 Substrates for Biohydrogen Production

9.7.1.1 Biohydrogen from Biomass

Wood substrates are the earliest scheme of energy employed by humankind. Wood biomass, agricultural crops and their residues, animal and municipal solid waste (MSW), food industrial wastes, aquatic plants, and algal species are the common potential biomass sources used for biohydrogen production. Biological as well as thermochemical approaches are the major processes in biohydrogen production. Hydrogen can be produced by thermochemical processes via gasification (supercritical water gasification (SCWG) and steam gasification), steam reforming of bio-oils and pyrolysis. The advantage of the thermochemical process is highly economic and highly efficient (up to 52%) (Zhou and Thomson 2009). Indeed, dark fermentation, photo-fermentation, biophotolysis of water by the aid of algal species, and developing hybrid reactor systems are the common biological hydrogen production processes. To establish biomass-based fuelling processes, the chemical, as well as organic composition of biomass employed in the fermentation process, should be scrutinized. Cellulose (40–50%), hemicelluloses (25–30%), lignin (15–20%), and extractives are the four primary components of all lignocellulose biomass and the estimated molecular weights of first three substrates are relatively high, whereas the last one is limited with minimal quantity (Mofijur et al. 2015).

9.7.1.2 Biohydrogen from MicroOrganisms

Investigations on biohydrogen producing anaerobic bacteria initiated in the 1980s and have been expanded due to its environmentally friendly characteristics. The widely known hydrogen producers are cyano-bacteria, anaerobic bacteria, and fermentative bacteria.Hydrogen generating microalgae include *Chlamydomonas reinhardtii, Chlorella fusca, Platymonas subcordiformis, Chlorococcum littorale,* and *Scenedesmus obliquus* were reported under direct biophotolysis method of biohydrogen production (Philipps et al. 2012; Mussgnug et al. 2010). So far, numerous studies were reported on the biological synthesis of hydrogen by the dark fermentation process using facultative (e.g., *Escherichia coli, Enterobacter cloacae, Enterobacter aerogenes,* and *Citrobacter intermedius*) and obligate anaerobic bacteria (e.g., *Ruminococcus albus, Clostridium beijerinckii,* and *C. paraputrificum*). There exists a substantial consent in employing mixed microbial consortia as a biocatalyst and feasible choice for scale-up of biohydrogen production chiefly with wastewater as the carbon energy source substrate (Sambusiti et al. 2015). This technique is widely approved because of the simple operation, stability, security, distinct biochemical

Table 9.2 Microbial strains used in biofuel production

Biofuel type	Microbial strain	Yield (g/g consumed feedstock)	References
Biodiesel	*Acinetobacter calcoaceticus*	0.69	Tiwari et al. (2007)
Bioethanol	*S. cerevisiae*	0.38	Shigechi et al. (2004)
Bioethanol	*E. coli*	3.87	Seelke and Yacobucci, (2007)
Biodiesel	*Zymomonas mobilis*	1.33	Waltermann and Steinbüchel (2010)
Biodiesel	*Clostridium thermocellum*	0.84	Surriya et al. (2015)
Biobutanol	*Clostridium acetobutylicum*	9.2	Solomon and Bailis (2014)
Bioethanol	*Clostridium beijerinckii*	1.8	Qureshi et al. (2001)
Bioethanol	*Bacillus coagulans*	0.33	Pathak et al. (2012)
Bioethanol	*Thermoanaerobacter mathranii*	3.48	Shigechi et al. (2004)
Bioethanol	*Coriolusversicolor*	2.96	Patni et al. (2011)
Biodiesel	*Mucor circinelloides*	6.45	Singh and Singh (2010)
Biobutanol	*Synechococcus elongatus*	0.45	Ramey (2004)
Biodiesel	*Botryococcus braunii*	1.2	Qureshi et al. (2001)
Bioethanol	*Chlamydomonas reinhardtii*	1.94	Philipps et al. (2012)
Biodiesel	*Scenedesmus dimorphus*	1.53	Prasad et al. (2007)
Biohydrogen	*Carboxydothermus hydrogenoformans*	1.32 ml H_2/L/h	Araújo et al. (2017)
Biohydrogen	*Nannochloropsis*	0.6 ml H_2/L/h	Schenk et al. (2008)
Biohydrogen	*Rhodopseudomonas faecalis*	2.76 ml H_2/L/h	Raju et al. (2009

functions, and the possibility of adopting an extensive range of substrates serving the dual purpose of biohydrogen generation as well as wastewater treatment. Various microbial strains involved in biofuel production are tabulated in Table 9.2.

9.8 Worldwide Biofuel Scenario

At present, Asia's best biofuel producers are Malaysia, Philippines, Indonesia, China, Thailand, and India. Malaysia produces biodiesel majorly from palm oil, after all, several researchers have been focusing on Jatropha for large-scale production. Malaysia and Thailand have established their first commercial plantation in the

1960s. Malaysia accounts for 0.5 million tons of waste cooking oil production annually and a mild refining and conversion process of this oil can simply be converted into high-value biodiesel. Thailand established around eight hundred gas stations marketing B-5 biodiesel in 2007. This even progress of Indonesia and Thailand were chiefly as long as the opportunity to utilize a different variety of feedstock. For bioethanol production in Thailand, the major feedstocks are cassava, molasses, and sugarcane and in Indonesia, sugarcane and cassava. Conversely, Malaysia focused on palm oil for biofuel production which made them more liable to the price fluctuations of petroleum as well as palm oil.

Poland is the only country favorable for cultivating oilseed rape among the newer producers due to the plentiful availability of agricultural lands and suitable climatic conditions. It is a net biofuel merchant (Kondili and Kaldellis 2007). In Lithuania, only two pilot-scale plants are in working in which one for the production of biodiesel and the other for the production of bioethanol and restricted their biofuel production for domestic purposes. Romania is recognized as a net sponsor of bioethanol by exploring excellent research in fuel processing as well as biofuel production using various feedstocks (Kondili and Kaldellis 2007).

Due to increased oil cost, Brazil started to develop sugarcane-based bioethanol and has become the most likely example of profitable utilization of biomass for bioenergy production. A great deal of experience in bioethanol production from sugarcane has driven Brazil the most leading producer worldwide. In 2001, South American countries like Peru and Colombia have enforced new law in order to promote the production and consumption of bioethanol derived from sugarcane which declared that the composition of gasoline should comprise 10% ethanol by 2009, with a progressive increment up to 25% in the next 20 years (IPS 2006). They are presently producing about 1,050 million liters of bioethanol per day and investigating diverse alternate substitute sources includes cassava and sugar beets for production of bioethanol. Their interest is not only focused on accomplishing the nation's demand for biofuel but also in the attainment of chances for biofuel export (Dufey 2006). Australia is performing a powerful position for bioethanol utilization within their transporting system (Dufey 2006). Colombia encouraged significant investment since 2005 in the biodiesel production by announcing an imperative demand of 5B biodiesel in their automotive fuel. In the United States of America, soybeans-based biodiesel production elevated from 284 million liters to 950 million liters in 2005–2006 (UNCTD 2008). In April 2006, Argentina endorsed the "Biofuels Act", which demands a 5% demand of biodiesel in petroleum by-products from January 2010 that require 60,000 tons of biodiesel annually for the indigenous market (IPS 2006).

The report of International Energy Agency "World Energy Outlook 2007" suggests that the global energy requirement would be 50% greater by the next 10 years than today. In this scenario, India and China were exclusively supposed for 45% of the increment in fuel demand. The Indian Ministry of Petroleum and Natural Gas has initiated the first stage of the Ethanol Blended Petrol (EBP) Program on 2003 that authorized 5% blending of ethanol in gasoline for 9 states out of 29 and 4 union territories out of 6 (Su et al. 2015; Khanna et al. 2013). As for India, the production of biodiesel was chiefly concentrated on nonedible crop oils such as Jatropha, Neem, Karanja, and Mahua. In China, currently, 80% fuel class ethanol was produced from corn, and remaining from wheat. They use the inferior quality corn for fuel-grade ethanol production to prevent the food stock. Sweet sorghum and cassava are used on an experimental basis and the fuel class ethanol production and marketing is reserved by state resident companies (Mofijur et al. 2015). There are six promising biofuel feedstock, i.e., corn-derived ethanol; sweet sorghum-derived ethanol; cassava-derived ethanol; Jatropha-derived biodiesel; soybean-derived biodiesel; and used cooking-oil-derived biodiesel. Chinese method of biodiesel production is marginal compared to the production of ethanol and their biodiesel production estimated to roughly 300,000 metric tons annually based on waste vegetable oils or animal fat (Elbehri et al. 2013). Various biofuel policies across the world are shown in Table 9.3.

Table 9.3 Biofuel policies across the world

Country	Timeline	Action	Economic measures	Impact
China	November 2018	10% blending mandates in some regions of the country	Tax exemption for biodiesel from animal or vegetable oil and Used cooking oil	Launched the world's first coal-to-ethanol production facility and signed a $100 million agreement of intent to jointly construct about 100 municipal solid waste-to-bioethanol plants by 2035
Japan	January 2017	Upper limits for blending are 3% (ethanol) and 5% (biodiesel)	Subsidies for bioethanol production and tax exemptions	Aim for 10,000–20,000 L of bio-jet fuel production in 2020
Indonesia	August 2016	Target for 30% blending in the transport fuel supply in 2025	Providing biofuels subsidies to producers and also support the domestic agricultural economy to mitigate climate change	The blending mandate B20 program was established domestically
Philippines	July 2006	Diesel: 1% coconut blend; 2% by 2009 Ethanol: 5% by 2008; 10% by 2010	Tax exemptions and priority in financing for biodiesel and bioethanol producers	Stop the sale of biofuels and biofuel-blended gasoline and diesel that are not in conformity with the specifications

(continued)

Table 9.3 (continued)

Country	Timeline	Action	Economic measures	Impact
Thailand	December 2018	Target to increase the current blend from 7% to 10 or 20%	No import tariff for biodiesel greater than B30 and up to and including B100	The government has raised the second large biodiesel plant in 2018, adding 210 million liters per annum to its current 450 million liters production capacity
India	December 2009	Blending 5% ethanol in gasoline in designated states in 2008, to increase to 20% by 2017	Ethanol and diesel: set minimum support prices for purchase by marketing companies	Conversion of surplus grains and agricultural biomass helps in price stabilization
Malaysia	December 2018	7% blending mandates	Plans to subsidize prices for 7% blended diesel	The use of palm oil would be subsequently reduced to zero by 2030

Sources Mofijur et al. (2015), Solomon and Bailis (2014), Pathak et al. (2012)

9.9 Challenges

The main challenges with first- and second-generation feedstocks include (i) threatening the food security, (ii) excess land requirement as well as farming inputs, (iii) high capital investment (Patni et al. 2011), (iv) little net energy benefits, (v) superior allegations over gaseous emission reductions (Solomon and Bailis 2014). The challenges regarding the land allotment for the cultivation of nonedible oil crops intend to be done on "wastelands" in the forest and nonforest areas but the definition of "wasteland" is not clear till now. However, according to few nations policy-makers and rulers, the term 'wasteland' means 'the uncultivated land that did not offer revenue to the government', i.e., semi-jungle lands, drylands, and wetlands (Zhou and Thomson 2009). In India, there is no agreement of mutual understanding among policy-makers regarding the vacancy of sufficient wasteland for the cultivation of biofuel crops to satisfy the future demand for driving fuel. The existing preferable crop was Jatropha. In the favor of reaching the aspiring target of B-10 biodiesel mandate, Indian government had assured to plant Jatropha on 11.2–13.4 million hectares area by 2012. Prominently, recommending suitable land allotments for Jatropha cultivation is one of the prime concerns in Indian biofuel production (Goswami and Choudhury 2015). Khanna et al. (2013) stated that no order available for the division of wasteland suitable for Jatropha cultivation for biofuel production in India. Further, it is concluded that policy-makers failed to consider farmers while framing decisions. Widely, Indian farmers cultivated Jatropha as a fence crop and certain farmers were objected for Jatropha plantation as a monoculture. Meanwhile, few farmers who grow Jatropha were extremely upset because of reduced productivity and profits.

Another major limitation includes the diverse tax structures, i.e., dissimilar state tax policies that vary from state to state. Raju et al. (2012) reported that though each state admits its own custom tax, biodiesel is excluded from 4% central excise duty as a marketing incentive. Researchers so far identified more than 400 species of nonedible oil seeds for biofuel production, but the feasible experiments affirmed a limited feedstock source. Various research communities are experimenting in the developing genetically improved eminent yielding nonedible plants and microbial species, but of limited success rate (Koçar and Civaş. 2013). Many National Policies on Biofuels did not establish their laws within the stipulated period likewise; very few voluntary institutions have been scheduled to accomplish the importing profits to farmers for gaining the carbon credits. Consequently, it is certain to focus the consequences through the country's traditional or novel mechanisms.

9.10 Conclusion and Future Prospects

In the future, biofuel will be the only possible option that plays a promising role in meeting the energy requirements of the world. To meet this large energy demand, the abundant raw material source is the typical need. Each generation of biofuel has its own pros and cons. Therefore, if a country has to evolve with satisfactory biofuel production, the dominant indigenous biofuel crops are essential to be planted within the country aside from influencing the food supply. Several standardization and promotional actions should be employed for the replacement of conventional fuels. It has been well approved globally that biofuel would serve as an energy source to meet the nation's energy security and it is solely a matter of time before they are added on the market than petroleum fuels. The expansion and application of biofuels still need progressive technological development, to extend its utility by upgrading the energy balance, lessening the noxious emissions, and manufacturing cost, so that the purpose of biofuels' future scheme as true alternatives will be accomplished.

Acknowledgements The authors gratefully acknowledge Vel Tech High Tech Dr. Rangarajan Dr. Sakunthala Engineering College, Avadi, Chennai for their continued support.

References

Abd-Aziz S (2002) Sago starch and its utilization. J Biosci Bioeng 94:526–529

Araújo K, Mahajan D, Kerr R, Silva M (2017) Global biofuels at the crossroads: an overview of technical, policy, and investment complexities in the sustainability of biofuel development. Agriculture 7(32):1–22

Bajpai D, Tyagi VK (2006) Biodiesel: source, production, composition, properties and its benefits. J Olio Sci 55:487–502
Balat M, Balat H, Oz C (2008) Progress in bioethanol processing. Prog Energy Combust Sci 34(5):551–573
Banerjee A, Sharma R, Chisti Y, Banerjee UC (2002) *Botryococcus braunii*: a renewable source of hydrocarbons and other chemicals. Crit Revs Biotechnol 22:245–279
Benson D, Kerry K, Malin G (2014) Algal biofuels: impact significance and implications for EU multi-level governance. J Clean Prod 72:4–13
Chen WH, Chen YC, Lin JG (2013) Evaluation of biobutanol production from nonpretreated rice straw hydrolysate under non-sterile environmental conditions. Bioresour Technol 135:262–268
Chisti Y (2008) Biodiesel from microalgae beats bioethanol. Trends Biotechnol 26:126
Chung BH, Nam JG (2002) Process for producing high concentration of ethanol using food wastes by fermentation. Patent KR20020072326
Demirbas A (2009) Progress and recent trends in biodiesel fuels. Energy Conserv Manag 50:14–34
Dufey A (2006) Biofuels production, trade and sustainable development: emerging issues. International Institute for Environment and Development, London
Dutta K, Daverey A, Lin J (2014) Evolution retrospective for alternative fuels: first to fourth generation. Renew Energy 69:114–122
Elbehri A, Liu A, Segerstedt A, Liu P, Babilonia ER, Hölldobler BW, Davies SJC, Stephen CLN, Andrew JF, Pérez H (2013) Biofuels and the sustainability challenge: a global assessment of sustainability issues, trends and policies for biofuels and related feedstocks. FAO, Rome, Italy
Ezeji TC, Blaschek HP (2008) Fermentation of dried distillers' grains and soluble (DDGS) hydrolysates to solvents and value-added products by solventogenic clostridia. Bioresour Technol 99:5232–5242
Food and Agriculture Organization (FAO) (2004) Global cassava market study. Business opportunities for the use of cassava. In: Proceedings of the validation forum on the global cassava development strategy, vol 6. FAO Rome
Ghosh P, Ghose TK (2003) Bioethanol in India: recent past and emerging future. Adv Biochem Eng/Biotechnol 85:1–27
Goswami K, Choudhury HK (2015) To grow or not to grow? Factors influencing the adoption of and continuation with Jatropha in North East India. Renew Energy 81:627–638
Gulati M, Kohlman K, Ladish MR, Hespell R, Bothast RJ (1996) Assessment of ethanol production options for corn products. Bioresour Technol 5:253–264
Haas MJ, Michalski PJ, Runyon S, Nunez A, Scott KM (2003) Production of FAME from acid oil, a by-product of vegetable oil refining. J Amer Oil Chem Soc 80:97–102
Hammond JB, Egg R, DigginsD CCG (1996) Alcohol from bananas. Bioresour Technol 56:125–130
Harun R, Singh M, Forde GM, Danquah MK (2010) Bioprocess engineering of microalgae to produce a variety of consumer products. Renew Sustain Energy Rev 14:1037–1047
IPS (2006) Biofuel boom sparks environmental fears. Inter Press Services News Agency
Jesse T, Ezeji TC, Qureshi N, Blaschek HP (2002) Production of butanol from starch-based waste packing peanuts and agricultural waste. J Ind Microbiol Biotechnol 29:117–123
Kaparaju P, Serrano M, Thomsen AB, Kongjan P, Angelidaki I (2009) Bioethanol, biohydrogen and biogas production from wheat straw in a biorefinery concept. BioresourTechnol 100(9):2562–2568
Khanna M, Önal H, Crago CL, Mino K (2013) Can India meet biofuel policy targets? Implications for food and fuel prices. Am J Agric Econ 95:296–302
Koçar G, Civaş N (2013) An overview of biofuels from energy crops: current status and future prospects. Renew Sustain Energy Rev 28:900–916
Kondili EM, Kaldellis JK (2007) Biofuel implementation in East Europe: current status and future prospects. Renew Sustain Energy Rev 11:2137–2151
Kour D, Rana KL, Yadav N, Yadav AN, Rastegari AA, Singh C et al. (2019) Technologies for biofuel production: current development, challenges, and future prospects. In: Rastegari AA, Yadav AN,

Gupta A (eds) Prospects of renewable bioprocessing in future energy systems, pp 1–50. Springer International Publishing, Cham. https://doi.org/10.1007/978-3-030-14463-0_1
Kumar S, Sharma S, Thakur S, Mishra T, Negi P, Mishra S et al (2019) Bioprospecting of microbes for biohydrogen production: Current status and future challenges. In: Molina G, Gupta VK, Singh BN, Gathergood N (eds) Bioprocessing for biomolecules production. Wiley, USA, pp 443–471
Maddox IS, Qureshi N, Gutierrez NA (1993) Utilization of whey and process technology by Clostridia. In: Woods DR (ed) The Clostridia and biotechnology. Butterworth Heinemann, MA, pp 343–369
Maity JP, Bundschuh J, Chen CY, Bhattacharya P (2014) Microalgae for third generation biofuel production, mitigation of greenhouse gas emissions and wastewater treatment: Present and future perspectives–a mini review. Energy 78:104–113
Manzanera M (2011) Biofuels from oily biomass. In: Muradov M, Veziroghu N (eds) Carbon-neutral fuels and energy carriers, pp 635–663. CRC, Taylor& Francis Group, Orlando. ISBN 978- 143–9818–57–2
Marchal R, Ropars M, Pourquie J, Fayolle F, Vandecasteele JP (1992) Large-scale enzymatic hydrolysis of agricultural lignocellulosic biomass. Part 2: conversion into acetonebutanol. Bioresour Technol 42:205–217
Mata TM, Martins AA, Caetano NS (2010) Microalgae for biodiesel production and other applications: a review. Renew Sustain Energy Rev 14:217–232
Mofijur M, Masjuki H, Kalam M, Rahman SA, Mahmudul H (2015) Energy scenario and biofuel policies and targets in ASEAN countries. Renew Sustain Energy Rev 46:51–61
Msangi S, Sulser T, Rosegrant M, Rowena V, Claudia R (2007) Global scenarios for biofuels: impacts and implications for food security and water use. Biofuels Glob Food Bal 1–20
Mussgnug JH, Klassen V, Schluter A, Kruse O (2010) Microalgae as substrates for fermentative biogas production in a combined biorefinery concept. J Biotechnol 150:51–56
Pandey A, Nigam P, Soccol CR, Soccol VT, Singh D, Mohan R (2000) Advances in microbial amylases. Biotechnol Appl Biochem 31:135–152
Pathak C, Mandalia HC, Rupala YM (2012) Biofuels: Indian energy scenario. Res J Recent Sci 1(4):88–90
Patil SG, Patil BG (1989) Chitin supplement speeds up the ethanol production in cane molasses fermentation. Enzyme Microb Technol 11:38–43
Patni N, Pillai SG, Dwivedi AH (2011) Analysis of current scenario of biofuels in India specifically bio-diesel and bio-ethanol. In: International conference on current trends in technology, Nuicone, pp 1–4
Philipps G, Happe T, Hemschemeier A (2012) Nitrogen deprivation results in photosynthetic hydrogen production in *Chlamydomonas reinhardtii*. Planta 235:729–745
Patzek TW, Pimentel D (2005) Ethanol production using corn, switchgrass, and wood; biodiesel production using soybean and sunflower. Nat Resour Res 14:65–76
Prasad S, Singh A, Jain N, Joshi HC (2007) Ethanol production from sweet sorghum syrup for utilization as automotive fuel in India. Energy Fuel 21:2415–2420
Qureshi N, Blaschek HP (2005) Butanol production from agricultural biomass. In: Shetty K, Qureshi N, Lolas A, Blaschek HP (2001) Soy molasses as fermentation substrate for production of butanol using Clostridium beijerinckii BA101. J Ind Microbiol Biotechnol 26:290–295
Raju SS, Shinoj P, Joshi PK (2009) Sustainable development of biofuels: Prospects and challenges. Econ Polit Wkly 26:65–72
Raju SS, Parappurathu S, Chand R, Joshi PK, Kumar P, Msangi S (2012) Biofuels in India: potential, policy and emerging paradigms. In: Policy Paper - National Centre for Agricultural Economics and Policy Research
Ramey D (2004) Butanol advances in biofuels, pp 105–118. The Light Party, Washington
Rastegari AA, Yadav AN, Gupta A (2019) Prospects of renewable bioprocessing in future energy systems. Springer International Publishing, Cham
Reyes JF, Sepulveda MA, Lo PM (2006) Emissions and power of a diesel engine fuelled with crude and refined biodiesel from salmon oil. Fuel 85:1714–1719

Sambusiti C, Bellucci M, Zabaniotou A, Beneduce L, Monlau F (2015) Algae as promising feedstocks for fermentative biohydrogen production according to a biorefinery approach: a comprehensive review. Renew Sustain Energy Rev 44:20–36

Schenk PM, Thomas-Hall SR, Stephans E, Mark VC (2008) Second generation biofuels: high efficiency microalgae for biodiesel production. Bioenergy 1:20–43

Seelke C, Yacobucci B (2007) Ethanol and other biofuels: potential for US-Brazil cooperation. CRS report RL34191, Environmental Protection Agency (EPA), Renewable Fuel Standard Program

Sharma YC, Singh B, Upadhay SN (2008) Advancements in development and characterization of biodiesel: a review. Fuel 87:2355–2373

Shigechi H, Fujita Y, Koh J, Ueda M, Fukuda H, Kondo A (2004) Energy-saving direct ethanol production from low-temperature cooked corn starch using a cell-surface engineered yeast strain codisplaying glucoamylase and a-amylase. Biochem Eng J 18:149–153

Sims R, Mabee W, Saddler JN, Taylor M (2010) An overview of second generation biofuel technologies. Bioresour Technol 101:1570–1580

Singh SP, Singh D (2010) Biodiesel production through the use of different sources and characterization of oils and their esters as the substitute of diesel: a review. Renew Sustain Energy Rev 14:200–216

Solomon B, Bailis R (2014) Sustainable development of biofuels in Latin America and the Caribbean. Springer, New York

Soni SK, Kaur A, Gupta JK (2003) A solid state fermentation based bacterial a-amylase and fungal glucoamylase system and its suitability for the hydrolysis of wheat starch. Process Biochem 39:185–192

Sorda G, Banse M, Kemfert C (2010) An overview of biofuel policies across the world. Energy Policy 38:6977–6988

Su Y, Zhang P, Su Y (2015) An overview of biofuels policies and industrialization in the major biofuel producing countries. Renew Sustain Energy Rev 50:991–1003

Suali E, Sarbatly R (2012) Conversion of microalgae to biofuel. Renew Sustain Energy Rev 16:4316–4342

Surriya O, Saleem SS, Waqar K, Kazi AG, Öztürk M (2015) Bio-fuels: a blessing in disguise. Phytoremediation for green energy. Springer, Dordrecht, pp 11–54

Tiwari AK, Kumar A, Rahamen H (2007) Biodiesel production from Jatropha oil (*Jatropha curcas*) with high free fatty acids: an optimized process. Biomass Bioenergy 31:569–578

United Nations Conference on Trade and Development (UNCTD) (2008) Biofuel production technologies: Status, prospects and implications for trade and development. New York and Geneva

Waltermann M, Steinbüchel A (2010) Neutral lipid bodies in prokaryotes: recent insights into structure, formation and relationship to eukaryotic lipid depots. J Bactriol 187:3607–3619

Wang S, Sosulski K, Sosulski F, Ingledew M (1997) Effect of sequential abrasion on starch composition of five cereals for ethanol fermentation. Food Res Inter 30:603–609

Wang S, Ingledew W, Thomas K, Sosulski K, Sosulski F (1999) Optimization of fermentation temperature and mash specific gravity for fuel alcohol production. Cereal Chem 76:82–86

Wilkie AC, Riedesel KJ, Owens JM (2000) Stillage characterization and anaerobic treatment of ethanol stillage from conventional and cellulosic feedstocks. Biomass Bioenergy 19:63–102

Wyman CE (1999) Opportunities and technological challenges of bioethanol. Presentation to the committee to review the R and D strategy for biomass-derived ethanol and biodiesel transportation fuels. Review for the research strategy for biomass-derived transportation fuels. National Research Council. National Academy, Washington, pp 1–48

Zhou A, Thomson E (2009) The development of biofuels in Asia. Appl Energy 86:11–20

Chapter 10
Bioconversion and Biorefineries: Recent Advances and Applications

José Francisco González-Álvarez, Judith González-Arias, Cristian B. Arenas, and Xiomar Gómez

Abstract The conversion of biomass is full of challenges requiring multiples steps for attaining high efficiencies in the transformation of this material for producing valuable goods and chemicals. There exist several biological processes capable of generating different fuels and green chemicals; however, their efficiency may be too low associated with the need of biomass pre-treatments or the maturity of these technologies may be at an early stage requiring for the development of pilot-scale experiences to get an insight on their performance under different conditions and for assessing their behaviour during extended periods. Some technical aspects are still in need of deep research to consider their implications in a global economic balance when the integration into multiple phases is proposed. Technologies for the production of fuels and the valorisation of the variety of side streams are reviewed in this chapter giving an approximation of the several possibilities of integrating these biological alternatives considering the production of ethanol, butanol, biodiesel and biogas along with the production of hydrogen. A cascade approach for applying a diversity of valorisation stages has been studied taking into account the use of different side streams for coupling biological and thermal processes in an attempt to increase process yields and reduce operating costs. The integration of anaerobic digestion and fermentative hydrogen production for the valorisation of cellulosic biomass into different processes as ethanol and biodiesel production has been assessed.

J. F. González-Álvarez
School of Industrial, Informatics and Aerospace Engineering, University of León, Campus de Vegazana, Leon, Spain

J. González-Arias · C. B. Arenas · X. Gómez (✉)
Chemical and Environmental Bioprocess Engineering Group, Natural Resources Institute (IRENA), University of León, Av. de Portugal 41, Leon, Spain
e-mail: xagomb@unileon.es

A. N. Yadav et al. (eds.), *Biofuels Production – Sustainability and Advances in Microbial Bioresources*, Biofuel and Biorefinery Technologies 11,
https://doi.org/10.1007/978-3-030-53933-7_10

10.1 Introduction

The use of biomass and its conversion into high value-added products is of great relevance when considering the development of a sustainable economy having as main aim the reduction or substitution of non-renewable sources. However, attaining this goal makes imperative the complete valorisation of biomass, reducing waste streams and developing processes characterised by a low energy demand. This concept irremediably leads to the concatenation of different biological and thermochemical technologies integrated in a way that allow for the maximisation of yields and economic revenues, otherwise their industrial application would be compromised. Therefore, attaining a green economy requires the development of processes similar to those already taking place in petroleum refinery, which allow the production of useful chemicals at a large scale, but in this case using renewable sources as raw material, thus the name of biorefineries (Fernando et al. 2006). The different value-added products that can be obtained from biomass in a biorefinery involves chemical building blocks, raw materials for different subsequent stages, biofuels and the production of energy (heat and power) (Aresta et al. 2012; Yadav et al. 2019). This strategy must be in compliance with physical conservation laws and it is to be achieved using the principles of Green Chemistry and Clean Technologies, where only pure substances are produced without waste and using by-products from another production step or conversion into energy, which also increases profitability (Kołtuniewicz and Dąbkowska 2016).

In recent years, the research activities have extensively reported on the valorisation of different types of biomass and the evaluation of microbiological processes capable of transforming these materials into a great variety of valuable products. However, there is still a need of extrapolating these results at larger scales and what it is of most relevance, to evaluate the global performance of coupling several types of technologies intended to maximise biomass conversion. The development of sustainable biorefineries calls for the suitable integration of innovative treatments to prove the technical and economic viability of the entire value chain (Aresta et al. 2012).

The experience and knowledge gained in the operation and management of conventional petroleum refineries can serve as a starting point to aid in making biorefineries a reality. The existence of petroleum refineries for over a century has allowed for a perfect control of thermal and catalytic processes. These processes have become increasingly sophisticated with the different products moving initially from a handful of fuels and lubricants to a full suite of chemical products (Mabee and Saddler 2006). However, one of the main factors that should be considered when comparing the technological development of refineries and that expected for its renewable counterpart is the dispersion of the feeding raw materials for the latter one. Dispersion and season availability will directly affect transportation costs and therefore will negatively influence carbon emissions in any type of energy efficiency balance assessed. Therefore, the future of these technologies is highly dependent of the management activities necessary for the supply of raw materials.

The lessons learned in petroleum refinery will serve as a wide knowledge base for the development of highly efficient biorefineries. Existing pulp and paper mills may be viewed as early examples of biorefineries, thus the integration of innovative processes in already operating industrial facilities would greatly help in developing complex conversion technologies that enable the production of value-added biomaterials and energy (Mabee and Saddler 2006). The biorefinery concept although offering several benefits to society and the environment is required to evolve in a way that allows for flexibility in the treatment of different feedstocks, increasing efficiency in the conversion of lignocellulosic materials and sustain production all year round in an attempt to avoid the low capital utilisation of several agro-industrial factories which depend on seasonal availability of feeding materials (Eggeman and Verser 2006; Kour et al. 2019a; Rana et al. 2019). The European Biobased Economy (based on an intensive agriculture) allows for a large production of materials from different types of biobased chains (food as well as non-food) along with the production of by-products that act as raw materials for a great variety of conversion techniques thus resembling a cascade approach of valorisation. The aim is to generate cyclic processes within which as many by-products as possible are valorised (Fava et al. 2015).

Different conversion platforms are available for transforming any kind of biomass into chemicals and/or fuels. Sugar and starch-based platforms were the first ones to be developed due to the relative low capital investment and the facility for controlling these types of fermentations. Yeast, specially *Saccharomyces cerevisiae* presents outstanding abilities for converting sugars to ethanol and it is part of one of the oldest human technologies being essential for many biotechnological processes (Dashko et al. 2014; Yadav et al. 2020). However, this yeast cannot utilise cellulosic materials, thus the extended use of starch in this fermentation processes, requires an additional pre-treatment in the form of hydrolysis to release glucose (Apiwatanapiwat et al. 2011) as main sugar with amylases being one of the most widely used family of enzymes capable of achieving the hydrolysis of starch and facilitating the subsequent fermentation stages.

The production of ethanol as a biofuel from the fermentation of sugars/or starchy materials was initially classified as first-generation biofuels, inside this same category were included fuels derived from vegetable oils or animal fats using conventional technologies (Cherubini and Jungmeier 2010). However, the competition created with food and feed agronomic production generated the need of transforming these processes into systems capable of treating more complex materials, leading to the so-called second- and third-generation biofuels.

The second-generation biofuels can be considered as the following reasonable step in producing biofuels from feedstock of lignocellulosic biomass and non-food materials along with the use of energy crops. On the other hand, the third-generation biofuels, which is an area currently under intensive research, are based on algal biomass production. This line of work requires a huge amount of experimental work to attain a significant improvement in biofuel yields and lower further the production costs (Aro et al. 2016). In an attempt to enhance the performance of these systems, the process has evolved into the so-called fourth-generation biofuels which can be defined as the combination of the third-generation biofuel with the enhancement of performance by means of genetic and metabolic engineering (Singh et al. 2017; Farrokh et al. 2019; Kour et al. 2019b).

Different bioconversion technologies have become available since the appearance of the first fermentation processes. Nowadays, bioconversion facilities are capable of integrating several technologies to attain the valorisation of lignocellulosic biomass and waste streams, in particular of agro food by-products, waste effluents and surplus materials, with the production of value-added fine chemicals, novel materials and biofuels (Fava et al. 2015). A biorefinery thus involves a multi-step valorisation approach starting with the collection of the raw material, followed by its transport to plant and selection, involving pre-treatment stages (Fig. 10.1). The development of the precursor containing biomass is a key step of the process along with the subsequent fractionation stage leading to the recovery of valuable products (FitzPatrick et al. 2010). Although there is great experience in the implementation of fermentation at large scales, there exists a great need for enhancing the yields of high-quality by-products and increasing the efficiency in energy and water use of these industrial systems, along with the optimisation of fractionation equipment given the intrinsic difficulty of operating with a great variety of components and the presence of organic compounds produced in the intermediary stages having the potential of interfering in the yields of separation and precipitation steps.

One of the main obstacles commonly reported in ethanol fermentation from lignocellulosic biomass is the insufficient separation of cellulose and lignin, the formation of by-products that inhibit ethanol fermentation, the high use of chemicals and/or energy and the considerable production of waste materials (Menon and Rao 2012; Rastegari et al. 2019). The present chapter deals with the different processes available for the conversion of biomass and integration approaches tested in an attempt to make of the biorefinery concept a reality, taking into consideration that the biorefinery concept is geared towards the production of both traditional and novel fuels and chemicals with a wider goal than simply imitating petroleum refineries but rather generating novel products, which are otherwise not obtainable from fossils (Amoah et al. 2019).

10.2 Biofuel production

The most widely used biofuels in the transport sector are bioethanol, biodiesel and biogas. Ethanol is mainly produced from sugarcane in Brazil and corn in the United States of America (USA), with these two countries being the main producers of this type of fuel worldwide (about 85% of the global production for ethanol production) (Sharma et al. 2019). Biodiesel, on the other hand, is mainly produced from plants oils and again USA and Brazil scaling the first position coping about 36% of the global production (Statista 2019). Biogas is mainly produced from wastes, energy crops and agricultural residues, although the valorisation of this gas is usually performed by means of combined heat and power units for electricity production and heat

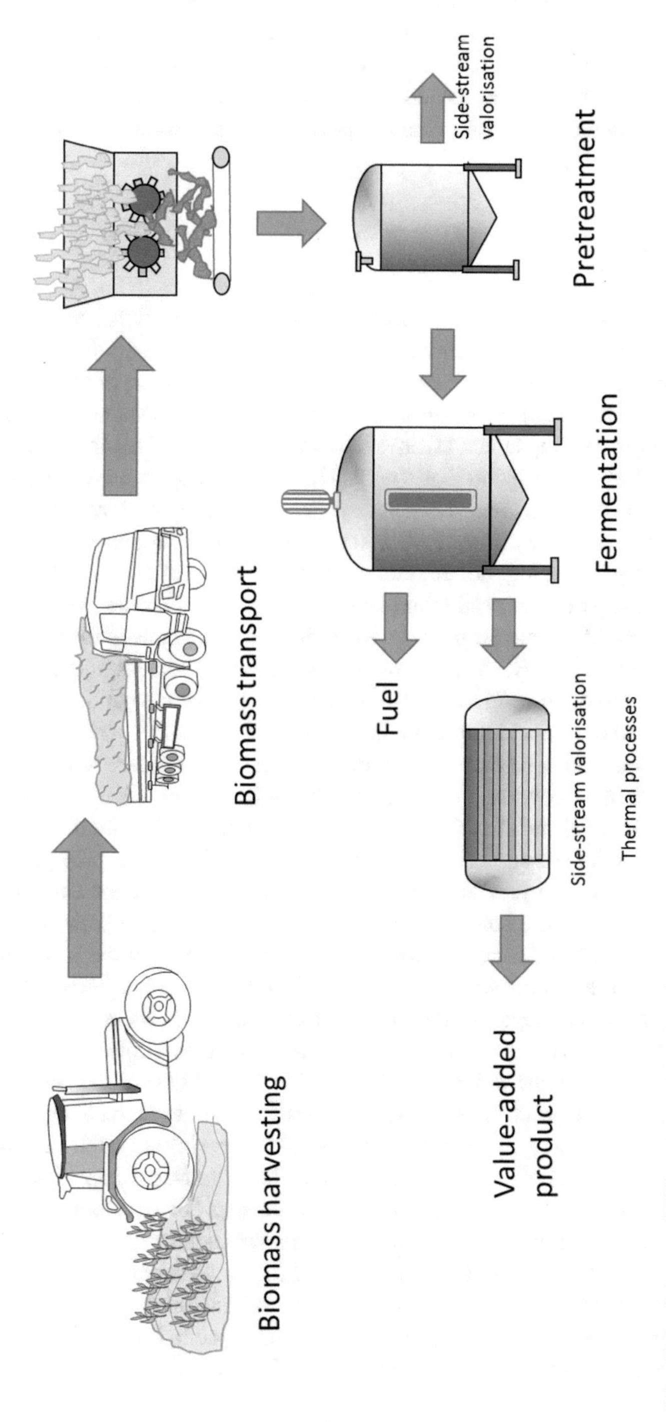

Fig. 10.1 Scheme of different stages in biorefinery concept

recovery, it also has an important share in the transport sector, with a high increase in the production of biomethane to an upgraded level compatible with injection to the natural gas grid. Germany is the leading country in the ranking of installed biogas producing plants and biomethane production in Europe (European biogas association report 2017), whereas the main producers of this gas are USA and European countries.

10.2.1 Bioethanol

The different bioconversion platforms for the production of biofuels usually involve a pre-treatment stage and the severity of this stage is associated with the type of substrate being treated. Bioethanol production was one of the first processes implemented at industrial scale to supplement gasolines with a renewable substitute. The fermentation for producing ethanol from sugarcane is capable of reaching extremely high yields (92–93% of the theoretical yield). The fermenter operates with high yeast cell densities (10–15% w/v) and fermentation volumes are as high as 0.5–3 billion litres (Amorim et al. 2011). Figure 10.2 shows a schematic representation of the Brazilian distillery technology for ethanol production. The fermentation from sugarcane takes advantage of the production of electricity from bagasse which favours the energy balance of the global process. Recycling of yeast cells is fundamental to achieve economic feasibility. In the Brazilian fermentation process, more than 90% of the yeast is reused from fermentation to the next one (Basso et al. 2008). It is also essential for the transformation of ethanol into ethyl tertiary butyl ether (ETBE) which is produced from a mediated catalytic reaction of isobutylene and ethanol.

The use of corn or cereals for producing ethanol involves additional steps for milling the raw material and the subsequent enzymatic hydrolysis at high temperature for liquefying starch type carbohydrates for the saccharification to take place. The fermentation in this case produces two types of stillage which are separated by centrifugation, leading to a solid fraction containing distiller wet grain and a thin stillage fraction which is further evaporated and then mixed with the wet grains to produce dried distiller grains with solubles (DDGS) (Eggeman and Verser 2006).

Due to the multiple stages in the ethanol fermentation process at large scale, the conversion of this type of plants into biorefinery centres allows for increasing the energy efficiency of the process. A remarkable case is that of the Bazancourt-Pomacle biorefinery which has developed a diversifying strategy to change its original nature of sugar factory and distillery to be transformed into a starch and glucose producing plant, and specialised research centre where start-up companies can test their demonstration and industrial pilot plants at its sites for developing lignocellulosic fractionation for ethanol and fine chemicals production (Schieb et al. 2015; Stadler and Chauvet 2018). A different story is that of Abengoa and its ethanol production plants, which due to different regulation constraints and unfavourable ethanol market prices was not able to keep industrial plants located in Spain and France, after being one of the main bioethanol producers in Europe.

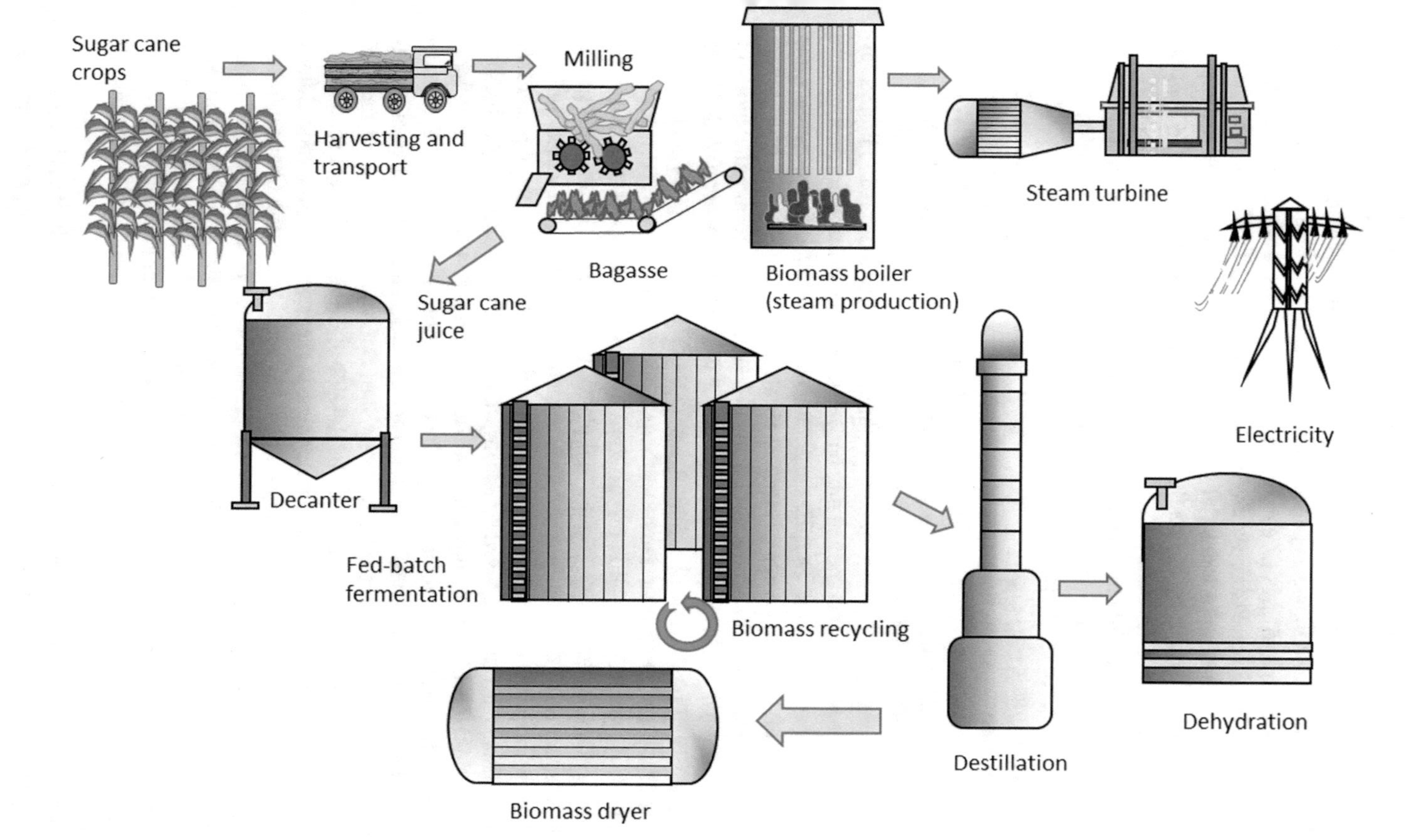

Fig. 10.2 Scheme of a typical plant for bioethanol production

The feasibility of these cereals processing plants is greatly dependent on the production of value-added by-products which can give higher revenues than the traditional sale of electricity, CO_2 and DDGS, with this later material presenting significant prices variations associated with the fluctuating price of cereals where it is produced from and thus affecting the final price of the resulting bioethanol (Pena et a1. 2012). Increasing quality (protein content) of DDGS is essential to gain profitability of ethanol industrial plants, but in any case the price reached of this by-product is limited by the market prices of animal feeding. This is the case of POET offering its DDGS improved product denominated Dakota Gold® HP™ (dakotagold.com) with higher protein content than that traditionally obtained from the standard DDGS. Increasing the protein content of this by-product has been subject of extensive research by several authors (Robinso et al. 2008; Singh et al. 2005) with Pena and co-workers (2012) proposing the use of a ligninolytic fungus selected from a screen of nine white-rot fungal strains. These authors reported a 32% increase in protein content after carrying out a secondary fermentation of DDGS. This process had the additional advantage of also producing an important ligninolytic enzyme with a variety of biotechnological applications.

The transformation of starch ethanol producing plants to those using lignocellulosic biomass is close to become a reality but yet, there is room for many intermediate stages needing optimisation. Despite the abundance of lignocellulosic biomass, the arrangement of its components presents a recalcitrant structure, requiring severe pre-treatments to allow the access of C-6 and C-5 sugars to the fermenting biomass, resulting also in a lignin fraction needing further valorisation (Amoah et al. 2019). The process and pre-treatment technologies for valorising this type of biomass are subject of a previous chapter. However, here it addressed the relevance in accomplishing high production yields on different by-products which would allow for attaining economic feasibility of industrial plants. One way of reaching this goal is by the integration of different fermentations capable of producing a variety of fuels from the valorisation of secondary streams from the multi-stage ethanol production process. This is the idea proposed by Ahring and Westermann (2007) for enhancing biofuel yields from biomass. The major fuels considered in this novel type of refinery are ethanol, hydrogen and methane from the use of corn- or grain-based bioethanol plants by the coupling of photofermentation and anaerobic digestion of volatile fatty acids, and the use of fuel cell systems along with catalytic reformation of methane for hydrogen production. This idea will be developed in the subsequent sections.

The different pre-treatment methods involved in the processing of lignocellulosic biomass are extrusion, steam explosion, liquid hot water, ammonia fibre explosion, supercritical CO_2 explosion and organosolv pre-treatment, other novel methods are ozonolysis pre-treatment, ionic liquids pre-treatment and biological pre-treatments along with enzymatic hydrolysis (Capolupo and Faraco 2016). Thus, the feasibility of refining lignocellulosic biomass to obtain either ethanol or any other class of fine chemicals is still in need of extensive research to make this process economically attractive.

Demonstration plants for the production of ethanol from cellulosic biomass are running at a pilot scale (some of them as large-scale pre-commercial prototypes) in an attempt to evaluate the technical feasibility of these technologies. It should be borne in mind that the process should confront several burdens associated with the collection and transport of a diffuse source of lignocellulosic material in addition to the set of high energy-intensive steps necessary for the complete turnover of this component into ethanol and by-products. Table 10.1 shows a source of different demonstration plants constructed for the production of cellulosic ethanol, the common feature of most of them is that after promising a prosperous production, many suffered from adverse financing and the lack of a favourable regulation leading to either the shut-down of the production line or selling the industrial plant in an attempt of refinancing. The DuPont cellulosic ethanol plant (Nevada, Iowa) which was inaugurated under the promise of becoming one of the largest commercial ethanol-producing plant using non-feed feedstock had to find a new investor to be reconverted into a different line of business to produce renewable natural gas.

The cellulosic production of ethanol at industrial scale is based on the PROESA® technology, developed by Chemtex. The process requires physical pre-treatment of the feedstock by steam explosion to release the cellulosic material. By means of enzymatic hydrolysis, (either Novozyme technology or DMS technology) the cellulose is transformed into simple sugars which can be fermented into ethanol and other types of fine chemicals. When comparing this process with the traditional fermentation from soluble sugars, it is obvious that several difficulties arise in this technology. Figure 10.3 represents a scheme of the basic approach of the patented PROESA® process for producing ethanol listing also some of the particular points needing optimisation at large-scale implementation, having special relevance the effect of inhibitory compounds and the need of adapting harvesting and pre-treatment stages of biomass to the specific lignocellulosic material (Green Car Congress 2019). What is considered an efficient and economical pre-treatment for one type of feedstock may not necessarily translate into an efficient process for another type of biomass (Menon and Rao 2012).

Understanding enzyme pre-treatment and the main characteristics of the solubilisation of biomass polysaccharides is the central core of the biomass-to-bioethanol process. Xyloglucan-active hydrolases are enzymes which carry out hydrolysis and transglucosylation. Xyloglucans cover and cross-link the cellulosic microfibrils in plant cell walls making cellulose inaccessible to saccharification by cellulases. This compound is the major hemicellulosic polysaccharide in plant biomass. Xyloglucan hydrolases which are known to act synergistically with cellulases and xylanases are vital enzymes to release the plant cell wall and attain a successful bioconversion process (Saritha et al. 2016). The further conversion of polymeric cellulose or hemicellulose into simple saccharides (sugars) is highly dependent on the use of another type of enzymes such as endo-1, 4-β-glucanases, cellobiohydrolases and β-glucosidases which act randomly breaking down cellulose by attacking the amorphous regions to produce more accessible new free chain ends for the action of cellobiohydrolases (Annamalai et al. 2016).

Based on the currently available ethanol production process, a classification of biorefineries was proposed by Kam and Kam (2004) considering 'phase I' biorefinery as those current dry-milling ethanol plants. These types of plants use grain

Table 10.1 Some examples of industrial plants built for the production of second-generation ethanol

Project	Characteristic	Location
COMETHA Project	Industrial-scale pre-commercial plant. Finalising construction	Porto Marghera, (Italy)
Beta Renewables	Industrial scale: The refinery was built and operated by Grupo M&G. It was the world's first commercial-scale refinery, but economic crisis and the need of restructuring effort forced the cease of operation	Crscentino (Italy)
Abengoa	Industrial scale using cereals as substrate and demonstration plant for valorisation of cellulosic biomass. Shut-down and sold to an investment group for refinancing the company along with other similar plants of Abengoa	Babilafuente, Spain
Granbio	The first commercial-scale cellulosic ethanol plant in the Southern Hemisphere. The biorefinery, named Bioflex 1, transforms sugarcane residue, straw and bagasse into 'second generation' ethanol	Alagoas, Brazil
Canergy	Primary feedstock will be energy cane which is an approved EPA cellulosic feedstock. Energy cane is a perennial highly fibrous form of sugarcane with high content of cellulose. The company is experiencing delays in getting the plant operating and running	Imperial county, California, USA
POET/DSM's Project Liberty	Corn stover is used as cellulosic substrate for the fermentation process. It took additional efforts to integrate the different multi-steps of the large-scale process to get the whole plant running. Optimisation of corn stover pre-treatment stage proved to be challenging	Emmetsburg, Iowa, USA

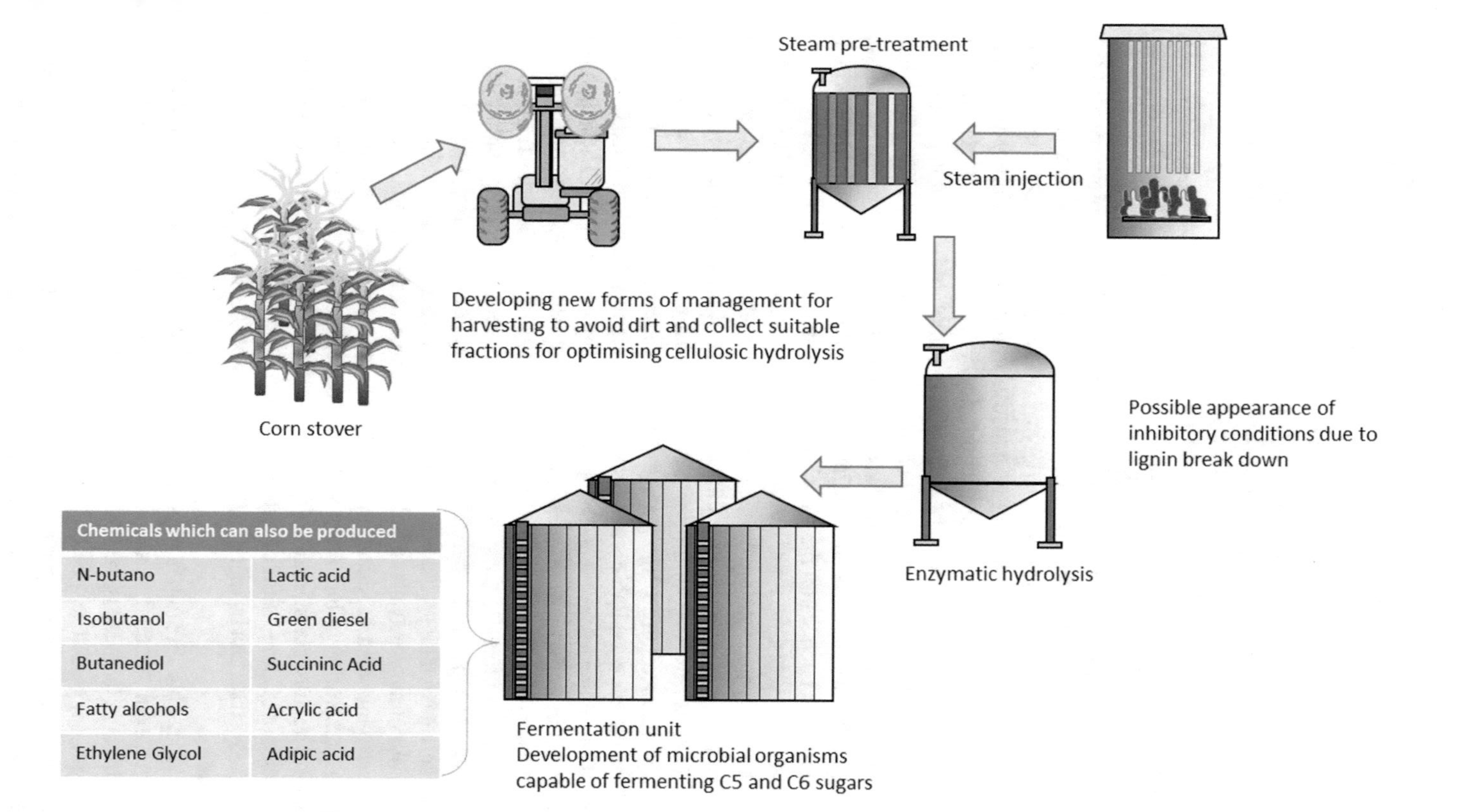

Fig. 10.3 Schematic representation of the production of ethanol from cellulosic biomass highlighting difficulties overcome at industrial scale to make the process a reality at running large-scale operating plants

as feedstock, have fixed processing capability producing a fixed amount of ethanol, feed co-products and carbon dioxide with no flexibility in processing. Phase II were considered by these same authors as those having current wet-milling technology. This technology uses grain feedstocks but it has the capability of producing various end products, depending on demand. Such products include starch, high fructose corn syrup, ethanol, corn oil and corn gluten feed and meal. Thus different fermentations can be connected to these biorefineries using the resulting stream of one previous stage as raw materials for different fermentation products, such as succinic acid, butanol and poly 3-hydroxybutyric acid among others (Du et al. 2007; Zverlov et al. 2006; Nonato et al. 2001; Rastegari et al. 2020).

On this same line, Kam and Kam (2004) classified a phase III biorefinery as the installation capable of producing not only a variety of chemicals, fuels and intermediates or end products, but also use a variety of feedstocks and processing methods to produce several types of goods for the industrial market. The flexibility on the use of different feedstock is the factor of first priority for adaptability towards changes in demand and supply feed, food and industrial commodities. The competition for the use of land for producing fuels and chemicals and that for producing food and feeds has led to an attempt for a new classification of biorefineries based on the use of feedstocks that would avoid interfering with traditional markets. This is why a great emphasis has been set on the valorisation of lignocellulosic biomass.

The products derived from the forest sector, and in particular the pulp and paper industry is currently undergoing a transitioning process where their traditional market may confront new opportunities for the development of novel products and market streams. Therefore, great opportunities exist for the cost-effective utilisation of wood components—hemicellulose, lignin and extractives (Kumar and Christopher 2017). Based on the different available ways for valorising lignocellulosic biomass, a classification of biorefineries has been developed by Dong et al. (2019). These authors denoted Type I biorefinery as those that attain a complete dissociation of lignin but keep it in the spent liquor. In this process, the majority of hemicelluloses remain in the fibre bundles so the structural sugars can be recovered after enzymatic hydrolysis. Solid–liquid separation operations are necessary for recovering lignin and the remaining solvents/catalysts are removed to avoid affecting bioconversion.

Type II biorefinery was classified by these authors as the complete removal of hemicelluloses intended to reduce the recalcitrance of lignocellulosic biomass but avoiding the generation of inhibitors to the subsequent saccharification and fermentation steps. The third classification (Type III) considers the decomposition of lignin and hemicelluloses along with the reduction of the crystallinity of cellulose. The de-crystallised substrate can be regenerated after washing and become easily accessible by cellulose.

The previous classification is important since the pre-treatment applied of lignocellulosic material selected as feedstock needs to be specifically designed based on their intrinsic characteristics. Wood hemicelluloses contain mainly five types of sugars (mannose, galactose, glucose, xylose, arabinose), which are partially acetylated and have some lateral groups like 4-O-methyl glucuronic acid. The major hemicelluloses found in softwoods are galactoglucomannans, whereas in hardwoods the dominant components are arabinoxylans (Bajpai 2018). Although one of the major features of a biorefinery is flexibility of treating different biomass materials, the complexity of each feedstock sets relevant technical impediments for optimising biomass fractionation and subsequent conversion.

In addition to lignocellulosic biomass, wastes are also important raw materials for obtaining biofuels and green chemicals. The initial concept of waste to energy (WtE) was defined by Villar and co-workers (2012) and refers to all technologies that convert, transport, manage and recover or reuse energy from any type of waste (solid, liquid, gas and heat) in a continuous industrial process. This concept associated with the transformation of any kind of waste stream is easily integrated into the biorefinery one, either by setting the conversion of biowastes into goods and energy or by valorising by-products derived from biomass fractionation technologies into the production of energy. Thus, the conversion of biomass (or wastes) needs to consider an integral approach of valorisation where all types of streams find an industrial use leading to zero emissions.

Table 10.2 presents different conversion alternatives for obtaining chemicals and energy. In addition to ethanol, butanol is also a short chain organic fuel compatible with gasoline which presents several advantages associated with the behaviour of butanol and gasoline mixtures. However, the production of this type of alcohol although being a well-known process presents several limitations associated with the low concentration levels tolerated by the fermentation broth and the multiple production of several solvents needing costly final refining stages.

10.2.2 Butanol

Butanol is produced in the so-called acetone-butanol-ethanol (ABE) fermentation. This fermentation was one of the main biological processes for producing chemicals having a scale of production similar to that of ethanol fermentation by yeast but its decline started after 1950 due to the increasing costs of substrate and the lower production price of chemical solvent synthesis by the petrochemical industry (Dürre 1998). However, the advances in the development of microbial processes for increasing product yield and new configurations of fermentation reactors have led to reviving the interest in solvent production in an attempt to decrease the high cost of butanol recovery stages associated with the low concentration attained in fermentation broth and the diversity of solvent product obtained (Qureshi and Blaschek 2001). Fed-batch reactor operation along with gas-stripping product recovery has led to increasing fermentation yields from 0.29 g/L h of total solvent productivity to 1.16 g/L h (Ezeji et al. 2004) which is a considerable success. The use of packed bed reactors under a continuous operation was evaluated by Wang et al. (2016) using immobilised *Clostridium acetobutylicum.* The continuous process was performed in the presence of oleyl alcohol used as extractant for in situ butanol recovery achieving high productivity (11 g/L h) while this value is significantly much lower when basic batch operating configuration was performed (0.2–0.4) (Formanek et al. 1997).

Table 10.2 Different non-conventional raw materials for producing ethanol and butanol

Fermentation product	Substrate	Characteristics	References
Ethanol	Municipal solid wastes	*Saccharomyces cerevisiae*	Li et al. (2007)
	Cotton gin waste pre-treated with organic acids	*Saccharomyces cerevisiae* and *Pichia stipitis* yeast strains	Sahu and Pramanik (2018)
	Lignocellulosic (agricultural wastes)	*Zymomonas mobilis*, *Candida tropicalis*	Patle and Lal (2007)
	newspaper waste	*Saccharomyces cerevisiae*	Bilal et al. (2017)
	Waste wheat straw	*Saccharomyces cerevisiae*	Han et al. (2015)
	Glycerol from biodiesel production	*Enterobacter aerogenes* HU-101, producing hydrogen and ethanol	Ito et al. (2005)
	Solka Floc, waste cardboard and paper sludge	*Kluyveromyces marxianus* (simultaneous saccharification and fermentation)	Kádár et al. (2004)
Butanol	Starch (cassava)	*B. subtilis* WD 161 and *C. butylicum* TISTR 1032	Tran et al. (2010)
	Cellulosic biomass	*C. thermocellum* and *C. saccharoperbutylacetonicum* N1–4	Nakayama et al. (2011)
	Alkali pre-treated rice straw	*C. thermocellum* NBRC 103,400 and *C. saccharoperbutylacetonicum* strain N1–4	Kiyoshi et al. (2015)
	Cheese whey	*K. marxianus* DSM 5422 and *S. cerevisiae* Ethanol Red	Díez-Antolínez et al. (2018)
	Food-industry wastes	*C. beijerinckii, C. acetobutylicum, C. saccharobutylicum* and *C. saccaroperbutylacetonicum*	Hijosa-Valsero et al. (2018)
	Orange peels	*Saccharomyces cerevisiae* NCIM 3495 and *C. acetobutylicum* NCIM 2877	Joshi et al. (2015)
	Paper mill sludge	*C. sporogenes* NCIM 2337	Gogoi et al. (2018)
	Grape pomace	*C. beijerinckii*	Jin et al. (2018)

Another interesting approach for by-passing the energy-intensive butanol recovery process is the use of biodiesel as the extractant. Fermentations of *Clostridium acetobutylicum* were evaluated using biodiesel as the in situ extractant by Li et al. (2010). Biodiesel added to the fermentation preferentially extracted butanol, minimising product inhibition, and increasing butanol production from 11.6 to 16.5 g/L. The fuel properties of the ABE-enriched biodiesel were also evaluated indicating that the key quality indicators of diesel fuel, such as the cetane number increased from 48 to 54 and the cold filter plugging point decreased significantly from 5.8 to 0.2 °C, resulting in an outstanding improvement of biodiesel characteristics.

However, the use of low-cost material for making feasible the industrial production of butanol is necessary and thus involves the use of starchy materials or lignocellulosic biomass. Solventogenic *Clostridium* sp. utilise starch ineffectively due to its inexpression of amylases. Therefore, hydrolysis of starch is required to obtain sugars for the ABE fermentation (Jiang et al. 2018a, b). In this line, a novel butanol fermentation process was developed by Qureshi et al. (2016) using as lignocellulosic biomass sweet sorghum bagasse pre-treated with liquid hot water (190 °C) followed by enzymatic hydrolysis. The hydrolysate was successfully fermented without inhibition, and an ABE productivity of 0.51 g/L h was achieved which was comparable to the 0.49 g/L h observed in the control fermentation using glucose as a feedstock. In this same line of research, the use of pre-treated corn stover as substrate was evaluated by Xue and co-workers (2016), in this case accompanying the fermentation process of butanol recovery by vapor stripping–vapor permeation (VSVP). The condensate produced from this separation technique contained butanol in a range from 212.0 to 232.0 g/L (306.6–356.1 g/L ABE) from a fermentation broth containing ~10 g/L butanol.

The high cost associated with pre-treatments of lignocellulosic and starchy materials along with the energy demand of these processes supposes an important obstacle to circumvent. Co-culturing systems are an ideal and simple way to achieve direct butanol production from starchy-based feedstocks, in which starch is firstly hydrolysed by amylolytic strains, and then released sugars are converted to butanol by mesophilic solventogenic organisms, such as *Clostridium beijerinckii* and *C. Acetobutylicum* (Jiang et al. 2018b). An increase in the efficiency of the whole process may be attained by coupling hydrolytic enzyme production, lignocellulose degradation and microbial fermentation in one single step. This microbial co-cultivation system was studied by Jiang and co-workers (2018b) consisting of *Thermoanaerobacterium* sp. M5 and *C. acetobutylicum* NJ4 achieving a butanol titer of 8.34 g/L from xylan.

The technical and economic feasibility of revitalising butanol production lies not only in the use of inexpensive lignocellulosic hydrolysates and high productivity bacteria, but also in the optimisation of techniques capable of detoxification and efficient continuous fermentation technologies along with in situ product recovery to avoid inhibitory conditions which are typical of this fermentation (Maiti et al. 2016). Life cycle assessment was performed by Pereira et al. (2015) to integrate the biobutanol production process in a sugarcane biorefinery in Brazil. This evaluation indicated that butanol derived from bagasse and straw pentoses using genetically modified microorganism presented the best environmental performance. The introduction of butanol and acetone to the product portfolio of biorefineries leads to an increase of revenues that should not be overestimated.

10.2.3 Biodiesel

Biodiesel, along with ethanol, is also a widely used biofuel compatible with diesel fuels consisting of a mixture of fatty acid methyl esters. Legal mandates for commercialising blends of petrol and diesel fuels with their compatible homologous sets the demand of these biofuels to be directly linked to the consumption of conventional transport fuels. Blends at 5 and 10% of biofuels are commercialised worldwide without the need of making changes in engines. These features have allowed the great expansion of bioethanol and biodiesel industry. The Brazilian transport sector has adapted to include flex-fuel motors capable of running on E0 to E100 (from zero to a hundred percent content in ethanol); thanks to the presence of sensors in the fuel system that automatically recognises the ethanol level in the fuel (Goldemberg 2008).

The production of biodiesel is also linked to the use of land, just as in the case of ethanol production, but regarding the harvesting of oil accumulating plant species. The fabrication of biodiesel is based on chemical reactions involving the transformation (transesterification) of lipids with alcohols (usually methanol or ethanol) in the presence of a catalyst for producing the methyl (or ethyl) esters. In this process, the transesterification reaction involves the separation of glycerine from the fatty acid by means of sodium or potassium hydroxide as catalysts (Refaat 2011). Glycerine is obtained as valuable by-product requiring neutralisation and further refining to be used in pharmaceutical and cosmetic industry. The alcohol used in excess is recovered in the final stage by distillation and returned to the fabrication process. A general scheme is presented in Fig. 10.4 where the main crops for obtaining lipids are also represented.

The great demand for the production of biodiesel worldwide has not been exempted of polemic. The substitution in the use of land traditionally dedicated to human and animal feeding is a risk that should be avoided. Another important burden for the further promotion of biodiesel and in general of any other type of biofuel is the price. High production costs make biofuels unprofitable without subsidies. Biodiesel in principle provides sufficient environmental advantages to merit subsidy in an attempt to lower the price of transportation biofuels, including also in this characteristic synfuel hydrocarbons and cellulosic ethanol (Hill et al. 2006). However, the environmental benefits may not be clear in all available production schemes, since a conscious emission study may result in negative outputs when all resources involved in the production of biofuels are considered. When compared to petroleum-derived fuels, it is usually assumed that biofuels derived from biomass feedstock provide substantial emission savings, due to the simple reasoning that emissions released from biofuel combustion are absorbed from the atmosphere throughout plant growth, thus resulting in a zero emissions footprint. The evaluation of the whole biofuel production process which should involve also the cultivation of the biomass feedstock, the effect on the increase in feedstock prices and economic incentives to acquire additional land to site plantations substituting the original use of land may result in a disappointing outcome where the released CO_2 with the use of biofuel

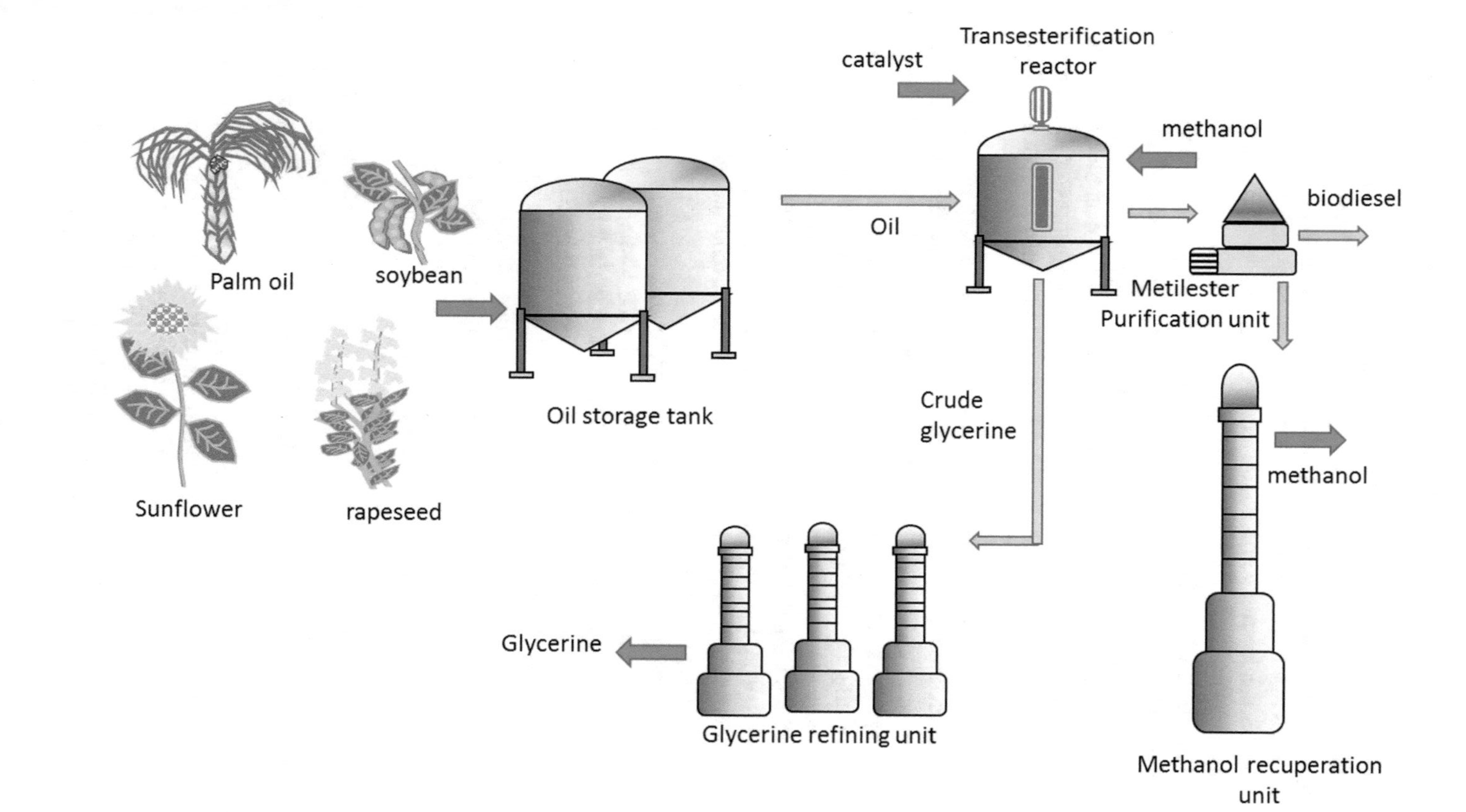

Fig. 10.4 General scheme of biodiesel fabrication process based on transesterification of lipid molecules

would be higher than if a traditional fossil fuel was combusted (Blakey et al. 2011). To all these previous facts, another point that should be taken into consideration is that the financing of governments is obtained to a great extent from taxes associated with conventional fuels sales. If the biofuel market becomes an important part of the transport sector, then the financing of governments should have to be derived from additional taxes associated with other industrial and social sectors.

There are tremendous opportunities for exploring alternative fuels especially with the growing importance of biodiesel and jet fuel in the trucking and aviation industries (Li and Mupondwa 2014). The need for these alternative fuels to be derived from biomass materials to keep a low greenhouse gas (GHG) emission balance causes an extra increase in the costs of production when low-input biomasses grown on agriculturally marginal lands or waste biomass are used as raw materials. However, this increase in production cost should counterbalance against the environmental benefits and the market distortions avoided against the use of food-based biofuels (Hill et al. 2006).

In this regard, crops such as jatropha and camelina are gaining attention as new feedstocks for biodiesel and jet fuel production based on the fact that nutrient needs of these crops are much lower than that of the traditional lipid crops. The study of Li and Mupondwa (2014) reported on GHG emissions from camelina derived biodiesel indicating that 1 MJ of energy contained in biodiesel derived from this source required a consumption ranging from 0.40 to 0.67 MJ/MJ non-renewable energy and for producing HRJ fuel ranged from −0.13 to 0.52 MJ/MJ. Camelina oil as a feedstock for fuel production accounted for the highest contribution to overall environmental performance, demonstrating the importance of reducing environmental burdens during the agricultural production process.

The interest in producing biofuels that are completely compatible with existing engines in all transportation sectors has set the focus on the development of processes for producing the so-called drop-in biofuels. The name is derived from the advantage these fuels offer for completely behaving in an equivalent manner to petroleum fuels. Currently, conventional/oleochemical feedstocks (lipids) can be easily upgraded and integrated into oil-refinery processes but the future interest is in developing thermochemical processes capable of using directly lingnocellulosic biomass to be transformed into drop-in biofuels (van Dyk et al. 2019). Thermal processes as pyrolysis and gasification allow for the conversion of lignocellulosic biomass for producing biocrude in the first case and mainly hydrogen and carbon monoxide (main constituents of syngas) in the later. Therefore, the two common alternative technologies for producing biodiesel fuel is the Fischer–Tropsch (FT) fuels to replace conventional kerosene and hydroprocessed renewable jet (HRJ) fuels made from hydroprocessed oils (Li and Mupondwa 2014).

Another source of lipid raw materials which is currently under intensive research is the culturing of microalgae for producing lipid feedstocks and also in a completely different but parallel novel line of research is the fermentation systems for producing single cell oils. Microalgae species can accumulate substantial contents of lipids based on the culturing conditions to which they are submitted. The oil content in microalgae may reach values as high as 75% (w/w of dry biomass) (Metzger and

Largeau 2005; Gonçalves and Silva 2018) but a relevant factor to take into account is the rate of lipid accumulation, since this parameter is crucial to set the volumetric productivity of the culturing pond. Most of the microalgae accumulate oils in a range between 20 and 50% (e.g. *Chlorella, Dunaliella, Isochrysis, Nannochloris, Nannochloropsis, Neochloris, Nitzschia, Phaeodactylum and Porphyridium* spp.) and the fact that many of these species can be grown on seawater makes of this option an interesting harvesting platform. However, some other factors besides high productivity should be carefully examined as it is the lipid profile of the microalgae cell since it will dictate the resulting characteristics of biodiesel (Amaro et al. 2011).

The principal investment for an algae biomass project may be split into the costs associated with the growth of these organisms, harvesting (steps as isolation of the biomass from the culture, dewatering and/or concentration of algae to facilitate further processing stages) and finally the extraction of algal oil (Singh and Gu 2010). The growth of microalga depends on a supply of carbon and light to carry out photosynthesis. Among the different types of metabolisms, controlled changes in environmental conditions can cause metabolic shifts affecting growth rate and lipid productivity (Amaro et al. 2011). Microalgae can grow photoautotrophically (in the light), heterotrophically (use of a substrate as carbon source) or photoheterotrophically (using simultaneously light and a substrate as carbon source), one organism capable of these three characteristic growths is *Spirulina* sp. (Chojnacka and Noworyta 2004).

The main advantages of producing biofuels from the culturing of microalgae systems are the high efficiency as it is evidenced from the high biomass yields per hectare when compared to lipid yields from conventional crops. The productivity of microalgae biomass can be estimated in 1.535 kg/m^3 d, if an average oil content as 30% (w/w dry biomass) is assumed, this yield would be 98.4 m^{3}per hectare, while this value for palm oil (which is the lipid producing crop with the highest productivity) is estimated in 4.8 m^{3}per hectare (Taparia et al. 2016). In addition, microalgae can be harvested all year round producing a continuous supply of oil, although harvesting and concentration stages may have a higher cost when compared with conventional oil-producing crops, the reliability of this process may counterbalance this disadvantage. Finally, the avoided use of freshwater resources in microalgae biofuel production is another feature which should not be disregarded (Schenk et al. 2008).

Farming is one of the largest commercial consumers of water, on average 20 mega litres of water/ha is required by the crop to fulfil evapotranspirational needs and account for losses during the course of irrigation (Shrivastava et al. 2011; Greenland et al. 2018). The requirements of water for producing ethanol as biofuel from sugarcane are estimated in 88 kg water/kg cane for a plant crop (Singh et al. 2018), which gives an idea of the relevance of this parameter when evaluating the efficiency in the use of resources. On this same line, Shi and co-workers (2017) evaluated the water needs for producing hydroprocessed ester and fatty acid (HEFA) jet fuel from rapeseed cultivation, indicating that the water footprint calculated for jet fuel production in North Dakota was 131–143 m^3 per GJ fuel. These data are strong arguments to favour the development of alternative production of lipid feedstock for biodiesel or drop-in biofuels. In fact, some experimental work is intended to the use of liquid digestate as culture medium form microalgae systems, this is the case of Montero and co-workers (2018) who evaluated the cultivation of *Chlorococcum* sp. obtaining biomass productivities of 23.4 mg/L d although a high dilution proportion was used (5.6% v/v). In a similar approach, anaerobic digestates were tested under batch cultivation of *Chlorella* sp. for oil production. Pig farm digestate was found most suitable as the growth medium generating 0.95 g/L medium (dry biomass) (Chaiprapat et al. 2017).

A schematic representation of the process for obtaining lipids from microalgae cultures is shown in Fig. 10.5. Different culturing ponds and reactors systems have been developed in an attempt to increase volumetric biomass productivity and counteract the effect of light shading of high-density cultures. A full description of factors affecting microalgae growth and types of the different reactors under development can be found elsewhere (Bajpai 2019; Grobbelaar 2010; Ugwu et al. 2008). Previous to lipid extraction, the removal of chlorophyll is necessary, since this compound makes the oil more susceptible to photo-oxidation and decreases its storage stability (Park et al. 2014). In addition, the presence of chlorophyll can decrease the efficiency of transesterification and interfere with biodiesel quality characteristics thus the relevance of its removal as a key step in the commercial production of microalgae oil (Li et al. 2016).

The residual microalgae biomass is a fraction needing further valorisation. The use of microalgae as input for different bioconversion processes has been studied by several authors, considering the anaerobic digestion as a feasible option due to its high biogas production potential (Sialve et al. 2009). Biogas potential of *S. platensis* was studied by Varol and Ugurlu (2016) showing high volatile solid removal in batch studies (about 89–93%) achieved under initial total solids concentrations of 0.6–5%. Another way of valorising this biomass is by thermal methods, either pyrolysis for producing biocrudes, hydrothermal liquefaction or co-combustion with conventional fuels (Coimbra et al. 2019; Eboibi 2019; Mohammed et al. 2018).

In addition to microalgae, many other microorganisms like yeast, bacteria and fungi, have the ability to accumulate oils under special culture conditions. These microbial oils might become one important raw material for the fabrication process of biodiesel once the reduction in fermentation costs is attained by the use of wastes as substrates and the avoidance of sterilisation stages which are crucial to increase economic feasibility (Martínez et al. 2015). Many yeast species, such as *Cryptococcus albidus*, *Lipomyces lipofera*, *Lipomyces starkeyi*, *Rhodosporidium toruloides*, *Rhodotorula glutinis*, *Trichosporon pullulan* and *Yarrowia lipolytica*, were found to be able to accumulate oils under some cultivation conditions where parameters such as C/N ratio, temperature, pH, oxygen and concentration of trace elements and inorganic salts have a significant influence on the yields of oil accumulation (Li et al. 2008).

Lipid accumulating microorganisms have the capacity to store lipids to a content greater than 20%, with a similar triacylglycerol (TAG) structure of that of oil derived from plants (Ratledge 1993). Fungi have been studied for producing specific polyunsaturated fatty acids (PUFA), whereas oleaginous moulds have been cultivated for producing high-value PUFA because the oil accumulated is characterised by a higher

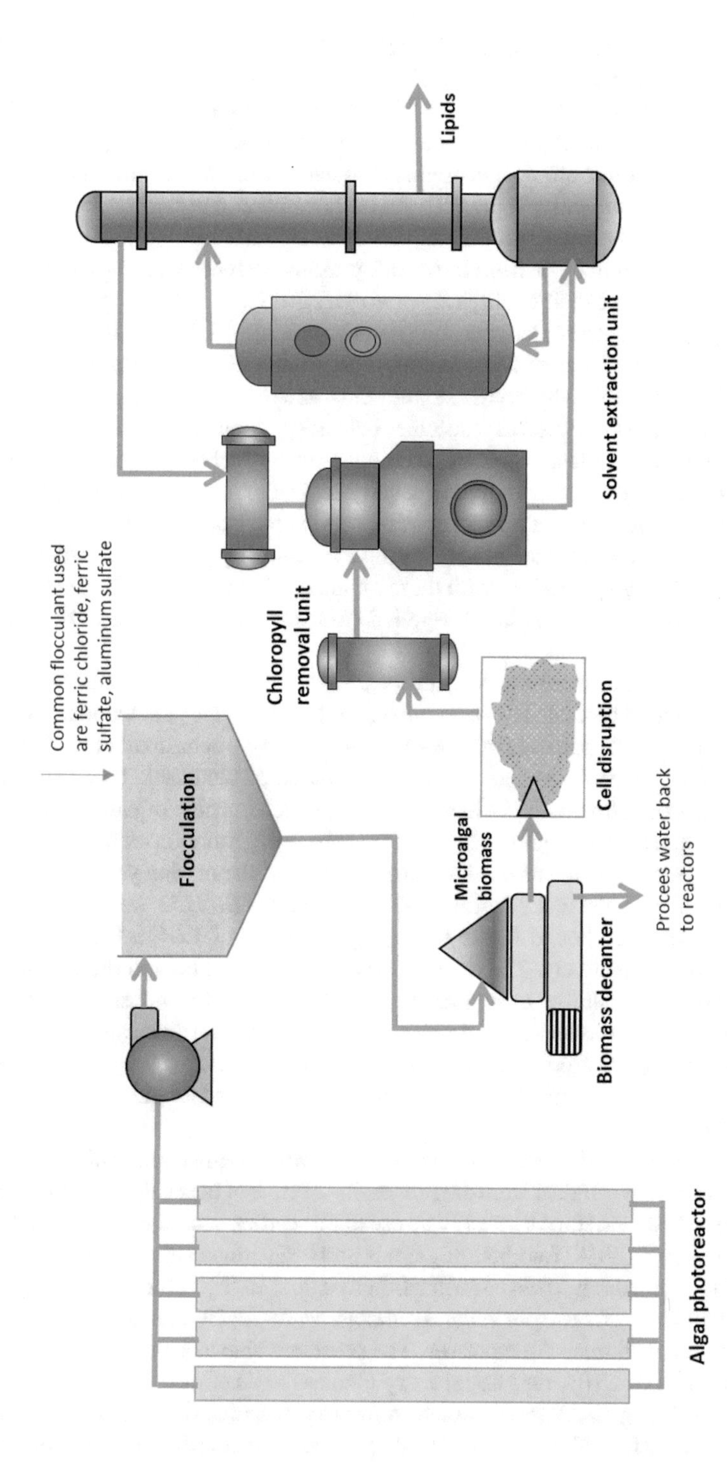

Fig. 10.5 Representation of microalgae lipid production process

level of unsaturation (Papanikolaou and Aggelis 2011). Yeast exhibit several advantages over other microbial sources, associated with high productivity because their duplication times are usually lower than 1 h, are much less affected than plants by season or climate conditions, and their cultures are more easily scaled up than those of microalgae since there are no constraints associated with the penetration of light into the reactor. Additionally, some oily yeasts have been reported to accumulate lipids up to 80% of their dry weight and can generate different types of oils from a variety of carbon sources or from low-quality lipids present in the culture media with the aim to increase their quality (Ageitos et al. 2011).

Some species of bacteria have the ability to accumulate oil, but the lipid composition is usually quite different from that of yeast strains. Bacteria usually produce complex lipids, such as polyhydroxyalkanoic acids, as a means of energy storage, and these compounds are deposited as insoluble inclusions in the cytoplasm. This accumulation process takes place when a carbon source is available in excess but there is also a deficiency of another nutrient (usually nitrogen) thus limiting the growth capacity. The accumulation of lipids for yeasts and some bacteria belonging to the actinomycetes group takes place mostly during the stationary phase of growth when proteins are not being synthesised with these organisms being highly affected by the type of carbon source and conditions applied (Martínez et al. 2015; Spiekermann et al. 1999).

In recent years, the search for valorising waste material and obtaining new sources for the production of biofuels has been intensive. In this regard, the valorisation of effluents obtained from palm oil mill has been studied by Louhasakul et al. (2016) using a novel approach for generating an extra source of biodiesel. These authors proposed the use of palm oil mill effluents (POME) which is a high organic (carbohydrate and proteins) content liquid stream also containing high amount of nutrients and mineral salts. This effluent was used as culture media of the marine yeast *Yarrowia lipolytica*. After the selection of strains, *Y. lipolytica TISTR 5151* was reported to produce lipids and cell-bound lipase at the highest levels of 1.64 ± 0.03 g/L and 3353 ± 27 U/L, respectively. The main relevance of the idea behind these type of experiments is the possibility of culturing a single organism for valorising a waste stream from palm oil production and perform the direct transesterification reaction using cell-bound lipase from the wet yeast cells and produce 40.9% of fatty acid methyl esters, without the need of costly procedures such as isolation, purification and immobilisation.

Another approach is the use of molasses for growing lipid accumulating organisms. Molasses is a by-product from the processing of cane or beet sugar and contains uncrystallised sugar and some sucrose. The use of this carbon source was proposed by Jiru and co-workers (2018) for obtaining lipids and evaluating also the quality of the resultant biodiesel from the transesterification reaction. In their study, *Rhodotorula kratochvilovae* (syn, *Rhodosporidium kratochvilovae*) SY89 was cultivated in a nitrogen-limited medium using molasses. The yeast was able to accumulate lipids to a content of 38.25 ± 1.10% on a cellular dry biomass basis corresponding to a lipid yield of 4.82 ± 0.27 g/L. Although these results are promising, they are still far from making the process feasible for an industrial application. Increasing concentration of titers in the reactor is vital to scale the process and lower production costs. In this line, Matsakas et al. (2015) reported on the use of sweet sorghum at high solid concentrations as a feedstock for single cell oil production by *Rhodosporidium toruloides*. Sweet sorghum is considered an excellent carbon source for this process because it possesses high photosynthetic activity yielding high amounts of soluble and insoluble carbohydrates, requiring low fertilisation inputs and irrigation rates. These authors obtained a fermentation yield of 13.77 g/L (content of lipid in the culturing media) when using sweet sorghum juice (20% w/w enzymatically liquefied sweet sorghum).

Pilot-scale tests have also been performed using a reactor volume of 1 m^3. This scale of process had allowed also to assess the economic evaluation of the lipid recovery stage and performance of biodiesel in diesel engines. This study was reported by Soccol et al. (2017) and attained a lipid concentration in the reactor of 20.5 g/L using as substrate sugarcane juice and *Rhodosporidium toruloides* DEBB 5533 as lipid accumulating organism. Under conditions tested, the estimated final cost of microbial biodiesel produced was US$ 0.76/L, considering in this assessment, energy and steam demands in addition to raw materials and fermentation costs.

Another interesting approach for increasing the efficiency in the utilisation of resources associated with the fabrication of biodiesel is the use of crude glycerol as carbon source for transforming this chemical into lipids. Thus, crude glycerol can be further valorised without the need of processing through costly fractionation and distillation stages (Ma and Hanna 1999). This is the approach tested by Dobrowolski et al. (2016) using *Yarrowia lipolytica* A101 in fermenter obtaining from batch cultivation in a bioreactor a lipid content of 4.72 g/L. Although productivity is not as high as in reports described above, these results allow for a cascade valorisation of raw materials for increasing the productivity of the global biorefinery performance.

10.2.4 Biogas

The energy and climate policies in the EU and the introduction of various support schemes intended to promote the use of renewable resources have encouraged the installation of industrial biogas plants. In Europe, most of the modern anaerobic digesters provide electricity and heat in electricity-only plants, heat only or combined heat and power (CHP) plants (Scarlat et al. 2018). However, in many European countries, the treatment of wastes for producing heat and electricity may not be economically feasible due to the low organic content of some waste streams or to the low biochemical methane potential of some of these materials. Many of these plants, dedicated to the treatment of animal manures struggle to find suitable co-substrates compatible with the process to increase the biogas productivity of the digester which has a direct impact on revenues derived from electricity and heat generation. Co-digestion with animal manures has become in many cases the most adequate alternative for attaining profitability. It has been extensively reported the increase in biogas production when co-substrates such as agricultural wastes (Cuetos et al. 2011, 2013), food wastes (Li et al. 2013; Ormaechea et al. 2017; Zhang et al.

2017) and industrial wastes (Gómez et al. 2007; Nordell et al. 2016) are treated along with animal manures and similar increments have also been reported in the case of sewage sludge treatment (Gómez et al. 2006; Martínez et al. 2012; Oliveira et al. 2018).

Another option to increase economic feasibility of digestion plants is to identify and explore alternative products/chemicals in addition to the production of energy by adopting the biorefinery approach. The integration of different processes intended for biomass conversion to produce fuels, power and chemicals seems an interesting configuration to increase the industrial efficiency in the production of biomass-derived products (Sawatdeenarunat et al. 2016). The production of biogas can be proposed as a last step valorisation in the biorefinery concept. This is the idea presented by Uellendahl and Ahring (2010) who proposed the valorisation of the effluent from the ethanol fermentation when using pre-treated lignocellulosic biomass. The anaerobic digestion of this effluent showed no signs of toxicity to the anaerobic microorganisms. This idea was materialised in a commercial strategy under the BioGasol company in the field of renewable energy for the sustainable production of bioethanol based on lignocellulosic biomasses. Conversion of straw and other agricultural residues into ethanol, biogas, hydrogen and solid fuel with reuse of process water is possible with this complete valorisation scheme.

There is a vast experience in anaerobic digestion processes with several reports indicating the successful digestion of agricultural, food-industry wastes and those derived as by-products from other processes conforming multiple valorisation approach of the biorefinery concept, as it would be, the digestion of vinasses which are side-stream effluents from ethanol fermentation (Buitrón et al. 2019; Cabrera-Díaz et al. 2017; Martínez et al. 2018a, b). The implementation of vinasse biodigestion in sugarcane biorefineries has been studied by Longati et al. (2019) who reported a positive impact when evaluating introduction of this technology into ethanol type biorefinery. The use of biogas from vinasse for a standard first-generation ethanol plant can increase in 9.20% the surplus of electric energy yielded to the grid, which has a significant impact on the global energy balance of the process. Estimated values of methane yield from vinasses are 0.234–0.300 m^3 $CH_4/COD_{removed}$ (Júnior et al. 2016; Fuess et al. 2016) which gives a clear idea of the high potential of this effluent for producing bioenergy. Despite the significant improvements in both scientific and technological aspects related to anaerobic digestion of vinasse, pilot- to full-scale experiences are still scarce even though biomethane production in ethanol processing plants results in outstanding performance regarding electricity generation (Fuess and Zaiat 2018).

In a similar line, of valorising side streams from conventional biofuel production processes, is the use of crude glycerol as co-substrate in digesters treating either sewage sludge or manures. Different authors report increments of methane yield from 35 to 50% in average with the increase in glycerine ratio (Lobato et al. 2010; Fierro et al. 2016). Crude glycerol provides high organic load to digesters allowing for a significant increase in the biogas performance of reactors. Valorisation of this side stream from biodiesel production process can thus be carried out without the need of further refining. However, there are limits to the use of glycerine as co-substrate since its presence in the digester alters the microbial flora and causes a preferential degradation of this readily biodegradable substrate leading to an incomplete stabilisation of the main feeding (animal manure) (Fierro et al. 2016; González et al. 2019). Another important issue is based on the fact that digestion is a process performed on sequential reactions, where the organic material is first acidified and then these short-chain fatty acids must be submitted to further degradation by the action or archaea microflora. Any unbalance associated with overloading of readily degradable material may cause fatty acid build-up leading to the decrease in biogas production by methanogenic inhibition.

Another approach where anaerobic digestion was proposed as the final valorisation step to be integrated into a biorefinery was that of Martínez et al. (2018a, b). The concept proposed by these authors considers the use of green biomass, where this material is first subjected to mechanical fractionation generating two fractions: one solid called press cake and another liquid known as green juice. The press cake is composed of lignocellulosic fibre material and residual proteins, which makes it a valuable feed, or it can either be used as lignocellulosic feedstock for biofuels and green chemical production. The green juice contains non-denatured proteins and free amino acids which can be valorised for producing protein concentrates, leading to a residual effluent called brown juice containing water-soluble carbohydrates, residual proteins and minerals which is suitable for anaerobic digestion. Therefore, the whole valorisation of biomass is attained producing a great diversity of green chemical products and energy.

It seems logical to consider conventional centralised waste treatment plants as centres for the transformation of organic materials and, therefore, grant these facilities greater status by converting them into biorefineries. Biomass such as lignocellulosic material and wastes can be valorised in conjunction. Wastes contain various high-value chemical substances and elements, including carbon sources in the form of carboxylic and other acids, carbohydrates, proteins and nutrients such as nitrogen (N), in the form of ammonium, phosphorus (P) and metals (Zacharof 2017). The use of recovered materials from waste is beneficial for the environment but also for the economy and the digestion process is capable of producing a valuable energy source and a stable form or organic matter suitable to be incorporated into different processes to benefit from the recovery of nutrients and organic compounds. For example, phosphate rock is a non-renewable natural resource with different applications including drinking water softening, feed and food additives and fertilisers. Mining phosphate rock is gradually becoming costlier (Zacharof 2017) and the depletion of this element is making imperative the search for the recovery of phosphorus from waste materials. This idea has been explored through the recovery of phosphorus from pig-slurry by a biological acidification step in the form of struvite (Daumer et al. 2010; Piveteau et al. 2017). By means of a lactic acid fermentation 60–90% of total phosphorus and total magnesium could be easily solubilised without interfering in a subsequent valorisation of the slurry by anaerobic digestion (Piveteau et al. 2017).

The production of biogas by means of anaerobic digestion also produces a digestate still needing final disposal. Digestates have been traditionally used as organic amendments for crops, but the great size of many of these installations sometimes makes unfeasible the spreading during some seasons or even the whole year round, thus becoming a problem. One alternative recently proposed was the thermal treatment by pyrolysis (Feng and Lin 2017) where organic compounds presenting high water content are first introduced in an anaerobic digester for biogas recovery and subsequent pyrolysis of the slurry is attained for producing gases and oil which can be valorised as fuels along with a char fraction having interesting properties in agronomic application and for improving the performance of fermentation systems due to its capacity to act as carbon conductive material (Gómez et al. 2018; González et al. 2018). Pyrolysis is a thermal process where organic compounds are decomposed under inert atmosphere, generating light gaseous products (short-chain hydrocarbons) along with hydrogen, carbon monoxide and carbon dioxide. The characteristics of products and yields obtained for the different fractions are highly dependent on process conditions (temperature and reactor operation) and heating ramp (Tripathi et al. 2016).

The use of char derived from pyrolysis has been studied for improving the performance of anaerobic digestion, thus reporting on a better stabilisation of the microbial system when inhibitors are present either by adsorbing onto the carbon surface the toxic compounds or by offering protecting sites to the microflora (Martínez et al. 2018b). In addition, the effect of char has been also evaluated for assessing the performance of high-loaded systems, indicating that the presence of this material accelerated the degradation rate of substrates up to 86% and favours the selective colonisation of functional microbes (*Methanosarcina* and *Methanosaeta*) (Luo et al. 2015).

Figure 10.6 shows a schematic representation of the valorisation of manures and lignocellulosic biomass integrated in a biorefinery concept for biomass conversion into biofuels. In this scheme, it also considered the treatment of pyrolysis water obtained from condensation reactions. The thermal conversion of biomass in a pyrolysis process yields in addition to oxygenated biooils, water derived from the initial content of the material and that formed through the thermal transformation process. Biooils and water form a miscible phase due to the oxygen content of the oily phase. The water thus obtained in this process needs further treatment. It is estimated that the water content of biooils may be as high as 52% of the total oil fraction (Abnisa et al. 2013; Mullen and Boateng 2011). The treatment of this aqueous phase by anaerobic digestion has been attempted by Hübner and Mumme (2015) reporting a removal of organic content of about 63.4% (measured as chemical oxygen demand, COD) and having a significant effect on the degradability of this liqueur, the temperature of the pyrolysis process. In addition to the use of char for enhancing the digestion process, this scheme is also considered the agronomic application of this material. Char addition to different crops has been reported (Rondon et al. 2007; Rosas et al. 2015) to favour carbon sequestration and enhance nutrient retention reducing thus run-off and the number of fertilisations needed because of the better use of fertilisers by plants (Van Zwieten et al. 2010) and causing a modification of the soil ecological niche in the long term (Hardy et al. 2019).

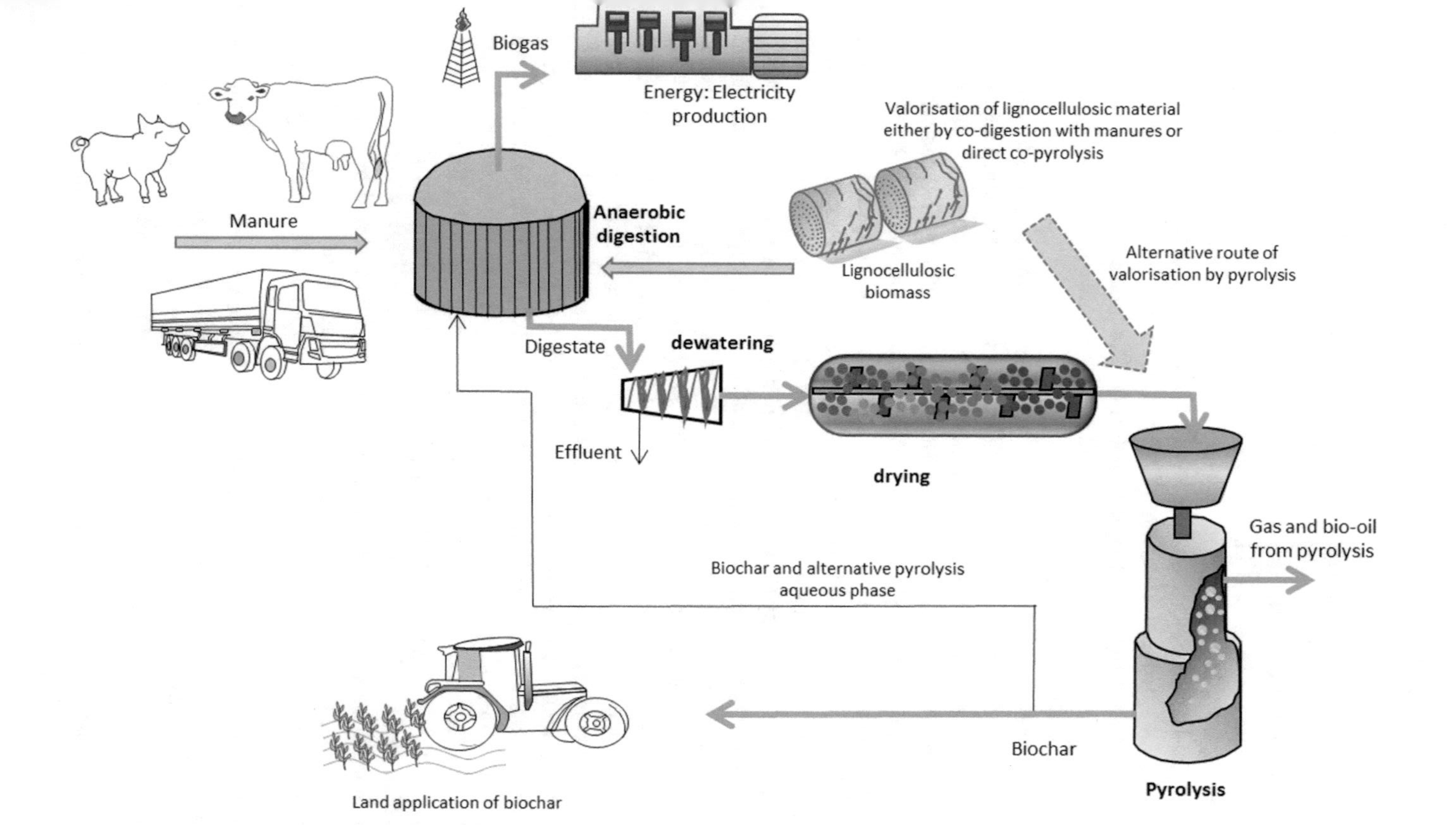

Fig. 10.6 Valorisation of manures and lignocellulosic biomass for producing biogas and pyrolysis valuable products from the combined pyrolytic treatment of digestates

10.2.5 Hydrogen

Biological production of hydrogen can be attained by different ways as it is direct and indirect biophotolysis and photofermentation (light-dependent methods), bioelectrochemical systems (BES) and dark fermentation process (no light-dependent methods) (Martínez et al. 2019a, b). Photofermentation is the biological process of converting organic molecules to H_2 and CO_2 in the presence of light, in anaerobic, nitrogen-limited conditions. Photosynthetic purple non-sulphur (PNS) bacteria as *Rhodobacter sphaeroides* and other PNS bacteria can produce hydrogen using a variety of organic compounds (Sagir et al. 2017). The limitation of nitrogen forces the bacteria to 'dump' the excess energy and reducing power through the production of hydrogen (Koku et al. 2002).

Two enzymes namely, nitrogenase and hydrogenase play an important role in biohydrogen production. Photofermentation by PNS bacteria can attain a significant increase in hydrogen yields of the biological process by optimisation of growth conditions and immobilisation of active cells (Basak and Das 2007). PNS bacteria have the ability to use light energy in a wide range of absorption spectra (522–860) nm without evolving oxygen which might cause inactivation of the system. Energy from light enables these organisms to overcome the thermodynamic barrier in the conversion of organic acids into hydrogen (Kumar et al. 2019; Miyake et al. 1982; Basak and Das 2009). Their ability to assimilate different types of carbon sources has led to the development of hydrogen-producing systems as a single stage, using glucose and sucrose, along with hydrocarbon-rich substrates as black strap, and beet molasses with hydrogen yields reported in a single-stage configuration in the range of 9–14 mol H_2/mol substrate (Abo-Hashesh et al. 2013; Keskin and Hallenbeck 2012; Sagir et al. 2017) using *Rhodobacter capsulatus.* Immobilised systems, on the other hand, can attain higher conversions of acid substrates, as it is the case of the use of lactic acid using a polyurethane foam reactor for the retention of *Rhodobacter sphaeroides* GL_{-1}. These organisms evolved hydrogen at a rate of 0.21 mL H_2/h mL foam and a conversion of 86% of lactic acid (Fedorov et al. 1998). These yields can be further improved by the alternation of light–dark periods. Sargsyan et al. (2015) reported on the effect of dark periods in hydrogen evolution when culturing *Rhodobacter sphaeroides* MDC 6522 from Armenian mineral springs. These authors reported that at inoculation of bacteria, illumination after 24 h dark period in comparison with continuous illumination can be used for enhancing H_2 yields, reporting values of 3–8 mmol H_2/g DW (cell dry weight) based on different alternating light–dark periods.

Another approach for increasing hydrogen productivity of reactors is the use of acid organic compounds as substrates for the photofermentative process. The acids can be derived from a previous fermentative hydrogen-producing system. Thus, in this two-stage configuration the fermentative hydrogen reactor evolves biogas composed mainly of hydrogen and CO_2, along with short-chain fatty acids (mainly C2 and C4 species) in the effluent stream which is subsequently treated in a second photofermentation stage. In general, the fermentative production process is considered to be ineffective due to the low-conversion rate of substrate into hydrogen. Pure cultures of *Enterobacter*, *Bacillus* and *Clostridium* are known for producing hydrogen from soluble sugars and starch (Hawkes et al. 2002). In the case of *Clostridium*, the fermentative process gives the higher yields due to the ability of these organisms for re-oxidising the NADH generated during glycolysis but even though this conversion only yields 33% of the theoretical value (12 mol H_2/mol glucose) (Hallenbeck 2009; Moreno and Gómez 2012). This process has similarities with those at industrial scale such as the acidogenic stage of anaerobic digestion and acetone–butanol (solvent) production by clostridia (Hawkes et al. 2002).

The combined approach for increasing the productivity of hydrogen is then based on the ability of the dark fermentative process (denomination based on the lack of needing light for evolving this gas) for assimilating not only carbohydrate but also cellulosic compounds along with the ability of operating using mixed cultures and wastewaters and solid wastes as substrates (Li and Fang 2007). Table 10.3 presents some results reported by different authors for enhancing hydrogen production using

Table 10.3 Hydrogen yields obtained from two-stage processes considering dark fermentation and subsequent photofermentation

Substrate	Dark fermentation	Photofermentation	Yield from two-stage	References
Sucrose	Mixed culture 3.67 mol H_2/mol sucrose (360 mL H_2/L h)	*R. sphaeroides* SH2C 4.06 mol H_2/mol sucrose	3.67 mol H_2/mol sucrose	Tao et al. (2007)
Olive mill wastewater (OMW)	Mixed culture	*R. sphaeroides* O.U.001	29 – 35 L H_2/L_{OMW}	Eroğlu et al. (2006)
Glucose	*Enterobacter cloacae* 1.86 mol H_2/mol glucose	*R. sphaeroides* O.U.001 1.5–1.72 mol H_2/mol acetic acid	2.78 mol H_2/mol glucose*	Nath et al. (2005)
Molasses	Thermophile *Caldicellulosiruptor saccharolyticus* (72 °C) 2.1 mol H_2/mol hexose	*R. capsulatus* (DSM1710) 3.71 mol H_2/mol hexose	5.81 mol H_2/mol hexose	Özgür et al. (2010)
Palm oil mill effluent (POME)	*Clostridium butyricum* LS2 0.784 mL H_2/mL POME (21 mL H_2/h)	*Rhodopseudomonas palustris* 26 mL H_2/h	3.064 mL H_2/mL POME	Mishra et al. (2016)
Glucose	Microaerobic dark fermentative process	*R. capsulatus* JP91	7.8 mol H_2/mol glucose	Sağır et al. (2018)

*Calculated based on data reported

a two-stage configuration. Other approaches include the combination of photofermentation by PNS bacteria as second stage of effluents derived from a thermophilic dark fermentation process of Miscanthus hydrolysate by *Thermotoga neapolitana*. However, in this case, the need of additional steps for coupling the two processes such as centrifugation, dilution, buffer addition, pH adjustment and sterilisation may lead to a significant increase in installation costs of this alternative when implemented at industrial scale (Uyar et al. 2009).

Other processes for producing hydrogen involve the utilisation of hydrogen protons and electrons derived from water photolysis. This feature is characteristic of green algae and cyanobacteria. The water photolysis process can be divided into indirect and direct pathways (Oey et al. 2016). In the indirect pathway, solar energy is first converted into carbohydrates which are then used as substrates for hydrogen production. This process is mediated by nitrogenases and hydrogenases enzymes depending on the Cyanobacteria species, whereas hydrogenases are exclusively used by microalgae (Dutta et al. 2005; Oncel et al. 2015). On the contrary, in direct photolysis, which has only been reported in microalgae, the process involves the use of electrons derived from the light-driven water splitting reaction of photosystem II to directly evolve hydrogen using hydrogenase as mediated enzyme (Melis et al. 2000). Many species of green algae have been reported to produce hydrogen by photolysis such as *Chlorella sorokiniana*, *Chlorella vulgaris*, *Scenedesmus obliquus* with *Chlamydomonas reinhardtii* one of the most studied organisms (Mortensen and Gislerød 2016; Rashid et al. 2013; Senger and Bishop 1979; Yadav et al. 2017).

The industrial feasibility of this process is, however, associated with its performance under the use of solar light, thus the relevance of carrying out studies under outdoor conditions which can be susceptible of contamination by other cultures. Because the hydrogen-producing hydrogenase is very sensitive to oxygen, the process of hydrogen production by microalgae must be performed in a two-stage configuration: under oxygenic photosynthesis for generation of the required algal biomass, followed by hydrogen biosynthesis under anaerobic conditions, this idea was explored by Geier et al. (2012) reporting 19.8–48.0 mL H_2/L reactor when light was set at 200 μmol photons/m^2 s but when increasing photosynthetically active radiation under outdoor cultivation only a maximum of 10% of the hydrogen amounts produced by cells grown under laboratory conditions was reached, indicating that further research will be required to investigate the effect of high irradiances and temperatures at midday along with carbon source content. A similar approach was tested by Xu et al. (2017) when adding a fermentative bacterium to the algae to enhance H_2 production without limiting electron resources, in this study *Chlamydomonas reinhardtii* cc849 was co-cultured with *Azotobacter chroococcum* to improve yields. Maximum production was in the range of 68–149 μmol H_2/mg Chl was reported in the co-culture at 100–200 μE/m^2 s of light intensity, values much higher to that of the pure algae culture (28 μmol H_2/mg Chl).

Another process which has been subject to intensive research in recent years is the production of hydrogen by bioelectrochemical systems (BES). This category includes microbial fuel cells (MFCs), microbial electrolysis cells (MECs) and microbial electrosynthesis cells. In these processes, electrochemically active bacteria grow attached on electrodes and degrade organic matter present in wastewater while producing either electricity, gas fuels and other value-added chemicals becoming a low energy-intensive technology capable of reducing the high energy demand of conventional waste treatment systems (Li et al. 2018; Khan et al. 2018).

MECs have been directly proposed for producing hydrogen from carbohydrate-rich effluent streams or as a second stage of the dark fermentative process to overcome the theoretical barrier associated with this process and improve its industrial feasibility by attaining the complete conversion of the organic compounds. In this line, the productivity of MECs has been evaluated using mixtures of volatile fatty acids as substrate in an attempt of coupling the biolectrochemical process to the fermentative one. The highest production rate reported by Rivera et al. (2015) was 81 mL H_2/L day when testing different acid concentrations ranging from 400 to 1200 mg/L measured as chemical oxygen demand. However, when lactic acid is also produced as a deviation of the dark fermentation process, a negative effect may be observed in the hydrogen yield of the MEC system. Moreno and co-workers (2015) reported on a decrease in hydrogen yield from 70 to 10 mL H_2/L day when the proportion of a dark fermentation effluent derived from the treatment of cheese whey (rich in lactic acid) was increased in the influent stream of the second stage MEC.

Because the dark fermentation process and anaerobic digestion are characterised by high organic loadings, additional research is needed to attain success in coupling any of these treatments to either BES or photofermentation systems. Many approaches consider the dilution of effluents obtained from dark fermentation to make it suitable for the subsequent stage, but these intermediary pre-treatments lead to additional costs which may suppose an excessive burden at large-scale implementation. Another important aspect is the negative effect on hydrogen yields that exert the presence of lactic acid bacteria due to the production of antimicrobial peptides, in this line Rosa et al. (2016) reported on maximum hydrogen yields of 1.7–2.1 L H_2/L day when using cheese whey as substrate but indicating also a severe decline when lactic acid bacterial proliferated in the reactor.

Some other attempts to gain stability on performance consider the coupling in a single reactor allowing for balance between different microbial cultures as it is the combination of MEC and anaerobic microflora or allowing the natural growth of anaerobic competitors in traditional MEC and dark fermentative systems. In the first case, the production of upgraded biogas can be obtained by coupling MEC and anaerobic digestion in a single chamber. Bo et al. (2014) reported on the enhancement of CH_4 yield (which was increased 2.3 times, whereas COD removal rate was tripled). The integrated process was capable of transforming the unwanted CO_2 component of biogas into CH_4 on the anode by the dominant microbes, hydrogenotrophic electromethanogens, using the hydrogen gas in situ generated. A similar idea was tested by Yin et al. (2016) in this case by co-culturing *Geobacter* with *Methanosarcina* in an AD–MEC coupled system, reporting a significant increase in organic matter removal thanks to the ability of co-existence of *Methanosarcina* and *Geobacter* in the biofilm, thus the first one obtaining electrons transferred from *Geobacter* and then reducing carbon dioxide into methane.

The production of biohythane follows a similar approach. This gas is a mixture of biogas enriched with hydrogen, either by the coupling of gases independently produced in biological transformation processes or by allowing the competition of methanogens in traditional MEC and dark fermentative processes. The presence of hydrogen in biogas allows for increasing the energy content of this mixture, and avoids the need of a separate installation for storing and upgrading a pure H_2 gaseous stream. An example of this approach is the integration of microalgae systems for producing hydrogen and the subsequent valorisation of the residual microalgae biomass through dark fermentation followed by conventional digestion process evaluated by Lunprom et al. (2019). These authors used *Chlorella* sp. biomass pre-treated by acid and thermal methods for obtaining yields of 12.5 mL H_2/g VS (volatile solid) and 81 mL CH_4/g VS.

A similar configuration with the same aim of producing hythane was studied by Farhat et al. (2018) using a standard H_2-CH_4 producing system in a two-phase configuration for treating waste materials, but operating in the acidification and H_2 fermentative phase as an anaerobic sequencing batch reactor, allowing for high microbial biomass retention but low hydraulic residence time and operating the subsequent methanogenic phase as standard continuously stirred tank reactor. The novelty in this case was based on the introduction of the gaseous stream generated in the first phase into the second methanogenic phase to enrich biogas, obtaining thus a fuel stream with 8% H_2, 28.5% CO_2 and 63.5% CH_4.

The conversion of H_2 and CO_2 into methane is greatly dependent on the predominant microflora present in the anaerobic reactor. The introduction of a H_2 stream into an anaerobic reactor digesting sewage sludge was evaluated by Martínez et al. (2019b) with the aim of calculating the efficiency of energy production from a MEC hydrogen-producing system for treating wastewaters and the enrichment of biogas derived from the conventional digester, when the hydrogen stream is introduced into the methanogenic reactor. These authors reported an increase in biogas production but not in methane content due to the enrichment of homoacetogenic groups along with other acetogenic microorganisms which produced acetate from hydrogen. Bacteria utilised hydrogen (transferred from the gas phase) and CO_2 to produce acetate, which was subsequently consumed by acetoclastic methanogens, thus the content of biogas was not modified, and CO_2 concentration was kept about 40% in average after hydrogen gas addition.

Several approaches in coupling different biological processes have been studied in an attempt of increasing conversion yields of organic materials, in particular wastes, for producing fuels, valuable products and energy. The most studied and used at an industrial scale is the anaerobic digestion processes for the production of methane. However, such bioconversion has limited net energy yields. The biorefinery concept is then based on the coupling of several steps for increasing the global efficiency. In recent years, a novel approach which is based on the multitude of studies regarding the ability of microorganisms for the direct transfer of electrons is electrofermentation. This technology has attracted much interest due to its ability to boost the microbial metabolism through extracellular electron transfer during fermentation. It has been studied on various acetogens and methanogens, where the enhancement in the biogas yield reached up to twofold (Kumar et al. 2018) probably becoming in the near future one of the alternatives for increasing economy feasibility of biorefineries.

10.3 Conclusion and Future Prospects

The biorefinery concept admits a diversity of technologies and biological transformations with the aim of producing green compounds. The conversion of biomass into chemicals and energy, however, possess several restrictions associated with the availability of sugars and cellulosic components. Pre-treatments favour the access of microbials to organic compounds but introduce a high demand of energy in the global process which should be carefully evaluated. As experience of performance, it can be used the one obtained from the installation and operation of conventional digestion processes and ethanol plants. These plants have not been a focus of success in all territories installed, and those dealing with the production of cellulosic ethanol have been through serious financial problems. Digestion plants, on the contrary, are well known for the high amount of subsidies or government incentives needed to attain economic feasibility. To favour the production of renewable energies and the treatment of wastes, digestion is at this moment the best environmental option and it is also considered as the best technological alternative regarding its energy demand. The lessons obtained from these two processes should be used as basis for evaluating future complex technologies if the biorefinery concept is to become a reality.

There is a great need on evaluating pilot-scale plants close to an industrial configuration in order to establish energy demands and costs of installation along with operation at a commercial scale. Several reports deal with laboratory scale with volumes of millilitres or litres, but there is an urgent need for obtaining reliable data at higher scale (m^3) for an extended time of evaluation to test microbial stability of the biological process and determining process conditions to avoid microbial shifts. Sterilisation needs and aseptic conditions (which are a common feature of biological processes operating with pure cultures and genetically modified microorganism) are usually against plant profitability, thus becoming an additional factor to be assessed if the aim is to transform the current economy into a green economy.

Acknowledgments Judith González-Arias would like to thank the Junta de Castilla y León (Consejería de Educación) fellowship, Orden EDU/1100/2017, co-financed by the European Social Fund. Cristian B. Arenas would like to thank Ministry of Economy and Competitiveness for fellowship BES-2016-078329

References

Abnisa F, Arami-Niya A, Daud WW, Sahu JN (2013) Characterization of bio-oil and bio-char from pyrolysis of palm oil wastes. BioEnergy Res 6:830–840

Abo-Hashesh M, Desaunay N, Hallenbeck PC (2013) High yield single step conversion of glucose to hydrogen by photofermentation with continuous cultures of Rhodobacter capsulatus JP91. Bioresour Technol 128:513–517

Ageitos JM, Vallejo JA, Veiga-Crespo P, Villa TG (2011) Oily yeasts as oleaginous cell factories. Appl Microbiol Biotechnol 90:1219–1227

Ahring BK, Westermann P (2007) Coproduction of bioethanol with other biofuels. In: Biofuels, pp 289–302. Springer, Berlin.

Amaro HM, Guedes AC, Malcata FX (2011) Advances and perspectives in using microalgae to produce biodiesel. Appl Energy 88:3402–3410

Amoah J, Kahar P, Ogino C, Kondo A (2019) Bioenergy and biorefinery: feedstock, biotechnological conversion and products. Biotechnol J 1800494

Amorim HV, Lopes ML, de Castro-Oliveira JV, Buckeridge MS, Goldman GH (2011) Scientific challenges of bioethanol production in Brazil. Appl Microbiol Biotechnol 91:1267

Annamalai N, Rajeswari MV, Balasubramanian T (2016) Endo-1, 4-β-glucanases: role, applications and recent developments. In: Microbial enzymes in bioconversions of biomass, pp 37–45. Springer, Cham.

Apiwatanapiwat W, Murata Y, Kosugi A, Yamada R, Kondo A, Arai T, Rugthaworn Ps, Mori Y (2011) Direct ethanol production from cassava pulp using a surface-engineered yeast strain co-displaying two amylases, two cellulases, and β-glucosidase. Appl Microbiol Biotechnol 90:377–384

Aresta M, Dibenedetto A, Dumeignil F (eds) (2012) Biorefinery: from biomass to chemicals and fuels. Walter de Gruyter

Aro EM (2016) From first generation biofuels to advanced solar biofuels. Ambio 45:24–31

Bajpai P (2018) Bioconversion of hemicelluloses. In: Biotechnology for pulp and paper processing, pp 545–560. Springer, Singapore

Bajpai P (2019) Cultivation of third generation biofuel. In: Third generation biofuels, pp 17–28. Springer, Singapore

Basak N, Das D (2007) The prospect of purple non-sulfur (PNS) photosynthetic bacteria for hydrogen production: the present state of the art. World J Microbiol Biotechnol 23:31–42

Basak N, Das D (2009) Photofermentative hydrogen production using purple non-sulfur bacteria *Rhodobacter sphaeroides* OU 001 in an annular photobioreactor: a case study. Biomass Bioenergy 33:911–919

Basso LC, De Amorim HV, De Oliveira AJ, Lopes ML (2008) Yeast selection for fuel ethanol production in Brazil. FEMS Yeast Res 8:1155–1163

Bilal M, Asgher M, Iqbal HM, Ramzan M (2017) Enhanced bio-ethanol production from old newspapers waste through alkali and enzymatic delignification. Waste Biomass Valori 8:2271–2281

Blakey S, Rye L, Wilson CW (2011) Aviation gas turbine alternative fuels: a review. Proc Combust Inst 33:2863–2885

Bo T, Zhu X, Zhang L, Tao Y, He X, Li D, Yan Z (2014) A new upgraded biogas production process: coupling microbial electrolysis cell and anaerobic digestion in single-chamber, barrel-shape stainless steel reactor. Electrochem Commun 45:67–70

Buitrón G, Martínez-Valdez FJ, Ojeda F (2019) Biogas production from a highly organic loaded winery effluent through a two-stage process. BioEnerg Res 1:8

Cabrera-Díaz A, Pereda-Reyes I, Oliva-Merencio D, Lebrero R, Zaiat M (2017) Anaerobic digestion of sugarcane vinasse through a methanogenic UASB reactor followed by a packed bed reactor. Appl Biochem Biotechnol 183:1127–1145

Capolupo L, Faraco V (2016) Green methods of lignocellulose pretreatment for biorefinery development. Appl Microbiol Biotechnol 100:9451–9467

Chaiprapat S, Sasibunyarat T, Charnnok B, Cheirsilp B (2017) Intensifying clean energy production through cultivating mixotrophic microalgae from digestates of biogas systems: effects of light intensity, medium dilution, and cultivating time. BioEnerg Res 10:103–114
Cherubini F, Jungmeier G (2010) LCA of a biorefinery concept producing bioethanol, bioenergy, and chemicals from switchgrass. Int J Life Cycle Ass 15:53–66
Chojnacka K, Noworyta A (2004) Evaluation of *Spirulina* sp. growth in photoautotrophic, heterotrophic and mixotrophic cultures. Enzyme Microb Technol 34:461–465
Coimbra RN, Escapa C, Otero M (2019) Comparative thermogravimetric assessment on the combustion of coal, Microalgae Biomass and Their Blend. Energies 12:2962
Cuetos MJ, Fernández C, Gómez X, Morán A (2011) Anaerobic co-digestion of swine manure with energy crop residues. Biotechnol Bioproc E 16:1044
Cuetos MJ, Gómez X, Martínez EJ, Fierro J, Otero M (2013) Feasibility of anaerobic co-digestion of poultry blood with maize residues. Bioresour Technol 144:513–520
Dakota. https://www.dakotagold.com/distillers-grains
Dashko S, Zhou N, Compagno C, Piškur J (2014) Why, when, and how did yeast evolve alcoholic fermentation? FEMS Yeast Res 14:826–832
Daumer ML, Picard S, Saint-Cast P, Dabert P (2010) Technical and economical assessment of formic acid to recycle phosphorus from pig slurry by a combined acidification–precipitation process. J Hazard Mater 180:361–365
Díez-Antolínez R, Hijosa-Valsero M, Paniagua-García AI, Garita-Cambronero J, Gómez X (2018) Yeast screening and cell immobilization on inert supports for ethanol production from cheese whey permeate with high lactose loads. PLoS ONE 13(12):e0210002. https://doi.org/10.1371/journal.pone.0210002
Dobrowolski A, Mituła P, Rymowicz W, Mirończuk AM (2016) Efficient conversion of crude glycerol from various industrial wastes into single cell oil by yeast Yarrowia lipolytica. Bioresour Technol 207:237–243
Dong C, Wang Y, Wang H, Lin CSK, Hsu HY, Leu SY (2019) New generation urban biorefinery toward complete utilization of waste derived lignocellulosic biomass for biofuels and value-added products. Energy Proc 158:918–925
Du C, Lin SKC, Koutinas A, Wang R, Webb C (2007) Succinic acid production from wheat using a biorefining strategy. Appl Microbiol Biotechnol 76:1263–1270
Dürre P (1998) New insights and novel developments in clostridial acetone/butanol/isopropanol fermentation. Appl Microbiol Biotechnol 49:639–648
Dutta D, De D, Chaudhuri S, Bhattacharya SK (2005) Hydrogen production by cyanobacteria. Microb Cell Fact 4:36
Eboibi BE (2019) Impact of time on yield and properties of biocrude during downstream processing of product mixture derived from hydrothermal liquefaction of microalga. Biomass Conv Bioref 9:379–387
Eggeman T, Verser D (2006) The importance of utility systems in today's biorefineries and a vision for tomorrow. Appl Microbiol Biotechnol 130:361–381
Eroğlu E, Eroğlu İ, Gündüz U, Türker L, Yücel M (2006) Biological hydrogen production from olive mill wastewater with two-stage processes. Int J Hydrog Energ 31:1527–1535
European biogas association report. https://european-biogas.eu/wp-content/uploads/2017/12/Statistical-report-of-the-European-Biogas-Association-web.pdf
Ezeji TC, Qureshi N, Blaschek HP (2004) Acetone butanol ethanol (ABE) production from concentrated substrate: reduction in substrate inhibition by fed-batch technique and product inhibition by gas stripping. Appl Microbiol Biotechnol 63:653–658
Farhat A, Miladi B, Hamdi M, Bouallagui H (2018) Fermentative hydrogen and methane co-production from anaerobic co-digestion of organic wastes at high loading rate coupling continuously and sequencing batch digesters. Environ Sci Pollut R 25:27945–27958
Farrokh P, Sheikhpour M, Kasaeian A, Asadi H, Bavandi R (2019) Cyanobacteria as an eco-friendly resource for biofuel production: a critical review. Biotechnol Progr. https://doi.org/10.1002/btpr.2835

Fava F, Totaro G, Diels L, Reis M, Duarte J, Carioca OB, Poggi-Varaldo HM, Ferreira BS (2015) Biowaste biorefinery in Europe: opportunities and research & development needs. New Biotechnol 32:100–108

Fedorov AS, Tsygankov AA, Rao KK, Hall DO (1998) Hydrogen photoproduction by *Rhodobacter sphaeroides* immobilised on polyurethane foam. Biotechnol Lett 20:1007–1009

Feng Q, Lin Y (2017) Integrated processes of anaerobic digestion and pyrolysis for higher bioenergy recovery from lignocellulosic biomass: a brief review. Renew Sustain Energy Rev 77:1272–1287

Fernando S, Adhikari S, Chandrapal C, Murali N (2006) Biorefineries: current status, challenges, and future direction. Energy Fuel 20:1727–1737

Fierro J, Martinez EJ, Rosas JG, Fernández RA, López R, Gómez X (2016) Co-digestion of swine manure and crude glycerine: Increasing glycerine ratio results in preferential degradation of labile compounds. Water Air Soil Pollut 227:78

FitzPatrick M, Champagne P, Cunningham MF, Whitney RA (2010) A biorefinery processing perspective: treatment of lignocellulosic materials for the production of value-added products. Bioresour Technol 101:8915–8922

Formanek J, Mackie R, Blaschek HP (1997) Enhanced butanol production by *Clostridium beijerinckii* BA101 grown in semidefined P2 medium containing 6 percent maltodextrin or glucose. Appl Environ Microbiol 63:2306–2310

Fuess LT, Zaiat M (2018) Economics of anaerobic digestion for processing sugarcane vinasse: applying sensitivity analysis to increase process profitability in diversified biogas applications. Process Saf Environ 115:27–37

Fuess LT, Kiyuna LSM, Garcia ML, Zaiat M (2016) Operational strategies for long-term biohydrogen production from sugarcane stillage in a continuous acidogenic packed-bed reactor. Int J Hydrog Energ 41:8132–8145

Geier SC, Huyer S, Praebst K, Husmann M, Walter C, Buchholz R (2012) Outdoor cultivation of *Chlamydomonas reinhardtii* for photobiological hydrogen production. J Appl Phycol 24:319–327

Gogoi H, Nirosha V, Jayakumar A, Prabhu K, Maitra M, Panjanathan R (2018) Paper mill sludge as a renewable substrate for the production of acetone-butanol-ethanol using Clostridium sporogenes NCIM 2337. Energy Source Part A 40:39–44

Goldemberg J (2008) The Brazilian biofuels industry. Biotechnol Biofuels 1:6

Gómez X, Cuetos MJ, Cara J, Moran A, García AI (2006) Anaerobic co-digestion of primary sludge and the fruit and vegetable fraction of the municipal solid wastes: conditions for mixing and evaluation of the organic loading rate. Renew Energy 31:2017–2024

Gómez X, Cuetos MJ, García AI, Morán A (2007) An evaluation of stability by thermogravimetric analysis of digestate obtained from different biowastes. J Hazard Mater 149:97–105

Gómez X, Meredith W, Fernández C, Sánchez-García M, Díez-Antolínez R, Garzón-Santos J, Snape CE (2018) Evaluating the effect of biochar addition on the anaerobic digestion of swine manure: application of Py-GC/MS. Environ Sci Pollut R 25:25600–25611

Gonçalves BC, Silva MB (2018) Green microalgae as substrate for producing biofuels and chlorophyll in biorefineries. In: Sustainable biotechnology-enzymatic resources of renewable energy, pp 439–461. Springer, Cham

González J, Sánchez M, Gómez X (2018) Enhancing anaerobic digestion: the effect of carbon conductive materials. C 4:59

González R, Smith R, Blanco D, Fierro J, Gómez X (2019) Application of thermal analysis for evaluating the effect of glycerine addition on the digestion of swine manure. J Therm Anal Calorim 135:2277–2286

Green Car Congress (2019). https://www.greencarcongress.com/2019/07/20190722-poet.html

Greenland SJ, Dalrymple J, Levin E, O'Mahony B (2018) Improving agricultural water sustainability: strategies for effective farm water management and encouraging the uptake of drip irrigation. In: The Goals of Sustainable Development, pp 111–123. Springer, Singapore

Grobbelaar JU (2010) Microalgal biomass production: challenges and realities. Photosynth Res 106:135–144

Hallenbeck PC (2009) Fermentative hydrogen production: principles, progress, and prognosis. Int J Hydrog Energy 34:7379–7389
Han Q, Jin Y, Jameel H, Chang HM, Phillips R, Park S (2015) Autohydrolysis pretreatment of waste wheat straw for cellulosic ethanol production in a co-located straw pulp mill. Appl Biochem Biotechnol 175:1193–1210
Hardy B, Sleutel S, Dufey JE, Cornelis JT (2019) The long-term effect of biochar on soil microbial abundance, activity and community structure is overwritten by land management. Front Environ Sci 7:110
Hawkes FR, Dinsdale R, Hawkes DL, Hussy I (2002) Sustainable fermentative hydrogen production: challenges for process optimisation. Int J Hydrog Energy 27:1339–1347
Hijosa-Valsero M, Paniagua-García AI, Díez-Antolínez R (2018) Industrial potato peel as a feedstock for biobutanol production. New Biotechnol 46:54–60
Hill J, Nelson E, Tilman D, Polasky S, Tiffany D (2006) Environmental, economic, and energetic costs and benefits of biodiesel and ethanol biofuels. P Natl Acad Sci USA 103:11206–11210
Hübner T, Mumme J (2015) Integration of pyrolysis and anaerobic digestion–use of aqueous liquor from digestate pyrolysis for biogas production. Bioresour Technol 183:86–92
Ito T, Nakashimada Y, Senba K, Matsui T, Nishio N (2005) Hydrogen and ethanol production from glycerol-containing wastes discharged after biodiesel manufacturing process. J Biosci Bioeng 100:260–265
Jiang Y, Guo D, Lu J, Dürre P, Dong W, Yan W, Zhang W, Ma J, Jiang M, Xin F (2018a) Consolidated bioprocessing of butanol production from xylan by a thermophilic and butanologenic *Thermoanaerobacterium* sp. M5. Biotechnol Biofuels 11:89
Jiang Y, Zhang T, Lu J, Dürre P, Zhang W, Dong W, Zhou J, Jiang M, Xin F (2018) Microbial co-culturing systems: butanol production from organic wastes through consolidated bioprocessing. Appl Microbiol Biotechnol 102:5419–5425
Jin Q, Neilson AP, Stewart AC, O'Keefe SF, Kim YT, McGuire M, Wilder G, Huang H (2018) Integrated approach for the valorization of red grape pomace: production of oil, polyphenols, and acetone–butanol–ethanol. ACS Sustain Chem Eng 6:16279–16286
Jiru TM, Steyn L, Pohl C, Abate D (2018) Production of single cell oil from cane molasses by *Rhodotorula kratochvilovae* (syn, *Rhodosporidium kratochvilovae*) SY89 as a biodiesel feedstock. Chem Cent J 12:91
Joshi SM, Waghmare JS, Sonawane KD, Waghmare SR (2015) Bio-ethanol and bio-butanol production from orange peel waste. Biofuels 6:55–61
Júnior ADNF, Koyama MH, de Araújo Júnior MM, Zaiat M (2016) Thermophilic anaerobic digestion of raw sugarcane vinasse. Renew Energy 89:245–252
Kádár Z, Szengyel Z, Réczey K (2004) Simultaneous saccharification and fermentation (SSF) of industrial wastes for the production of ethanol. Ind Crop Prod 20:103–110
Kamm B, Kamm M (2004) Principles of biorefineries. Appl Microbiol Biotechnol 64:137–145
Keskin T, Hallenbeck PC (2012) Hydrogen production from sugar industry wastes using single-stage photofermentation. Bioresour Technol 112:131–136
Khan N, Khan MD, Sultana S, Khan MZ, Ahmad A (2018) Bioelectrochemical systems for transforming waste to energy. In: Modern age environmental problems and their remediation, pp 111–128. Springer, Cham
Kiyoshi K, Furukawa M, Seyama T, Kadokura T, Nakazato A, Nakayama S (2015) Butanol production from alkali-pretreated rice straw by co-culture of Clostridium thermocellum, and Clostridium saccharoperbutylacetonicum. Bioresour Technol 186:325–328
Koku H, Eroğlu I, Gündüz U, Yücel M, Türker L (2002) Aspects of the metabolism of hydrogen production by *Rhodobacter sphaeroides*. Int J Hydrog Energy 27:1315–1329
Kołtuniewicz AB, Dąbkowska K (2016) Biorefineries–factories of the future. Chem Process Eng 37:109–119
Kour D, Rana KL, Yadav N, Yadav AN, Rastegari AA, Singh C et al (2019a) Technologies for biofuel production: current development, challenges, and future prospects. In: Rastegari AA,

Yadav AN, Gupta A (eds) Prospects of renewable bioprocessing in future energy systems, pp 1–50. Springer International Publishing, Cham. https://doi.org/10.1007/978-3-030-14463-0_1
Kour D, Rana KL, Yadav N, Yadav AN, Singh J, Rastegari AA et al. (2019b) Agriculturally and industrially important fungi: current developments and potential biotechnological applications. In: Yadav AN, Singh S, Mishra S, Gupta A (eds) Recent advancement in white biotechnology through fungi. Perspective for value-added products and environments, vol 2, pp 1–64. Springer International Publishing, Cham. https://doi.org/10.1007/978-3-030-14846-1_1
Kumar H, Christopher LP (2017) Recent trends and developments in dissolving pulp production and application. Cellulose 24:2347–2365
Kumar P, Chandrasekhar K, Kumari A, Sathiyamoorthi E, Kim BS (2018) Electro-fermentation in aid of bioenergy and biopolymers. Energies 11:343
Kumar S, Sharma S, Thakur S, Mishra T, Negi P, Mishra S et al (2019) Bioprospecting of microbes for biohydrogen production: current status and future challenges. In: Molina G, Gupta VK, Singh BN, Gathergood N (eds) Bioprocessing for biomolecules production. Wiley, USA, pp 443–471
Li C, Fang HH (2007) Fermentative hydrogen production from wastewater and solid wastes by mixed cultures. Crit Rev Environ Sci Technol 37:1–39
Li X, Mupondwa E (2014) Life cycle assessment of camelina oil derived biodiesel and jet fuel in the Canadian Prairies. Sci Total Environ 481:17–26
Li A, Antizar-Ladislao B, Khraisheh M (2007) Bioconversion of municipal solid waste to glucose for bio-ethanol production. Bioproc Biosyst Eng 30:189–196
Li Q, Du W, Liu D (2008) Perspectives of microbial oils for biodiesel production. Appl Microbiol Biotechnol 80:749–756
Li Q, Cai H, Hao B, Zhang C, Yu Z, Zhou S, Chenjuan L (2010) Enhancing clostridial acetone-butanol-ethanol (ABE) production and improving fuel properties of ABE-enriched biodiesel by extractive fermentation with biodiesel. Appl Biochem Biotechnol 162:2381–2386
Li Y, Zhang R, Liu X, Chen C, Xiao X, Feng L, He Y, Liu G (2013) Evaluating methane production from anaerobic mono-and co-digestion of kitchen waste, corn stover, and chicken manure. Energy Fuel 27:2085–2091
Li T, Xu J, Wu H, Wang G, Dai S, Fan J, He H, Xiang W (2016) A saponification method for chlorophyll removal from microalgae biomass as oil feedstock. Mar Drugs 14:162
Li Z, Fu Q, Kobayashi H, Xiao S (2018) Biofuel production from bioelectrochemical systems. In: Bioreactors for microbial biomass and energy conversion, pp 435–461. Springer, Singapore
Lobato A, Cuetos MJ, Gómez X, Morán A (2010) Improvement of biogas production by co-digestion of swine manure and residual glycerine. Biofuels 1:59–68
Longati AA, Lino AR, Giordano RC, Furlan FF, Cruz AJ (2019) Biogas production from anaerobic digestion of vinasse in sugarcane biorefinery: a techno-economic and environmental analysis. Waste Biomass Valori 1:19
Louhasakul Y, Cheirsilp B, Prasertsan P (2016) Valorization of palm oil mill effluent into lipid and cell-bound lipase by marine yeast *Yarrowia lipolytica* and their application in biodiesel production. Waste Biomass Valori 7:417–426
Lunprom S, Phanduang O, Salakkam A, Liao Q, Imai T, Reungsang A (2019) Bio-hythane production from residual biomass of *Chlorella* sp. biomass through a two-stage anaerobic digestion. Int J Hydrog Energy 44:3339–3346
Luo C, Lü F, Shao L, He P (2015) Application of eco-compatible biochar in anaerobic digestion to relieve acid stress and promote the selective colonization of functional microbes. Water Res 68:710–718
Ma F, Hanna MA (1999) Biodiesel production: a review1. Bioresour Technol 70:1–15
Mabee WE, Saddler JN (2006) The potential of bioconversion to produce fuels and chemicals. In Annual meeting-pulp and paper technical association of Canada, June 2006, vol 92, No C, p 185. Pulp and Paper Technical Association of Canada, 1999
Maiti S, Gallastegui G, Sarma SJ, Brar SK, Le Bihan Y, Drogui P, Buelna G, Verma M (2016) A re-look at the biochemical strategies to enhance butanol production. Biomass Bioenergy 94:187–200

Martínez EJ, Fierro J, Sánchez ME, Gómez X (2012) Anaerobic co-digestion of FOG and sewage sludge: study of the process by Fourier transform infrared spectroscopy. Int Biodeterior Biodegrad 75:1–6

Martínez E, Raghavan V, González-Andrés F, Gómez X (2015) New biofuel alternatives: integrating waste management and single cell oil production. Int J Mol Sci 16:9385–9405

Martínez EJ, Rosas JG, González R, García D, Gómez X (2018) Treatment of vinasse by electrochemical oxidation: evaluating the performance of boron-doped diamond (BDD)-based and dimensionally stable anodes (DSAs). Int J Environ Sci Technol 15:1159–1168

Martínez EJ, Rosas JG, Sotres A, Morán A, Cara J, Sánchez ME, Gómez X (2018) Codigestion of sludge and citrus peel wastes: evaluating the effect of biochar addition on microbial communities. Biochem Eng J 137:314–325

Martínez EJ, Blanco D, Gómez X (2019a) Two-stage process to enhance bio-hydrogen production. In: Improving biogas production, pp 149–179. Springer, Cham

Martínez EJ, Sotres A, Arenas CB, Blanco D, Martínez O, Gómez X (2019) Improving anaerobic digestion of sewage sludge by hydrogen addition: analysis of microbial populations and process performance. Energies 12:1228

Matsakas L, Bonturi N, Miranda EA, Rova U, Christakopoulos P (2015) High concentrations of dried sorghum stalks as a biomass feedstock for single cell oil production by *Rhodosporidium toruloides*. Biotechnol Biofuels 8:6

Melis A, Zhang LP, Forestier M, Ghirardi ML, Seibert M (2000) Sustained photobiological hydrogen gas production upon reversible inactivation of oxygen evolution in the green alga *Chlamydomonas reinhardtii*. Plant Physiol 122:127–135

Menon V, Rao M (2012) Trends in bioconversion of lignocellulose: biofuels, platform chemicals & biorefinery concept. Prog Energy Combust 38:522–550

Metzger P, Largeau C (2005) *Botryococcus braunii*: a rich source for hydrocarbons and related ether lipids. Appl Microbiol Biotechnol 66:486–496

Mishra P, Thakur S, Singh L, Ab Wahid Z, Sakinah M (2016) Enhanced hydrogen production from palm oil mill effluent using two stage sequential dark and photo fermentation. Int J Hydrog Energy 41:18431–18440

Miyake J, Tomizuk N, Kamibayashi A (1982) Prolonged photo-hydrogen production by *Rhodospirillum rubrum*. J Ferment Technol 60:199–203

Mohammed IY, Abba Z, Matias-Peralta HM, Abakr YA, Fuzi SFZM (2018) Thermogravimetric study and evolved gas analysis of new microalga using TGA-GC-MS. Biomass Conv Bioref 8:669–678

Montero E, Olguín EJ, De Philippis R, Reverchon F (2018) Mixotrophic cultivation of *Chlorococcum* sp. under non-controlled conditions using a digestate from pig manure within a biorefinery. J Appl Phycol 30:2847–2857

Moreno R, Gómez X (2012) Dark fermentative H2 production from wastes: effect of operating conditions. J Environ Sci Eng A 1:936

Moreno R, Escapa A, Cara J, Carracedo B, Gómez X (2015) A two-stage process for hydrogen production from cheese whey: integration of dark fermentation and biocatalyzed electrolysis. Int J Hydrog Energy 40:168–175

Mortensen LM, Gislerød HR (2016) The growth of *Chlorella sorokiniana* as influenced by CO 2, light, and flue gases. J Appl Phycol 28:813–820

Mullen CA, Boateng AA (2011) Production and analysis of fast pyrolysis oils from proteinaceous biomass. BioEnergy Res 4:303–311

Nakayama S, Kiyoshi K, Kadokura T, Nakazato A (2011) Butanol production from crystalline cellulose by cocultured *Clostridium thermocellum* and *Clostridium saccharoperbutylacetonicum* N1-4. Appl Environ Microbiol 77:6470–6475

Nath K, Kumar A, Das D (2005) Hydrogen production by *Rhodobacter sphaeroides* strain OU 001 using spent media of *Enterobacter cloacae* strain DM11. Appl Microbiol Biot 68:533–541

Nonato R, Mantelatto P, Rossell C (2001) Integrated production of biodegradable plastic, sugar and ethanol. Appl Microbiol Biot 57:1–5

Nordell E, Nilsson B, Påledal SN, Karisalmi K, Moestedt J (2016) Co-digestion of manure and industrial waste–the effects of trace element addition. Waste Manag 47:21–27
Oey M, Sawyer AL, Ross IL, Hankamer B (2016) Challenges and opportunities for hydrogen production from microalgae. Plant Biotechnol J 14:1487–1499
Oliveira JV, Duarte T, Costa JC, Cavaleiro AJ, Pereira MA, Alves MM (2018) Improvement of biomethane production from sewage sludge in co-digestion with glycerol and waste frying oil, using a design of experiments. BioEnergy Res 11:763–771
Oncel SS, Kose A, Faraloni C (2015) Genetic optimization of microalgae for biohydrogen production (Chap. 25). In: Kim S-K (ed) Handbook of marine microalgae. Academic Press, Boston, pp 383–404
Ormaechea P, Castrillón L, Marañón E, Fernández-Nava Y, Negral L, Megido L (2017) Influence of the ultrasound pretreatment on anaerobic digestion of cattle manure, food waste and crude glycerine. Environ Technol 38:682–686
Özgür E, Afsar N, de Vrije T, Yücel M, Gündüz U, Claassen PA, Eroglu I (2010) Potential use of thermophilic dark fermentation effluents in photofermentative hydrogen production by *Rhodobacter capsulatus*. J Clean Prod 18:S23–S28
Papanikolaou S, Aggelis G (2011) Lipids of oleaginous yeasts. Part I: biochemistry of single cell oil production. Eur J Lipid Sci Technol 113:1031–1051
Park JY, Choi SA, Jeong MJ, Nam B, Oh YK, Lee JS (2014) Changes in fatty acid composition of *Chlorella vulgaris* by hypochlorous acid. Bioresour Technol 162:379–383
Patle S, Lal B (2007) Ethanol production from hydrolysed agricultural wastes using mixed culture of *Zymomonas mobilis* and *Candida tropicalis*. Biotechnol Lett 29:1839–1843
Pena R, Lú-Chau TA, Lema JM (2012) Use of white-rot fungi for valorization of stillage from bioethanol production. Waste Biomass Valori 3:295–303
Pereira LG, Chagas MF, Dias MO, Cavalett O, Bonomi A (2015) Life cycle assessment of butanol production in sugarcane biorefineries in Brazil. J Clean Prod 96:557–568
Piveteau S, Picard S, Dabert P, Daumer ML (2017) Dissolution of particulate phosphorus in pig slurry through biological acidification: a critical step for maximum phosphorus recovery as struvite. Water Res 124:693–701
Qureshi N, Blaschek HP (2001) Recent advances in ABE fermentation: hyper-butanol producing *Clostridium beijerinckii* BA101. J Ind Microbiol Biotechnol 27:287–291
Qureshi N, Liu S, Hughes S, Palmquist D, Dien B, Saha B (2016) Cellulosic butanol (ABE) biofuel production from sweet sorghum bagasse (SSB): impact of hot water pretreatment and solid loadings on fermentation employing *Clostridium beijerinckii* P260. BioEnergy Res 9:1167–1179
Rana KL, Kour D, Sheikh I, Yadav N, Yadav AN, Kumar V et al (2019) Biodiversity of endophytic fungi from diverse niches and their biotechnological applications. In: Singh BP (ed) Advances in endophytic fungal research: present status and future challenges, pp 105–144. Springer International Publishing, Cham. https://doi.org/10.1007/978-3-030-03589-1_6
Rashid N, Lee K, Han JI, Gross M (2013) Hydrogen production by immobilized *Chlorella vulgaris*: optimizing pH, carbon source and light. Bioproc Biosyst Eng 36:867–872
Rastegari AA, Yadav AN, Yadav N (2020) New and Future Developments in Microbial Biotechnology and Bioengineering: Trends of Microbial Biotechnology for Sustainable Agriculture and Biomedicine Systems: Diversity and Functional Perspectives. Elsevier, Amsterdam
Rastegari AA, Yadav AN, Gupta A (2019) Prospects of renewable bioprocessing in future energy systems. Springer International Publishing, Cham
Ratledge C (1993) Single cell oils—have they a biotechnological future? Trends Biotechnol 11:278–284
Refaat AA (2011) Biodiesel production using solid metal oxide catalysts. Int J Environ Sci Technol 8:203–221
Rivera I, Buitrón G, Bakonyi P, Nemestóthy N, Bélafi-Bakó K (2015) Hydrogen production in a microbial electrolysis cell fed with a dark fermentation effluent. J Appl Electrochem 45:1223–1229

Robinson PH, Karges K, Gibson ML (2008) Nutritional evaluation of four co-product feedstuffs from the motor fuel ethanol distillation industry in the Midwestern USA. Anim Feed Sci Technol 146:345–352

Rondon MA, Lehmann J, Ramírez J, Hurtado M (2007) Biological nitrogen fixation by common beans (*Phaseolus vulgaris* L.) increases with bio-char additions. Biol Fert Soils 43:699–708

Rosa PRF, Gomes BC, Varesche MBA, Silva EL (2016) Characterization and antimicrobial activity of lactic acid bacteria from fermentative bioreactors during hydrogen production using cassava processing wastewater. Chem Eng J 284:1–9

Rosas JG, Gómez N, Cara J, Ubalde J, Sort X, Sánchez ME (2015) Assessment of sustainable biochar production for carbon abatement from vineyard residues. J Anal Appl Pyrol 113:239–247

Sagir E, Ozgur E, Gunduz U, Eroglu I, Yucel M (2017) Single-stage photofermentative biohydrogen production from sugar beet molasses by different purple non-sulfur bacteria. Bioproc Biosyst Eng 40:1589–1601

Sağır E, Yucel M, Hallenbeck PC (2018) Demonstration and optimization of sequential microaerobic dark-and photo-fermentation biohydrogen production by immobilized *Rhodobacter capsulatus* JP91. Bioresour Technol 250:43–52

Sahu S, Pramanik K (2018) Evaluation and optimization of organic acid pretreatment of cotton gin waste for enzymatic hydrolysis and bioethanol production. Appl Biochem Biotechnol 186:1047–1060

Sargsyan H, Gabrielyan L, Hakobyan L, Trchounian A (2015) Light–dark duration alternation effects on Rhodobacter sphaeroides growth, membrane properties and bio-hydrogen production in batch culture. Int J Hydrog Energy 40:4084–4091

Saritha M, Arora A, Choudhary J, Rani V, Singh S, Sharma A, Sharma S, Nain L (2016) The role and applications of xyloglucan hydrolase in biomass degradation/bioconversion. In: Microbial enzymes in bioconversions of biomass, pp 231–248. Springer, Cham

Sawatdeenarunat C, Nguyen D, Surendra KC, Shrestha S, Rajendran K, Oechsner H, Xie L, Khanal SK (2016) Anaerobic biorefinery: current status, challenges, and opportunities. Bioresour Technol 215:304–313

Scarlat N, Dallemand JF, Fahl F (2018) Biogas: developments and perspectives in Europe. Renew Energy 129:457–472

Schenk PM, Thomas-Hall SR, Stephens E, Marx UC, Mussgnug JH, Posten C, Kruse O, Hankamer B (2008) Second generation biofuels: high-efficiency microalgae for biodiesel production. BioEnergy Res 1:20–43

Schieb PA, Lescieux-Katir H, Thénot M, Clément-Larosière B (2015) Prospects for the Bazancourt-Pomacle biorefinery between now and 2030. In: Biorefinery 2030, pp 81–100. Springer, Berlin

Senger H, Bishop NI (1979) Observations on the photohydrogen producing activity during the synchronous cell cycle of *Scenedesmus obliquus*. Planta 145:53–62

Sharma HK, Xu C, Qin W (2019) Biological pretreatment of lignocellulosic biomass for biofuels and bioproducts: an overview. Waste Biomass Valori 10:235–251

Shi R, Ukaew S, Archer DW, Lee JH, Pearlson MN, Lewis KC, Shonnard DR (2017) Life cycle water footprint analysis for rapeseed derived jet fuel in North Dakota. ACS Sustain Chem Eng 5:3845–3854

Shrivastava AK, Srivastava Arun K, Soloman S (2011) Sustaining sugarcane productivity under depleting water resources. Curr Sci India 101:748–754

Sialve B, Bernet N, Bernard O (2009) Anaerobic digestion of microalgae as a necessary step to make microalgal biodiesel sustainable. Biotechnol Adv 27:409–416

Singh J, Gu S (2010) Commercialization potential of microalgae for biofuels production. Renew Sustain Energy Rev 14:2596–2610

Singh V, Johnston DB, Naidu K, Rausch KD, Belyea RL, Tumbleson ME (2005) Comparison of modified dry-grind corn processes for fermentation characteristics and DDGS composition. Cereal Chem 82:187–190

Singh SP, Pathak J, Sinha RP (2017) Cyanobacterial factories for the production of green energy and value-added products: An integrated approach for economic viability. Renew Sustain Energy Rev 69:578–595

Singh AK, Kumari VV, Gupta R, Singh P, Solomon S (2018) Efficient irrigation water management in sugarcane through alteration of field application parameters under subtropical India. Sugar Tech 20:21–28

Soccol CR, Neto CJD, Soccol VT, Sydney EB, da Costa ESF, Medeiros ABP, de Souza Vandenberghe LP (2017) Pilot scale biodiesel production from microbial oil of *Rhodosporidium toruloides* DEBB 5533 using sugarcane juice: performance in diesel engine and preliminary economic study. Bioresour Technol 223:259–268

Spiekermann P, Rehm BHA, Kalscheuer R, Baumeister D, Steinbüchel AA (1999) Sensitive, viable-colony staining method using Nile red for direct screening of bacteria that accumulate polyhydroxyalkanoic acids and other lipid storage compounds. Arch Microbiol 171:73–80

Stadler T, Chauvet JM (2018) New innovative ecosystems in France to develop the bioeconomy. New Biotechnol 40:113–118

Statista. https://www.statista.com/statistics/271472/biodiesel-production-in-selected-countries/

Tao Y, Chen Y, Wu Y, He Y, Zhou Z (2007) High hydrogen yield from a two-step process of dark-and photo-fermentation of sucrose. Int J Hydrog Energy 32:200–206

Taparia T, MVSS M, Mehrotra R, Shukla P, Mehrotra S (2016) Developments and challenges in biodiesel production from microalgae: a review. Biotechnol Appl Biochem 63:715–726

Tran HTM, Cheirsilp B, Hodgson B, Umsakul K (2010) Potential use of *Bacillus subtilis*, in a co-culture with *Clostridium butylicum*, for acetone–butanol–ethanol production from cassava starch. Biochem Eng J 48(2):260–267

Tripathi M, Sahu JN, Ganesan P (2016) Effect of process parameters on production of biochar from biomass waste through pyrolysis: a review. Renew Sustain Energy Rev 55:467–481

Uellendahl H, Ahring BK (2010) Anaerobic digestion as final step of a cellulosic ethanol biorefinery: biogas production from fermentation effluent in a UASB reactor—pilot-scale results. Biotechnol Bioeng 107:59–64

Ugwu CU, Aoyagi H, Uchiyama H (2008) Photobioreactors for mass cultivation of algae. Bioresour Technol 99:4021–4028

Uyar B, Schumacher M, Gebicki J, Modigell M (2009) Photoproduction of hydrogen by *Rhodobacter capsulatus* from thermophilic fermentation effluent. Bioproc Biosyst Eng 32:603–606

van Dyk S, Su J, Mcmillan JD, Saddler J (2019) Potential synergies of drop-in biofuel production with further co-processing at oil refineries. Biofuel Bioprod Bior 13:760–775

Van Zwieten L, Kimber S, Morris S, Chan KY, Downie A, Rust J, Joseph S, Cowie A (2010) Effects of biochar from slow pyrolysis of papermill waste on agronomic performance and soil fertility. Plant Soil 327:235–246

Varol A, Ugurlu A (2016) Biogas production from microalgae (*Spirulina platensis*) in a two stage anaerobic system. Waste Biomass Valori 7:193–200

Villar A, Arribas JJ, Parrondo J (2012) Waste-to-energy technologies in continuous process industries. Clean Technol Envir 14:29–39

Wang YR, Chiang YS, Chuang PJ, Chao YP, Li SY (2016) Direct in situ butanol recovery inside the packed bed during continuous acetone-butanol-ethanol (ABE) fermentation. Appl Microbiol Biotechnol 100:7449–7456

Xu L, Cheng X, Wu S, Wang Q (2017) Co-cultivation of *Chlamydomonas reinhardtii* with *Azotobacter chroococcum* improved H_2 production. Biotechnol Lett 39:731–738

Xue C, Wang Z, Wang S, Zhang X, Chen L, Mu Y, Bai F (2016) The vital role of citrate buffer in acetone–butanol–ethanol (ABE) fermentation using corn stover and high-efficient product recovery by vapor stripping–vapor permeation (VSVP) process. Biotechnol Biofuels 9:146

Yadav AN, Kumar R, Kumar S, Kumar V, Sugitha T, Singh B et al (2017) Beneficial microbiomes: biodiversity and potential biotechnological applications for sustainable agriculture and human health. J Appl Biol Biotechnol 5:45–57

Yadav AN, Singh S, Mishra S, Gupta A (2019) Recent advancement in white biotechnology through fungi. Perspective for value-added products and environments, vol 2. Springer International Publishing, Cham

Yadav AN, Rastegari AA, Yadav N (2020) Microbiomes of extreme environments: biodiversity and biotechnological applications. CRC Press, Taylor & Francis, Boca Raton, USA

Yin Q, Zhu X, Zhan G, Bo T, Yang Y, Tao Y et al (2016) Enhanced methane production in an anaerobic digestion and microbial electrolysis cell coupled system with co-cultivation of *Geobacter* and *Methanosarcina*. J Environ Sci 42:210–214

Zacharof MP (2017) Grape winery waste as feedstock for bioconversions: applying the biorefinery concept. Waste Biomass Valori 8:1011–1025

Zhang J, Loh KC, Lee J, Wang CH, Dai Y, Tong YW (2017) Three-stage anaerobic co-digestion of food waste and horse manure. Sci Rep-UK 7:1269

Zverlov VV, Berezina O, Velikodvorskaya GA, Schwarz WH (2006) Bacterial acetone and butanol production by industrial fermentation in the Soviet Union: use of hydrolyzed agricultural waste for biorefinery. Appl Microbiol Biotechnol 71:587–597

Chapter 11
Microbial Technologies for Biorefineries: Current Research and Future Applications

Deepika Goyal, Sushma Mishra, and Prem Kumar Dantu

Abstract Conventional resources becoming limited due to the increase in population and energy demand. This rise in energy demand has increased consumer prices and pressure on the environment. This prompted researchers to take care of sustainable energy resources. In this case, biomass is only environmentally friendly renewable resource which is used for the production of chemicals and fuels. A system similar to a petroleum refinery is required to produce fuels and useful chemicals from biomass and is known as a biorefinery. Biorefineries have been subdivided into various categories on the basis of technology and biomass used. In this chapter, types of biorefineries and microbes which are used for the production of valuable products are discussed.

11.1 Introduction

International Energy Agency (IEA) Bioenergy Task 42 has defined biorefinery as the sustainable processing of biomass into a variety of marketable products (food, feed, materials, chemicals) and energy (fuels, power, heat) (de Jong and Jungmeier 2015). The National Renewable Energy Laboratory (NREL) defined biorefinery as a facility that facilitates conversion of biomass into fuels, power, and chemicals. A biorefinery can utilize all types of biomass and producing agricultural by-products (wheat bran, rapeseed meal, straw, corn stover, bagasse), waste from the food industry (including kitchen and household waste), grains/cereals (wheat, maize, corn, soybean), starch and sugars, aquatic biomass (algae and seaweeds), as well as wood and lignocellulosic materials. A biorefinery is not a completely new concept.

According to Berntsson et al., biorefinery promotes industrial trades, economic, and environmental sustainability. Biorefineries are found helpful in generating added-value products, bio-based products, and bioenergy utilizing sustainable biomass (de Jong and Jungmeier 2015). As per the increasing energy demand nowadays, interest

---

D. Goyal (✉) · S. Mishra · P. K. Dantu
Department of Botany, Dayalbagh Educational Institute, Dayalbagh, Agra, Uttar Pradesh, India
e-mail: deepikagoyal1307@gmail.com

A. N. Yadav et al. (eds.), *Biofuels Production – Sustainability and Advances in Microbial Bioresources*, Biofuel and Biorefinery Technologies 11,
https://doi.org/10.1007/978-3-030-53933-7_11

of scientists is increasing in renewable and sustainable biotechnological processes for energy, biofuels, and chemicals. Use of microorganisms in chemical industries is to derive the same product; using biological materials is an alternative sustainable and economical approach. It is estimated that by 2025, 15% of chemical products will be bioformulated (Vijayendran 2010). Thus, the development of biorefineries is an alternative to diesel and petroleum-based products. Biorefineries can be defined as processing of biomass (mainly lignocelluloses) into marketable and commercial products (food, feed, material, and chemicals) and energy (fuels, power, and heat) mediated by physical, chemical, or biological materials (IEA 2010).

The biorefinery concept is eye-catching because it facilitates production of high added-value products at lesser price and reducing waste disposal and maintaining ecological harmony. Few biorefineries have established, for instance, the pulp- and paper-based biorefinery, Borregaard, in Norway (Borregaard 2014), but attempts are required to establish such biorefineries in several other countries aswell. Microorganisms are the basis of biorefineries and backbone of industrial bioprocesses; they either produce desired chemical or produce intermediate required for the process. Most of the industries in world utilize the potential of microorganisms for the production of food additives, medicines, antibiotics, enzymes, bioethanol, biodiesel, and other chemicals. Lignocellulosic biomass is the most abundant biomass on earth obtained as agricultural by-product and renewable source of sugars, and is an advisable feedstock for the production of biodiesel, biogas, biohydrogen, and chemical products through the biorefinery processes (Menon and Rao 2012). In biorefinery processes, lignocellulosic biomass is firstly pre-treated, and then cellulosic and hemicellulosic are decomposed into simple sugars mediated by enzymes (Rastegari et al. 2019a). Microbes metabolize and ferment these simple sugars producing chemical products such as alcohols, fatty acids, organic acids, and amino acids. Bioethanol is a more preferred alternative over conventional petroleum-based transport fuels. However, complex structure of lignocellulosic biomass is a challenge in its bioconversion than simple starch and sugar materials (Mussatto et al. 2010; Yadav et al. 2020). Cellulose, hemicellulose, and lignin are building blocks of lignocellulosic biomass.

Biorefineries have led new opportunities to the industrial application of microorganisms. Potential of unexplored or new microbe for desired product can be checked. New substrates may be added, and along with these industrial processes can be optimized to achieve maximum conversion processes. In addition, we highlight and exemplify general strategies to develop microorganisms that are able to produce fuels and chemicals from renewable feedstocks. All types of biomass from forestry, aquaculture, agriculture, organic and forest residues, and aquatic biomass (algae and seaweeds) are converted into valuable products of humankind. Many of the industries converting sugar, starch, pulp, and paper industries are considered as biorefineries. There are many differences between refineries and biorefineries (Table 11.1).

Table 11.1 Comparison of refineries and biorefineries regarding feedstocks, building block composition, processes, and chemical intermediates produced at commercial scale

Sources	Refinery	Biorefinery
Feedstock	Feedstock relatively homogeneous	Feedstock heterogeneous regarding bulk components e.g., carbohydrates, lignin, proteins, oils, extractives, and/or ash Most of the starting material present in polymeric form (cellulose, starch, proteins, lignin)
	Low in oxygen content	High in oxygen content
	The weight of the product (mole/mole) generally increases with processing	The weight of the product (mole/mole) generally decreases with processing. It is important to perceive the functionality in the starting material
	Sometimes high in sulfur	Sometimes high in inorganics, especially silica
Building block composition	Main building blocks: Ethylene, propylene, methane, benzene, toluene, xylene isomers	Main building blocks: Glucose, xylose, fatty acids (e.g., oleic, stearic, sebacic)
(Bio)chemical processes	Introduction of heteroatoms (O, N, S)	Removal of oxygen
	Relative homogeneous processes to arrive at building blocks: Steam cracking,	Relative heterogeneous processes to arrive building blocks
Chemical intermediates produced at commercial scale	Many	Few but increasing (e.g., ethanol, furfural, biodiesel, mono-ethanol glycol, lactic acid, succinic acid)

11.2 Classification of Biorefineries

Biorefineries have been classified in different categories on the basis of different criteria (de Jong and Jungmeier 2015). On the basis of technologies used, biorefineries are divided into conventional and advanced biorefineries: first-, second-, and third-generation biorefineries. On the basis of raw material used, biorefineries are divided into whole crop biorefineries, oleochemical biorefineries, lignocellulosic feedstock biorefineries, green biorefineries, and marine biorefineries. On the basis of conversion process used, biorefineries are divided into thermochemical biorefineries, biochemical biorefineries, and two-platform concept biorefineries. On the basis of intermediate produced, biorefineries are syngas platform biorefineries and sugar platform biorefineries. On the basis of availability of biomass, biorefineries have been classified into six types (Lange 2017). Yellow biorefinery utilize straw, corn stover, and wood. Green biorefinery utilizes fresh green biomass, grass for protein-rich feed. Blue biorefineries use fish by-catch/cut-offs, fish discards and innards, mussels as biomass, brown seaweed, red and green algae, and invertebrates such as sea cucumber. Red biorefinery utilizes slaughterhouse waste. White biorefinery uses agro-industry-side streams.

11.3 Microbial Fermentation Processes for the Development of Biorefineries

Due to large consumption of fuels and foods, sustainable way to produce new foods and fuels from agro-residues is required. Sustainable production is an effective technology utilizing raw materials, agro-waste to produce new, commercial, and valuable products. Solid-state fermentation is an alternative and long term used approach for the production of biotechnology-based commercial products. Fermentation technology of microbes has been used in East for the manufacture of fermented foods and for manufacture of mold-ripened cheese in West. In fermentation technology, microbes are allowed to grow on solid material with low moisture content. Fermentation is an economical, large-scale process of bioconversion and biodegradation process. With the aid of this technology food, enzymes, chemicals, cosmetics, and pharmaceutical compounds have been produced (Kour et al. 2019a; Kumar et al. 2019). This fermentation technology is driving attention of researchers widely nowadays. Various alternative terms are currently being used as synonyms of solid-state fermentation likewise solid-state fermentation, surface cultivation, surface culture, solid-state digestion, and solid-state fermentation.

Botella et al. (2009) used a new term "particulate bioprocessing", in order to define solid-state fermentation. Particulate bioprocessing defines growth of microorganism in moist condition in a particulate solid medium. Amore and Faraco (2012) used the term consolidated bioprocessing (CBP) defining fungi as alternative microbe for the degradation of lignocellulosic materials. Cellulose degrading fungi produce saccharolytic enzymes for the digestion of lignocellulose and converting sugars to ethanol. These technologies reduce the cost of production of ethanol and show that the fungi have all the pathways required for conversion of lignocellulose to bioethanol. Viniegra-Gonzàlez (1997) defined solid-state fermentation as a process where microbes grow on the surface of solid material without the addition of nutrients. Pandey et al. (2000) defined solid-state fermentation, a technology, where microbes are grown on moist solid support, either on inert carriers or on insoluble substrates that can also be used as carbon and energy source.

Rahardjo et al. (2006) defined solid-state fermentation as the growth of microorganisms on moistened solid substrate with enough moisture is to maintain microbial growth and metabolism. Adopting the technology of solid-state fermentation, microbes have been used in biorefineries for conversion of sugar containing polymers such as cellulose and hemicellulose in commercial products. Biofuels, bioethanol, biomethanol, biogas, pharmaceutical products, and biodegradable products have been produced using microbes (Koutinas et al. 2007). Webb et al. proposed a model for wheat-based biorefining strategy in economical way using microbial fermentation (Fig. 11.1).

11.4 Genetic Improvement of Microorganisms for Development of Biorefinery Products

Microbial strains are required which can result in high yield and productivity of compounds tolerating several stresses (Rastegari et al. 2019b, c). For the same, microbes are genetically modified. *S. cerevisiae* has been used in bio-industries since last 30 years, each year with an improved version. Different strategies have been adopted for this genetic engineering likewise (i) driving carbon flux, (ii) increase tolerance to toxic compounds, (iii) increase of substrate uptake range, and (iv) generation of new products (Fig. 11.2).

11.4.1 Driving Carbon Flux

Naturally, microbes have capability to produce desired chemical compounds, and they are optimized for maximal growth. But the production of bioactive compounds is hindered due to expense of carbon, energy, and by-product formation. Thus, modifications in microorganisms which lead to higher production are driving carbon flux. Microbes of different groups such as bacteria, fungi, and yeast have been genetically modified to enhance production of biofuel and desired compounds. Microbial strains which are able to produce 90% m/m of desired chemical compound are available (Table 11.2). There are many steps where microbes have been modified such as modification in microbial metabolism by overexpression or knockout of enzymes (Jiang et al. 2009; Mojzita et al. 2010), modification in transcription and change in redox reactions (Alper and Stephanopoulos 2007; Almeida et al. 2009; Nissen et al. 2000). For instance, *S. cerevisiae* is modified to produce ethanol from sugars present in lignocellulosic biomass (Hahn-Hägerdal et al. 2007).

11.4.2 Increased Tolerance to the Substrate

Low tolerance to end product also hampers product formation by microbes. Fermentation medium also causes a harsh environment for the microorganism. In case of unavailability of tolerant strains, genetic engineering approaches have been used to

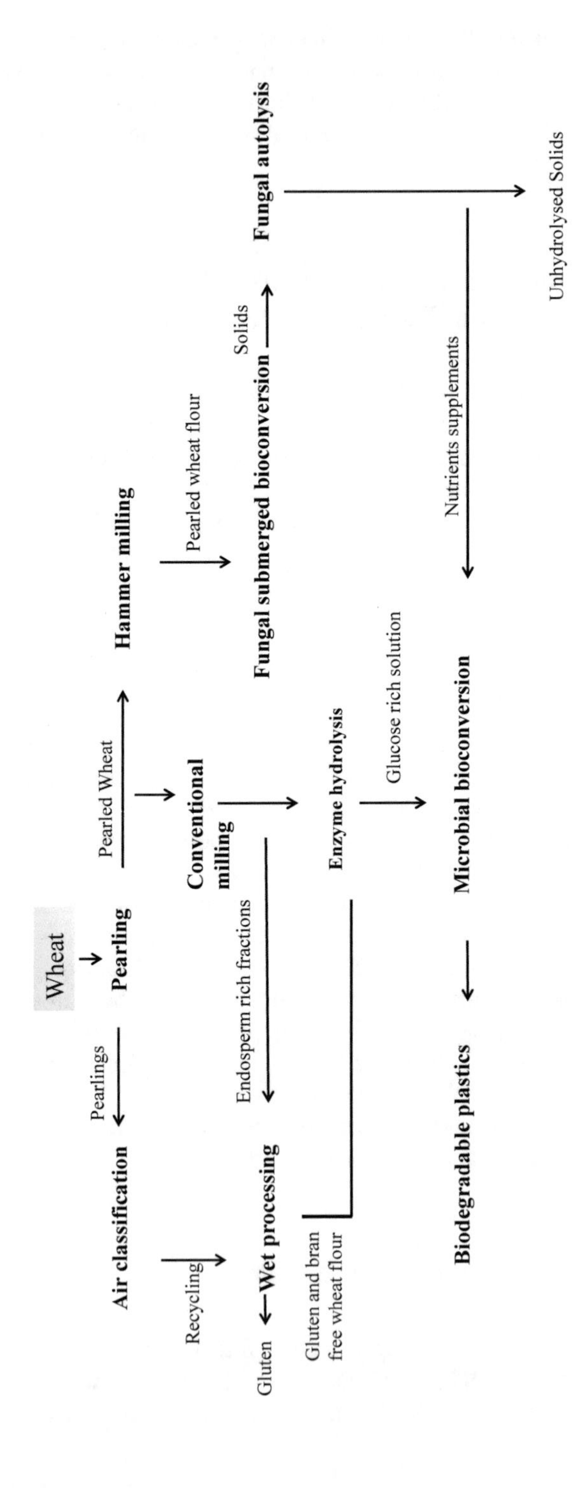

Fig. 11.1 Schematic diagram of microbial fermentations proposed in a possible biorefinery utilizing wheat for the production of poly-hydroxyl butyrate and succinic acid

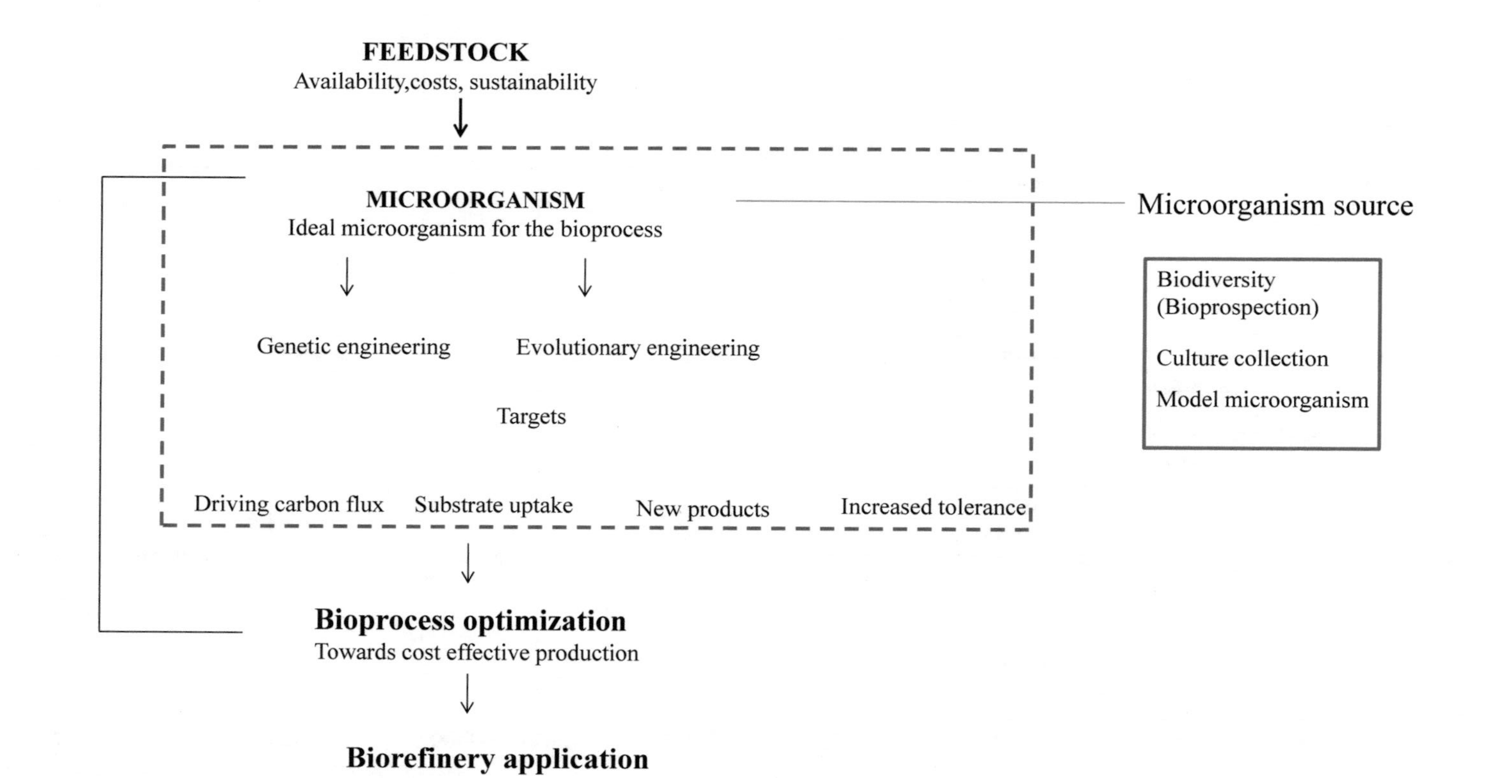

Fig. 11.2 Main steps for the development of a new bioprocess integrated to a biorefinery

Table 11.2 Microbial bioresources and biofuel production

Organism	Product	Main substrate	Yield*	Productivity	Concentration	Outcomes	Main genetic modifications	References
Driving carbon flux toward the desired pathway								
E. coli SY4	Ethanol	Glycerol	0.42 g g^{-1}	0.15 g L^{-1} h^{-1}	7.8 g L^{-1}	Yield improved 69-fold. Engineered strains efficiently utilized glycerol in a minimal medium without rich supplements	Deletion of genes to minimize the synthesis of by-products	Durnin et al. (2009)
E. coli LA02Δdld	Lactic acid	Glycerol	0.80 g g^{-1}	1.25 g g^{-1} h^{-1}	32 g L^{-1}	Low-value glycerol streams to a higher value product like D-lactate. Yield improved sevenfold	Overexpression of pathways involved in the conversion of glycerol to lactic acid and blocking those leading to the synthesis of competing by-products	Mazumdar et al. (2010)
E. coli	Acetate	Glucose	0.456 g g^{-1}	1.38 g g^{-1} h^{-1}	53 g L^{-1}	Reduction of the fermentation by products concentration by 1, 25 (succinate) to 33 fold (lactate). Yield improved over sevenfold	Deletion of genes involved in the succinate formation as fermentation product	Causey et al. (2003)
Y. lipolytica	Succinic acid	Glycerol	0.45 g g^{-1}	Not determined	45 g L^{-1}	Succinic acid production yield increased over 20 fold	Deletion in the gene coding one of succinate dehydrogenase subunits	Blankschien et al. (2010)

(continued)

Table 11.2 (continued)

Organism	Product	Main substrate	Yield*	Productivity	Concentration	Outcomes	Main genetic modifications	References
Y-3314 Mannheimia succiniciproducens	Succinic Acid	Glucose	0.76 g g^{-1}	1.8 g g^{-1} h^{-1}	52.4 g L^{-1}	Nearly complete elimination of fermentation by-products, (acetic, formic, and lactic acids) and carbon recovery increased to 58–77% by fed-batch culture	Disruption of genes responsible for by-product formation (ldhA, pflB, pta, and ackA)	Lee et al. (2006)
Increasing of tolerance to toxic compounds								
C. acetobutylicum	Butanol	Glucose	Not determined	Not determined		Increased tolerance and extendedmetabolism response to butanol stress	Overexpression of spo0A, responsible for the transcription of solvent formation genes	Alsaker et al. (2004)
C. acetobutylicum	Butanol	Glucose	70.8%	Not determined	13.6 g L^{-1}	Reduction of acetone production from 2,83 g L^{-1} to 0,21 g L^{-1} and enhanced butanol yield from 57 to 70.8%	Disruption of the acetoacetate decarboxylase gene (adc) avoiding acetone production and optimization of medium	Jiang et al. (2009)
S. cerevisiae	Ethanol	Glucose plus HMF (inhibitor)	0.43 g g^{-1}	0.61 g g^{-1} h^{-1}	Not determined	Four times higher specific uptake rate of HMF and 20% higher specific ethanol productivity	Overexpression of alcohol dehydrogenases ADH6 or ADH1-mutated	Almeida et al. (2008)

(continued)

Table 11.2 (continued)

Organism	Product	Main substrate	Yield*	Productivity	Concentration	Outcomes	Main genetic modifications	References
S. cerevisiae	Ethanol	Spruce hydrolysate	Not determined	0.39 g g^{-1} h^{-1}	Not determined	HMF conversion rate and ethanol productivity for the engineered strains four to five times and 25% higher than for the control strain	Overexpression of alcohol dehydrogenases ADH6 or ADH1-mutated	Almeida et al. (2008)
E. coli XW068(pLOI4319)	Lactate	Xylose plus HMF	85% of the theoretical maximum	Not determined	Not determined	Furfural tolerance increased by 50%. Minimal growth and lactate production occurred after 120 h for the control strain	Overexpression of NADH-dependent propanediol oxidoreductase (FucO)	Wang et al. (2011)
Increasing substrate uptake range								
E. coli	Ethanol	Xylose	0.48 g g^{-1}	2.00 g g^{-1} h^{-1}	43 g L^{-1}	Rapid co-fermentation due to reduced repression of xylose metabolism by glucose, and 60% less time required for fermentation of 5-sugar mix to ethanol	Deletion of methylglyoxal synthase gene (mgsA), involved in sugar metabolism	Yomano et al. (2009)

(continued)

Table 11.2 (continued)

Organism	Product	Main substrate	Yield*	Productivity	Concentration	Outcomes	Main genetic modifications	References
Lactobacillus plantarum	Lactic Acid	Corn starch	0.89 g g^{-1}	4.51 g g^{-1} h^{-1}	86 g L^{-1}	First direct and efficient fermentation of optically pure D-lactic acid from raw corn starch reported	Deletion of L-lactate dehydrogenase gene (ldhL1) and expression of Streptococcus bovis 148 α-amylase (AmyA)	Okano et al. (2009)
S. cerevisiae	Ethanol	Xylose	0.43 g g^{-1}	0.02 g g^{-1} h^{-1}	7.3 g L^{-1}	Higher ethanol yields than XR/XDH carrying strains	Overexpression of Piromyces sp. xylose isomerase (XI)	Kuyper et al. (2003)
S. cerevisiae	Ethanol	Xylose	0.33 g g^{-1}	0.04 g g^{-1} h^{-1}	13.3 g L^{-1}	Higher specific ethanol productivity and final ethanol concentration than XI carrying strains	Overexpression of xylose reductase (XR) and xylitol dehydrogenase (XDH) enzymes from Scheffersomyces stipitis	Karhumaa et al. (2007)
E. coli	Butanol	Glucose	6.1%	0.02 g g^{-1} h^{-1}	1.2 g L^{-1}	Anaerobic production of butanol by a microorganism expressing genes from a strict aerobic organism	Expression of *C. acetobutylicum* butanol pathway synthetic genes in E. coli	Inui et al. (2008)

(continued)

Table 11.2 (continued)

Organism	Product	Main substrate	Yield*	Productivity	Concentration	Outcomes	Main genetic modifications	References
Generation of new products								
E. coli	Fatty acid ethyl esters (FAEEs)	Glucose	7%	Not determined	30.7 g L^{-1}	Tailored fatty ester (biodiesel) production	Heterologous expression of a "FAEE pathway" engineered in E. coli	Steen et al. (2010)
S. cerevisiae	Butanol	Galactose	Not determined	Not determined	2.5 mg L^{-1}	First demonstration of n-butanol production in S. cerevisiae	N-butanol biosynthetic pathway engineered in S. cerevisiae	Steen et al. (2008)
E. coli K12	1,3-propanediol	Glycerol	90.2%	2.61 g g^{-1} h^{-1}	104.4 g L^{-1}	Substantially high yield and productivity efficiency of 1,3-PD with glycerol as the sole source of carbon	Heterologous overexpression of genes from natural producers of 1,3-PDO	Tang et al. (2009a, b

improve strain response for toxic and end product. Strains have been improved to produce biofuels from lignocellulosic hydrolysate. Lignocellulose is composed of cellulose, hemicellulose, and lignin (Hahn-Hägerdal et al. 2007). Prior to fermentation, this hydrolysate is allowed for pretreatment to reduce its recalcitrance. Later, it is allowed for hydrolysis where sugar monomers have been formed from cellulose and hemicellulose. These sugar monomers form biofuels. During this pretreatment and hydrolysis, many toxic compounds are produced which inhibit microbial processes, microbial metabolism, and microbial growth as well. Compounds like furaldehyde, organic acids (acetic, levulinic, and furoic), and phenolic derivatives are found in lignocellulose. These compounds inhibit microbial growth, cause lowering in product yield, and reduce cellular viability (Almeida et al. 2007, 2011). Metabolic engineering and genetic engineering have been applied to make these strains tolerant. *S. passalidarum, S. cerevisiae,* and *P. stipites* have been evolutionary engineered to ferment lignocellulose more than the native strains (Heer and Sauer 2008; Hughes et al. 2012; Liu et al. 2004; Kour et al. 2019b). Yeast tolerance to lignocellulose has been improved by genetic engineering (Almeida et al. 2011) (Table 11.2). Genes having resistance to inhibitors are transferred in microbial strain for providing tolerance to end product.

11.4.3 Increase of Substrate Uptake Range

Genetic engineering of microbes has been done to increase substrate and its better utilization in product formation. Utilization of lignocellulosic biomass requires xylose utilization. Xylose is the second most abundant pentose sugar present in sugarcane bagasse (30%) (Ferreira-Leitão et al. 2010). Naturally, *S. cerevisiae* does not utilize pentose sugars; it is genetically modified to use this pentose sugar (Table 11.2).

11.4.4 New Products

Genetically modified microorganisms are able to produce compounds that are not possible by natural pathways. For this, enzymes and pathways from one organism have been transferred in an organism of choice. Nowadays, many new compounds have been reported by microbes rather than bioethanol which increase economy and can be produced in lesser time (Table 11.2). Acids produced from this lignocellulose serve as precursors of plastics (Werpy et al. 2004). *Acetobacter*, *Aerobacter*, *Pseudomonas, Gluconobacter,* and *Erwinia* produce a five-carbon acid xylonic acid, derived from xylose. Obviously, wild-type bacteria are able to produce this xylonic acid; however, this yield was very low. *E. coli, S. cerevisiae, Kluyveromyces lactis,* and *Pichia kudriavzevii* have been produced by genetic recombination to enhance yield of this xylonic acid (Toivari et al. 2010; Nygård et al. 2011; Liu et al. 2012).

11.5 Microbial Technologies for Biodiesel-Based Biorefineries

Production of biofuels from renewable feedstocks is demanded in the period of crisis of energy where petrol fuels are becoming limited and expensive (Rastegari et al. 2020; Yadav et al. 2019). Production of biofuels is a costly process, and various residues are produced; however, this cost can be reduced if residues can be converted into valuable coproducts (Zhang 2011; Yazdani and Gonzalez 2007). Biodiesel is an alternative biofuel obtained by the transesterification of fat and vegetable oils and reduces net greenhouse effect (O'Connor 2011). Many plants such as sunflower, soybean, rape, and palm oils are used to produce biodiesel. In Brazil, soybean oil was the source of 80% of biodiesel in 2010. Pies and glycerol are produced as residues in the production of biodiesel. Pies are used as animal feed or fertilizers, whereas glycerol is used as crude sample in biorefineries and many valuable products are formed (Fig. 11.3).

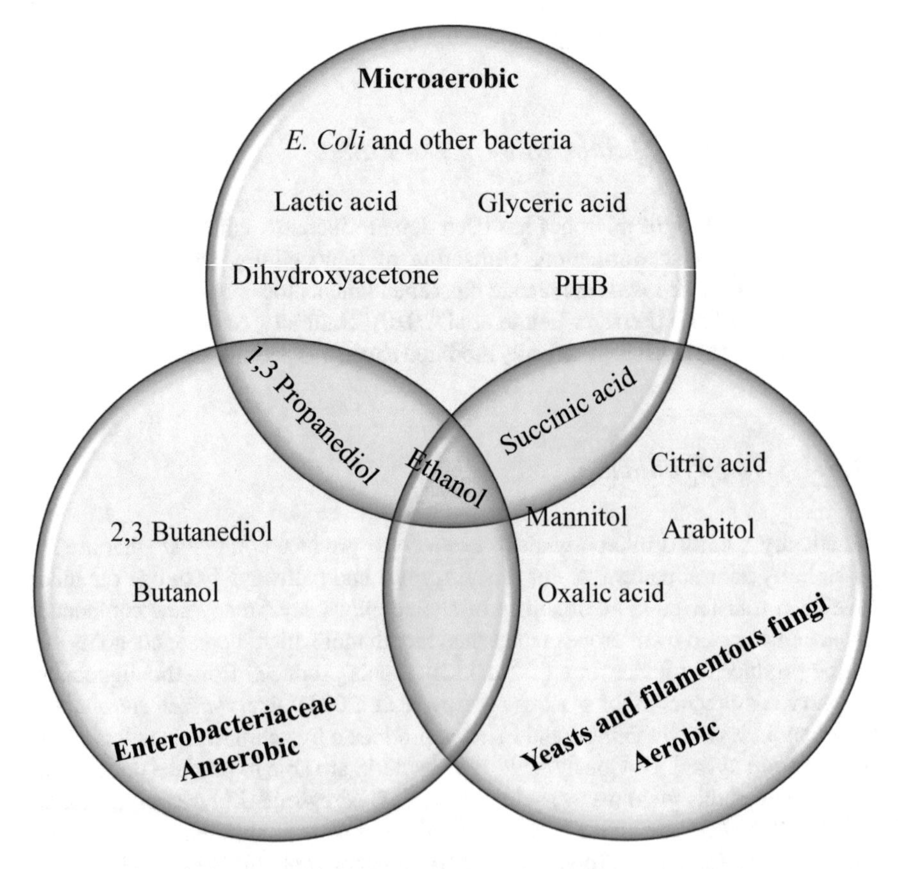

Fig. 11.3 List of chemicals produced by microbes by the fermentation of glycerol

Many microbes such as *Klebsiella, Enterobacter, Clostridium,* Yeasts, and filamentous fungi are used for the production of organic acids, polyols, 1,3-propanediol, 2,3-butanediol, butanol, and ethanol (Yadav et al. 2017). 1,3-propanediol (1,3-PDO) can be produced by *Klebsiella* spp. and *Clostridium* spp. from glycerol (Celinska 2010). *K. pneumoniae* G31 also produces 2,3-Butanediol (BDO) from the fermentation of glycerol (Petrov and Petrova 2009). This BDO can be used in the preparation of synthetic rubber, plastics, and as a precursor of pharmaceutical drugs and medicine (Syu 2001; Ji et al. 2011). Ethanol is a widely used fuel and solvent in industries, produced from lignocellulose by yeasts. However, there are many reports where glycerol also acts as a source of ethanol (Liu et al. 2007; Petrov and Petrova 2009). *E. coli* can convert glycerol to ethanol aerobically and anaerobically (Dharmadi et al. 2006; Durnin et al. 2009). *Hansenula polymorpha,* a methylotrophic yeast, possesses potential to produce ethanol from glycerol (Hong et al. 2010). Genes encoding for pyruvate decarboxylase and aldehyde dehydrogenase II, from *Zymomonas mobilis,* are transferred into *H. polymorpha,* and increase in ethanol production was found (Hong et al. 2010). Butanol is an alternative fuel which is used in the manufacturing of plastics, paints, resin formulation, and lacquers (Harvey and Meylemans 2011). *C. pasteurianum* has been found to produce butanol from glycerol (Taconi et al. 2009). Apart from these, glycerol has been used to produce mannitol, arabitol, erythritol, succinic acid, lactic acid, oxalic acid, citric acid, and glyceric acid (Table 11.3).

11.6 Conclusion

Plant cell wall is composed of cellulose and lignin, which are very complex and poorly understood. Utilization of this for bioenergy needs more understanding and research inputs. In biorefineries, a consortium of microbes is used, where microbe–microbe interaction takes place. Attention should be paid toward population dynamics, interrelationship between species for scale-up of a process. It is possible to optimize microbial processes with the aid of computer simulations. Application of biotechnological aspects such as CRISPR/Cas, genome shuffling, transcription, and translational machinery in microbes can make them more potent for biorefineries

Table 11.3 Chemicals produced at high yield and/or high concentration by microbial fermentation of glycerol

Product	Organism	Fermentation mode	Oxygen availability	Yield (product/glycerol)	Productivity	Product concentration	References
1,3-Propanediol	*K. pneumoniae* DSM 2026	Fed-batch	Microaerobic	0.52 mol/mol	1.57 g/L/h	59.50 g/L	Chen et al. (2003)
	K. pneumoniae LDH 526	Fed-batch	Aerobic	0.52 mol/mol	2.13 g/L/h	102.1 g/L	Xu et al. (2009)
	C. butyricum F2b	Batch	Anaerobic	0.53 g/g	1.05 g/L/ha	47.1 g/L	Papanikolaou et al. (2008)
	E. coli K12	Fed-batch	Anaerobic	90.2%	2.61 g/L/h	104.4 g/L	Tang et al. (2009b
	K. pneumoniae	Fed-batch 1 m^3	Anaerobic	61 mol/mol	2.2 g/L/h	75 g/L	Liu et al. (2010)
2,3-Butanediol	*K. pneumoniae* G31	Fed-batch	Microaerobic	0.36 mol/mol	0.18 g/L/h	49.2 g/L	Petrov and Petrova (2009)
	K. pneumoniae G31	Fed-batch	Aerobic	0.39 g/g	0.47 g/L/h	70.0 g/L	Petrov and Petrova (2009)
Ethanol	*E. coli* SY 4	Batch	Microaerobic	85%	0.15 g/L/h	7.8 g/L	Durnin et al. (2009)
Butanol	*C. pasteurianum*	Batch	Anaerobic	0.36 g/g	Not determined	1.8 g/La	Taconi et al. (2009)
Dihydroxyacetone	*G. oxydans* ZJB09112	Fed-batch	Aerobic	88.7%	Not determined	161.9 g/L	Hu et al. (2010)
Glyceric acid	*G. frateurii* NBRC103465	Fed-batch	Aerobic	0.76 g/g	0.81 g/L/ha	136.5 g/Lc	Habe et al. (2009)
	A. tropicalis NBRC16470	Fed-batch	Aerobic	0.46 g/g	0.71 g/L/ha	101.8 g/Ld	Habe et al. (2009)

(continued)

Table 11.3 (continued)

Product	Organism	Fermentation mode	Oxygen availability	Yield (product/glycerol)	Productivity	Product concentration	References
Lactic acid	*E. coli* AC-521	Fed-batch	Aerobic	0.9 mol/mol	0.49 g/g/ha	85.8 g/L	Hong et al. (2009)
	E. coli LA02Δdld	Batch	Microaerobic	0.83 g/g	1.25 g/g//h	32 g/L	Mazumdar et al. (2010)
Succinic acid	engineered *E. coli*	Batch	Microaerobic	0.69 g/g	~4 g/g/h	14 g/L	Blankschien et al. (2010)
	Y. lipolytica Y-3314	Batch	Oxygen limited	0.45 g/g	Not determined	45 g/L	Yuzbashev et al. (2010)
Citric acid	*Y. lipolytica*	Repeated batch	Aerobic	0.77 g/g	0.85 g/L/h	124.2 g/L	Rymowicz et al. (2010)
Oxalic acid	*A. niger*	Batch	Aerobic	0.62 g/g	Not determined	21 g/L	Andre et al. (2010)
Mannitol	*C. magnoliae*	Batch	Aerobic	0.51 g/g	0.53 g/L/h	51 g/L	Khan et al. (2009)
Erythritol	*Y. lipolytica* Wratislavia K1	Fed-batch	Aerobic	0.56 g/g	1.0 g/L/h	170 g/L	Rymowicz et al. (2009)
Arabitol	*D. hansenii* SBP1	Batch	Aerobic	0.50 g/g	0.12 g/L/h	14 g/L	Koganti et al. (2011)
PHB	*E. coli* Arc2	Fed-batch	Microaerobic		0.18 g/L/h	10.81 g/L	Nikel et al. (2008)
	Z. denitrificans MW1	Fed-batch	Aerobic	0.25 g/g	1.09 g/L/h	54.3 g/L	Ibrahim and Steinbuchel (2009)

Acknowledgments The authors would like to thank Director, DEI, for his continuous support and encouragement. SM is grateful to Dayalbagh Educational Institute, Deemed University, Agra, for sanctioning the Research Project, DEI/Minor Project/2017-18 (iv), as a start-up grant. DG is thankful to DST-INSPIRE for providing the fellowship.

References

Almeida JR, Modig T, Petersson A, Hähn-Hägerdal B, Lidén G, Gorwa-Grauslund MF (2007) Increased tolerance and conversion of inhibitors in lignocellulosic hydrolysatesby *Saccharomyces cerevisiae*. J Chem Technol Biotechnol 82:340–349

Almeida JRM, Bertilsson M, Hahn-Hägerdal B, Lidén G, Gorwa-Grauslund M-F (2009) Carbon fluxes of xylose-consuming Saccharomyces cerevisiae strains are affected differently by NADH and NADPH usage in HMF reduction. Appl Microbiol Biotechnol 84:751–761

Almeida JRM, Röder A, Modig T, Laadan B, Lidén G, Gorwa-Grauslund M-F (2008) NADH- vs NADPH-coupled reduction of 5-hydroxymethyl furfural (HMF) and its implications on product distribution in Saccharomyces cerevisiae. Appl Microbiol Biotechnol 78:939–945

Almeida JRM, Runquist D, Sànchez i Nogué V, Lidén G, Gorwa-Grauslund MF (2011) Stress-related challenges in pentose fermentation to ethanol by the yeast Saccharomyces cerevisiae. Biotechnol J 6:286–299

Alper H, Stephanopoulos G (2007) Global transcription machinery engineering: a new approach for improving cellular phenotype. Metab Eng 9:258–267

Alsaker K, Spitzer T, Papoutsakis E (2004) Transcriptional analysis of spo0A overexpression in clostridium acetobutylicum and its effect on the cell's response to butanol stress. J Bacteriol 186:1959–1971

Amore A, Faraco V (2012) Potential of fungi as category I Consolidated BioProcessing organisms for cellulosic ethanol production. Renew Sustain Energy Rev 16(5):3286–3301

Andre A, Diamantopoulou P, Philippoussis A, Sarris D, Komaitis M, Papanikolaou S (2010) Biotechnological conversions of bio-diesel derived waste glycerol into added-value compounds by higher fungi: production of biomass, single cell oil and oxalic acid. Ind Crop Prod 31:407–416

Blankschien MD, Clomburg JM, Gonzalez R (2010) Metabolic engineering of *Escherichia coli* for the production of succinate from glycerol. Metab Eng 12:409–419

Botella C, Diaz AB, Wang R, Koutinas A, Webb C (2009) Particulate bioprocessing: a novel process strategy for biorefineries. Process Biochem 44(5):546–555

Causey TB, Zhou S, Shanmugam KT, Ingram LO (2003) Engineering the metabolism of *Escherichia coli* W3110 for the conversion of sugar to redox-neutral and oxidized products: homoacetate production. Proc Natl Acad Sci USA 100:825–832

Celinska E (2010) Debottlenecking the 1,3-propanediol pathway by metabolic engineering. Biotechnol Adv 28:519–530

Chen X, Zhang DJ, Qi WT, Gao SJ, Xiu ZL, Xu P (2003) Microbial fed-batc production of 1,3-propanediol by *Klebsiella pneumoniae* under micro-aerobic conditions. Appl Microbiol Biotechnol 63:143–146

de Jong E, Jungmeier G (2015) Biorefinery concepts in comparison to petrochemical refineries. In: Industrial biorefineries & white biotechnology. Elsevier, pp 3–33

Dharmadi Y, Murarka A, Gonzalez R (2006) Anaerobic fermentation of glycerol by *Escherichia coli*: a new platform for metabolic engineering. Biotechnol Bioeng 94(5):821–829

Durnin G, Clomburg J, Yeates Z, Alvarez PJJ, Zygourakis K, Campbell P, Gonzalez R (2009) Understanding and harnessing the microaerobic metabolism of glycerol in *Escherichia coli*. Biotechnol Bioeng 103:148–161

Ferreira-Leitão V, Perrone CC, Rodrigues J, Franke APM, Macrelli S, Zacchi G (2010) An approach to the utilisation of CO_2 as impregnating agent in steam pretreatment of sugar cane bagasse and leaves for ethanol production. Biotechnol Biofuels 3:7

Habe H, Shimada Y, Yakushi T, Hattori H, Ano Y, Fukuoka T, Kitamoto D, Itagaki M, Watanabe K, Yanagishita H (2009) Microbial production of glyceric acid, an organic acid that can be mass produced from glycerol. Appl Environ Microb 75:7760–7766

Hahn-Hägerdal B, Karhumaa K, Fonseca C, Spencer-Martins I, Gorwa-Grauslund MF (2007) Towards industrial pentose-fermenting yeast strains. Appl Microbiol Biotechnol 74:937–953

Harvey BG, Meylemans HA (2011) The role of butanol in the development of sustainable fuel technologies. J Chem Technol Biotechnol 86(1):2–9

Heer D, Sauer U (2008) Identification of furfural as a key toxin in lignocellulosichydrolysates and evolution of a tolerant yeast strain. Microb Biotechnol 1:497–506

Hong AA, Cheng KK, Peng F, Zhou S, Sun Y, Liu CM, Liu DH (2009) Strain isolation and optimization of process parameters for bioconversion of glycerol to lactic acid. J Chem Technol Biot 84:1576–1581

Hong WK, Kim CH, Heo SY, Luo LH, Oh BR, Seo JW (2010) Enhanced production of ethanol from glycerol by engineered *Hansenula polymorpha* expressing pyruvate decarboxylase and aldehyde dehydrogenase genes from *Zymomonas mobilis*. Biotechnol Lett 32(8):1077–1082

Hu ZC, Liu ZQ, Zheng YG, Shen YC (2010) Production of 1,3-Dihydroxyacetone from Glycerol by *Gluconobacter oxydans* ZJB09112. J Microbiol Biotechnol 20:340–345

Hughes SR, Gibbons WR, Bang SS, Pinkelman R, Bischoff KM, Slininger PJ, Qureshi N, Kurtzman CP, Liu S, Saha BC, Jackson JS, Cotta M, Rich JO, Javers JE (2012) Random UV-C mutagenesis of *Scheffersomyces* (formerly *Pichia*) *stipitis* NRRL Y-7124 to improve anaerobic growth on lignocellulosic sugars. J Ind Microbiol Biotechnol 39:163–173

Ibrahim MHA, Steinbuchel A (2009) Poly(3-Hydroxybutyrate) Production from Glycerol by *Zobellella denitrificans* MW1 via High-Cell-Density Fed-Batch fermentation and simplified solvent extraction. Appl Environ Microb 75:6222–6231

Inui M, Suda M, Kimura S, Yasuda K, Suzuki H, Toda H, Yamamoto S, Okino S, Suzuki N, Yukawa H (2008) Expression of *Clostridium acetobutylicumbutanol* synthetic genes in *Escherichia coli*. Appl Microbiol Biotechnol 77:1305–1316

Ji XJ, Huang H, Ouyang PK (2011) Microbial 2,3-butanediol production: A state-of-the -art review. Biotechnol Adv 29:351–364

Jiang Y, Xu C, Dong F, Yang Y, Jiang W, Yang S (2009) Disruption of the acetoacetate decarboxylase gene in solvent-producing *Clostridium acetobutylicum* increases the butanol ratio. Metab Eng 11:284–291

Karhumaa K, Garcia Sanchez R, Hahn-Hägerdal B, Gorwa-Grauslund M-F (2007) Comparison of the xylose reductase-xylitol dehydrogenase and the xylose isomerase pathways for xylose fermentation by recombinant *Saccharomyces cerevisiae*. Microb Cell Fact 6:5

Khan A, Bhide A, Gadre R (2009) Mannitol production from glycerol by resting cells of *Candida magnoliae*. Bioresour Technol 100:4911–4913

Koganti S, Kuo TM, Kurtzman CP, Smith N, Ju LK (2011) Production of arabitol from glycerol: strain screening and study of factors affecting production yield. Appl Microbiol Biotechnol 90:257–267

Kour D, Rana KL, Yadav N, Yadav AN, Rastegari AA, Singh C et al (2019a) Technologies for biofuel production: current development, challenges, and future prospects. In: Rastegari AA, Yadav AN, Gupta A (eds) Prospects of renewable bioprocessing in future energy systems. Springer International Publishing, Cham, pp 1–50. https://doi.org/10.1007/978-3-030-14463-0_1

Kour D, Rana KL, Yadav N, Yadav AN, Singh J, Rastegari AA et al (2019b) Agriculturally and industrially important fungi: current developments and potential biotechnological applications. In: Yadav AN, Singh S, Mishra S, Gupta A (eds) Recent advancement in white biotechnology through fungi, Volume 2: Perspective for Value-Added Products and Environments. Springer International Publishing, Cham, pp 1–64. https://doi.org/10.1007/978-3-030-14846-1_1

Koutinas AA, Xu Y, Wang R, Webb C (2007) Polyhydroxybutyrate production from a novel feedstock derived from a wheat-based biorefinery. Enzyme Microb Technol 40(5):1035–1044

Kumar S, Sharma S, Thakur S, Mishra T, Negi P, Mishra S et al (2019) Bioprospecting of microbes for biohydrogen production: current status and future challenges. In: Molina G, Gupta VK, Singh BN, Gathergood N (eds) Bioprocessing for biomolecules production. Wiley, USA, pp 443–471

Kuyper M, Harhangi H, Stave A, Winkler A, Jetten M, Delaat W, Denridder J, Opdencamp H, Vandijken J, Pronk J (2003) High-level functional expression of a fungal xylose isomerase: the key to efficient ethanolic fermentation of xylose by? FEMS Yeast Res 4:69–78

Lange L (2017) Fungal enzymes and yeasts for conversion of plant biomass to bioenergy and high-value products. In: The Fungal Kingdom, pp 1027–1048

Lee SJ, Song H, Lee SY (2006) Genome-based metabolic engineering of mannheimia succiniciproducens for succinic acid production. Appl Environ Microbiol 72:1939–1948

Liu H, Valdehuesa KNG, Nisola GM, Ramos KRM, Chung W-J (2012) High yield production of D-xylonic acid from D-xylose using engineered Escherichia coli. Bioresour Technol 115:244–248

Liu HJ, Xu YZ, Zheng ZM, Liu DH (2010) 1,3-Propanediol and its copolymers: research, development and industrialization. Biotechnol J 5:1137–1148

Liu ZL, Slininger PJ, Dien BS, Berhow MA, Kurtzman CP, Gorsich SW (2004) Adaptive response of yeasts to furfural and 5-hydroxymethylfurfural and new chemical evidence for HMF conversion to 2,5-bis-hydroxymethylfuran. J Ind Microbiol Biotechnol 31:345–352

Liu HJ, Zhang DJ, Xu YH, Mu Y, Sun YQ, Xiu ZL (2007) Microbial production of 1, 3-propanediol from glycerol by *Klebsiella pneumoniae* under micro-aerobic conditions up to a pilot scale. Biotechnol Lett 29(8):1281–1285

Mazumdar S, Clomburg JM, Gonzalez R (2010) *Escherichia coli* strains engineered for homofermentative production of D-lactic acid from glycerol. Appl Environ Microbiol 76:4327–4336

Menon V, Rao M (2012) Trends in bioconversion of lignocellulose: biofuels, platform chemicals & biorefinery concept. Prog Energy Combust Sci 38(4): 522–550

Mojzita D, Wiebe M, Hilditch S, Boer H, Penttilä M, Richard P (2010) Metabolic engineering of fungal strains for conversion of D-galacturonate to meso-galactarate. Appl Environ Microbiol 76:169–175

Mussatto SI, Dragone G, Guimarães PM, Silva JPA, Carneiro LM, Roberto IC, Teixeira JA (2010) Technological trends, global market, and challenges of bio-ethanol production. Biotechnol Adv 28(6): 817–830

Nikel PI, Pettinari MJ, Galvagno MA, Mendez BS (2008) Poly(3 hydroxybutyrate) synthesis from glycerol by a recombinan *Escherichia coli* arcA mutant in fed-batch microaerobic cultures. Appl Microbiol Biotechnol 77:1337–1343

Nissen TL, Kielland-Brandt MC, Nielsen J, Villadsen J (2000) Optimization of ethanol production in *Saccharomyces cerevisiae* by metabolic engineering of the ammonium assimilation. Metab Eng 2:69–77

Nygård Y, Toivari MH, Penttilä M, Ruohonen L, Wiebe MG (2011) Bioconversion of D-xylose to D-xylonate with *Kluyveromyces lactis*. Metab Eng 13:383–391

O'Connor D (2011) Report T39-T3. Biodiesel GHG emissions, pas, present, and future. A report to IEA Bioenergy Task 39. In: Commercializing liquid biofuels from biomass.: International Energy Agency (IEA); www.ieabioenergy.com/Liblinks.aspx

Okano K, Zhang Q, Shinkawa S, Yoshida S, Tanaka T, Fukuda H, Kondo A (2009) Efficient production of optically pure D-lactic acid from raw corn starch by using a genetically modified L-lactate dehydrogenase gene-deficient and alpha-amylase-secreting *Lactobacillus plantarum* strain. Appl Environ Microbiol 75:462–467

Pandey A, Soccol CR, Mitchell D (2000) New developments in solid state fermentation: I-bioprocesses and products. Process Biochem 35(10):1153–1169

Papanikolaou S, Fakas S, Fick M, Chevalot I, Galiotou-Panayotou M, Komaitis M, Marc I, Aggelis G (2008) Biotechnological valorisation of raw glycerol discharged after bio-diesel (fatty acid methyl esters) manufacturing process: production of 1,3-propanediol, citric acid and single cell oil. Biomass Bioenerg 32:60–71

Petrov K, Petrova P (2009) High production of 2, 3-butanediol from glycerol by *Klebsiella pneumoniae* G31. Appl Microbiol Biotechnol 84(4):659–665

Rahardjo YS, Tramper J, Rinzema A (2006) Modeling conversion and transport phenomena in solid-state fermentation: a review and perspectives. Biotechnol Adv 24(2):161–179

Rastegari AA, Yadav AN, Yadav N (2020) New and future developments in microbial biotechnology and bioengineering: Trends of microbial biotechnology for sustainable agriculture and biomedicine systems: diversity and functional perspectives. Elsevier, Amsterdam

Rastegari AA, Yadav AN, Gupta A (2019a) Prospects of renewable bioprocessing in future energy systems. Springer International Publishing, Cham

Rastegari AA, Yadav AN, Yadav N (2019b) Genetic manipulation of secondary metabolites producers. In: Gupta VK, Pandey A (eds) New and future developments in microbial biotechnology and bioengineering. Elsevier, Amsterdam, pp 13–29. https://doi.org/10.1016/B978-0-444-63504-4.00002-5

Rastegari AA, Yadav AN, Yadav N, Tataei Sarshari N (2019c) Bioengineering of secondary metabolites. In: Gupta VK, Pandey A (eds) New and future developments in microbial biotechnology and bioengineering. Elsevier, Amsterdam, pp 55–68. https://doi.org/10.1016/B978-0-444-63504-4.00004-9

Rymowicz W, Fatykhova AR, Kamzolova SV, Rywinska A, Morgunov IG (2010) Citric acid production from glycerol-containing waste of biodiesel industry by *Yarrowia lipolytica* in batch, repeated batch, and cell recycle regimes. Appl Microbiol Biotechnol 87:971–979

Rymowicz W, Rywinska A, Marcinkiewicz M (2009) High-yield production of erythritol from raw glycerol in fed-batch cultures of Yarrowia lipolytica. Biotechnol Lett 31:377–380

Steen EJ, Chan R, Prasad N, Myers S, Petzold CJ, Redding A, Ouellet M, Keasling JD (2008) Metabolic engineering of Saccharomyces cerevisiae for the production of n-butanol. Microb Cell Fact 7:36

Steen EJ, Kang Y, Bokinsky G, Hu Z, Schirmer A, McClure A, Del Cardayre SB, Keasling JD (2010) Microbial production of fatty-acid-derived fuels and chemicals from plant biomass. Nature 463:559–562

Syu MJ (2001) Biological production of 2,3-butanediol. Appl Microbiol Biotechnol 55:10–18

Taconi KA, Venkataramanan KP, Johnson DT (2009) Growth and solvent production by Clostridium pasteurianum ATCC (R) 6013 (TM) utilizing biodiesel-derived crude glycerol as the sole carbon source. Environ Prog Sustain Energy 28:100–110

Tang X, Tan Y, Zhu H, Zhao K, Shen W (2009a) Microbial conversion of glycerol to 1,3-propanediol by an engineered strain of *Escherichia coli*. Appl Environ Microbiol 75:1628–1634

Tang XM, Tan YS, Zhu H, Zhao K, Shen W (2009b) Microbial conversion of glycerol to 1,3-Propanediol by an engineered strain of *Escherichia coli*. Appl Environ Microb 75:1628–1634

Toivari MH, Ruohonen L, Richard P, Penttilä M, Wiebe MG (2010) *Saccharomyces cerevisiae* engineered to produce D-xylonate. Appl Microbiol Biotechnol 88:751–760

Vijayendran B (2010) Bio products from bio refineries-trends, challenges and opportunities. J Bus Chem 7(3)

Viniegra-Gonzàlez G (1997) Solid state fermentation: definition, characteristics, limitations and monitoring. In: Advances in solid state fermentation. Springer, Dordrecht, pp 5–22

Wang X, Miller EN, Yomano LP, Zhang X, Shanmugam KT, Ingram LO (2011) Increased furfural tolerance due to overexpression of NADH-dependent oxidoreductase FucO in *Escherichia coli* strains engineered for the production of ethanol and lactate. Appl Environ Microbiol 77:5132–5140

Werpy T, Petersen G, Aden A, Bozell J (2004) Top value added chemicals from biomass. Volume 1-Results of screening for potential candidates from sugars and synthesis gas

Xu YZ, Guo NN, Zheng ZM, Ou XJ, Liu HJ, Liu DH (2009) Metabolism in 1,3- propanediol fed-batch fermentation by a D-lactate deficient mutant of *Klebsiella pneumoniae*. Biotechnol Bioeng 104:965–972

Yadav AN, Kumar R, Kumar S, Kumar V, Sugitha T, Singh B et al (2017) Beneficial microbiomes: biodiversity and potential biotechnological applications for sustainable agriculture and human health. J Appl Biol Biotechnol 5:45–57

Yadav AN, Rastegari AA, Yadav N (2020) Microbiomes of extreme environments: biodiversity and biotechnological applications. CRC Press, Taylor & Francis, Boca Raton, USA

Yadav AN, Singh S, Mishra S, Gupta A (2019) Recent advancement in white biotechnology through fungi. In: Perspective for value-added products and environments, vol 2. Springer International Publishing, Cham

Yazdani SS, Gonzalez R (2007) Anaerobic fermentation of glycerol: a path to economic viability for the biofuels industry. Curr Opin Biotech 18:213–219

Yomano LP, York SW, Shanmugam KT, Ingram LO (2009) Deletion of methylglyoxal synthase gene (mgsA) increased sugar co-metabolism in ethanol-producing Escherichia coli. Biotechnol Lett 31:1389–1398

Yuzbashev TV, Yuzbasheva EY, Sobolevskaya TI, Laptev IA, Vybornaya TV, Larina AS, Matsui K, Fukui K, Sineoky SP (2010) Production of succinic acid at low pH by a recombinant strain of the aerobic yeast *Yarrowia lipolytica*. Biotechnol Bioeng 107:673–682

Zhang Y-HP (2011) What is vital (and not vital) to advance economically competitive biofuels production. Process Biochem 46(11):2091–2110

Chapter 12
Microbial Bioresources and Their Potential Applications for Bioenergy Production for Sustainable Development

N. K. Ismail, M. A. Amer, M. E. Egela, and A. G. Saad

Abstract There are many inexhaustible resources in the natural environment that can be used for the production of bioenergy. There are also many ways to produce such energy, depending on your requirements. The production and utilization of different forms of bioenergy, such as bioelectric and different biofuels, helps to preserve the environment.

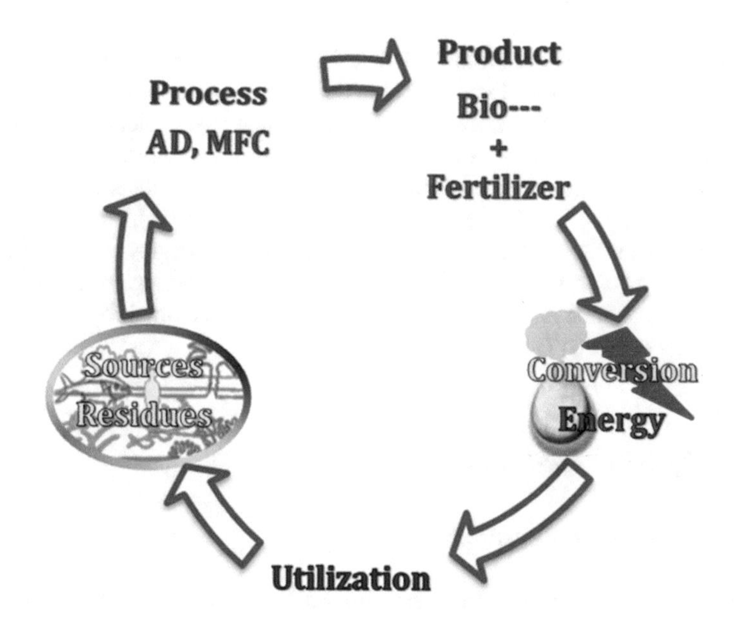

N. K. Ismail · M. A. Amer · M. E. Egela · A. G. Saad (✉)
Bio-System Engineering Department, Agricultural Engineering Research Institute (AEnRI), Agricultural Research Center (ARC), Giza, Egypt
e-mail: Dr.abdelgawad2012@gmail.com; en_gawad2000@yahoo.com

A. N. Yadav et al. (eds.), *Biofuels Production – Sustainability and Advances in Microbial Bioresources*, Biofuel and Biorefinery Technologies 11,
https://doi.org/10.1007/978-3-030-53933-7_12

12.1 Introduction

One concept behind the provision of energy is to ensure that there is not a reliance on any one form of energy production, thereby avoiding energy shortages should one energy source be depleted. Having many energy sources also eases the economic pressure associated with a reliance on any one form. Therefore, we must make good use of all the raw materials available that can be used for energy production.

In general, microbes can be produced and grown naturally when conditions are suitable in terms of moisture, temperature, and nutrients. Environments associated with agricultural processes using plants, animals, and food residues; farms, including poultry, other livestock, and fisheries; and wastewater, are considered suitable for microbe production because of their levels of organic matter, moisture, etc.

12.2 Bioresources

Bioresources are biomass or biological material from living or recently living organisms that can decompose under aerobic and anaerobic conditions using processes of burning, gasification, or fermentation to produce bioenergy. Protecting the environment and improving standards of living are the most important factors driving the management of bioresources, in addition to integrating them with energy-producing technologies (Rasool and Hemalatha 2016; Bhatia et al. 2018). Bioresources can be classified according to their origin and the different strategies required for their pretreatment and conversion into bioenergy. Sources include legume plants, algae, monocot plants, edible and non-edible vegetable oils, and animal fats (Bhatia et al. 2018; Gaurav et al. 2017).

12.2.1 Types of Bioresources

12.2.1.1 Agricultural By-Products

The production of bioenergy from agricultural biomass, such as oil palm shells, pineapple residue, forest (logging) residue, coir pith, sugarcane bagasse, empty fruit palm bunches, oil palm fronds, coconut husks, soybean hulls, corn stover, wheat straw, oil palm fibers, oil palm trunks, silk cotton, rice husks, banana residue, paddy straw, reeds, and rapeseed, is linked to microbial action on lignocellulose. Such sources are well known and considered ecofriendly (Gaurav et al. 2017; Rastegari et al. 2020; Yadav et al. 2019).

12.2.1.2 Food Processing Residue

Food processing residue comes from the manufacture of vegetable oils and the processing of meat and can be divided into liquid and solid waste (Kumar et al. 2017; Ravindran and Jaiswal 2016). Liquid waste comes from meat, vegetables, and fruits that have been washed to remove solid organic matter, starch, and sugar. However, processing fruits or vegetables produces solid waste residue from peeling and pulping. Such residue often lacks quality control standards (Bhatia et al. 2018).

12.2.1.3 Energy from Plant Biomass

Plant biomass comes from dedicated crops that are regularly replanted after harvesting. Use of this biomass resource depends on crop availability and required biomass product (Najafi et al. 2009a, b; Balat et al. 2008).

12.2.1.4 Animal and Poultry Residue

Animal residue is the perfect raw material for biogas production because it already contains most of the microbes used in this technology (biowaste-to-bioenergy). Animal residue exists in abundance as organic matter such as feathers, bones, skin, hair, and meat (Mathias 2014; Gebrezgabher et al. 2010).

12.2.1.5 Algal Biomass

Algal biomass has been used, through the process of anaerobic digestion, to produce methane. Its low level of lignin favors biofuel production. Using algae to produce biofuel has no requirement for pesticides, freshwater, or fertilizers for growth. In addition, the growth rates of algae are found to be higher than plants. Moreover, the land requirement for cultivation is lower than for agricultural plants (Bruton et al. 2009; Gaurav et al. 2017; Panjiar et al. 2017). Algae utilize enormous amounts of CO_2 for their growth, remove CO_2 from the atmosphere (some of which originates from power plant emissions), convert biomass via photosynthesis, and liberate oxygen to the atmosphere. Algal biomass can be transformed into different types of biofuel according to three types of production processes: thermochemical processes, biological processes, and chemical reactions (Figs. 12.1 and 12.2) (Dalena et al. 2017).

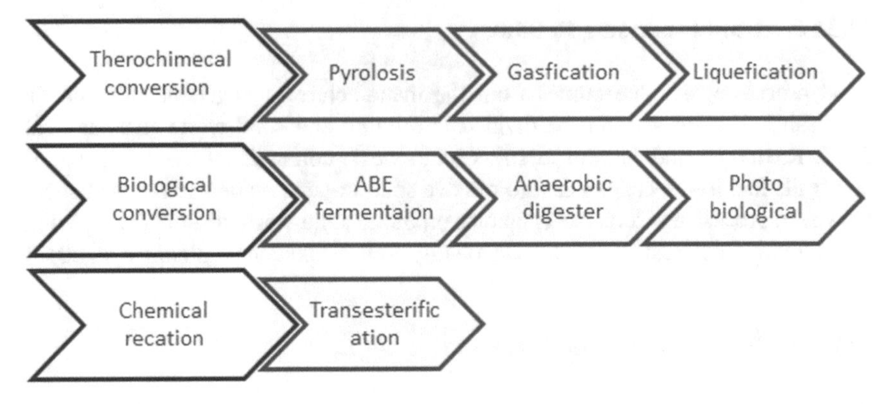

Fig. 12.1 Production processes

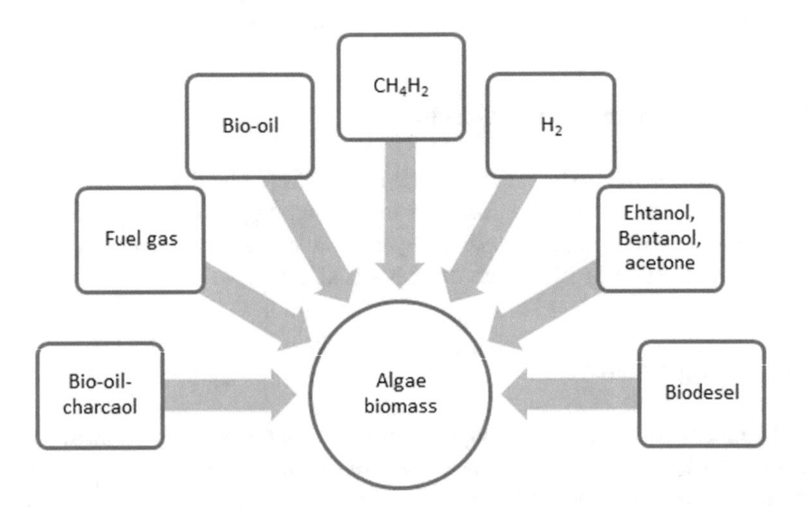

Fig. 12.2 Types of biofuel

12.2.2 *Bioresource Strategies for Bioenergy Technology*

12.2.2.1 Anaerobic Digestion

Anaerobic digestion (AD) is a biological process that transforms residue into energy. Anaerobic digestion is the disintegration of complex organic matter by microorganisms, in the absence of oxygen, into simpler chemical components (Chen et al. 2018; Li et al. 2019; Momayez et al. 2019; Pramanik et al. 2019; Timonen et al. 2019). The AD process is a multi-step biochemical process; four processes occur simultaneously, namely, hydrolysis, acidogenic fermentation, hydrogen-producing acetogenesis, and methanogenesis (Zhang et al. 2014; Feng and Lin 2017; Gould 2015; Li et al. 2019; Kainthol et al. 2019; Pramanik et al. 2019). AD, a gas that is often referred to as biogas, is comprised of methane and carbon dioxide as well as small volumes of other gases such as hydrogen sulphide (H_2S), ammonia (NH_3), nitrogen, hydrogen, and water vapor (Monnet 2003; Abbasi et al. 2012). Different microorganisms are important to the production of AD, with several types of bacteria degrading constantly and other bacteria producing the gas irregularly (Wang et al. 2018).

For bacteria responsible for the degradation of biowaste there is a relationship between microbial structure and process stability (Li et al. 2015). In the process of hydrolysis, carbohydrates, proteins, lipids, and other organics that are contained within insoluble complex polymers are broken down by hydrolases, produced by microbes, into simple, smaller soluble molecules such as sugars, amino acids, and fatty acids. This phase is a comparatively slow process (Ostrem 2004; Kothari et al. 2014; Zhang et al. 2014, 2015; Leung and Wang 2016). The next phase is the fermentation of molecules such as sugars, amino acids, and fatty acids which are converted into different volatile fatty acids (VFAs) and gaseous components (H_2 and CO_2) by acetogenic bacteria which also reduce these components to acetic acid. This is called the acidogenic phase (Ostrem 2004; Kothari et al. 2014; Zhang et al. 2015; Amer et al. 2019). The final stage in AD is the methanogenic process, where methane gas is produced from acetic acid, hydrogen, and carbon dioxide by bacteria on the intermediate products of the previous steps and fermentation process. A suitable pH for methanogenic bacteria is between 6.5 and 7.5 (Leung and Wang 2016). Figure 12.3 shows the four phases of anaerobic biodegradation.

Operational Conditions in the Anaerobic Digestion Process

Environmental factors affect the stability of the AD process as well as the equilibrium of microorganisms when producing biogas from biomass. Factors include temperature (Gerardi 2003; Khalid et al. 2011), pH (Appels et al. 2008; Leung and Wang 2016), VFAs (Xu et al. 2014; Shi et al. 2018), carbon and nitrogen ratio (C/N ratio) (Yadvika et al. 2004; Krishna and Kalamdhad 2014), retention time (Deepanraj et al. 2014; Mao et al. 2015), and organic loading rate (Kothari et al. 2014). The process of digestion can be wet (Deepanraj et al. 2014; Kothari et al. 2014) or dry (Kothari et al. 2014; Yi et al. 2014).

12.2.2.2 Transesterification

Transesterification is also called alcoholysis. In this process, non-edible oil is allowed to chemically react with alcohols, such as methanol and ethanol, according to their availability and cost. Another organic reaction is where an ester is transformed into another through an interchange of the alkoxy moiety. This process is used to reduce the viscosity of non-edible oil and convert triglycerides into esters (Atabania et al. 2013; Azad et al. 2017). The transesterification reaction is outlined in the following equation (Gerpen 2005; Romano et al. 2006):

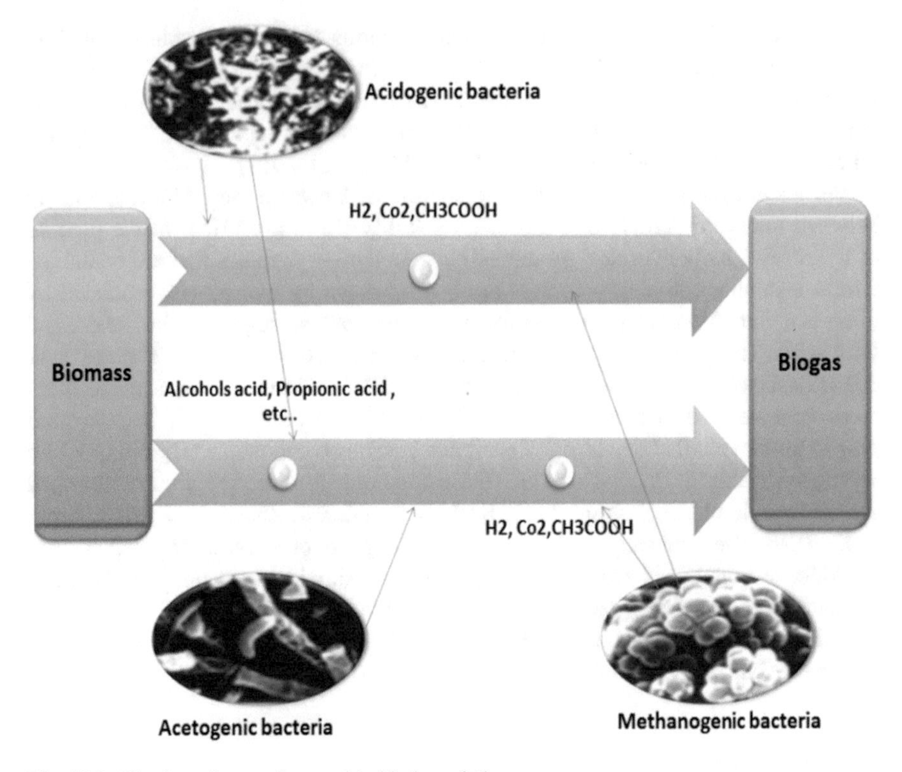

Fig. 12.3 The four phases of anaerobic biodegradation

$$\mathrm{RCOOR}' + \mathrm{R}''\mathrm{OH} \overset{\mathrm{cat}}{\Longleftrightarrow} \mathrm{R}'\mathrm{OH} + \mathrm{RCOOR}''$$

where RCOOR′ is an ester; R″OH is an alcohol; R′OH is another alcohol (glycerol); RCOO R″ is an ester mixture; and "cat" represents a catalyst.

The drawback related to this process is the length of time needed for the separation of the oil, alcohol, catalyst, and saponified impurity mixture from the biodiesel (Azad 2017). Transesterification can be basic, acidic, or enzymatic.

Base-Catalyzed Transesterification

Base-catalyzed transesterification is the most economical and commonly used technique because it demands only low temperatures and pressures. Base-catalyzed transesterification produces a conversion yield of over 98% when the starting oil is low in moisture and free fatty acid (FFAs) content—a high FFA content causes the formation of soap which reduces catalyst efficiency, causes increased viscosity, leads to gel formation, and makes the separation of glycerol difficult (Singh et al. 2006; Leung and Guo 2006).

Acid-Catalyzed Transesterification

Acid catalysts can be used to produce biodiesel from low-cost lipid feedstock with FFA contents greater than 1%. In this process, residue cooking oil was found overall to be the most economically feasible, providing a lower total manufacturing cost and a lower biodiesel break-even price (Zhang et al. 2003; Lotero et al. 2005).

12.2.2.3 Microbial Fuel Cells

Microbial fuel cell (MFC) technology converts biomass or biowaste directly to electricity using microbial catalyzed "anodic" and microbial, enzymatic, abiotic "cathodic" electrochemical reactions (Santoro et al. 2017; Kumar et al. 2019; Rastegari et al. 2019). In other words, this technology combines classic abiotic electrochemical reactions and physics with biological catalytic redox activity (Logan et al. 2006; Rinaldi et al. 2008). The most important advantages of MFC are considered as an energy-saving technology. Because it reduces the energy used for aerating. Moreover, this technology can be used for the removal of pollutants, retrieval of nutrients, and generation of electrical energy from wastewater (Oh et al. 2010; He et al. 2015; Palanisamy et al. 2019). MFCs are categorized according to electrolyte nature and alignment: (1) single-chambered MFCs (SCMFCs), (2) double-chambered MFCs (DCMFCs), (3) stacked MFCs, and (4) up-flow mode MFCs (Ou et al. 2016; Wu et al. 2017).

Microbial Fuel Cell Operation

Initially, substrate oxidation occurs inside an anode chamber. This leads to the generation and transportation of electrons and protons (He et al. 2005; Palanisamy et al. 2019). At the same time, through an external circuit, electrons are moved from the anode to the cathode and protons are transported via a polymer electrolyte membrane (Rabaey and Verstraete 2005). In the last step of the process water molecules are produced in the cathode chamber where electrons and protons integrate with oxygen (Sharma and Li 2010). Microorganisms such as *Clostridium, Geobacter, Shewanella,* and *Pseudomonas* act as biocatalysts, oxidizing the substrate and moving electrons to the anode through substrate oxidation thereby generating bioelectricity (Yadav et al. 2017, 2020). Sometimes, microorganisms perform this process without an exogenous electron mediator (Nimje et al. 2012; Zhi et al. 2014). An MFC is shown in Fig. 12.4. Operational conditions in MFCs are associated with pH (He et al. 2006 and Huang et al. 2012) and temperature (Amend and Shock 2001; Logan 2004; Oh et al. 2010; Patil et al. 2011; Tang et al. 2015).

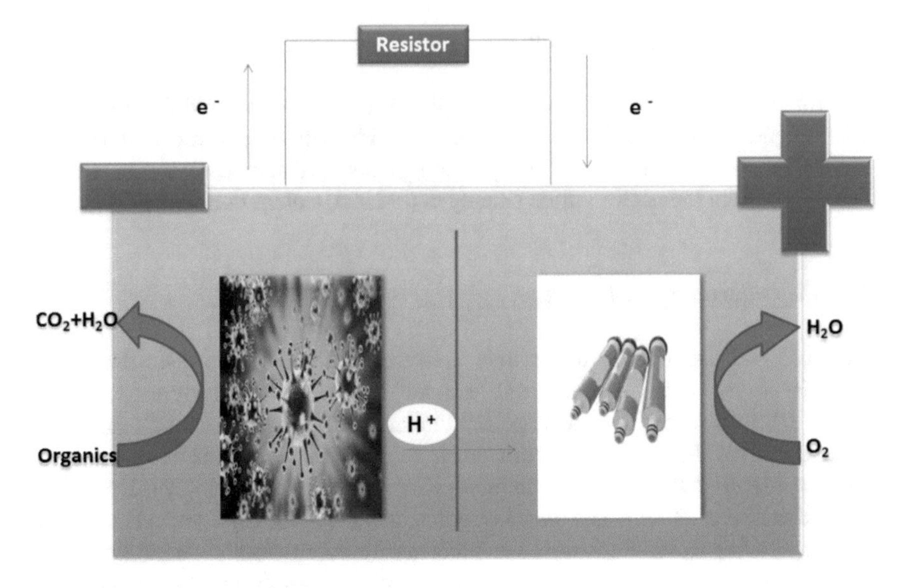

Fig. 12.4 Schematic diagram of a microbial fuel cell

12.3 Potential Applications

The form of bioenergy produced mainly depends on microbial activation (Milano et al. 2016). Bioenergy forms include bioelectricity (Moqsud et al. 2013; El-Chakhtoura et al. 2014; Mekawy et al. 2015; Rahimnejad et al. 2015) and biofuels such as bioethanol (Ballesteros et al. 2002; Najafi et al. 2009a, b; Gelfand et al. 2013; Nitsos et al. 2016, 2017; Achinas and Euverink 2016; Matsakas et al. 2018), biobutanol (Raganati et al. 2012; Jang and Choi 2018), biodiesel, and biohydrogen (Ibrahim 2012; Alavijeh and Yaghmaei 2016).

12.3.1 Bioelectricity

Fermentation processes used to produce bioelectricity (Moqsud et al. 2013) have obtained about 350 mV from MFCs, being significantly influenced by volatile ash, cell tissues, and electrode design. The MFC method is affected by chemical oxygen demand and bioresource loading rate (Jia et al. 2013). Using mixed of organic residues, from paddy or rice, compost and soil the maximum obtained voltage was 700 mV (Moqsud et al. 2015), from stream of wastewater or animal manure the maximum power density were (MFCs 116 mWm^{-2} and 123 mWm^{-2}) respectively (El-Chakhtoura et al. 2014).

12.3.2 *Biofuel*

The merits of any form of bioenergy include a reduction in greenhouse gas emissions compared with fossil fuels, the ease with which large volumes of bioresources are fermented as biofuels, and from the social point of view the generation of employment (Lin and Tanaka 2006; Kour et al. 2019). Wen et al. (2016) reported the generation of about 12 g m^{-2} per day of biomass using 10 L of high-lipid microalgae like *Graesiella sp.* WBG-1, as well as 5.4 g m^{-2} per day of lipid with 15 mol m^{-2} per day irradiation of artificial light at an optimum temperature and level of natural solar radiation. Also, Schnürer (2016) explained that methane production is the important stage in terms of biogas as a biofuel. Microbial growths with other basic treatments mainly affect the amount of energy obtained from methane.

12.4 Sustainable Development

Sustainable bioenergy mainly depends on crop and food residues. Environmental, social, and economic requirements influence the sustainability of bioenergy. Consequently, bioenergy must be carefully managed (Uwe et al. 2006; Srivastava 2019). Sustainable bioenergy fuels such as biodiesel, biogas, bioethanol, and biohydrogen can be generated from different types of biomass, such as plant and food residue, wastewater, and other waste materials, as well as microalgae grown using advanced techniques (Tan et al. 2015). Saxena et al. (2009) reported the likelihood of there being about 220×10^9 Mega-g of available dry biomass globally. Hall and Rosillo-Calle (1998) and Gaurav et al. (2017) calculated available biomass production, with high lignocellulose content, to be about 200×10^9 Mega-g per year, of which only about $8–20 \times 10^9$ Mega-g per year can be converted to energy.

12.4.1 *Bioenergy from Sustainable Residues*

12.4.1.1 Sustainable Bioelectricity

In sustainable bioelectricity systems the preferred source for the anode is any carbon material, like bamboo charcoal. However, the cathode is made from synthesized fiber to ensure its good design and maximize its bioelectrical power generation (Moqsud et al. 2013). Bioelectricity systems utilize food and agricultural wastewater as bioresources (Mekawy et al. 2015). In addition, there are some innovative technologies that can process bio-residues from food and wastewater to produce bioenergy. These technologies include treatment by means of bioelectrochemistry. The effectiveness electrode of anode which can make from the phyla Firmicutes (67%) in electricity generation (El-Chakhtoura et al. 2014), In addition, Khater et al. (2017) found the bio-film and microbial fuel-cell at act as the anode are effectively showed a high coulombic efficiency of about 65%. Anti-clockwise, they practiced the ability utilize of microbial fuel cell "MFCs" as anode or cathode in biosensor. Moqsud et al. (2015) reported the use of plants as MFCs—producing bioelectricity via soil, compost, or some other organic components. Such a system is considered truly green energy.

12.4.1.2 Sustainable Biofuels

The main bioresources used to produce bioenergy are materials that are rich in lignocellulose (Rashid and Altaf 2008). Therefore, Sun et al. (2016), in a trial using cellulosic agricultural plants, found it difficult to produce biogas especially when using raw materials from wheat and rice—which affected the cells of microorganisms.

12.4.2 Bioenergy from Microbial Substrate

12.4.2.1 Sustainable Bioelectricity

Jia et al. (2013) identified that the more durable the MFC the more effective the electrical power production. Such systems use exoelectrogenic species of *Geobacter* along with organic components in their fermentation cycles. Electrons flowing from anode to cathode can be obtained using different species of bacteria such as *Geobacter*, *Bacteroides*, *Clostridium* (Karluval et al. 2015), and *Clostridium cellulolyticum* (Sun et al. 2016). Helder et al. (2010) used the membrane from *S. anglica* as the surface for their plant associated microbial fuel cell (P-MFC)—it generated a maximum power density of about 222 mW m^{-2}.

12.4.2.2 Sustainable Biofuel

Wang et al. (2017) observed that in many studies there are some obstacles facing high efficiency methane production, such as pH or pectin type of bacteria to help activate the fermentation processes where it was found that CH_4 reduced in minimization, about 37.12% at used H group as, Thermovirga, Soehngenia and Actinomyces, to methane generation. Wirth et al. (2012) cleared that to produce the hydrogen as a biofuel the main importance bacteria in metabolism in biogases synthesizing is Closteria.

12.5 Conclusion

When producing bioenergy it should be noted that a sustainable source of biomaterial is essential, whether terrestrial or marine. Environmental, social, and economic aspects must also be considered at all stages of production and utilization.

References

Abbasi T, Tauseef SM, Abbasi SA (2012) Anaerobic digestion for global warming control and energy generation - an overview. Renew Sustain Energy Rev 16:3228–3242

Achinas S, Euverink GJW (2016) Consolidated briefing of biochemical ethanol production from lignocellulosic biomass. Electron Biotechnol 23:44–53

Alavijeh MK, Yaghmaei S (2016) Biochemical production of bioenergy from agricultural crops and residue in Iran. Waste Manag 52:375–394

Amend JP, Shock EL (2001) Energetics of overall metabolic reactions of thermophilic and hyperthermophilic Archaea and Bacteria. FEMS Microbiol Rev 25:175–243

Amer M, Saad A, Ismail NK (2019) Biofuels from microorganisms. In: Srivastava N, Srivastava M, Mishra PK, Upadhyay SN, Ramteke PW, Gupta (eds) "Sustainable approaches for biofuels production technologies" - from current status to practical implementation. biofuel and biorefinery technologies, Spring, 7th edn, pp 93–110

Appels L, Baeyens J, Degrève J, Dewil R (2008) Principles and potential of the anaerobic digestion of waste-activated sludge. Prog Energy Combust Sci 34:755–781

Atabania AE, Silitongaab AS, Onga HC, Mahliac TMI, Masjukia HH, BadruddinaI A et al (2013) Nonedible vegetable oils: a critical evaluation of oil extraction, fatty acid compositions, biodiesel production, characteristics, engine performance and emissions production. Renew Sustain Energy Rev 18:211–45

Azad AK (2017) Biodiesel from mandarin seed oil: a surprising source of alternative fuel. Energies 10:1689

Azad AK, Rasul MG, Khan MMK, Sharma SC (2017) Macadamia biodiesel as a sustainable and alternative transport fuel in Australia. Energy Proc 110:543–548

Balat M, Balata H and Cahide OZ (2008) Progress in bioethanol processing. Prog Energy Combust Sci 34:551–73

Ballesteros I, Oliva JM, Negro MJ, Manzanares P, Ballesteros M (2002) Ethanol production from olive oil extraction residue pretreated with hot water. Appl Biochem Biotechnol 98:717–732

Bhatia SK, Joo HS, Yang HY (2018) Biowaste-to-bioenergy using biological methods – a mini-review. Energy Convers Manag 177:640–660

Bruton T, Lyons H, Lerat Y, Stanley M, Rasmussen MB (2009) A review of the potential of marine algae as a source of biofuel in Ireland. Sustainable energy Ireland report, pp 1–88. https://www.seai.ie/Publications/RenewablesPublications/Bioenergy/Algaereport.pdf/

Chen Y, Ho S, Nagarajan D, Ren N, Chang J (2018) Waste biorefineries - integrating anaerobic digestion and microalgae cultivation for bioenergy production. Curr Opin Biotechnol 50:101–110

Deepanraj B, Sivasubramanian V, Jayaraj S (2014) Biogas generation through anaerobic digestion process-an overview. Res J Chem Environ 18:80–93

El-Chakhtoura J, El-Fadel M, Rao HA, Li D, Ghanimeh S, Saikaly PE (2014) Electricity generation and microbial community structure of air-cathode microbial fuel cells powered with the organic fraction of municipal solid waste and inoculated with different seeds. Biomass Bioenergy 67:24–31

Feng Q, Lin Y (2017) Integrated processes of anaerobic digestion and pyrolysis for higher bioenergy recovery from lignocellulosic biomass: a brief review. Renew Sustain Energy Rev 77:1272–1287

Gaurav N, Sivasankari S, Kiran GS, Ninawe A, Selvin J (2017) Utilization of bioresources for sustainable biofuels: a review. Renew Sustain Energy Rev 73:205–214

Gebrezgabher SA, Meuwissen MPM, Prins BAM, Lansink AGJMO (2010) Economic analysis of anaerobic digestion - a case of Green power biogas plant in The Netherlands. NJAS-Wagen J Life Sci 57:109–115

Gelfand I, Sahajpal R, Zhang X, Izaurralde RC, Gross KL, Robertson GP (2013) Sustainable bioenergy production from marginal lands in the US Midwest. Nature 493(7433):514–517

Gerardi MH (2003) The microbiology of anaerobic digesters. Wiley.

Gerpen JV (2005) Biodiesel processing and production. Fuel Proc Technol 86:1097–1107

Gould MC (2015) Bioenergy and anaerobic digestion. In: Bioenergy, pp 297–317. Academic Press

Hall D, Rosillo-Calle F (1998) The role of bioenergy in developing countries. In: 10th European conference and technology exhibition on biomass energy and industry, pp 52–55

He Y, Caporaso JG, Jiang XT, Sheng HF, Huse SM, Rideout JR et al (2015) Stability of operational taxonomic units: an important but neglected property for analyzing microbial diversity. Microbiome 3:20

He Z, Minteer SD, Angenent LT (2005) Electricity generation from artificial wastewater using an upflow microbial fuel cell. Environ Sci Technol 39:5262–5267

He Z, Wagner N, Minteer SD, Angenent LT (2006) The upflow microbial fuel cell with an interior cathode: assessment of the internal resistance by impedance spectroscopy. Environ Sci Technol 40:5212–5217

Helder M, Strik DPBTB, Hamelers HVM, Kuhn AJ, Blok C, Buisman CJN (2010) Concurrent bioelectricity and biomass production in three Plant-Microbial Fuel Cells using *Spartinaanglica, Arundinellaanomala* and *Arundodonax*. Bioresour Technol 101:3541–3547

Huang L, Chai X, Quan X, Logan BE, Chen G (2012) Reductive dechlorination and mineralization of pentachlorophenol in biocathode microbial fuel cells. Bioresour Technol 111:167–174

Ibrahim HAH (2012) Pretreatment of straw for bioethanol production. Energy Proc 14:542–551

Jang MO, Choi G (2018) Techno-economic analysis of butanol production from lignocellulosic biomass by concentrated acid pretreatment and hydrolysis plus continuous fermentation. Biochem Eng J 134: 30–43

Jia J, Tang Y, Liu B, Wu D, Ren N, Xing D (2013) Electricity generation from food wastes and microbial community structure in microbial fuel cells. Bioresour Technol 144:94–99

Kainthol J, Kalamdhad AS, Goud VV (2019) A review on enhanced biogas production from anaerobic digestion of lignocellulosic biomass by different enhancement techniques. Proc Biochem 84:81–90

Karluval A, Köroğlu EO, Manav N, Çetinkaya AY, Özkaya B (2015) Electricity generation from organic fraction of municipal solid wastes in tubular microbial fuel cell. Sep Purif Technol 156:502–511

Khalid A, Arshad M, Anjum M, Mahmood T, Dawson L (2011) The anaerobic digestion of solid organic waste. Waste Manag 31:1737–1744

Khater DZ, El-Khatib KM, Hassan HM (2017) Microbial diversity structure in acetate single chamber microbial fuel cell for electricity generation. J Genet Eng Biotechnol 15:127–137

Kothari R, Pandey AK, Kumar S, Tyagi VV, Tyagi SK (2014) Different aspects of dry anaerobic digestion for bio-energy: an overview. Renew Sustain Energy Rev 39:174–195

Kour D, Rana KL, Yadav N, Yadav AN, Rastegari AA, Singh C et al. (2019) Technologies for biofuel production: current development, challenges, and future prospects. In: Rastegari AA, Yadav AN, Gupta A (eds) Prospects of renewable bioprocessing in future energy systems, pp 1–50. Springer International Publishing, Cham. https://doi.org/10.1007/978-3-030-14463-0_1

Krishna D, Kalamdhad AS (2014) Pre-treatment and anaerobic digestion of food waste for high rate methane production - a review. J Environ Chem Eng 2:1821–1830

Kumar K, Yadav AN, Kumar V, Vyas P, Dhaliwal HS (2017) Food waste: a potential bioresource for extraction of nutraceuticals and bioactive compounds. Bioresour Bioprocess 4:18. https://doi.org/10.1186/s40643-017-0148-6

Kumar S, Sharma S, Thakur S, Mishra T, Negi P, Mishra S et al (2019) Bioprospecting of microbes for biohydrogen production: current status and future challenges. In: Molina G, Gupta VK, Singh BN, Gathergood N (eds) Bioprocessing for biomolecules production. Wiley, USA, pp 443–471

Leung DYC, Guo Y (2006) Transesterification of neat and used frying oil: Optimization for biodiesel production. Fuel Process Tech 87:883–890

Leung DYC, Wang J (2016) An overview on biogas generation from anaerobic digestion of food waste. Int J Green Energy 13:119–131

Li L, He Q, Ma Y, WangX PX (2015) Dynamics of microbial community in a mesophilic anaerobic digester treating food waste: relationship between community structure and process stability. Bioresour Technol 189:113–120

Li Y, Chen Y, Wu J (2019) Enhancement of methane production in anaerobic digestion process: a review . Appl Energy 240:120–137

Lin Y, Tanaka S (2006) Ethanol fermentation from biomass resources: current state and prospects. Appl Microbiol Biotechnol 69:627–642

Logan BE (2004) Extracting hydrogen and electricity from renewable resources. Environ Sci Technol 38:160–167

Logan BE, Aelterman P, Hamelers B, Rozendal R, Schrödeder U, Keller J et al (2006) Microbial fuel cells: methodology and technology. Environ Sci Technol 40:5181–5192

Lotero E, Liu Y, Lopez DE, Suwannakarn K, Bruce DA, Goodwin JG (2005) Synthesis of biodiesel via acid catalysis. Ind Eng Chem Res 44:5353–5363

Mao C, Feng Y, Wang X, Ren G (2015) Review on research achievements of biogas from anaerobic digestion. Renew Sustain Energy Rev 45:540–555

Mathias JFCM (2014) Manure as a resource: livestock waste management from anaerobic digestion, opportunities and challenges for Brazil. Int Agribus Manag Rev 7:87–109

Matsakas L, Nitsos C, Raghavendran V, Yakimenko O, Persson G, Olsson E et al (2018) A novel hybrid organosolv: steam explosion method for the efficient fractionation and pretreatment of birch biomass. Biotechnol Biofuels 11:160

Mekawy AE, Srikanth S, Bajracharya S, Hegab HM, Nigam PS, Singh A et al (2015) Food and agricultural wastes as substrates for bioelectrochemical system (BES): the synchronized recovery of sustainable energy and waste treatment. Food Res Int 73:213–225

Milano J, Ong HC, Masjuki HH, Chong WT, Lam MK, Loh PK, Vellayan V (2016) Microalgae biofuels as an alternative to fossil fuel for power generation. Renew Sustain Energy Rev 58:180–197

Momayez F, Karimi K, Taherzadeh MJ (2019) Energy recovery from industrial crop wastes by dry anaerobic digestion: a review. Ind Crop Prod 129:673–687

Monnet F (2003) An introduction to anaerobic digestion of organic wastes. Remade Scotland report, pp 1–48

Moqsud MA, Omine K, Yasufuku N, Hyodo M, Nakata Y (2013) Microbial fuel cell (MFC) for bioelectricity generation from organic wastes. Waste Manag 33:2465–2469

Moqsud MA, Yoshitake J, Bushra QS, Hyodo M, Omine K, Strik D (2015) Compost in plant microbial fuel cell for bioelectricity generation. Waste Manag 36:63–69

Najafi G, Ghobadian B, Tavakoli T, Yusaf T (2009a) Potential of bioethanol production from agricultural wastes in Iran. Renew Sustain Energy Rev 13:1418–1427

Najafi G, Ghobadian B, Tavakoli T, Yusaf T (2009b) Potential of bioethanol production from agricultural wastes in Iran. Renew Sustain Energy Rev 13:1418–1427

Nimje VR, Chen CY, Chen HR, Chen CC, Huang YM, Tseng MJ et al (2012) Comparative bioelectricity production from various wastewaters in microbial fuel cells using mixed cultures and a pure strain of *Shewanella oneidensis*. BioresourceTechnol 104:315–323

Nitsos CK, Choli-Papadopoulou T, Matis KA, Triantafyllidis KS (2016) Optimization of hydrothermal pretreatment of hardwood and softwood lignocellulosic residues for selective hemicellulose recovery and improved cellulose enzymatic hydrolysis. ACS Sustain Chem Eng 4:4529–4544

Nitsos C, Matsakas L, Triantafyllidis K, Rova U, Christakopoulos P (2017) Investigation of different pretreatment methods of Mediterranean-type ecosystem agricultural residues: characterization of pretreatment products, high-solids enzymatic hydrolysis and bioethanol production. Biofuels 65:1–14

Oh S, Kim J, Premier G, Lee T, Changwon K, Sloan W (2010) Sustainable wastewater treatment: how might microbial fuel cells contribute. Biotechnol Adv 28:871–881

Ostrem K (2004) Greening waste: anaerobic digestion for treating the organic fraction of municipal solid waste. Department of Earth and Environmental Engineering, Fu Foundation School of Engineering and Applied Science, Columbia University, pp 1–59

Ou S, Zhao Y, Aaron DS, Regan JM, Mench MM (2016) Modeling and validation of single-chamber microbial fuel cell cathode biofilm growth and response to oxidant gas composition. J Power Sources 328:385–396

Palanisamy G, Jung HY, Sadhasivam T, Kurkuri MD, Kim SC, Roh SH (2019) A comprehensive review on microbial fuel cell technologies: processes, utilization, and advanced developments in electrodes and membranes. J Clean Prod 221:598–662

Panjiar N, Mishra S, Yadav AN, Verma P (2017) Functional foods from cyanobacteria: an emerging source for functional food products of pharmaceutical importance. In: Gupta VK, Treichel H, Shapaval VO, Oliveira LAd, Tuohy MG (eds) Microbial functional foods and nutraceuticals, pp 21–37. Wiley, USA. https://doi.org/10.1002/9781119048961.ch2

Patil SA, Harnisch F, Koch C, Hubschmann T, Fetzer I, Carmona-Martinez AA et al (2011) Electroactive mixed culture derived biofilms in microbial bioelectrochemical systems: the role of pH on biofilm formation, performance and composition. Bioresour Technol 102:9683–9690

Pramanik SK, Suja FB, Zain SM, Pramanik BK (2019) The anaerobic digestion process of biogas production from food waste: prospects and constraints. Bioresour Technol Rep 8:100–310

Rabaey K, Verstraete W (2005) Microbial fuel cells: novel biotechnology for energy generation. Trends Biotechnol 23:291–298

Raganati F, Curth S, Götz P, Olivieri G, Marzocchella A (2012) Butanol production from Lignocellulosic-based Hexoses and Pentoses by Fermentation of *Clostridium acetobutylicum.* Chem Eng Trans 27:91–96

Rahimnejad M, Adhamia A, Darvari S, Zirepour A, Oh SE (2015) Microbial fuel cell as new technology for bioelectricity generation: a review. Alex Eng J 4:745–756

Rashid MT, Altaf Z (2008) Potential and environmental concerns of ethanol production from sugarcane molasses in Pakistan. Nat Preced 1499:1–13

Rasool U, Hemalatha S (2016) A review on bioenergy and biofuels: sources and their production. Braz J Biol Sci 3:3–21. ISSN 2358-2731

Rastegari AA, Yadav AN, Gupta A (2019) Prospects of renewable bioprocessing in future energy systems. Springer International Publishing, Cham

Rastegari AA, Yadav AN, Yadav N (2020) New and future developments in microbial biotechnology and bioengineering: Trends of microbial biotechnology for sustainable agriculture and biomedicine systems: diversity and functional perspectives. Elsevier, Amsterdam

Ravindran R, Jaiswal AK (2016) Exploitation of food industry waste for high-value products. Review, Trends Biotechnol 34:58–69

Rinaldi A, Mecheri B, Garavaglia V, Licoccia S, Nardo PD, Traversa E (2008) Engineering materials and biology to boost performance of microbial fuel cells: a critical review. Energy Environ Sci 1:417–429

Romano SD, González SE, Laborde MA (2006) Biodiesel. In: Combustibles alternativos, 2nd edn. Ediciones Cooperativas, Buenos Aires

Santoro C, Arbizzani C, Erable B, Ieropoulos I (2017) Microbial fuel cells: from fundamentals to applications a review. J Power Sources 356:225–244

Saxena RC, Adhikari DK, Goyal HB (2009) Biomass-based energy fuel through biochemical routes: a review. Renew Sustain Energy Rev 13:167–178

Schnürer A (2016) Biogas production: microbiology and technology. Adv Biochem Eng Biotechnol 156: 195–234

Sharma Y, Li B (2010) The variation of power generation with organic substrates in singlechamber microbial fuel cells (SCMFCs). Bioresour Technol 101:1844–1850
Shi X, Guo X, Zuo J, Wang Y, Zhang M (2018) A comparative study of thermophilic and mesophilic anaerobic co-digestion of food waste and wheat straw: process stability and microbial community structure shifts. Waste Manag 75:261–269
Singh A, He B, Thompson J, van Gerpen J (2006) Process optimization of biodiesel production using different alkaline catalysts. Appl Eng Agric 22:597–600
Srivastava RK (2019) Bio-energy production by contribution of effective and suitable microbial system. Mat Sci Energy Technol 2:308–318
Sun L, Liu T, Müller B, Schnürer A (2016) The microbial community structure in industrial biogas plants influences the degradation rate of straw and cellulose in batch tests. Biotechnol Biofuels 9:128
Tan CH, Show PL, Chang JS, Ling TC, Lan JCW (2015) Novel approaches of producing bioenergies from microalgae: a recent review. Biotechnol Adv 33:1219–1227
Tang J, Chen S, Yuan Y, Cai X, Zhou S (2015) In situ formation of graphene layers on graphite surfaces for efficient anodes of microbial fuel cells. Biosens Bioelectron 71:387–395
Timonen K, Sinkko T, Luostarinen S, Tampio E, Joensuu K (2019) LCA of anaerobic digestion: emission allocation for energy and digestate. J. Clean Prod 235:1567–1579
Uwe RF, Katja H, Andreas H, Falk SW (2006) Sustainability standards for bioenergy. WWF Germany, Frankfurt am Main
Wang P, Wang H, Qiu Y, Ren L, Jiang B (2018) Microbial characteristics in anaerobic digestion process of food waste for methane production-a review. Bioresour Technol 248:29–36
Wang S, Hou X, Su H (2017) Exploration of the relationship between biogas production and microbial community under high salinity conditions. Sci Rep 7:1149
Wen X, Du K, Wang Z, Peng X, Luo L, Tao H et al (2016) Effective cultivation of microalgae for biofuel production: a pilot-scale evaluation of a novel oleaginous microalga *Graesiella*sp. WBG-1, Biotechnol Biofuels 9:123
Wirth R, Kovács E, Maráti G, Bagi Z, Rákhely G, Kovács KL (2012) Characterization of a biogas-producing microbial community by short-read next generation DNA sequencing. Biotechnol Biofuels 5:41
Wu LC, Tsai TH, Liu MH, Kuo JL, Chang YC, Chung YC (2017) A Green microbial fuel cell-based biosensor for in situ chromium (VI) measurement in electroplating wastewater. Sensors 17:2461
Xu Z, Zhao M, Miao H, Huang Z, Gao S, Ruan W (2014) In situ volatile fatty acidsinfluence biogas generation from kitchen wastes by anaerobic digestion. Bioresour Technol 163:186–192
Yadav AN, Kumar R, Kumar S, Kumar V, Sugitha T, Singh B et al (2017) Beneficial microbiomes: biodiversity and potential biotechnological applications for sustainable agriculture and human health. J Appl Biol Biotechnol 5:45–57
Yadav AN, Rastegari AA, Yadav N (2020) Microbiomes of extreme environments: biodiversity and biotechnological applications. CRC Press, Taylor & Francis, Boca Raton, USA
Yadav AN, Singh S, Mishra S, Gupta A (2019) Recent advancement in white biotechnology through fungi. Perspective for value-added products and environments, vol 2. Springer International Publishing, Cham
Yadvika S, Sreekrishnan TR, Kohli S, Rana V (2004) Enhancement of biogas production from solid substrates using different techniques - a review. Bioresour Technol 95:1–10
Yi J, Dong B, Jin J, Dai X (2014) Effect of increasing total solids contents on anaerobic digestion of food waste under mesophilic conditions: performance and microbial characteristics analysis. PLoS ONE 9(7):e102548
Zhang A, Shen J, Ni Y (2015) Anaerobic digestion for use in the pulp and paper industry and other sectors: an introductory mini-review. BioResources 10:8750–8769
Zhang C, Su H, Baeyens J, Tan T (2014) Reviewing the anaerobic digestion of food waste for biogas production. Renew Sustain Energy Rev 38:383–392

Zhang Y, Dube MA, McLean DD, Kates M (2003) Biodiesel production from waste cooking oil: 2 economic assessment and sensitivity analysis. Bioresour Technol 90:229–240
Zhi W, Ge Z, He Z, Zhang H (2014) Methods for understanding microbial community structures and functions in microbial fuel cells: a review. Bioresource Technol 171:461–468

Chapter 13
Organic Waste for Biofuel Production: Energy Conversion Pathways and Applications

Vinayak Vandan Pathak, Meena Kapahi, Roopa Rani, Jaya Tuteja, Sangita Banga, and Versha Pandey

Abstract Global energy supply is predominantly dependent on fossil fuels, which are not only limited in availability but also harm the environment. Fossil fuel-based carbon dioxide generation has been identified as one of the main causes of the global warming phenomenon, which leads to various adverse effects. In recent years, both developed and developing countries have enhanced the share of renewable energy in overall energy scenario. Renewable energy sources are considered as a potential, reliable and environment-friendly way to substitute fossil fuels. Biomass energy has long been used for cooking and heating application; however, traditional biomass application has several drawbacks such as low energy efficiency and emission of harmful gases. Therefore, the transformation of biomass through proper technology is crucial for the development of sustainable and environmentally safe energy resources. Globally, biomass energy shares around 56.2 EJ out of the total energy supply of 560 EJ. The estimated bioenergy potential of India is around 18,000 MW. If this estimated amount of energy is achieved, the country will get rid of the energy crisis problem. This chapter provides an overview of the significance of bioenergy at the national and global levels with possible energy conversion technologies.

13.1 Introduction

Global primary energy consumption has been on its rise at an alarming rate. In 2018, it became double as compared to its last 10 years of average despite moderate global GDPgrowth (World Energy Council 2019). Industrial sectors consume about half of the global energy and feedstock fuels, while transport and residential sectors consume 29% and 21%, respectively. Globally, the fossil fuels are still playing a dominant role

---

V. V. Pathak · M. Kapahi (✉) · R. Rani · J. Tuteja · S. Banga
Department of Chemistry, Manav Rachna University, Faridabad, Haryana, India
e-mail: meenakapahi@mru.edu.in

V. Pandey
Agronomy and Soil Science, Central Institute of Medicinal and Aromatic Plants, Lucknow, Uttar Pradesh, India

A. N. Yadav et al. (eds.), *Biofuels Production – Sustainability and Advances in Microbial Bioresources*, Biofuel and Biorefinery Technologies 11,
https://doi.org/10.1007/978-3-030-53933-7_13

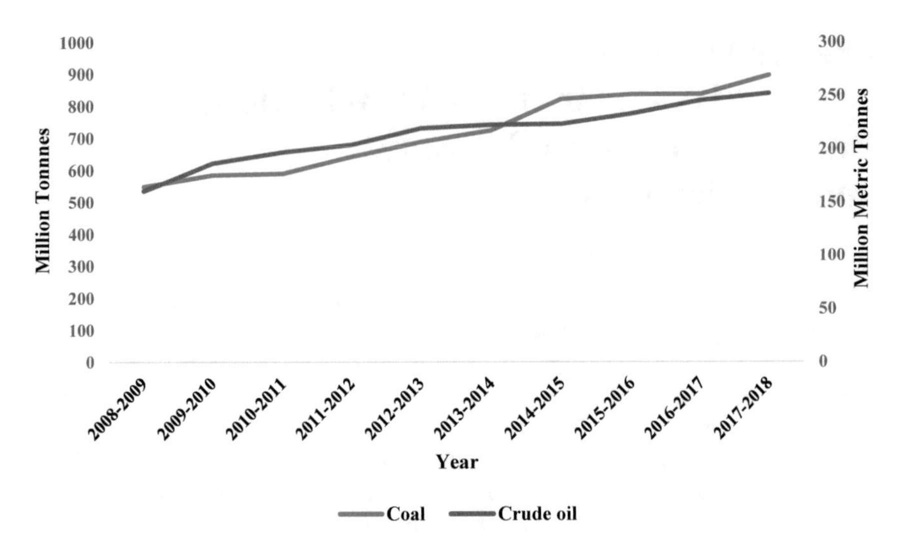

Fig. 13.1 Trends in consumption of coal and crude oil in India

as the primary source of energy consumption (80%), while consumption of non-fossil energy resources reports for only a lesser share (Chen et al. 2019). Out of the total contribution from fossil fuels, the transport sector demands 58% (Escobar et al. 2009).

World's major developing countries like India reflect consistent consumption of fossil fuels to meet the demand for electricity generation. The estimated compound annual growth rate (CAGR) for consumption of coal and crude oil in India remained 5.01% and 4.59%, during the period 2008–09 to 2017–18, respectively (Energy Statistics 2019). Figure 13.1 indicates a steady increase in coal and crude oil consumption during this period (Energy Statistics 2019). Global consumption of natural gas is growing much faster than other fossil fuels. In this context, natural gas consumption has increased by 2.6% per year from 2013 to 2018. In India, import of natural gas has risen from 8.06 billion cubic metres (BCM) in 2008–09 to 19.87 BCM in 2017–18, with a CAGR of 9.44%. Figure 13.2 indicates the consumption of energy resources in India in terms of petajoules which indicates the highest share of coal followed by crude oil, electricity, natural gas and lignite (Energy Statistics 2019).

Emission of harmful gases and limited availability of fossil fuels are the key challenges of the global energy sector. Combustion of fossil fuels emits harmful gases like oxides of nitrogen, sulphur, carbon and particulate matter, resulting in climate change and consequent biodiversity loss and sea-level rise (Agrawal 2007). The global average temperature continues to rise due to higher than ever carbon dioxide emissions from various sources. Global fossil fuel-based CO_2 emission have grown 3 years consecutively, i.e. +1.5% in 2017, +2.1% in 2018 and +0.6% in 2019. However, the European Union and the United States have declined their CO_2

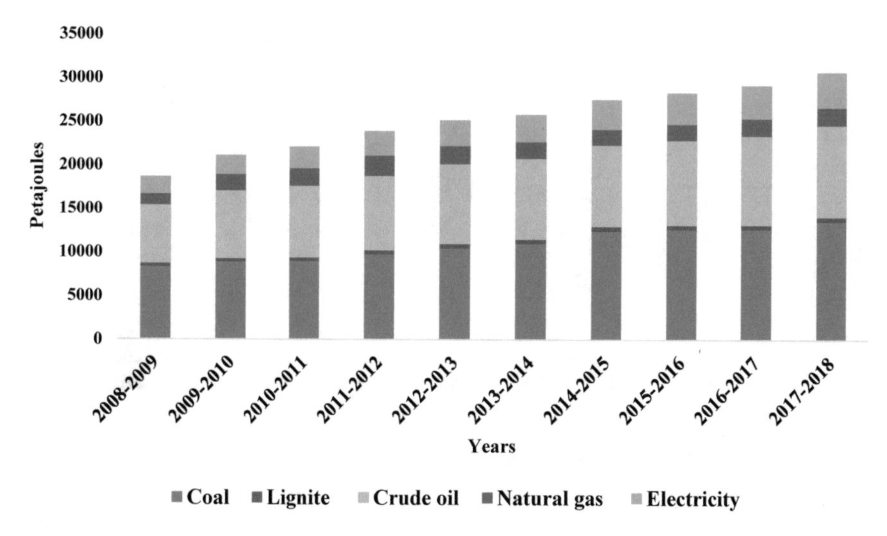

Fig. 13.2 Consumption of energy sources in India

emission. On the other hand, CO_2 emission continues to increase in India with a rate of 1.8% in the year 2019 (Jackson et al. 2019) as shown in Figs. 13.3 and 13.4.

Energy demand has escalated with an increase in industrialization, transportation and living standards. To meet this growing energy demand, environmentally safe and sustainable energy sources are being sought. In this regard, researchers are working on alternative energy sources like solar energy, wind energy, small hydropower, geothermal energy and biomass energy. Because of the high and volatile fossil fuel prices, national energy security and negative environmental impacts, biomass-based energy generation is considered an alternative solution to resolve these challenges. Biomass is any organic material derived from tree, plant (including crop), organic waste such as animal dung, crop residues and municipal solid wastes, and can be used

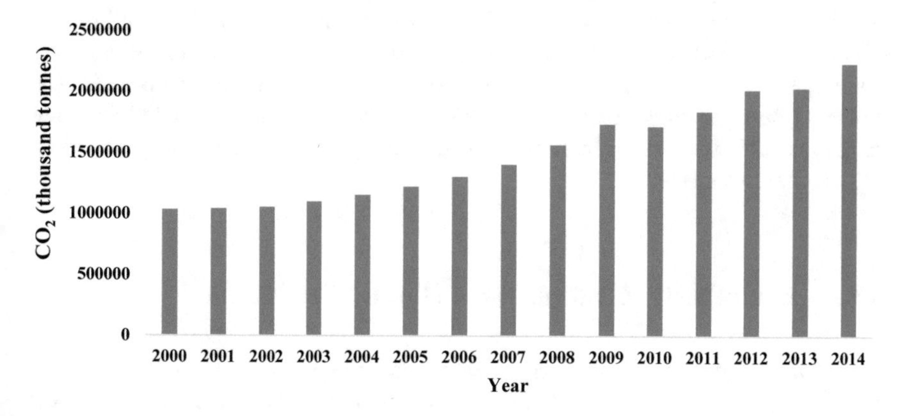

Fig. 13.3 CO_2 emissions per year

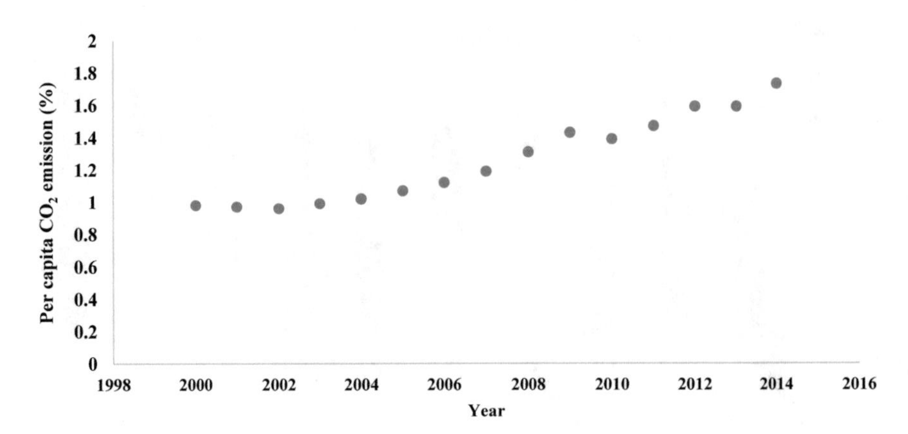

Fig. 13.4 Trend of CO_2 emissions in India

as energy sources, nutrient supplements and other industrial applications. Biomass has been reported to be the fourth largest energy resource available globally (Acma and Yaman 2010). It provides a natural and inexpensive storage device for energy that can be utilized at any time (Arthe et al. 2008) and can be converted into desired energy products as a potential feedstock employing various conversion routes such as combustion, gasification, thermochemical and biochemical.

Researchers are continuously working for the development of biofuels using sustainable sources as it is considered an effective alternative to fuel from non-renewable sources (Weldemichael and Assefa 2016). Annual biomass availability in India is estimated at around 500 million metric ones, which involves surplus biomass around 120–150 million metric tonnes per annum. The potential of biomass-based power generation estimated in India is around 18,000 MW (Gaurav et al. 2017).

Biomass-based energy generation mainly relies on terrestrial biomass which has a constraint of land availability for biomass cultivation. Of lately, biomass from the aquatic ecosystems is emerging as a new source for biofuel feedstock. Aquatic species such as algal biomass show higher productivity than the terrestrial planets (Gaurav et al. 2017). Waste to energy generation is another sustainable approach for bioenergy production, which is being explored by various researchers and stakeholders (Rastegari et al. 2019a). The present chapter provides a detailed overview of the available sources for biofuel production and the current technologies used for the bioenergy conversion.

13.2 Availability of Biomass for Biofuel Production

Land availability to cultivate biomass, its productivity and the cost incurred are the major issues in countries like India holding the second largest population in the world. More than 77% of domestic consumption in rural areas in the country has been reported to be dependent on wood. Biofuels govern the rural energy sector and are responsible for 80% of energy consumption. Fuelwood, an important source of combustion fuel, holds for 54% of biomass fuel in India. Some of the other biomass-based fuels being used in India include agricultural residues (rice husk and bagasse) and animal dung. Fuel head loading is the largest source of employment employing about 2–3 million people in the Indian energy sector. The major sources of fuelwood are the tree growing farmlands and common lands. Forestry, occupying 22% of the land, becomes the second largest land use in India after agriculture. Approximately, 275 million poor rural people in India earn their livelihoods from fuelwood, fodder and various Non-Timber Forest Products (NTFP) like medicinal plants, fruits and flowers. Due to paucity of fuelwood in rural areas, timber residues are usually sold freely in India. In Jammu and Kashmir, there is plenty of biomass potential for bioenergy production due to the presence of enormous agricultural and forest resources.

The agricultural residues contribute around 480 metric tonnes, while the residues from food grains account for approximately 100 metric tonnes. Residue production largely depends on the type of crop such as straw (a low-density residue) which is the dominant residue. The straw to grain ratio of the cereals differs from crop to crop. Rice husk, rice-milling by-product, forms 20% of the paddy crop. The crops like cotton, red gram, mustard, mulberry and plantation crops yield woody residues. The total crop residue production in India during 1996–97 is estimated to be 626 Mt of air-dry weight (Rashad 2013; Devi et al. 2017). The dominant residues are those of rice, wheat, sugarcane and cotton accounting for 66% of the total residue production. Sugarcane and cotton residue productions are 110 and 50 Mt, respectively (Ravindra 2005).

Various issues need to be taken into account while evaluating the biomass production potential from various sources, e.g. those associated with land (availability and cost, ownership status, biomass production techniques) and the sustainable plantation forestry strategies, to name a few.

13.3 Biomass to Energy Conversion Pathways

Bioenergy conversion is one of the major steps in biofuel production that depends on the composition of biomass feedstock. Direct application of biomass for heat generation has been practised for a long time; however, it was found inefficient and unsafe to the environment due to harmful emissions. Biomass to bioenergy generation can be achieved by following three major conversion pathways such as physicochemical, thermochemical and biochemical (Fig. 13.5).

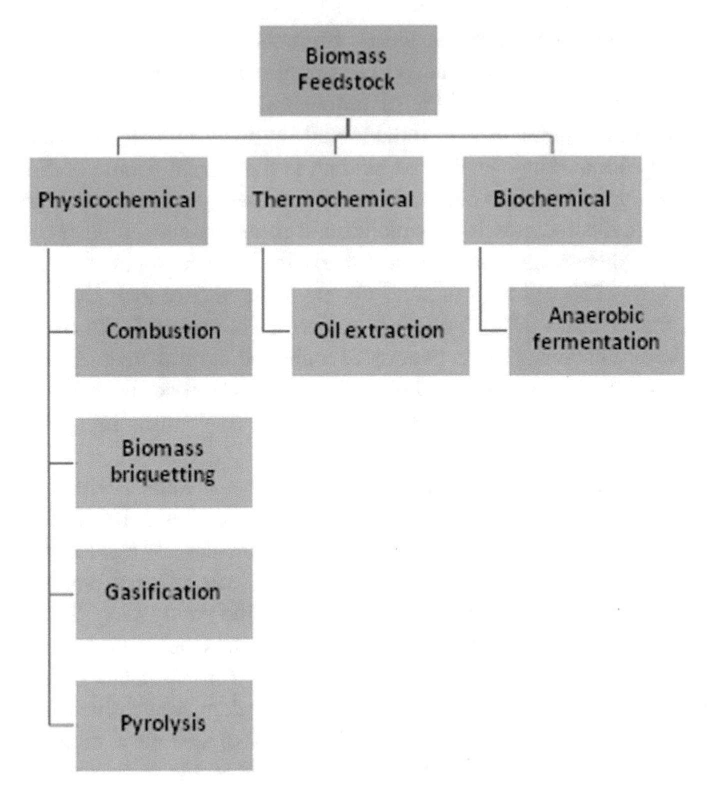

Fig. 13.5 Biomass to energy conversion pathways (Singh et al. 2014)

13.3.1 Direct Combustion

It is the most commonly used technology for conversion of biomass to heat. The combustible fraction of biomass can be converted into heat by the direct combustion process. Further, the produced can be converted into electricity generation. Biomass quality varies with its sources which causes various challenges in the combustion process. Generally, biomass has higher moisture and oxygen, lesser carbon and lower heating value in comparison to coal affecting the efficiency of the combustion process. The blending of biomass rich in sulphur with biomass containing high chorine and potassium resolves the problem of fouling and corrosion in the combustion device (Hupa et al. 2017). Operational problems such as unwanted emissions and ash-induced corrosion are caused by minor constituents of the biomass which can be controlled by an understanding of the chemistry of minor constituents of biomass during the combustion biomass.

13.3.2 Biomass Gasification

Biomass gasification is also known as "Creating valuable gases through environment-friendly manner". The process is a thermochemical process which involves the oxidation of renewable biomass into valuable gases such as CO, H_2, CH_4, CO_2, C_2H_6, C_3H_8 and other hydrocarbons as tars in the presence of gasifying agents (Ruiz et al. 2013). The factors which determine the quality of produced gas include quality of feedstock used (the type of biomass), reactor design (moving or fixed bed gasifier), gasifying agent (air, oxygen, steam, CO_2 or mixture of these), the presence or absence of a catalyst and other operational conditions of the reactor (Parthasarathy and Narayanan 2014).

The first report on electricity production from gasification was reported in 1792, followed by the installation of the first gasifier unit in 1861 by Siemens. After almost 60 years, a revolutionary change has occurred with the development of fluidized bed gasifier in 1926 as the first commercial coal gasification plant in the USA (Somayeh et al. 2016). As we are progressing towards globalization, we have observed that our demands for energy have increased a lot while the sources of energy are very limited. It requires hundreds of years to produce fossil fuels, and their consumption rate is way higher than their production. There is a strong need to balance the production and consumption rate of energy sources to meet future demands with the help of renewable energy sources. Biomass is the only renewable source which has C energy stored in the form of carbohydrates, as it converts environmental CO_2 to carbohydrates via photosynthesis (Tuteja et al. 2012). Biomass includes agriculture and forestry residues, municipality waste, woody biomass and biological materials (Tuteja et al. 2014).

Biomass gasification is one of the best ways to produce bioenergy from biomass via thermochemical or biochemical conversion processes. Thermochemical conversion of biomass has attained much attention in the twenty-first century, as it can be utilized to produce syngas, electricity, heat, hydrogen, light hydrocarbons and methanol as a replacement of fossil fuel sources (Lapuerta et al. 2008). The biggest challenge which scientists are facing is the production of tars and high cost associated with cleaning of products from tar. Many improvements in the gasification procedure have been done by various researchers to maximize product formation and minimize the tar formation.

13.3.2.1 Gasification Technologies

Biomass gasification involves a combination of four major procedures that are oxidation, drying, pyrolysis and reduction (Kumar et al. 2009). It is a method to valorize the waste biomass to important feedstock such as CH_4, CO, H_2 and petrochemicals. The conventional route for gasification technologies uses fixed bed, fluidized bed and flow reactors. In the latest technology, the system has been developed with plasma gasification and gasification using supercritical water to utilize a variety of biomass to produce gas. The supercritical water shows the properties of liquids as well as gases to maximize the product (Heidenreich and Foscolo 2015; Sikarwar et al. 2016).

The gasification process is endothermic and requires an amount of energy coming from the oxidation maintaining the required temperature (Ahmad et al. 2016). Oxidation is generally carried out in anaerobic conditions as the biomass contains high oxygen content. If we observe the structure of petrochemical products, they are mostly hydrocarbons or contain fewer amounts of oxygen species (Petrus and Noordermeer 2006). Thus, the procedure requires the removal of oxygen from the carbohydrates obtained from biomass to produce such products which further can be transformed into hydrocarbons by catalysis (Fischer–Tropsch process) (Whitty et al. 2008). The major disadvantage of gasification is the high cost associated with cleaning of products from tar. Therefore, the current technologies involve the integration of gasification and cleaning products from the tar.

The new emerging technologies are (i) multi-stage gasification step that has the advantages of improved efficiency and high-quality syngas; (ii) plasma gasification with the benefit of decomposition of any organic matter and treatment of hazardous waste but suffers from the disadvantage of high cost, low efficiency and high-power requirement; (iii) supercritical water gasification where liquid and high moisture biomass are treated directly without any pretreatment but this process also requires high cost and power; and (iv) Fischer–Tropsch process coupled with gasification with the advantage of the production of clean and carbon–neutral liquid biofuels with a disadvantage of complexity in process design. Many other technologies have been discovered along with above mentioned like a combination of gasification and gas cleaner in one reactor, distributed pyrolysis plants with central gasification plant, co-generation of thermal energy with power (Somayeh et al. 2016).

13.3.2.2 Type of Gasifiers

The classification of gasifiers has been done on the basis of their bed and flow. Based on the type of bed, the gasifiers are of two types: fixed bed and fluidized bed gasifiers. Fixed bed is further categorized into two types on the basis of flow, i.e. updraft and downdraft fixed bed gasifiers. In updraft gasifier, the feed material is introduced from top, and air known as gasifying agents enters from below and flows upwards. The highest temperature is at the bottom where combustion takes place, while the produced product gas exits from the top which is a lower temperature region containing a high amount of tar. In the downdraft gasifier, the feed or biomass material is introduced top to bottom along with the flow of gasifying agents. Both feed material and air travel to bottom to combustion zone, and product gas exits from the bottom (the highest temperature region). In downdraft gasifier, the tar production is low. Due to combustion in high temperature, the tar is broken down and hence, cleaner gases are obtained in the product.

Though both updraft and downdraft gasifiers have been used extensively in the past, they have their advantages and disadvantages. The updraft gasifier is used because of its simple design, the flexibility of using high moisture content and a large amount of feed. The downdraft gasifier has the advantage of producing cleaner gas, which can be used in gas engines and production of low tar.

Fluidized bed gasifiers product gas exits from the top which includes air/steam N_2 and the reactant of feed is introduced at the bottom (Lv et al. 2004). The advantage of fluidized bed over fixed bed gasifier is the ease in heat transfer to each particle, high reaction rate, uniform temperature and hence high conversions (Narváez et al. 1996). Fluidized bed gasifier with the use of catalysts represents many advanced techniques to improve reaction rate and high product gas (producer gas). Considering all the advantages of fluidized bed gasifier, it represents a good technique for producing the producer gas. The disadvantage being that the biomass with particle size 0.1–1 cm only can be introduced.

Thus, it can be concluded that biomass gasification is a potential technique to produce valuable products such as heat, electricity and power (fuel precursors, syngas). The products obtained are dependent on applied technology, temperature, pressure and gasifying agents (Maschio et al. 1994). Thus, there is a flexibility of altering the parameters as per the requirements to obtain the best results.

13.3.2.3 Biomass Densification

Direct application of biomass for energy generation is associated with various challenges such as agglomeration due to low ash melting point, fouling, hazardous emissions, slagging, and corrosion. Most of these problems are due to low particle density, a large proportion of unburnt content and high volatile matter (Balatinecz 1983). Researchers (Chen et al. 2009; Werther et al. 2000) have designed suitable chambers and furnaces for the efficient and controlled combustion of biomass; however, such process is uneconomic on a large scale. Therefore, upgradation of biomass through the densification process is considered to be the most economic and sustainable process for the efficient combustion process. Biomass densification process can be achieved through mechanical or thermochemical densification which provides several benefits such as improvement in the rate of combustion of biomass (almost equivalent to coal), reduced particulate emission, uniform combustion, easier transportation and storage of biomass. Biomass densification is defined as any process which causes lower physical density and higher energy density in biomass. Generally, after the densification process, the bulk density of biomass ranges from 40 to 800 kgm^3- and biomass appears in the form of a pellet, briquette, cubes, bales, pucks and wood chips.

Mechanical densification (Table 13.1) of biomass is achieved through exerting pressure mechanically on biomass while in thermochemical densification, biomass is heated in the absence of oxygen. Selection of densification method mainly depends on the type of residue and availability of technology at the local level (Nalladurai and Vance Morey 2009).

Table 13.1 Common products of mechanical densification (Nalladurai and Vance Morey 2009)

Mechanical densification
Bales:
• Compressed chopped biomass • Produced by using a farm machinery known as Baler • Low production cost
Pellets:
• Most commonly used product of mechanical densification • Highest biomass density among all products of mechanical densification • Pellets are uniform cylindrical shape fuel with a length smaller than 38 mm and a diameter of around 7 mm
Cubes:
• Cube size ranges from 13 to 38 mm • Chopped biomass is produced under high pressure exerted through heavy press wheel
Briquettes:
• These are produced using a piston press system • The product diameter is around 25 mm or greater • Screw extrusion process is also used alternatively which produces efficient biomass briquettes that contain higher storability and energy density
Pucks:
• It is a disc-shaped biomass fuel with a 75 mm of diameter • Techniques for production of pucks are similar to that of the briquettes • It has similar density with pellets with lower production cost
Wood chips:
• These are small pieces of woods produced by shredder • Wood chips are consumed in various operations such as household application to power plant industries • Wood chips used for boiler range from 5 to 50 mm

13.3.3 Pyrolysis

Pyrolysis is defined as the thermal decomposition of biomass in the absence of oxygen or with a limited supply of oxygen, which results in a combination of products such as hydrocarbon-rich gas mixture, bio-oil and carbon-rich solid residues (Demirbas2004). Based on the rate of heating and residence time, biomass pyrolysis technology is classified as slow, fast and flash pyrolysis. Slow pyrolysis is carried out at temperature range of 300–700 °C with a biomass particle size of 5–50 mm, while in fast pyrolysis rate of heating varies from 10 to 200 °C/s with a residence time of 0.5–10 s. In the case of flash pyrolysis, the heating rate ranges from 10^3 to 10^4 °C/s with a residence time of less than 0.5 s. Temperature and residence time are the major determinants of the pyrolysis product. Low residence time and moderate temperature favour the formation of liquid products, while low residence time and low temperature convert biomass into charcoal (Bridgwater 2012). Biomass pyrolysis yields three major products which are bio-oil, char and pyrolytic gas.

Table 13.2 Production of bio-oil from micro-algae via biomass pyrolysis

Biomass type	Type of pyrolysis	Temperature (°C)	Bio-oil yield (wt%)	References
Spirulina platensis	Catalytic pyrolysis	400	49.71	Xu et al. (2019)
Chlorella vulgaris	Fast pyrolysis	400	42.2	Belotti et al. (2014)
Chlorella protothecoides	Fast pyrolysis	600	57.9	Miao and Wu (2004)

Bio-oil possesses high thermal instability and low heating value due to high content of oxygenated compounds. Therefore, bio-oil cannot be used as an efficient fuel in a diesel engine. In the process of formation of bio-oil, fragmentation and depolymerization of lignin, hemicellulose and cellulose are carried out in the fast pyrolysis (Rastegari et al. 2020; Yadav et al. 2020). Hasty heating of biomass with fast slaking of vapour produced in fast pyrolysis results in the formation of bio-oil (Isahak et al. 2012). Bio-oil production from fast pyrolysis process has mainly been investigated with lignocellulosic biomass such as pinewood, straw and stalk of crops like cotton, maize, rice, wheat and tobacco (Demirbas 2002; Gercel 2002; Putun 2002). In recent years, many authors have investigated bio-oil production using micro-algal biomass through the fast pyrolysis process (Table 13.2). The quality of bio-oil produced from micro-algal biomass not only depends on the reaction conditions but also the biochemical composition of selected algal biomass. It has been observed that the algal biomass grown under nitrogen-starved conditions tends to produce higher lipid content, which improves its yield and calorific value (Belotti et al. 2014).

13.4 Organic Waste for Biofuel Production

Rapid modernization and industrialization have resulted in the generation of an enormous amount of solid waste which is difficult to handle and creates a nuisance. The solid waste generated through household and other activities (termed as Municipal solid waste, MSW) is generally collected by the Municipal Co-operation of each district and is dumped at a particular place. This impacts the environment in a negative way (Goel 2008; Pandey et al. 2007). Municipal solid waste contains a significant amount of organic waste which mainly comes from food waste. A survey conducted on the characterization of MSW reveals that India produces a larger amount of food waste (approx. 31.9%) as compared to other wastes like paper, glass, leather, metal or plastic waste (Srivastava et al. 2014). Food wastes are produced from food and meat processing industries, kitchens of residential societies, canteens, hostels and restaurants (Kumar et al. 2017). Many times the organic or food waste generated from different locations is thrown or dumped in landfills resulting in the formation of greenhouse gases (GHG) like methane and carbon dioxide and leachates in contact with water (Karmee and Lin 2014; Karmee 2016). Leachates so produced may contaminate the underlying soil and water bodies. Organic waste has several adverse effects on biodiversity, land and water bodies leading to different social and environmental issues across the globe.

Reprocessing of food wastes for biodiesel production is a concept to create fuel, thereby reducing energy crisis. Different food or organic wastes can be utilized for the generation of different biogas or biodiesel products through various routes. Fatty acids present in the lipids extracted through organic waste are used to produce biodiesel. The processes included in biodiesel production are thermal cracking, micro-emulsions, pyrolysis and transesterification. Transesterification, being the most significant step towards biodiesel formation, where oil content of organic waste is treated with alcohol in the presence of a catalyst to produce alkyl esters and glycerol.

Organic waste generated through meat rendering and aquaculture may be hydrogenated, esterified and digested to produce biodiesel or biogas. Most of the agricultural wastes or residues are treated by thermal gasification or hydrolysis and fermentation to produce renewable diesel or ethanol. Waste generated from food services and wholesale can be converted into biodiesel or renewable diesel through the process of hydrogenation and esterification. Household organic waste can result in the formation of biogas and ethanol through anaerobic digestion, fermentation and hydrogenation. Biomass from MSW can be thermally gasified followed by anaerobic digestion which can lead to the production of ethanol, biodiesel and biogas (Nordic Energy Research 2019).

Biodiesel production from organic waste is now commercializing at an alarming rate. Even though there are several biodiesel production plant setups across the globe which utilize most of the organic wastes, the problem of landfills persists to a greater extent. One of the biodiesel production plants, established in Loosening Denmark, utilizes animal fats obtained through slaughterhouse or used cooking oil (not suitable for cooking anymore) for the production of biodiesel (Nordic Energy Research 2019). Two-step transesterification methodology is adopted for the production of biodiesel which also generates the by-products of glycerin and potassium sulphate having other applications too (Nordic Energy Research 2019). Similarly, the joint venture of Labio Oy and Gasum Oy is responsible for the production of biogas (through the process of anaerobic digestion) and its distribution taking the biowaste feedstock from food industries, household or sewage sludge (Nordic Energy Research 2019). Another renewable diesel production plant Neste has been set up in Porvoo (Finland) employing NEXBTL hydrogenation technology that can replace fossil diesel in the transportation sector. The feedstock utilized in the plant is organic waste, vegetable oil or used cooking oil (Nordic Energy Research 2019).

In today's scenario, biofuel production is largely carried out through edible wastes across the globe (Pimentel and Patzek 2005). Organic wastes (or food waste) can also be composted, recycled and incinerated to generate energy, fuel and other different value-added products requiring the development of advanced technology (Luque and Clark 2013). Biodiesel can be produced through numerous edible plant oils like rapeseed, canola and soybean, by the process of transesterification (Pimentel and Patzek 2005; Karmee and Chadha 2005). The process of transesterification involves the reaction of mono-, di- and triglycerides present in edible oils with methanol in the presence of a catalyst to produce biodiesel. Edible plant oils can be used for different biofuels' production (Mathews 2008). The biofuel like bioethanol production from edible oils involves the process of pretreatment followed by fermentation, enzymatic hydrolysis and distillation steps. Several non-edible plant sources like Pongamia and jatropha can be used for the generation of biofuels (Karmee and Chadha 2005; Sun and Cheng 2002).

The production of biodiesel and bioethanol is through lipids, amino acids, phosphates and carbohydrate treatment and degradation processes which are present in a surplus amount in food wastes, bakery products and other organic wastes originated from soil or land (Pleissner et al. 2013, 2014). Some other food products can also be treated or hydrolyzed enzymatically to produce lipids and carbohydrates useful for the production of biofuels (Pleissner et al. 2014). Research also reveals that noodles waste can also be utilized for the production of biofuels by extracting oil from its waste material using non-polar solvent hexane (Yang et al. 2014a, b). This oil can be converted into bioethanol following different steps like separation of starch from oil, saccharification of starch followed by its fermentation (Yang et al. 2014a, b). However, the production of biodiesel can be carried out through the separated oil content by the process of treatment with methanol with a catalyst. Bioethanol production from organic waste was studied through various processes using different food wastes or other organic wastes. Yan et al. (2011) utilized the kitchen waste to obtain hydrolysates which could be further reduced to bioethanol. Potato peels from potato product producing industries were taken to produce bioethanol through biocatalytic conversion methods treating with the process of liquefaction, saccharification and fermentation (Yan et al. 2011; Arapoglou et al. 2010; Matsakas et al. 2014). The conversion of organic waste (old newspaper waste) and food waste was carried out to produce bioethanol through microbial or acid hydrolysis that yielded fermented sugar wort. The obtained sugar wort was then fermented with *Saccharomyces cerevisiae* to yield ethanol (Uduak et al. 2008).

Microorganisms like algae, bacteria, fungi and yeast can also be used to produce energy since these organisms can accrue the main components—lipids (oil), proteins and carbohydrates (Martinez et al. 2015; Yadav et al. 2017, 2019). The natural oil or lipid content of microorganisms is in the form of triacylglycerol which is the effective component for biodiesel production (Wen and Johnson 2009).

Biogas may be produced from a variety of organic wastes including kitchen wastes, paper wastes and other municipal solid wastes having organic nature. The process of conversion of waste to biogas proceeds through anaerobic digestion which mainly produces methane and carbon dioxide along with a solid residue (Papacz 2011). The pretreatment of organic content of municipal solid waste also yields biogas through anaerobic digestion (Nasir et al. 2012). Shrestha et al. (2017) utilized canteen's kitchen waste to carry out anaerobic digestion to produce biogas. The study was carried out with the support of Solid Waste Management Technical Support Centre, Lalitpur, and it was found that each kilogram of waste was capable to generate 22.03

Lof biogas (Shrestha et al. 2017). Organic wastes also include grass wastes. Their decomposition can produce approx. 50–110 m^3 of CO_2 and 90–140 m^3 of CH_4 in the atmosphere (Yu et al. 2002).

The production of biofuel from different types of organic wastes depends upon the factors—availability of organic waste, the distance between the landfill site of organic waste and their processing sites, the competence of chemical process, lipid, protein or carbohydrate content present in organic waste and efficiency of transesterification process to convert the organic waste to biodiesel and the ability of microorganisms to transform organic waste to degradable chemicals useful to produce biofuel (Karmee and Lin 2014; Rastegari et al. 2019b, c).

13.5 Key issues—The Road Ahead

Replacing fossil fuels with biofuels has the potential to generate low carbon economy. However, for complete sustainable development, bioenergy generation process must meet three major criteria, i.e. economic viability, environmental performance and social acceptability (Elghali et al. 2007). The utilization of competition for agricultural land for food or biofuel has been an issue of controversy and is considered to be responsible for an increase in agricultural food crops from 2006 to 2008 and 2010 to 2011 (Thompson 2012; Oladosu and Msangi 2013; Tomei and Helliwell 2016). During these periods, there were simultaneous increases in biofuel production leading to a direct relationship between the two. It is reported that 20 to 40% of the increase in food prices was because of biofuel growth, which leads to larger concerns like increasing food prices impacting the poor and land use for fuel crops instead of food crops and availability of markets for such crops.

With the ever-increasing population, land usage can be expected to increase because of food, societal expansion and biofuels. As per the UN FAO estimates, the cultivated areas globally have been increased by approximately 12% since 1961 (UNFAO report 2016) along with the extensive usage of fertilizers and pesticides aided with IT and genetics (Fischer et al. 2002). Only 3% of the agricultural area is under biofuel plantations (Borras and Franco 2012; Edenhofer et al. 2011; FAO report 2011; Popp et al. 2014).

There have been two schools of thoughts regarding the emissions generated by biofuels. Depending upon the process of production and the type of feedstock, there may a net increase in the GHG emissions from such fuels as compared to the conventional fuels (Searchinger et al. 2008; Yang and Chen 2013; Kahn Ribeiro et al. 2012). Fertilizer usage emits nitrous oxide, a potential GHG (Yu 2008). The biorefineries involved in the process employ fossil fuels. As per the other reports, biofuels have been reported to reduce GHG emissions in the range of 60–94% as compared to conventional ones (Holma et al. 2013; Highina et al. 2014). Biodiesel has been reported to decrease particulate matter by approximately 88% but releases more amount of nitrogen oxides as compared to conventional fuel (Xue et al. 2011). The emissions from biodiesel depend largely on its source and mixing ratio with conventional fuels. As reported, unblended biodiesel (100%) can reduce hydrocarbons by 70%, particulates and carbon monoxide pollution by 50%, while increasing NO_x by 10% (USEPA 2002). SO_2 pollution is also reduced to a greater extent (Cowie et al. 2016; USEPA 2002). It further reduces the ozone formation as compared to that by conventional diesel. Second- and third-generation biofuels, produced on wasteland, can decrease GHG pollution as compared to fossil fuels.

Raising biofuel crops may impact water requirement in terms of quality and quantity which is a matter of serious spatial/temporal concern. It aggravates the debate to utilize this scarce resource for food or biofuel crops. The existing production technology coupled with the resources like availability of local skills and capacity to handle the biofuel plant is required for the bioenergy plant operation (Kour et al. 2019; Kumar et al. 2019). Apart from this, the economy and commercialization of the biofuel industry depend on the investment by the public and private sectors and the demands raised by the related market.

As compared to the exhaustible fossil fuels, biofuels are derived or produced from local renewable sources, thus reducing our dependence on foreign suppliers. The other factors like feedstock production, availability of better varieties and water, local skill enhancement through training and the infrastructure like roads and water also contribute to the production cost of the biofuels. Increasing demands for biofuels positively affects the farmer's income and boosts the local economy. To promote the biofuel economy, it may require subsidies and other favourable market policies.

13.6 Conclusion and Future Prospects

Biomass-based energy generation has several advantages over the fossil fuel-based energy generation. Biomass as a source of energy can be used as decentralized energy generation and can be implemented in rural areas. Biomass for energy generation can be exploited from different types of biomass such as forest biomass, crop residues, aquatic biomass and organic wastes. Commercialization of biomass-based energy generation is still under development. However, various demonstration projects are running across the country to spread awareness regarding bioenergy generation. The supply chain of biomass for consistent bioenergy production is one of the major challenges in India. The statistical data of its spatial distribution on a local scale is required for their effective utilization in the industrial sector. Attention must be diverted towards woody biomass due to its largely untapped potential and flexibility to be used with different conversion technologies. Bioenergy application of aquatic biomass requires more research and development for its commercialization.

References

Acma H, Yaman S (2010) Interaction between biomass and different rank coals during co-pyrolysis. Renew Energy 35:288–292
Agrawal AK (2007) Biofuels (alcohols and biodiesel) applications as fuels for internal combustion engines. Prog Energy Combust Sci 33:233–271
Ahmad AA, Zawawi NA, Kasim FH, Inayat A, Khasri A (2016) Assessing the gasification performance of biomass: a review on biomass gasification process conditions, optimization and economicevaluation.Renew Sust Energy Rev 53:1333–1347.
Arapoglou D, Varzakas T, Vlyssides A, Israilides C (2010) Ethanol production from potato peel waste (PPW). Waste Manag 30:1898–1902
Arthe R, Rajesh R, Rajesh EM, Rajendran S, Jeyachandran S (2008) Production of bioethanol from cellulosic cotton waste through microbial extracellular enzymatic hydrolysis and fermentation. EJEAF Che 7:2984–2992
Balatinecz JJ (1983) The potential of densification in biomass utilization. In: Cote WA (eds) Biomass utilization. Nato Advanced Science Institutes Series, Vol 76, Plenum Press, Springer, Boston, MA, pp 181–190.
Belotti G, de Caprariis B, de Filippis P, Scarsella M, Verdone N (2014) Effect of *Chlorella vulgaris* growing conditions on bio-oil production via fast pyrolysis. Biomass Bioenergy 61:187–195
Borras S, Franco J (2012) Global land grabbing and trajectories of agrarian change: a preliminary analysis. J Agrar Chang 12:34–59
Bridgwater AV (2012) Review of fast pyrolysis of biomass and product upgrading. Biomass Bioenergy 38:68–94
Chen J, Wu Y, Xu C, Song M, Liu X (2019) Global non-fossil fuel consumption: driving factors, disparities, and trends. Manag Decis 57(4):791–810
Chen L, Xing L, Han L (2009) Renewable energy from agro-residues in China: solid biofuels and biomass briquetting technology. Renew Sust Energy Rev 13(9):2689–2695
Cowie A, Soimakallio S, Brandao M (2016) Environmental risks and opportunities of biofuels. In: Bouthillier YL, Cowie A, Martin P, McLeod-Kilmurray H (eds) The law and policy of biofuels. Edward Elgar, Cheltenham, United Kingdom, pp 3–29
Demirbas A (2002) Analysis of liquid products from biomass via flash pyrolysis. Energy Source 24:337–345
Demirbas A (2004) Combustion characteristics of different biomass fuels. Prog Energy Combust Sci 30:219–230
Devi S, Gupta C, Jat SL, Parmar MS (2017) Crop residue recycling for economic and environment sustainability: the case of India. Open Agri 2. https://doi.org/10.1515/opag-2017-0053
Edenhofer O Pichs-Madruga R, SokonaY,Seyboth K, MatschossP,Kadner S, Zwickel T, Eickemeier P, Hansen G, Schlomer S (2011) Renewable Energy Sources and Climate Change Mitigation, Special Report of the IPCC, IPCC. https://www.ipcc.ch/report/srren/
Elghali L, Clift R, Sinclair P, Panoutsou C, Bauen A (2007) Developing a sustainability framework for the assessment of bioenergy system. Energy Policy 35(12):6075–6083
Energy Statistics (2019) Central Statistics Office Ministry of Statistics and Programme Implementation Government of India. www.mospi.gov.in
Escobar JC, Lora ES, Venturini OJ, Yanez EE, Castillo EF, Almazan O (2009) Biofuels: environment, technology and food security. Renew Sust Energy Rev 13:1275–1287
FAO report (2011) The state of the world's land and water resources for food and agriculture. FAO and Earthscan/Routledge, Abingdon, UK
Fischer G, Van Velthuizen H, Shah M, Nachtergaele F (2002) Global agro-ecological assessment for agriculture in the 21st century. International Institute of Applied System Analysis, Laxenburg, Austria, Rome, Italy
Gaurav N, Sivasankari S, Kiran GS, Ninawe A, Selvin J (2017) Utilization of bioresources for sustainable biofuels: a Review. Renew Sust Energy Rev 73:205–214

Gercel HF (2002) The effect of a sweeping gas flow rate on the fast pyrolysis of biomass. Energy Sour 24:633–642
Goel S (2008) Municipal solid waste management in India: a critical review. J Environ Sci Eng 50:319–328
Heidenreich S, Foscolo PU (2015) New concepts in biomass gasification. Prog Energy Combust Sci 46:72–95
Highina B, Bugaje I, Umar B (2014) A review of second generation biofuel: a comparison of its carbon footprints. Eur J EngTechnol 2:117–125
Holma A, Koponen K, Antikainen R, Lardon L, Leskinen P, Roux P (2013) Current limits of life cycle assessment framework in evaluating environmental sustainability—case of two evolving biofuel technologies. J Clean Prod 54:215–228
Hupa M, Karlstrom O, Vainio E (2017) Biomass combustion technology development- It is all about chemical details. P Combust Inst 36(1):113–134
Isahak WRW, Hisham MWM, Yarmo MA, Yun TYH (2012) A review on bio-oil production from biomass using pyrolysis method. Renew Sust Energy Rev 16:5910–5923
Jackson RB, Friedlingstein P, Andrew RM, Canadell JG, Le Quere C, Peters GP (2019) Persistent fossil fuel growth threatens the Paris agreement and planetary health. Environ Res Lett 14(12):1–8
Kahn Ribeiro S, Figueroa M, Creutzig F, Dubeux C, Hupe J, Kobayashi S (2012) Energy end-use: transport. In: Global energy assessment—Toward a sustainable future. Cambridge University Press, pp 575–648. https://doi.org/10.1017/CBO9780511793677.015
Karmee SK (2016) Liquid biofuels from food waste: Current trends, prospect and limitation. Renew Sust Energy Rev 53:945–953
Karmee SK, Chadha A (2005) Preparation of biodiesel from crude oil of *Pongamiapinnata*. BioresourTechnol 96:1425–1429
Karmee SK, Lin CSK (2014) Valorization of food waste to biofuel: Current trends and technological challenges. Sustain Chem Process 2(22):2–4
Kour D, Rana KL, Yadav N, Yadav AN, Rastegari AA, Singh C et al (2019) Technologies for biofuel production: current development, challenges, and future prospects. In: Rastegari AA, Yadav AN, Gupta A (eds) Prospects of renewable bioprocessing in future energy systems. Springer International Publishing, Cham, pp 1–50. https://doi.org/10.1007/978-3-030-14463-0_1
Kumar K, Yadav AN, Kumar V, Vyas P, Dhaliwal HS (2017) Food waste: a potential bioresource for extraction of nutraceuticals and bioactive compounds. Bioresour Bioprocess 4:18. https://doi.org/10.1186/s40643-017-0148-6
Kumar S, Sharma S, Thakur S, Mishra T, Negi P, Mishra S et al (2019) Bioprospecting of microbes for biohydrogen production: Current status and future challenges. In: Molina G, Gupta VK, Singh BN, Gathergood N (eds) Bioprocessing for biomolecules production. Wiley, USA, pp 443–471
Kumar A, Jones DD, Hanna MA (2009) Thermochemical biomass gasification: a review of the current status of the technology. Energies 2:556–581
Lapuerta M, Hernández JJ, Pazo A, López J (2008) Gasification and co-gasification of biomass wastes: effect of the biomass origin and the gasifier operating conditions. Fuel Process Technol 89:828–837
Luque R, Clark JH (2013) Valorization of food residues: waste to wealth using green chemical technologies. Sustain Chem Process 1:10
Lv PM, Xiong ZH, Chang J, Wu CZ, Chen Y, Zhu JX (2004) An experimental study on biomass air-steam gasification in a fluidized bed. BioresourTechnol 95:95–101
Martinez EJ, Raghavan V, Gonzalez-Andres F, Gomez X (2015) New biofuel alternatives: Integrating waste management and single cell oil production. Int J Mol Sci 16:9385–9405
Maschio G, Lucchesi A, Stoppato G (1994) Production of syngas from biomass. Bioresour Technol 48:119–126
Mathews JA (2008) Is growing biofuel crops a crime against humanity? Biofuel Bioprod Biorefin 2(2):97–99
Matsakas L, Kekos D, Loizidou M, Christakopoulos P (2014) Utilization of household food waste for the production of ethanol at high dry material content. Biotechnol Biofuels 7(1):4–14

Miao X, Wu Q (2004) High yield bio-oil production from fast pyrolysis by metabolic controlling of *Chlorellaprotothecoides*. J Biotechnol 110:85–93
Nalladurai K, Vance Morey R (2009) Factors affecting strength and durability of densified biomass products. Biomass Bioenergy 33(3):337–359
Narváez I, Orío A, Aznar MP, Corella J (1996) Biomass gasification with air in an atmospheric bubbling fluidized bed effect of six operational variables on the quality of the produced raw gas. Ind Eng Chem Res 35(7):2110–2120
Nasir IM, Ghazi TIM, Omar R (2012) Production of biogas from solid organic wastes through anaerobic digestion: a review. Appl Microbiol Biotechnol 95:321–329
Nordic Energy Research (2019) Report on Food waste to Biofuels. 28 Feb 2019
Oladosu G, Msangi S (2013) Biofuel-food market interactions: a review of modeling approaches and findings. Agriculture 3:53–71
Pandey SK, Tyagi P, Gupta AK (2007) Municipal solid waste management in Ghazipur city- a case study. J Agric Biol Sci 2:41–43
Papacz W (2011) Biogas as vehicle fuel: a European overview. J KONES Powertrain Transp 18(1):7–49
Parthasarathy P, Narayanan KS (2014) Hydrogen production from steam gasification of biomass: influence of process parameters on hydrogen yield-a review. Renew Energy 66:570–579
Petrus L, Noordermeer MA (2006) Biomass to biofuels, a chemical perspective. Green Chem 8:861–867
Pimentel D, Patzek TW (2005) Ethanol production using corn, switchgrass, and wood; biodiesel production using soybean and sunflower. Nat Resour Res 14:65–76
Pleissner D, Kwan TH, Lin CSK (2014) Fungal hydrolysis in submerged fermentation for food waste treatment and fermentation feedstock preparation. Bioresour Technol 158:48–54
Pleissner D, Lam WC, Sun Z, Lin CSK (2013) Food waste as nutrient source in heterotrophic microalgae cultivation. BioresourTechnol 137:139–146
Popp J, Lakner Z, Harangi-Rakos M, Fari M (2014) The effect of bioenergy expansion: food, energy and environment. Renew Sust Energy Rev 32:559–578
Putun AE (2002) Biomass to bio-oil via fast pyrolysis of cotton straw and stalk. Energy Sour 24:275–285
Rashad AH (2013) Biomass production for energy in India: review. J Technol Innov Ren Energy 2:366–375
Rastegari AA, Yadav AN, Yadav N (2020) New and future developments in microbial biotechnology and bioengineering: trends of microbial biotechnology for sustainable agriculture and biomedicine systems: diversity and functional perspectives. Elsevier, Amsterdam
Rastegari AA, Yadav AN, Gupta A (2019) Prospects of renewable bioprocessing in future energy systems. Springer International Publishing, Cham
Rastegari AA, Yadav AN, Yadav N (2019b) Genetic manipulation of secondary metabolites producers. In: Gupta VK, Pandey A (eds) New and future developments in microbial biotechnology and bioengineering. Elsevier, Amsterdam, pp 13–29. https://doi.org/10.1016/B978-0-444-63504-4.00002-5
Rastegari AA, Yadav AN, Yadav N, Tataei Sarshari N (2019c) Bioengineering of secondary metabolites. In: Gupta VK, Pandey A (eds) New and future developments in microbial biotechnology and bioengineering. Elsevier, Amsterdam, pp 55–68. https://doi.org/10.1016/B978-0-444-63504-4.00004-9
Ravindra NH (2005) Assessment of sustainable non-plantation biomass resource potential for energy in India. Centre for sustainable Technologies Indian Institute of Science, Bangalore
Ruiz JA, Juárez MC, Morales MP, Muñoz P, Mendívil MA (2013) Biomass gasification for electricity generation: review of current technology barriers. Renew Sust Energy Rev 18:174–183
Searchinger T, Heimlich R, Houghton RA, Dong F, Elobeid A, Fabiosa J, Tokgoz S, Hayes D, Yu TH (2008) Use of U.S. croplands for biofuels increases greenhouse gases through emissions from land-use change. Science 319:1238–1240

Shrestha S, Chaulagain NP, Shrestha KR (2017) Biogas production for organic waste management: a case study of canteen's organic waste in solid waste management technical support centre, Lalitpur, Nepal. J Environ Sci Technol 5. https://doi.org/10.3126/njes.v5i0.22714

Sikarwar VS, Zhao M, Clough P, Yao J, Zhong X, Memon MZ, Shah N, Anthony E, Fennell P (2016) An overview of advances in biomass gasification. Energy Environ Sci 9:2939–2977

Singh NB, Kumar A, Rai S (2014) Potential production of bioenergy from biomass in an Indian perspective. Renew Sust Energy Rev 39:65–78

Somayeh F, Mohsen AM, Johann FG (2016) A critical review on biomass gasification, co-gasification, and their environmental assessments. Biofuel Res J 12:483–495

Srivastava R, Krishna V, Sonkar I (2014) Characterization and management of municipal solid waste: a case study of Vanaras city, India. Int J Curr Res Acad Rev 2:10–16

Sun Y, Cheng J (2002) Hydrolysis of lignocellulosic materials for ethanol production: a review. BioresourTechnol 83:1–11

Thompson BP (2012) The agricultural ethics of biofuels: the food vs Fuel Debate. Agriculture 2:339–358

Tomei J, Helliwell R (2016) Food versus fuel? Going beyond biofuels. Land Use Policy 56:320–326

Tuteja J, Choudhary H, Nishimura S, Ebitani K (2014) Direct synthesis of 1,6-Hexanediol from HMFover a heterogeneous Pd/ZrP catalyst using formic acid as hydrogen source. Chem Sus Chem 7:96–100

Tuteja J, Nishimura S, Ebitani K (2012) One-pot synthesis of furans from various saccharides using a combination of solid acid and base catalysts. Bull Chem Soc Jpn 85:275–281

Uduak GA, Adamu AA, Udeme JJI (2008) Production of ethanol fuel from Organic and food wastes. Leonardo El J PractTechnol 13:1–11

UN Food and Agriculture Organization (FAO) (2016) Food outlook, biennial report on global food markets. FAO, Rome, Italy

USEPA (2002) A comprehensive analysis of biodiesel impacts on exhaust emissions. US Environmental Protection Agency (EPA), Washington DC, USA, p 2002

Weldemichael Y, Assefa G (2016) Assessing the energy production and GHG (greenhouse gas) emissions mitigation potential of biomass resources for Alberta. J Clean Prod 112:4257–4264

Wen Z, Johnson MB (2009) Microalgae as a feedstock for Biofuel production, Virginia Cooperative extension, Publication No. 442–886

Werther J, Saenger M, Hartge E-U, Ogadab T, Siagi Z (2000) Combustion of agricultural residues. Prog Energy Combust Sci 26:1–27

Whitty KJ, Zhang HR, Eddings EG (2008) Emissions from syngas combustion. Comb Sci Technol 180:1117–1136

World Energy Council (2019) World energy scenario: exploring innovation pathways to 20140 Cornhills London, United Kingdom, pp 62–64

Xu Y, Hu Y, Peng Y, Yao L, Dong Y, Yang B, Song R (2019) Catalytic pyrolysis and liquefaction behaviour of micro-algae for bio oil production.BioresTechnol. https://doi.org/10.1016/j.biortech.2019.122665

Xue J, Grift T, Hansen A (2011) Effect of biodiesel on engine performance and emissions. Renew Sust Energy Rev 15:1098–1116

Yadav AN, Kumar R, Kumar S, Kumar V, Sugitha T, Singh B et al (2017) Beneficial microbiomes: biodiversity and potential biotechnological applications for sustainable agriculture and human health. J Appl Biol Biotechnol 5:45–57

Yadav AN, Rastegari AA, Yadav N (2020) Microbiomes of extreme environments: biodiversity and biotechnological applications. CRC Press, Taylor & Francis, Boca Raton, USA

Yadav AN, Singh S, Mishra S, Gupta A (2019) Recent advancement in white biotechnology through fungi. Volume 2: Perspective for value-added products and environments. Springer International Publishing, Cham

Yan S, Li J, Chen X, Wu J, Wang P, Ye J, Yao J (2011) Enzymatical hydrolysis of food waste and ethanol production from the hydrolysate. Renew Energy 36:1259–1265

Yang Q, Chen GQ (2013) Greenhouse gas emissions of corn—Ethanol production in China. Ecol Model 252:176–184

Yang X, Lee JH, Yoo HY, Shin HY, Thapa LP, Park C, Kim SW (2014) Production of bioethanol and biodiesel using instant noodle waste. Bioproc Biosyst Eng. https://doi.org/10.1007/s00449-014-1135-3

Yang X, Lee SJ, Yoo HY, Choi HS, Park C, Kim SW (2014) Biorefinery of instant noodle waste to biofuels. Bioresour Technol 159:17–23

Yu HW, Samini Z, Hanson A, Smith G (2002) Energy recovery from grass using two-phase anaerobic digestion. Waste Manage 22:1–5

Yu TH (2008) Use of U.S. croplands for biofuels increases greenhouse gases through emissions from land-use change. Science 319:1238–1240

Chapter 14
Lignocellulosic Biofuel Production Technologies and Their Applications for Bioenergy Systems

Hamideh Bakhshayeshan-Agdam, Seyed Yahya Salehi-Lisar, and Gholamreza Zarrini

Abstract The use of energy by humans is increasing day by day and there is a need for an infinite energy source. The increasing use of fossil fuels along with limitations on these resources such as their finite nature, geopolitical instability, and deleterious global effects have led scientists to seek and discover alternative renewable resources for energy. Biofuels are one of the most important renewable energy resources that lessen the dependence on fossil fuels. Energy-enriched chemicals produced chiefly by photosynthetic organisms such as photosynthetic bacteria, microalgae, macroalgae, and plants are biofuel resources. Lignocellulosic biomass refers to a plant's dry matter and is an energy-enriched chemical that can be used as a renewable fuel resource. There are numerous groups of raw materials that contain lignocellulosic biomass the most important of which are woody feedstocks, agricultural residues, municipal solid wastes, and marine algae. Biofuel production from lignocellulosic biomass is dependent on the yield of fermentable sugars available. The various steps involved in the production of biofuels from lignocellulosic materials include pretreatment, hydrolysis, fermentation, and product separation. Lignocellulosic biomass can potentially be converted into biofuels such as bioethanol, biodiesel, and biogas. Despite the many challenges that have had to be overcome lignocellulosic biomass is likely to become *the* renewable resource for the economical production of biofuels in the near future because the raw materials that make up such a biomass are available, cheap, and contain high levels of carbohydrate that can be used for biofuel production.

H. Bakhshayeshan-Agdam · S. Y. Salehi-Lisar (✉)
Department of Plant Sciences, Faculty of Natural Sciences, University of Tabriz, Tabriz, Iran
e-mail: y_salehi@tabrizu.ac.ir

H. Bakhshayeshan-Agdam
e-mail: h_bakhshayeshan@tabrizu.ac.ir

G. Zarrini
Department of Animal Sciences, Faculty of Natural Sciences, University of Tabriz, Tabriz, Iran

A. N. Yadav et al. (eds.), *Biofuels Production – Sustainability and Advances in Microbial Bioresources*, Biofuel and Biorefinery Technologies 11,
https://doi.org/10.1007/978-3-030-53933-7_14

14.1 Introduction

Providing sustainable energy sources for the modern industrialized world is becoming more difficult day by day. Today up to 80% of the world's energy is provided by fossil fuels (Lund 2007). Recently there have been alterations in rainfall patterns in many regions because of global climate change brought about by increases in temperature and atmospheric CO_2 levels. Global climate change is the main factor triggering worldwide droughts and will represent a significant challenge to the communities of the world in the near future. Studies have revealed there is a direct relationship between atmospheric CO_2 levels and global warming (Salehi-lisar and Bakhshayeshan-agdam 2016). Although technologies, especially motorization, have facilitated human life at many levels, they have taken their toll on the natural ecosystem. One of the most important effects has been the intensive climatic change widely believed to be due to high CO_2 generation by fossil fuel consumption (Voloshin et al. 2016; Zabed et al. 2016).

CO_2 generation can be lessened by capturing and sequestering CO_2 during the consumption of fossil fuels and utilizing renewable energy sources such as wind, solar, nuclear, and geothermal, as well as various biomass sources (Demirbas 2010; Nakagawa et al. 2007). In order to reduce their energy dependence on fossil fuels many countries are today focusing on alternative energy production strategies. This has led to biofuels becoming one of the most important fuel sources since they produce cheap energy, release low amounts of (or even no) greenhouse gases, and bring about foreign exchange savings related to socioeconomic benefits (Azad et al. 2015; Bahadar and Khan 2013; Sarkar et al. 2012). Lignocellulosic biomass consisting of cellulose, hemicellulose, and lignin is the most abundant raw material for biofuel production, especially bioethanol. The combustion of bioethanol produced from lignocellulosic biomass adds no net carbon dioxide to the earth's atmosphere. It is a green fuel. The CO_2 released due to the combustion of bioethanol is in fact that captured by photosynthetic plants from the atmosphere. This chapter presents an overview of lignocellulosic biomass, its sources, and lignocellulosic biofuel production technologies.

14.2 Biofuel: Future Energy

Liquid fuel and other components produced from biomass (organisms like plants) are called biofuels. All biofuels are renewable in that they involve photosynthetic conversion of solar energy to chemical energy thus setting them apart from fossil fuels. In recent years different arguments have been raised against the use of biofuels as future energy supply despite their contributing to a reduction in carbon dioxide emissions (Kour et al. 2019b; Kumar et al. 2019). There is increasing pressure today to reduce the use of fossil fuels. They are not only the main source of energy but also the main source of CO_2 emissions (Alalwan et al. 2019). Biofuel is classified into two types: primary and secondary. Primary biofuels are fuels used essentially in their natural forms and come from organic material such as firewood, wood chips, and pellets. Although secondary biofuels also come from organic materials, they do not naturally exist in nature. Such biofuels include charcoal, ethanol, biodiesel, biooil, biogas, synthesis gas (syngas), and hydrogen and can be used for a wider range of applications such as transport and high-temperature industrial processes (Azad et al. 2015; Doshi et al. 2016; Yadav et al. 2019). Secondary biofuels are today based on primary sources and production techniques and classified in a number of categories (Fig. 14.1) such as first-, second-, third-, and even fourth-generation biofuels (Alalwan et al. 2019).

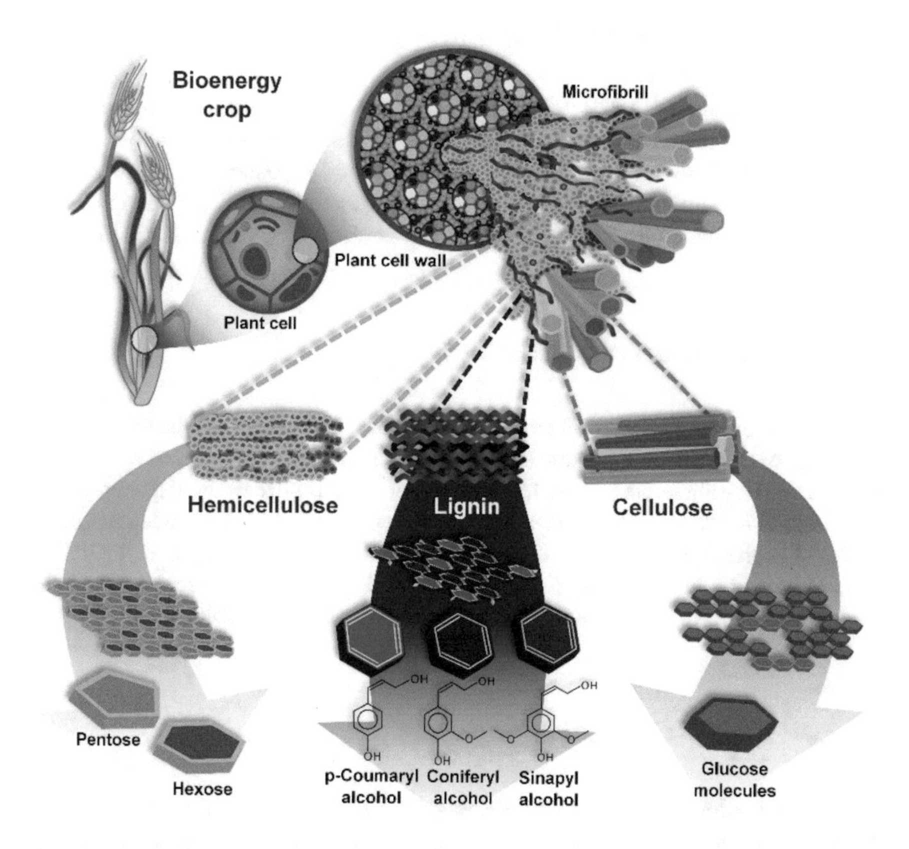

Fig. 14.1 Structure of lignocellulosic biomass and its biopolymers: cellulose, hemicellulose, and lignin (Hernández-Beltrán et al. 2019)

14.2.1 First-Generation Biofuels

First-generation biofuels are conventional biofuels produced from food crops. Such biofuels include bio-ethanol or butanol that are produced by the fermentation of starch (from crop residues of barley, corn, potato, wheat, etc.) or sugar (from sugar beet and sugarcane). Biodiesel is another first-generation biofuel produced directly from vegetable oils of oleaginous crops (such as rapeseed, palm, soybean, coconut, and sunflower) by transesterification (Alalwan et al. 2019; Jorgensen 2011).

14.2.2 Second-Generation Biofuels

Second-generation biofuels are fuels produced from various types of biomass, especially lignocellulosic biomass. Biomass means any source of organic carbon rapidly renewed by the carbon cycle of plants. Biomass used to produce second-generation biofuels is more efficient than that used for first-generation biofuels due to the low cost of feed biomass (Alalwan et al. 2019; Dar et al. 2018; Doshi et al. 2016).

14.2.3 Third-Generation Biofuels

Microalgal biomass is the material used for the production of third-generation biofuels. Ever since 1978 aquatic species of algae have been introduced as a biofuel source. Oil-rich algae can be used for biofuel production and their dried residue can be reprocessed to create ethanol. The microalgae typically targeted include *Dunaliella salina*, *Chlorella vulgaris*, and *Chlamydomonas reinhardtii* because of their high lipid content (Alalwan et al. 2019; Bahadar and Khan 2013; Carere et al. 2008; Demirbas 2010; Kong et al. 2010; Gouveia and Oliveira 2009; Voloshin et al. 2016).

14.2.4 Fourth-Generation Biofuels

Fourth-generation biofuels are the result of the biotechnological manipulation of algae and cyanobacteria. This is a young but strongly evolving research field (Heimann 2016). Unlike the first three generations of biofuels, fourth-generation biofuels do not require the biomass used to be destroyed. This class of biofuels includes electrobiofuels and photobiological solar fuels. Fourth-generation biofuels are today produced by photosynthetic microorganisms designed to produce photobiological solar fuels (Alalwan et al. 2019).

14.3 Lignocellulosic Biomass: Renewable Resource for Liquid Biofuels

14.3.1 Concept of Lignocellulosic Biomass

Lignocellulosic biomass is precisely defined as the dry matter of plants (biomass) such as cellulose, hemicellulose, and lignin that are the most abundant raw materials on earth for biofuel production (Fig. 14.2). Cellulose is a linear homopolysaccharide consisting of glucose units (500–15,000) linked by β(1–4) glycosidic bonds. Hydrogen bonds make the cellulose very rough and crystalline and provide protection from enzyme activity. Hemicellulose is an amorphous and variable polymer that originates from heteropolymers including hexoses (D-glucose, D-galactose, and D-mannose) as well as pentose (D-xylose and L-arabinose) and might contain sugar acids (uronic acids).

Hemicellulose plays a key role in the linkage between lignin and cellulose. Lignin is an aromatic polymer and is the second major constituent of biomass after cellulosic matter. When burnt lignin produces a lot of energy and could even be used as a better source for the production of heat and power than the cost-effective yield of bioethanol. Since the carbohydrate polymers of a plant's cell wall are tightly bound to lignin the biomass obtained from the cell walls is called lignocellulosic biomass (Hernández-Beltrán et al. 2019; Limayem and Ricke 2012). Lignocellulosic biomass is classified into three major categories that consist of (1) intact biomass that includes all terrestrial plants such as trees, shrubs, bushes, and grasses; (2) waste biomass that is produced as a low-value by-product of various industrial sectors such as agriculture (corn stover, sugarcane bagasse, straw, etc.) and forestry; and (3) energy crops defined as those crops that produce a high yield of lignocellulosic biomass and serve as a

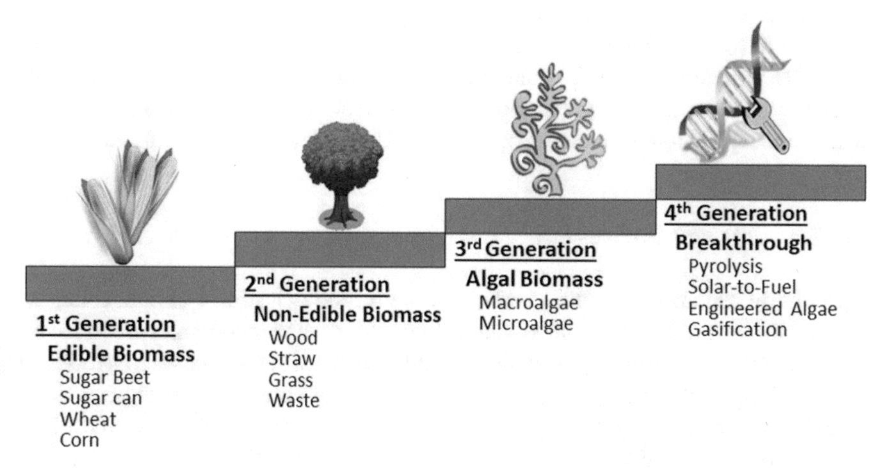

Fig. 14.2 The four generations of secondary biofuels (Alalwan et al. 2019)

raw material for the production of second-generation biofuels (Jorgensen 2011; Kour et al. 2019c; Limayem and Ricke 2012).

Lignocellulosic biomass is the raw material used in the pulp and paper industry where the focus is on separating the lignin and cellulose of the biomass of plants. The fermentation of lignocellulosic biomass to ethanol is an attractive alternative to fossil fuels for the production of fuel. The combustion of lignocellulosic ethanol adds no net carbon dioxide to the earth's atmosphere because the CO_2 involved is photosynthetically fixed (Nakagawa et al. 2007).

14.3.2 Lignocellulosic Biomass Sources

Numerous groups of raw materials are classified by their origin, composition, and structure the most important of which are woody feedstock, agricultural residue, municipal solid waste, and marine algae (Fig. 14.3).

14.3.2.1 Woody Feedstock

Primary and secondary industries take wood and convert it into products such as furniture, kitchen cabinets, flooring, building products, pallets, containers, and paper products. The high volume of woody residue produced during such manufacture is called woody feedstock (Organization 2015). Secondary manufacturing industries use products manufactured by primary industry. The residue produced by secondary industries is less than that of primary industries (Rooney 1998; McKeever 1998) because any residue produced is used to meet their energy needs for heat, especially in the winter. Wood residues have long been considered significant energy sources worldwide. Woody feedstock consists of sawdust, wood chips, and wood bark all of which have been found to be favorable for bioethanol production (Huzir et al. 2018). Such materials have the potential to be an additional resource for biological-based products for bioenergy. This potential should be properly managed using forest management techniques for bioenergy to be a sustainable source of fuel for the future.

14.3.2.2 Agricultural Residue

Agricultural residue represents a large, renewable, and rich source of lignocellulose biomass for the production of bioethanol. In countries whose economies are based on agriculture large quantities of agricultural field residues are available such as residues from growing beet, corn, fruit, sugarcane, and cobs of corn. Such residues include corn stover, leaves, orchard trimmings, rice husk, rice straw, and stalks (Sims 2004). Even weeds are available in large quantities. Although usually ultimately burned or left in the fields, such weeds can be better used for biofuel production (Kim and Dale 2004; Kumar et al. 2017; Sarkar et al. 2012). Agricultural biomass typically

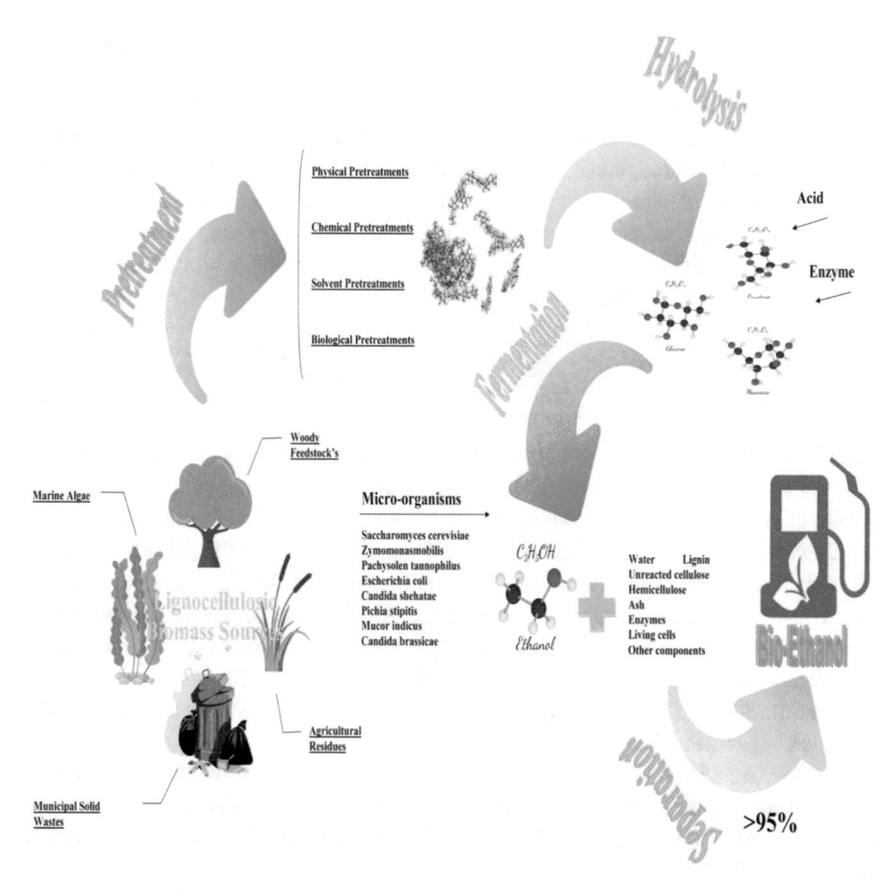

Fig. 14.3 Lignocellulosic biomass sources and bioethanol production technologies that come from them

comprises ash, cellulose, hemicellulose, lignin, and protein. Moreover, crop residues offer a cheap and sustainable resource that can be used to produce biofuels worldwide (Bhatia and Paliwal 2011; Braide et al. 2016; Cheng et al. 2012; Voloshin et al. 2016).

14.3.2.3 Urban Residue

The two principal sources of urban lignocellulosic biomass are municipal solid waste and construction or demolition debris (Ahmed and Ahmaruzzaman 2016). Municipal solid waste such as sewage, any industrial leftover that is organic, and waste of modern urban life are low-cost sources of lignocellulosic biomass containing significant amounts of CO_2 (Nakagawa et al. 2007). Construction and demolition debris make up the other principal source of urban residue and are considered separate from municipal solid waste because they come from different sources. The amount of such residue is highly dependent on economic activity, population, demolition activity, and recycling programs. Construction debris is potentially usable, while demolition debris is not since it tends to be contaminated. There is not enough data currently available about the volume of urban wood residue that could be used. Industrial residues have excellent potential to be recycled as cellulosic materials irrespective of whether they come from residential or nonresidential sources. However, there has been limited research into the utilization of municipal solid waste to generate cost-efficient biofuels profitably and on a large scale worldwide (Ahmed and Ahmaruzzaman 2016).

14.3.2.4 Marine Algae

Algae are aquatic organisms capable of converting solar energy into energy-rich chemicals such as starch and lipids. Consisting of macroalgae and microalgae (Falkowski and Raven 2004; Ho et al. 2013; Koutra et al. 2018) marine algae are highly cost-effective, abundant, and sustainable raw materials for the production of biofuels such as alcohol, diesel, methane, and hydrogen. Marine algae biomass is getting a lot of attention today as a third-generation biofuel as a result of the setting up of a number of quick biorefineries. Algal species such as *Chlorococcum* spp., *Chlamydomonas reinhardtii* (Choi et al. 2010; Kong et al. 2010), *Schizocytrium* spp. (Kim et al. 2012), *Dictyochloropsis splendida* (Abd El-Moneim et al. 2010), *Spirulina* spp. (Markou et al. 2013), *Stichococcus bacillaris* (Olivieri et al. 2011), and *Chlorella vulgaris* (Lee et al. 2011) are the best candidates for biofuel production. Algal biomass could also be used as a raw material for the production of aircraft fuel and rocket fuel, biocrude oils, bioplastics, and improved livestock co-products (Bahadar and Khan 2013).

14.4 Lignocellulosic Biomass-Based Biofuel

Lignocellulosic biomass holds out the greatest potential when it comes to production of biofuels such as biohydrogen, bioethanol, biomethanol, biobutanol, biodiesel, and biogas in an eco-friendly manner. Bioethanol is particularly popular and the technology behind its production is well-known (Azad et al. 2015; Almodares and Hadi 2009; Banerjee et al. 2010).

14.4.1 Bioethanol

Bioethanol is the principal petrol substitute for transport vehicles. It is largely produced from the fermentation of sugar, although it can also be produced when ethylene reacts with steam. The basic sources of sugar required to produce ethanol come from energy crops such as corn, maize, and wheat and their residue such as straw and stover. Other sources include willow and poplar, sawdust, reed canary grass, cord grasses, Jerusalem artichoke, *Miscanthus* spp., and sorghum (Jorgensen 2011; Weijde et al. 2013). Ethanol or ethyl alcohol (C_2H_5OH) is a clear and colorless liquid that is biodegradable with low toxicity and causes little environmental pollution. Ethanol is a chemical that can be used for a number of different purposes such as anti-freeze, beverages, solvents, depressants, germicides, and fuel (Braide et al. 2016).

Bioethanol has a number of advantages over traditional fuels the most important of which are: (1) resources are renewable in that energy crops are used for its production rather than finite resources; (2) greenhouse gas emissions are reduced; (3) blending bioethanol with petrol will help extend the life of oil supplies; (4) widescale production of bioethanol would give the rural economy a boost by growing the necessary crops; (5) bioethanol is not only biodegradable but also far less toxic than fossil fuels; (6) bioethanol can be easily inserted into the existing road transport fuel system; and (7) bioethanol will eventually be produced using well-known methods such as fermentation. Looked at collectively these advantages make the production and use of bioethanol eco-friendly (Perlack et al. 2005). Bioethanol can be produced from lignocellulosic biomass using hydrolysis and sugar fermentation processes described in detail in the following sections.

14.5 Lignocellulosic Biofuel Production Technologies

The first challenge to overcome in producing fuels from lignocellulosic biomass is releasing the fermentable sugars trapped inside the biomass. The fermentable sugars can be extracted by first disconnecting the celluloses from the lignin and then using acid or enzymatic methods to hydrolyze the celluloses to break them down into simple monosaccharides (Rastegari et al. 2019a). Another challenge that needs to be overcome associated with biofuel production is the high percentage of pentoses in the hemicellulose such as xylose. Unlike hexoses such as glucose and mannose, pentoses are difficult to ferment (Alvira et al. 2010). Synthesizing biofuels from lignocellolusic biomass is generally dependent on the yield of fermentable sugars available and on the effectiveness of the various steps involved in the production of biofuels from lignocellulosic materials such as pretreatment, hydrolysis, fermentation, and product separation or distillation (Banerjee et al. 2010; Sarkar et al. 2012) (Fig. 14.3).

14.5.1 Pretreatment

In order to hydrolyze lignocellulosic biomass into fermentable sugars a standard pretreatment method capable of removing lignin is required. Pretreatment is the most important, costly, and complex step in the biofuel production process. The cellulose–hemicellulose complex acts as a chemical barrier and affects the biofuel production process by restricting cellulase enzyme activity. Lignin physically encapsulates the cellulose-hemicellulose complex and is an important barrier to cellulase enzyme activity (Alvira et al. 2010; Procentese et al. 2017). The pretreatment processes that have been used can be classified into four categories: physical, chemical, solvent, and biological.

14.5.1.1 Physical Pretreatments

Physical pretreatment methods do not involve using chemicals of any kind for biomass transformation. Alterations in the biomass material that take place during physical pretreatment include increasing the surface area of the material to facilitate enzyme penetration and action, reducing the degree of crystallinity and polymerization of the cellulose content, hydrolyzing hemicelluloses, and disrupting the lignin structure albeit incompletely. Physical pretreatment methods include chipping, milling, grinding, and even freezing. Lignocellulosic biomass pretreatment using radiation such as microwaves also fits into this category.

However, the major drawback with physical pretreatment is its limitations when it comes to large quantities of feedstock as a result of the high amount of power requited for the radiation process (Kumari and Singh 2018; Sheikh et al. 2015). This has led to other physical pretreatment methods being developed such as hydrothermal processes including steam explosion and liquid hot water treatment. Biomass material in such pretreatment methods is hydrolyzed by applying high temperature (160–290 °C) and pressure (20–50 MPa) over a short period of time. Although a number of degrading compounds that inhibit microbial growth and are detrimental to ethanol fermentation such as furfural and carboxylic acid are generated in these methods, high xylose recovery (up to 90%) and lack of acid or any chemical requirement makes this method very economic (das Neves et al. 2007).

14.5.1.2 Chemical Pretreatments

One of the most important chemical pretreatment techniques is acidic pretreatment in which H_2SO_4 or HCl is used to extract sugar (Kumar et al. 2009). Concentrated and dilute acids can both be used for this purpose. Acid recovery is the biggest challenge facing application of this process on a commercial scale. This means dilute acid with a concentration lower than 2% is favored since it can easily be neutralized by alkaline compounds such as ammonium and lime. Despite the economic advantages of dilute acid pretreatment, there are some limitations to this technique that have led to new alternative pretreatment process techniques being developed such as alkaline pretreatment (Harun et al. 2011). Alkaline pretreatment using a number of alkalis such as sodium hydroxide, calcium hydroxide, potassium hydroxide, lime, and aqueous ammonia has some advantages such as bringing about delignification, decreasing cellulose crystallinity, and facilitating enzyme action on cellulose by increasing its surface area. In addition, alkaline pretreatment can be done under ambient conditions and does not require high temperature and pressure.

Alkaline pretreatment is typically chosen for lignocellolusic materials that contain low amounts of lignin. There are a number of other unusual chemical pretreatment methods such as ozone pretreatment (ozonolysis), ionic liquid pretreatment, CO_2 explosion pretreatment, liquid hot-water pretreatment, wet oxidation pretreatment, steam explosion pretreatment, and ultrasonication (Kumar et al. 2009; Kumari and Singh 2018). Chemical pretreatment processes also suffer a number of limitations such as the time required for process completion can vary from hours to weeks and salt production during pretreatment not only inhibits microorganism growth but also affects the fermentation process raising concerns related to the environment. In addition, according to the high cost of this pretreatment method there is little chance of it being applied on a commercial scale (Procentese et al. 2017).

14.5.1.3 Solvent Pretreatments

Solvent pretreatment is a fractionation technique in which aqueous organic solvents with or without catalysts are used to bring about lignocellolusic material delignification. Methanol, ethanol, trimethyleneglycol, tetrahydrofurfuryl alcohol, ethylene glycol, glycerol, acetone, phenol, and n-butanol have been used in this method for lignin extraction (Kumar et al. 2009; Zhao et al. 2011). One of the advantages solvent pretreatment has over other pretreatments is that it recovers lignin as a by-product. However, organic solvents are expensive and difficulties in solvent recovery make this technique costly and impracticable commercially (Procentese et al. 2017; Zhao et al. 2011).

14.5.1.4 Biological Pretreatments

Biological pretreatment offers a solution to the disadvantages that plague physical and chemical pretreatment methods such as their need for expensive equipment, chemicals, and high energy usage for biomaterial processing (Kumar et al. 2009; Xu et al. 2016). Biological pretreatments are generally carried out by growing microorganisms directly on feedstocks or by using the microorganism's enzymes. Microorganisms are typically chosen to hydrolyze lignocellulosic biomass under common conditions since there is no need for specific equipment. Bacteria and fungi can both be used for this purpose. Rot fungi in particular are rich in lignin-degrading enzymes such as lignin peroxidase, laccases, and manganese peroxidase. They are considered the best candidate for biological pretreatment (Bak et al. 2009; Kirk and Moore 2007; Zhang et al. 2007). Although cellulase plays the most important role in biomass hydrolysis, there are other enzymes such as hemicellulase, ligninase, and pectinase (Binod et al. 2010). Accessory enzymes are also a crucial part of the biological pretreatment process. Important accessory enzymes involved in the hydrolysis of lignocellulosic biomass are α-arabinofuranidase, endoxylanases, exoxylanases, and β-xylosidases (Sindhu et al., 2016). Major challenges facing biological pretreatment in commercializing the use of lignocellulosic biomass in bioethanol production are the cost of hydrolyzing enzymes and such cocktails (Rodionova et al. 2017; Sassner et al. 2008).

Although the biological method is very energy efficient, it also has several disadvantages such as (1) it is extremely slow; (2) a significant amount of biomass is lost; and (3) much of the fermentable sugar available for bioethanol production is used by the microbes themselves for their own growth. Nevertheless, biological pretreatment is necessary because it brings about an increase in digestibility and fermentation rates (Steffen et al. 2000). When it comes to biofuel production this means it needs to be merged with other pretreatment technologies and that novel strains of microorganisms with rapid and effective hydrolysis capabilities and low growth rates need to be screened in order to make this method commercially viable.

14.5.1.5 Combined Pretreatments

No single pretreatment method to degrade lignocellulosic biomass provides suitable results on its own because of the influence of many factors such as lignin content, cellulose crystallinity, linkages between lignin and cellulose, and even intrinsic disadvantages. Studies show that incorporating two or more pretreatments from different categories, called the combined pretreatment method, can be more effective than single pretreatment processes. The combined pretreatment method can include a variety of combinations such as alkali and electron beam irradiation, alkali and ionic liquid, alkali and photocatalysis, biological and dilute acid, biological and steam explosion, dilute acid and microwave, dilute acid and steam explosion, enzyme hydrolysis and superfine grinding with steam, ionic liquid and ultrasonic, organosolvent and biological, SO_2 and steam explosion, supercritical CO_2 and steam explosion, microwave-assisted acid, and microwave-assisted alkali (Kumari and Singh 2018; Procentese et al. 2017).

14.5.1.6 Hydrolysis

Hydrolysis is a stage in which complex carbohydrate is degraded to monomeric sugars that are usually called fermentable sugars. Hydrolysis can lead to the complete breakdown of carbohydrates into simple monomeric sugars and ethanol or incomplete breakdown into oligosaccharides that require further hydrolysis before fermentation takes place (Alvira et al. 2010). Fermentable sugars produced during hydrolysis include mannose in softwood; xylose, arabinose, and galactose in hardwood and agricultural residues; glucose; and even fructose (Taherzadeh and Karimi 2008). Lignocellulosic biomass hydrolysis into fermentable sugars is generally carried out using either acids or enzymes.

Sulfuric acid (H_2SO_4) is the acid currently used for hydrolysis. However, a number of other acids have also been used for hydrolysis such as hydrochloric acid (HCl), nitric acid (HNO_3), trifluoroacetic acid (TFA), and phosphoric acid (H_3PO_4). Hydrolysis using acids is carried out either as dilute acid (< 1%) or as concentrated acid (30–70%). High temperature and pressure that have a low reaction time and low temperature that has a high reaction time (up to several hours) are required for dilute acid and concentrated acid treatment, respectively (Gírio et al. 2010).

Carbohydrates (hemicellulose and cellulose) in lignocellulosic biomass can be converted into fermentable sugars by enzymatic hydrolysis. This can be done either using degrading enzymes produced by microorganisms during their growth in media or using commercial enzymes. Enzymatic hydrolysis of cellulose into fermentable sugars is carried out by cellulase enzymes that consist of different enzymes including endoglucanase, exoglucanase, cellobiohydrolase, β-glucosidase, acetylesterase, glucuronidase, xy-lanase, β-xylosidase, galactomannanase, and glucomannanase (Kour et al. 2019a; Nigam and Singh 2011; Yadav et al. 2016; Zabed et al. 2016). Many microorganisms are capable of cellulase production such as *Clostridium*, *Bacillus*, *Cellulomonas*, *Ruminococcus*, *Bacteroides*, *Erwinia*, *Thermomonospora*, *Acetovibrio*, *Streptomyces*, *Microbispora*, *Sclerotium-rolfsii*, *Phanerochaete*, *Trichoderma*, *Schizophyllum*, *Aspergillus*, and *Penicillium* (Alvira et al. 2010; Kour et al. 2019c; Yadav et al. 2018). Several factors influence the enzymatic hydrolysis of lignocellulosic biomass the most important of which are temperature, pH and mixing rate, substrate concentration, cellulase loading, surfactant addition, and even pretreatment approach (Rastegari et al. 2019b, c; Sarkar et al. 2012; Taherzadeh and Karimi 2008).

Low amounts of energy and moderate conditions are generally required for enzymatic hydrolysis making it advantageous over acid hydrolysis. Enzymatic hydrolysis is one of few methods available that are advantageous as a result of it being less toxic, very cost-effective, and not generating any inhibitory by-products.

14.5.1.7 Fermentation

Fermentation is the final step of lignocellulosic biomass alteration in which microorganisms convert six-carbon sugars such as glucose, galactose, and mannose into ethanol (Harun et al. 2010). Finding microbial strains available in sufficient numbers that have ideal traits such as holding the potential for broad substrate utilization, high ethanol generation capacity, ability to tolerate high ethanol concentration and heat, and resistance to inhibitors is the main limitation to bioethanol production on an industrial scale (Singh et al. 2010). *Saccharomyces cerevisiae*, *Zymomonas mobilis*, *Pachysolen tannophilus*, *Escherichia coli*, *Candida shehatae*, *Pichia stipitis*, *Mucor indicus*, and *Candida brassicae* are the most frequently used microorganisms in the fermentation process (Lee et al. 2011; Sarkar et al. 2012).

14.5.1.8 Product Separation

The separation of biofuel, especially ethanol recovery from fermentation broth, is traditionally conducted using distillation (alone or in combination with adsorption). Ethanol existing in the fermentation broth can be concentrated depending on membrane selectivity via hydrophobic pervaporation before transferring it to distillation thus reducing the energy load on distillation (Sushil et al. 2013). During distillation the fermentation broth is distilled by separating ethanol from water in order to reach an ethanol concentration above 95%. Lignin, unreacted cellulose, hemicellulose, ash, enzymes, living cells, and other components are leftovers of this process that remain in the waste water. All remaining components can be concentrated and used either to provide energy or to be transformed into other co-products (Sarkar et al. 2012). Although most types of lignocellulosic biomass result in the ethanol produced being highly concentrated, this leads to a couple of problems: (1) the concentrations of inhibitors such as acetic acid and furfural are increased thus suppressing the performance of yeast and enzymes; and (2) high viscosity leads to the fermentor consuming more power and to a decline in mixing and heat transfer efficiency (Georgieva et al. 2007).

14.6 Lignocellulosic Biomass: Sustainable Renewable Resource for the Future

Lignocellulosic biomass has been getting a lot of attention in recent years as a renewable resource for the economical production of biofuel in the near future. This is because such raw materials are widely available, cheap, and have a high carbohydrate content (Lund 2007; Rastegari et al. 2020; Yadav et al. 2020a). Despite all these advantages, there are a number of challenges that need to be overcome to utilize lignocellulosic biomass for fuel production: (1) ethanol production from lignocellulosic biomass is not commercially viable because of low yields that result from the production of less fermentable sugars from different biomasses and technical limitations; (2) the contents of lignocellulosic biomasses differ one from another and depend on the source and type of raw material; (3) incomplete fermentation of pentose and hexose sugars present in the hydrolysate; (4) technological barriers existing in this technology such as high viscosity, inhibitor production, reaction temperature, and sugar availability in hydrolysate; and (5) a number of other factors such as lignocellulosic biomass nature, pretreatment methods, enzyme type, enzyme source and amount, microorganisms used, process conditions and reactor type influence biofuel production (Hernández-Beltrán et al. 2019; Rodionova et al. 2017). Biofuel production, especially ethanol from lignocellulosic biomass, is considered so complex that its utilization on a commercial scale is limited. However, as a result of biofuel science developing and technology improving there are new approaches showing great promise for fuel generation in the future.

Although lignocellulosic biomass is abundant, it is largely unutilizable as a resource for biofuels. However, biotechnology and gene engineering should come up with smarter strategies to facilitate the production of renewable raw materials and secure the future of the biofuel industry worldwide. Lignocellulosic biomass is mainly derived from crop residues and from the cultivation of perennial energy crops. The challenges facing biotechnology at the moment are hence increasing crop yield sustainably and developing crops with a suitable set of chemical and physical traits for biofuel production. Plant growth can be improved by increasing photosynthesis. The most successful approaches to improving plant growth include: (1) transferring genes from photosynthetic bacteria into plants without affecting the activity of plant-specific genes (traditional breeding techniques here are unsuitable for the development of crops for biofuel production); (2) manipulating genes involved in the metabolism of nitrogen (an essential element in proteins) and DNA has been successful as shown by the overexpression of a glutamine synthesis gene (GS1) in plants; and (3) extending the growth phase of plants by reducing seed dormancy or delaying flowering such that plants appropriate much of their energy in vegetative growth (Welker et al. 2015).

Abiotic and biotic stresses are the main cause of crop loss worldwide (Yadav et al. 2020b). Accordingly, developing crops with higher resistance to stresses is the prime focus of crop yield improvement either by traditional breeding methods or by biotechnological approaches. BT cotton has been engineered (genetically modified) with an insecticidal gene from the soil bacterium *Bacillus thurengiensis* and represents a very successful example of the use of biotechnology to develop crops with higher resistance to stresses (Welker et al. 2015).

Furthermore, the switch to renewable biomass sources will require the development of energy crops with desirable chemical (especially) and physical traits. The biosynthesis of cellulose and lignin is co-regulated; hence reducing the proportion of lignin will also increase the proportion of cellulose. Moreover, using techniques that can alter the properties of the cell wall could be key to facilitating sugar accessibility in the fermentation process (Alalwan et al. 2019; Jorgensen 2011; Welker et al. 2015).

Consideration should also be given to a number of concerns about utilizing lignocellulosic biomass derived from crops for biofuel production: (1) socioeconomic concerns about field management and choice of biomass source that should be carefully considered to ensure biofuel production does not negatively impact food production or biodiversity; (2) environmental concerns about the effects of fertilizers and herbicides used during the production of energy crops, especially in terms of human health; and (3) concerns about the environmental impact the combustion of specific types of biofuels could have in terms of emissions (Hernández-Beltrán et al. 2019).

14.7 Conclusion and Future Prospects

Lignocellulosic biomass has a lot of potential to be converted into biofuels such as bioethanol, biodiesel, and biogas. It has shown itself to be a good candidate to provide a solution to the world energy crisis in an ecofriendly manner, especially since the biofuels it produces are widely available as indigenous resources. Although extensive research has been carried out into the development of production technologies for biofuels from lignocellulosic biomass worldwide, there are still many limitations in such technologies to be overcome for future biofuel production to be commercially viable. Moreover, understanding the effects of unexpected events such as climate change on biofuel production from lignocellulosic biomass and its management is vital to the availability of sustainable biofuels in the future.

References

Abd El-Moneim MR Afify, Emad A Shalaby, Sanaa MM Shanab (2010) Enhancement of biodiesel production from different species of algae. Grasas y Aceites 61(4):416–422

Ahmed M-J-K, Ahmaruzzaman M (2016) A review on potential usage of industrial waste materials for binding heavy metal ions from aqueous solutions. J Water Process Eng 10:39–47

Alalwan H-A, Alminshid A-H, Aljaafari H-A-S (2019) Promising evolution of biofuel generations. Subject review. Renew Energy Focus 28:127–139

Almodares A, Hadi M (2009) Production of bioethanol from sweet sorghum: a review. Afr J Agric Res 4:772–780

Alvira P, Tomás-Pejó E, Ballesteros M, Negro M (2010) Pretreatment technologies for an efficient bioethanol production process based on enzymatic hydrolysis: a review. Bioresour Technol 101:4851–4861

Azad AK, Rasul M, Khan MMK, Sharma SC, Hazrat M (2015) Prospect of biofuels as an alternative transport fuel in Australia. Renew Sust Energ Rev 43:331–351

Bahadar A, Khan MB (2013) Progress in energy from microalgae: a review. Renew Sust Energ Rev 27:128–148

Bak JS, Ko JK, Choi IG, Park YC, Seo JH, Kim KH (2009) Fungal pretreatment of lignocellulose by *Phanerochaete chrysosporium* to produce ethanol from rice straw. Biotechnol Bioeng 104:471-482

Banerjee S, Mudliar S, Sen R, Giri B, Satpute D, Chakrabarti T, Pandey R (2010) Commercializing lignocellulosic bioethanol: technology bottlenecks and possible remedies. Biofuel Bioprod Biorefin 4:77–93

Bhatia L, Paliwal S (2011) Ethanol producing potential of *Pachysolen tannophilus* from sugarcane bagasse. Int J Biotechnol Bioeng Res 2(2):271–276

Binod P, Sindhu R, Singhania RR, Vikram S, Devi L, Nagalakshmi S, Kurien N, Sukumaran RK, Pandey A (2010) Bioethanol production from rice straw: an overview. Bioresour Technol 101:4767–4774

Braide W, Kanu I, Oranusi U, Adeleye S (2016) Production of bioethanol from agricultural waste. J Fund Appl Sci 8:372–386

Carere CR, Sparling R, Cicek N, Levin DB (2008) Third generation biofuels via direct cellulose fermentation. Int J Mol Sci 9:1342–1360

Cheng C-L, Che P-Y, Chen B-Y, Lee W-J, Lin C-Y, Chang J-S (2012) Biobutanol production from agricultural waste by an acclimated mixed bacterial microflora. Appl Energy 100:3–9

Choi SP, Nguyen MT, Sim SJ (2010) Enzymatic pretreatment of *Chlamydomonas reinhardtii* biomass for ethanol production. Bioresour Technol 101:5330–5336
Dar RA, Dar EA, Kaur A, Phutela UG (2018) Sweet sorghum-a promising alternative feedstock for biofuel production. Renew Sust Energ Rev 82:4070–4090
das Neves M-A, Kimura T, Shimizu N, Nakajima M (2007) State of the art and future trends of bioethanol production. Dyn Biochem Process Biotechnol Mol Biol 1(1):1–14
Demirbas A (2010) Social, economic, environmental and policy aspects of biofuels. Energ Edu Sci Technol Part B-Soc Edu Stud 2:75–109
Doshi A, Pascoe S, Coglan L, Rainey T-J (2016) Economic and policy issues in the production of algae-based biofuels: a review. Renew Sustain Energy Rev 64:329–337
Falkowski PG, Katz ME, Knoll AH, Quigg A, Raven JA, Schofield O, Taylor F (2004) The evolution of modern eukaryotic phytoplankton. Science 305:354–360
Georgieva T, Ahring B-K (2007) Evaluation of continuous ethanol fermentation of dilute-acid corn stover hydrolysate using thermophilic anaerobic bacterium *Thermoanaerobacter* BG1L1. Appl Microbiol Biotechnol 77:61–68
Gírio F-M, Fonseca C, Carvalheiro F, Duarte L-C, Marques S, Bogel-Łukasik R (2010) Hemicelluloses for fuel ethanol: a review. Bioresour Technol 101(13):4775–4800
Gouveia L, Oliveira AC (2009) Microalgae as a raw material for biofuels production. J Ind Microbiol Biotechnol 36:269–274
Harun R, Danquah MK, Forde GM (2010) Microalgal biomass as a fermentation feedstock for bioethanol production. J Chem Technol Biotechnol 85:199–203
Harun R, Jason W, Cherrington T, Danquah MK (2011) Exploring alkaline pretreatment of microalgal biomass for bioethanol production. Appl Energy 88:3464–3467
Heimann K (2016) Novel approaches to microalgal and cyanobacterial cultivation for bioenergy and biofuel production. Curr Opin Biotechnol 38:183–189
Hernández-Beltrán J-U, Hernández-De Lira I-O, Cruz-Santos M-M, Saucedo-Luevanos A, Hernández-Terán F, Balagurusamy N (2019) Insight into pretreatment methods of lignocellulosic biomass to increase biogas yield: current state, challenges, and opportunities. Appl Sci 9:3721
Ho S-H, Huang S-W, Chen C-Y, Hasunuma T, Kondo A, Chang J-S (2013) Bioethanol production using carbohydrate-rich microalgae biomass as feedstock. Bioresour Technol 135:191–198
Huzir NM, Aziz MMA, Ismail S, Abdullah B, Mahmood NAN, Umor N, Muhammad SAFaS (2018) Agro-industrial waste to biobutanol production: eco-friendly biofuels for next generation. Renew Sust Energ Rev 94:476–485
Jorgensen U (2011) Benefits versus risks of growing biofuel crops: the case of *Miscanthus*. Curr Opin Environ Sust 3:24–30
Kim JK, Um B-H, Kim TH (2012) Bioethanol production from micro-algae, *Schizocytrium* sp., using hydrothermal treatment and biological conversion. Kor J Chem Eng 29:209–214
Kim S, Dale BE (2004) Global potential bioethanol production from wasted crops and crop residues. Biomass Bioenergy 26:361–375
Kirk T-K, Moore WE (2007) Removing lignin from wood with white-rot fungi and digestibility of resulting wood. Wood Fiber Sci 4:72–79
Kong Q-x, Li L, Martinez B, Chen P, Ruan R (2010) Culture of microalgae *Chlamydomonas reinhardtii* in wastewater for biomass feedstock production. Appl Biochem Biotechnol 160:9
Kour D, Rana KL, Kaur T, Singh B, Chauhan VS, Kumar A et al. (2019a) Extremophiles for hydrolytic enzymes productions: biodiversity and potential biotechnological applications. In: Molina G, Gupta VK, Singh B, Gathergood N (eds) Bioprocessing for biomolecules production, pp 321–372. https://doi.org/10.1002/9781119434436.ch16
Kour D, Rana KL, Yadav N, Yadav AN, Rastegari AA, Singh C et al. (2019b) Technologies for biofuel production: current development, challenges, and future prospects. In: Rastegari AA, Yadav AN, Gupta A (eds) Prospects of renewable bioprocessing in future energy systems. Springer International Publishing, Cham, pp 1–50. https://doi.org/10.1007/978-3-030-14463-0_1

Kour D, Rana KL, Yadav N, Yadav AN, Singh J, Rastegari AA et al. (2019c) Agriculturally and industrially important fungi: current developments and potential biotechnological applications. In: Yadav AN, Singh S, Mishra S, Gupta A (eds) Recent advancement in white biotechnology through fungi, Volume 2: Perspective for value-added products and environments. Springer International Publishing, Cham, pp 1–64. https://doi.org/10.1007/978-3-030-14846-1_1

Koutra E, Economou CN, Tsafrakidou P, Kornaros M (2018) Bio-Based Products from Microalgae Cultivated in Digestates. Trends Biotechnol 36(8):819–833

Kumar K, Yadav AN, Kumar V, Vyas P, Dhaliwal HS (2017) Food waste: a potential bioresource for extraction of nutraceuticals and bioactive compounds. Bioresour Bioprocess 4:18. https://doi.org/10.1186/s40643-017-0148-6

Kumar P, Barrett DM, Delwiche MJ, Stroeve P (2009) Methods for pretreatment of lignocellulosic biomass for efficient hydrolysis and biofuel production. Ind Eng Chem Res 48:3713–3729

Kumar S, Sharma S, Thakur S, Mishra T, Negi P, Mishra S et al (2019) Bioprospecting of microbes for biohydrogen production: current status and future challenges. In: Molina G, Gupta VK, Singh BN, Gathergood N (eds) Bioprocessing for biomolecules production. Wiley, USA, pp 443–471

Kumari D, Singh R (2018) Pretreatment of lignocellulosic wastes for biofuel production: a critical review. Renew Sust Energ Rev 90:877–891

Lee S, Oh Y, Kim D, Kwon D, Lee C, Lee J (2011) Converting carbohydrates extracted from marine algae into ethanol using various ethanolic Escherichia coli strains. Appl Biochem Biotechnol 164:878–888

Limayem A, Ricke S-C (2012) Lignocellulosic biomass for bioethanol production: current perspectives, potential issues and future prospects. Prog Energy Combust Sci 38(4):449–467

Lund H (2007) Renewable energy strategies for sustainable development. Energy 32:912–919

Markou G, Angelidaki I, Nerantzis E, Georgakakis D (2013) Bioethanol production by carbohydrate-enriched biomass of *Arthrospira* (*Spirulina*) *platensis*. Energies 6:3937–3950

McKeever D-B (1998) Wood residual quantities in the United States. BioCycle J Composting Recycl 39(1):65–68

Nakagawa H, Harada T, Ichinose T, Takeno K, Matsumoto S, Kobayashi M, Sakai M (2007) Biomethanol production and CO_2 emission reduction from forage grasses, trees, and crop residues. Japan Agric Res 41:173–180

Nigam P-S, Singh A (2011) Production of liquid biofuels from renewable resources. Prog Energy Combust Sci 37(1):52–68

Olivieri G, Marzocchella A, Andreozzi R, Pinto G, Pollio A (2011) Biodiesel production from Stichococcus strains at laboratory scale. J Chem Technol Biotechnol 86(6):776–783

Organization W-H (2015) Food and agriculture organization of the united nations. Probiotics in food: health and nutritional properties and guidelines for evaluation. Fao food and nutrition, Rome

Perlack R-D, Wright L-L, Turhollow A-F, Graham R-L, Stokes B-J, Erbach D-C (2005) Biomass as feedstock for a bioenergy and bioproducts industry: the technical feasibility of a billionton annual supply. Oak Ridge National Lab TN

Procentese A, Raganati F, Olivieri G, Russo ME, Marzocchella A (2017) Pretreatment and enzymatic hydrolysis of lettuce residues as feedstock for bio-butanol production. Biomass Bioenergy 96:172–179

Rastegari AA, Yadav AN, Yadav N (2020) New and future developments in microbial biotechnology and bioengineering: Trends of microbial biotechnology for sustainable agriculture and biomedicine systems: diversity and functional perspectives. Elsevier, Amsterdam

Rastegari AA, Yadav AN, Gupta A (2019a) Prospects of renewable bioprocessing in future energy systems. Springer International Publishing, Cham

Rastegari AA, Yadav AN, Yadav N (2019b) Genetic manipulation of secondary metabolites producers. In: Gupta VK, Pandey A (eds) New and future developments in microbial biotechnology and bioengineering. Elsevier, Amsterdam, pp 13–29. https://doi.org/10.1016/B978-0-444-63504-4.00002-5

Rastegari AA, Yadav AN, Yadav N, Tataei Sarshari N (2019c) Bioengineering of secondary metabolites. In: Gupta VK, Pandey A (eds) New and future developments in microbial biotechnology and bioengineering. Elsevier, Amsterdam, pp 55–68. https://doi.org/10.1016/B978-0-444-63504-4.00004-9

Rodionova M, Poudyal R, Tiwari I, Voloshin R, Zharmukhamedov S, Nam H, Zayadan B, Bruce B, Hou H, Allakhverdiev S (2017) Biofuel production: challenges and opportunities. Int J Hydrogen Energy 42:8450–8461

Rooney T (1998) Lignocellulosic feedstock resource assessment. NREL Report SR-580-24189, 123

Salehi-lisar S-Y, Bakhshayeshan-agdam H (2016) Drought stress in plants: Causes, consequences, and tolerance. In: Hossain MA et al (eds) Drought stress tolerance in plants, vol 1, pp 1–16. Springer Press, New York, USA

Sarkar N, Ghosh SK, Bannerjee S, Aikat K (2012) Bioethanol production from agricultural wastes: an overview. Renew Energy 37:19–27

Sassner P, Mårtensson C-G, Galbe M, Zacchi G (2008) Steam pretreatment of H2SO4-impregnated Salix for the production of bioethanol. Bioresour Technol 99(1):137–145

Sheikh MMI, Kim CH, Park HH, Nam HG, Lee GS, Jo HS, Lee JY, Kim JW (2015) A synergistic effect of pretreatment on cell wall structural changes in barley straw (*Hordeum vulgare* L.) for efficient bioethanol production. J Sci Food Agric 95:843–850

Sims RE (2004) Biomass, bioenergy and biomaterials: future prospects. Biomass and agriculture–sustainability markets and policies. OECD, Paris, pp 37–61

Sindhu R, Binod P, Pandey A (2016) Biological pretreatment of lignocellulosic biomass–an overview. Bioresour Technol 199:76–82

Singh A, Pant D, Korres NE, Nizami A-S, Prasad S, Murphy JD (2010) Key issues in life cycle assessment of ethanol production from lignocellulosic biomass: challenges and perspectives. Bioresour Technol 101:5003–5012

Steffen K, Hofrichter M, Hatakka A (2000) Mineralisation of 14 C-labelled synthetic lignin and ligninolytic enzyme activities of litter-decomposing basidiomycetous fungi. Appl Microbiol Biotechnol 54(6):819–825

Sushil S Gaykawad, Ying Zha, Peter J Punt, Johan W van Groenestijn, Luuk AM van der Wielen, Adrie JJ Straathof (2013) Pervaporation of ethanol from lignocellulosic fermentation broth. Bioresour Technol 129:469–476

Taherzadeh M-J, Karimi K (2008) Pretreatment of lignocellulosic wastes to improve ethanol and biogas production: a review. Int J Mol Sci 9(9):1621–1651

Voloshin RA, Rodionova MV, Zharmukhamedov SK, Veziroglu TN, Allakhverdiev SI (2016) Biofuel production from plant and algal biomass. Int J Hydrogen Energy 41:17257–17273

Weijde TVD, Alvim Kamei CL, Torres AF, Vermerris W, Dolstra O, Visser RGF, Trindade LM (2013) The potential of C4 grasses for cellulosic biofuel production. Front Plant Sci 4:107

Welker C, Balasubramanian V, Petti C, Rai K, DeBolt S, Mendu V (2015) Engineering plant biomass lignin content and composition for biofuels and bioproducts. Energies 8:7654–7676

Xu G-C, Ding J-C, Han R-Z, Dong J-J, Ni Y (2016) Enhancing cellulose accessibility of corn stover by deep eutectic solvent pretreatment for butanol fermentation. Bioresour Technol 203:364–369

Yadav AN, Rastegari AA, Yadav N (2020a) Microbiomes of extreme environments: biodiversity and biotechnological applications. CRC Press, Taylor & Francis, Boca Raton, USA

Yadav AN, Sachan SG, Verma P, Kaushik R, Saxena AK (2016) Cold active hydrolytic enzymes production by psychrotrophic Bacilli isolated from three sub-glacial lakes of NW Indian Himalayas. J Basic Microbiol 56:294–307

Yadav AN, Singh J, Rastegari AA, Yadav N (2020b) Plant microbiomes for sustainable agriculture. Springer International Publishing, Cham

Yadav AN, Singh S, Mishra S, Gupta A (2019) Recent advancement in white biotechnology through fungi. Volume 2: Perspective for value-added products and environments. Springer International Publishing, Cham

Yadav AN, Verma P, Kumar V, Sangwan P, Mishra S, Panjiar N et al. (2018) Biodiversity of the Genus *Penicillium* in Different Habitats. In: Gupta VK, Rodriguez-Couto S (eds) New and future developments in microbial biotechnology and bioengineering, *Penicillium* system properties and applications. Elsevier, Amsterdam, pp 3–18. https://doi.org/10.1016/b978-0-444-63501-3.00001-6
Zabed H, Sahu J, Boyce A, Faruq G (2016) Fuel ethanol production from lignocellulosic biomass: an overview on feedstocks and technological approaches. Renew Sustain Energy Rev 66:751–774
Zhang X, Yu H, Huang H, Liu Y (2007) Evaluation of biological pretreatment with white rot fungi for the enzymatic hydrolysis of bamboo culms. Int Biodeteriorat Biodegrad 60:159–164
Zhao X-Q, Zi L-H, Bai F-W, Lin H-L, Hao X-M, Yue G-J, Ho N-W (2011) Bioethanol from lignocellulosic biomass. In: Biotechnology in China III: Biofuels Bioenergy. Springer, pp 25–51

Chapter 15
Jatropha: A Potential Bioresource for Biofuel Production

Archita Sharma and Shailendra Kumar Arya

Abstract There has been an increased urgency in the demands of energy worldwide because of (a) the exhaustion of fossil fuels, (b) extended growth of global population, and (c) the economy from industrialization. Considering various countries, India has outperformed Japan and Russia and evolved as the third best consumer of oil, universally. Apart from the high demands of oil fuels, environmental problems like global warming, pollution, etc. have great consequences, and thus, there is a dire need for the development of an alternate form of energy in the R&D domain. An alternate form like production of energy from biomass is considered as a sustainable form of energy and has also gained positive responses from various sectors such as public sector, industrial sector, and policies of the government. Another alternate form of energy which is the talk of the talk from the recent past is jatropha. Jatropha is considered as a novel and a promising plant which results in the amplification of a renewable source of energy. Because of numerous advantages, it is one of the exclusive nominees with appreciable and ethereal merits toward ecology and the environment. The majority of the plantations are done on reduced wastelands globally. There is dearth awareness about jatropha in order to understand the contribution to the societies and toward the environment. Currently, jatropha has grabbed much of the attention of researchers due to its enormous performance in the production of biodiesel, an environment-friendly fuel, which is biodegradable and renewable in nature with no toxicity in the environment compared to petroleum oil, diesel, etc. There is an utmost requirement for some blueprint or plan to sort the issues of the crisis associated with the energy and to make use of jatropha as a substitute for the fossil fuels and other sources of energy. This chapter deals with the use, strengths and weaknesses, and toxicity of jatropha and its associated issues. Also, the dire need for alternative fuels has also been discussed following capital investment, cost of production, processing technologies, and some examples.

A. Sharma · S. K. Arya (✉)
Department of Biotechnology, University Institute of Engineering and Technology (UIET), Panjab University (PU), Chandigarh, Punjab, India
e-mail: skarya_kr@yahoo.co.in

A. N. Yadav et al. (eds.), *Biofuels Production – Sustainability and Advances in Microbial Bioresources*, Biofuel and Biorefinery Technologies 11,
https://doi.org/10.1007/978-3-030-53933-7_15

15.1 Introduction

There is an utmost requirement for bulk research regarding the development and production of alternative fuels along with meeting the demand of the energy in a sustainable way and with least effects on the environment as there has been rapid expenditure of fossil fuels. Considering the report of IPCC (Intergovernmental Panel on Climate Change—an intergovernmental body of the UN), the significant concern toward the environment is that with the increase in the concentrations of the anthropogenic greenhouse gas (GHG), there has been an increase in the range of the average temperatures worldwide during the mid-20th century (Report of Working Group of the Intergovernmental Panel on Climate Change). The increase in the concentrations of the greenhouse gases (GHGs) because of the combustion of fossil fuels has results in global warming. Yet another grave concern toward the development and production of alternative fuels is the increase in the prices of fossil fuels because of the extreme scarcity of the underground reserves of petroleum due to increased consumption of energy. It has been assumed that there will be a condition of an extreme catastrophe during the years 2010–2020 due to an increase in the demand for energy, as compared to the supply, which will keep on increasing with time. As there is no persistence of the scarcity of the energy till now, this situation can be postponed by the efforts made in the direction of the development and production of alternative fuels (Fusco 2013). Last but not the least, as it is well known that maximum amount of the reserves of petroleum are situated in Middle East Asia, hence, there is a casual notion with the increase in the scarcity of the fuel globally will result in the state of agitation in Middle East Asia. This may even lead to further conflict and war (Datta and Mandal 2014).

An alternative or non-conventional fuel (biodiesel, biofuels, hydrogen, fuels as a derivative of biomass) is any entity or element which is the source of the fuel apart from conventional fuels which consist of fossil fuels namely reserves of petroleum oil, coal, etc. and certain fuels with nuclear origin such as uranium in some cases (Datta and Mandal 2014). In the recent past, researchers are acknowledging jatropha for the production of the biofuel. It is a versatile small tree with large shrubs which is found in the entire tropical area (Mexico, Central America). It is broadly and widely allocated in wild/semi-cultivated areas of Latin America, Africa, India, and South-East Asia. In India, it is mostly dispersed in the Andaman Islands (in majority) as a live fence. It is well suitable for arid and semi-arid environments. Jatropha is a dynamic and drought receptive plant and with unsavory taste and hence not eaten by animals (Francis et al. 2005; Openshaw 2000b).

Jatropha is categorized as a novel contestant in a group of renewable sources of energy because of its distinguishing properties such as (a) drought tolerance (Openshaw 2000a), (b) quick growth, (c) effortless propagation, (d) high amount of oil content as compared to the other crops that produce oil (Achten et al. 2008), (e) small time duration of gestation, (f) broad range of adaptation in the environment, and (g) favorable size and architecture of the plant. The life of the productivity of jatropha is around 30 years or so with a yield of the seed in the range 0.5–12 ton year^{-1} ha^{-1} relying on certain factors like soil, amount of the nutrients, rainfall pattern, etc. (Francis et al. 2005; Openshaw 2000b). The seeds of jatropha consist of 30–35% of oil which can later be transformed into biodiesel with favorable via transesterification production process (Foidl et al. 1996). However, the actual potential of jatropha is still under the covers but there has been a lot of improvement in the conditions in the recent past for taking or using its advantages because of the increase in the prices of crude oil. Nowadays, many venture capitalists, policymakers, and project developers are showing keen interest in jatropha on the basis of its distinctive features, promising potentials, etc. engaged to handle and sort out the issues of supply of energy and reduce the emission of greenhouse gases (GHGs) (Pratap Singh et al. 2015). This chapter gives the insights of multifunctional jatropha, its benefits to the environment, and how jatropha is considered as a suitable candidate for the production of biofuel from the renewable source of energy with appropriate examples (Pandey et al. 2012).

15.2 Why Alternative Fuels?

It has been reported that in the year 2005, there was a consumption of approximately thirty million tons of oil in India in the domain of transportation. Out of these thirty million tons, 29% consists of gasoline and 71% consists of diesel. Taking into consideration the demands of energy in India only, there has been anticipation that there will be a hike in the demands of energy at an annual rate of 4.8% within the coming decades. The projection score was assumed to be double of the present consumption of the oil, latest by the year 2030 (Datta and Mandal 2014).

Hence, alternative fuels (biofuels) will have significant performance in this context. Various strategies for the development of biofuels and respective legal actions have been devised in countries in order to produce biofuels, such as a compulsory 5% blending of ethanol in gasoline with their respective trials. With respect to the Indian government, a goal was fixed to enhance the blending of the biofuels with diesel and gasoline, respectively, to nearly about 20% by the year 2017 (Report of the Committee on Development of Biofuel, Planning Commission, Government of India, 2003). It has been reported that crude oil production (domestic) in 2003–2004 was about 33.38 million tons, when in fact the imported amount of crude oil production was 90.43 million tons. This accounts for approximately 73% of total oil which is consumed. From the reports already published, it has been acknowledged that the production of crude oil is increasing in a brisk manner with static rates of the production process (Report of United States Energy Information Administration). With prompt increase in the importing of the fuel, the economy of the fuel in the forthcoming era will be depending too much on the countries which are considered as the main and significant countries for the production of the fuel. The economy of the domestic areas of countries that imports fuels has devastating effects with the increase in the prices of the fuels (Datta and Mandal 2014).

All these issues and concerns have urged researchers, governments, implementing institutes to think and plan toward the development and production of alternative fuels for the betterment of the future and economy of the countries and also to reduce the import of the crude oil. Alternative fuels from the endemic sources are recently the talk of the town and hence are explored at a higher extent. These endemic alternative fuels will help in the reduction in the amount of the bill regarding import of the oil which will alter help in improving the economy of the domestic region of the country (Datta and Mandal 2014).

15.3 Biofuels

15.3.1 Overview

The primary source of energy is fossil fuels which have a contribution of approximately 80% in which the share of the transportation domain is 58% (Escobar et al. 2009). There is a rapid depletion of fossil fuels and their reserves which are considered as a major contributor toward the emission of harmful gases. These harmful gases have a negative impact on the environment such as (a) lessening of glaciers, (b) biodiversity loss, (c) changes in the climate, and (d) sea level rise. With an increase in the demand for fossil fuels, the worldwide economy is also getting affected because of the high costs of crude oil. Industrialization and transportation are the two very basic amenities for this existing high-paced modernized world. This modernized world is the sole cause of the erratic demand for fuel (Agarwal 2007). There are many alternate sources of energy like biofuels which are available already in the market (Kour et al. 2019; Kumar et al. 2019). Continuous research work is going on for the production of the biofuel from renewable resources in order to replace the nonrenewable fuels (Weldu and Wondimagegnehu 2015). The growing research in the sector of biofuels over fuels produced from the petroleum reserves is because of certain merits like easy and clean extraction from the biomass, biodegradable nature, combustion process on the basis of carbon dioxide (CO_2) cycle, and eco-friendly nature. These particular merits will endow a rapid increase in the shares of the automobile market in the coming decades, active and durable growth of the agricultural sector, etc. (Demirbas 2008; Kim and Dale 2005).

There are basically three categories (Fig. 15.1) of the biofuels and they are first (biodiesel, vegetable oil), second (bio-ethanol, bio-hydrogen), and third (biogas) which are categorized in the chemical nature and complexity of the biomass (Rastegari et al. 2019, 2020). The production of the first-generation biofuels is from the crop plants, the production of the second-generation biofuels is from the byproducts of agriculture and from the energy plants which need a fertile land for the growth, and the production of the third generation biofuels is from the biomass in bulk in a particular time duration and these also do not need fertile land to grow (Kang et al. 2014). It has been reported in the papers that biomass is considered as the fourth

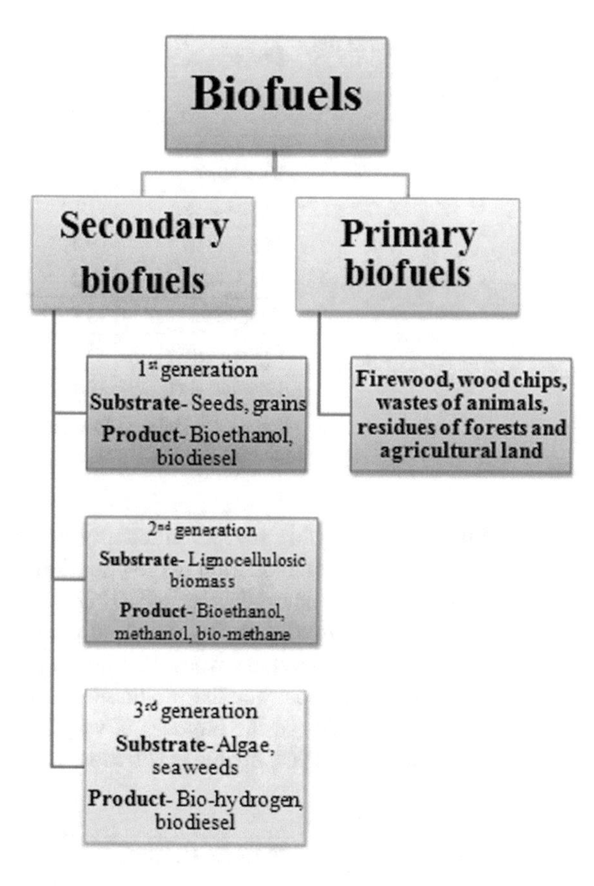

Fig. 15.1 Classification of the biofuels (Reddy 2013)

largest source of energy around the globe (Haykiri-Acma and Yaman 2010) due to its natural existence, an economical device for its storage with respect to energy. This particular energy can later be exploited at any time (Raajendiran et al. 2008). Annually, the fresh availability of the biomass in India has an estimation of approximately 500 million metric tons, which consists of the biomass of agricultural and forest areas. Also, the estimated potential for energy is approximately 18000 MW (MNRE 2006). Extensive research work is going on, all over the globe, to find an efficient way for grasping the energy potential hidden inside the biomass via some energy transfer mechanism which is still under investigation. Generation of biomass in bulk from various industries, forests, agricultural sector, and marine areas is ultimately decaying with unrestrained mechanism thus resulting in the damage to the environment due to the emission of the toxic/harmful gases (Taherzadeh and Karimi 2007a, 2007b). Table 15.1 depicts the generation of the biomass from the significant agricultural crops of India in million tons during the years 2009–2015 (Gaurav et al. 2017).

Table 15.1 Generation of the biomass from significant agricultural crops of India

Crops	Residues of biomass (Million tons-2009–2015)	Average biomass production (Million tons-2009–2015)
Rice husk	303	50.5
Wheat straw	719.4	119.9
Maize cobs	39.1	6.517
Jute Sticks	13.2	2.642
Groundnut shell	13.8	2.3
Coconut fiber	15.1	3.02
Coconut shell	12.5	2.5

Source Agricultural Research Data Book 2016 (http://www.iasri.res.in)

Utilizing these renewable sources efficiently will result in the reduction of the greenhouse gases (GHGs), reduction in the pollutions of the environment, improvement in the economy of the rural areas, etc. (Ragauskas et al. 2006). The generation of biofuels depends primarily on the terrestrial plants (with limited exploitation of the cultivable land) and biomass from the marine plants like seaweeds which are the fresh source of feedstock for the production of biofuel and thus giving a high percentage of productivity of the biomass in very less time. Table 15.2 depicts the sources of renewable and sustainable production of energy (Gaurav et al. 2017).

Table 15.2 Sources of renewable and sustainable production of energy

Terrestrial Sources	Marine Sources (A) Macroalga seaweeds	Marine Sources (B) Microalgae
Jatropha	*Acrosiphonia orientalis*	*Scenedesmus obliquus*
Switchgrass	*Ulva fasciata*	*Cyanobacteria*
Bermuda grass	*Enteromorpha compressa*	*Phormidium sp.*
Silver grass	*Bryopsis pennata*	*Spirogyra sp.*
Trailblazer	*Dictyota adnata*	*Euglena sp.*
	Gelidium pusillum	*Chalmydomonas reinhardtii*
	Gracilaria corticata	*Dunaliella tertiolecta*

Source Gaurav et al. (2017)

15.3.2 Universal View

It has been recognized worldwide that there will be a high estimate of the biofuel production process with respect to the energy systems in the near future. The availability of the bioresources for the production of biofuels will be a critical factor when the shares of biofuels in the automobile sector will touch their maximum peak market value. It is possible to attain the transition of the economy of the hydrocarbons (HCNs) into the economy of the carbohydrates via biomass for the production of the bio-methanol and bio-ethanol which will replace the fuels based on the oil (Yadav et al. 2020). With the production of the biofuels, there has been a reduction in the imports of the oil, growth in Gross Domestic Product (GDP), decrease in the emissions of the carbon dioxide, employment opportunities, etc. (Ghosh 2016; Hassan and Kalam 2013).

15.3.3 Economy of Biofuels

Till date, there is no such compromise between the prices of the biofuels with that of the conventional fuels even though there are higher prices of the oil. There has been a surge in the biofuel economy ever since the twenty-first century. In the past, the forces that have shaped the economy of the hydrocarbons are now being used to shape the economy of the biofuels and their respective bio-refineries. It has been assumed that the respective targets can be achieved by utilizing the biofuels in the public transport sector. The major and significant approaches to accomplish the task are the obligation of the biofuel and decrease in the tax (Demirbas 2008).

To develop the economy of the agricultural sector, the production of the energy has a major role to play. Hence, it is advised to adopt developmental programs for the communities sanctioning the development of a socioeconomic attitude of the country. Presently, producing methanol from natural gas and synthesizing ethanol from ethylene is a very costly affair. Another approach to produce bio-ethanol and bio-methanol simultaneously from the juice of sugar is pretty much attractive with respect to the aspects of the economy in particular areas where electricity from water is available at cheaper rates. The studies and investigations have already warned the exhaustion of the energy from conventional sources such as petroleum reserves, natural gas, coal, etc. and have projected that the eco-friendly biomass will be the apt alternative for meeting the demands of the energy (Demirbas 2008).

15.4 Jatropha

Jatropha is a tree from Central America (Fairless 2007) and Brazil. It is listed as a weed in countries like Australia, South Africa, India, Brazil, parts of Caribbean, etc. Jatropha is a plant which can tolerate drought and can grow on wastelands. The cultivation of jatropha is very easy and can be done by farmers with low incomes. The flowering stage of jatropha occurs within one year of the plantation after activities like irrigation, fertilization, and tilling of the soil and after flowering there is production of abundant crops. Additionally, oil cake formed from jatropha is not appropriate for nourishing livestock due to the presence of toxic substances like phorbol-ester, curcin, etc. (Menezes et al. 2006). A very minute amount of information is available over the globe about the germplasm of jatropha. The germplasm studies will help in improving the quality of the crops (Carvalho et al. 2008).

The chemical composition of jatropha is shown in Fig. 15.2. The content of oil present in the seed is within the range of 25–30% (Table 15.3). The oil consists of saturated (21%) and unsaturated (79%) fatty acids. These are certain chemical substances present in the seed namely curcin, which is considered as a poisonous substance and thus makes oil not suitable for consumption by humans (Raja et al. 2011).

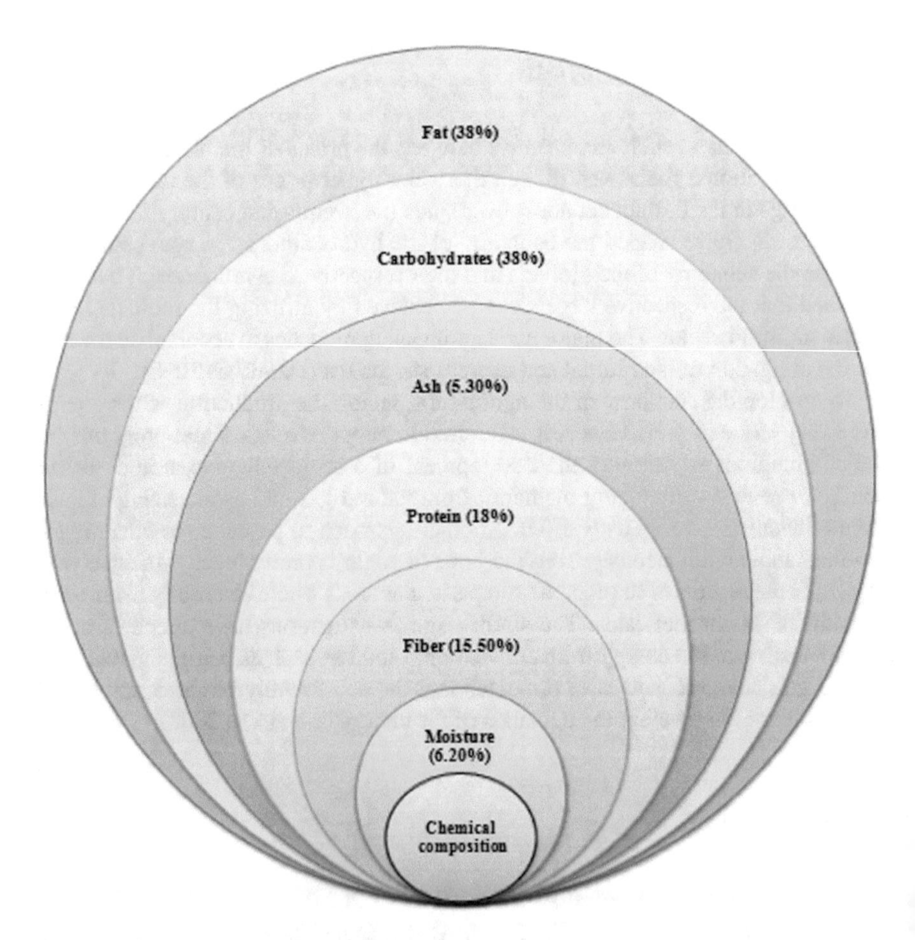

Fig. 15.2 Chemical composition of Jatropha seeds (Raja et al. 2011)

Table 15.3 Per hectare production of oil from biofuel crops

Biofuel crops	Liters of oil per hectare
Oil palm	2400
Jatropha	1300
Rapeseed (Canola)	1100
Sunflower	690
Soya bean	400

Source United Nations Development Programme/World Bank
*Indian Planning Commission

15.4.1 Toxicity

There is a formation of seed pods within the seeds of jatropha. Each and every seed pod consists of three seeds. Apart from a priceless oil source, these seeds are a rich source of protein as well. The composition of the protein in the seeds of jatropha is in alignment with the composition with the soybean meal (Makkar et al. 1998) which includes a favorable equilibrium of essential amino acids, except lysine. Mostly, the seeds are unfit to consume and are resistant to the treatments of heat-inactivation which are used in the processing of seed-meal (Heller et al. 1996). As a result, the seed rich in proteins is not suitable for feeding animals. There are varieties of toxins and anti-nutrients present in the seeds of jatropha. The basic toxic nature of the seeds of jatropha is due to the occupancy of curcin (which is a protein) and phorbol-esters (which are diterpenoids: a classification of terpenes) (King et al. 2009).

15.4.1.1 Curcin

It is frequently categorized as lectin and is similar to ricin which is a product of castor beans. Both curcin and ricin have the same levels of toxicity. Precisely, both curcin and ricin are called ribosome inactivating proteins (RIPs), which results in depurination of ribosome RNA which arrests the synthesis of proteins (King et al. 2009).

Curcin is considered as a type-I ribosome inactivating protein (Juan et al. 2002; Qin et al. 2005) whereas ricin is designated as a type-II ribosome inactivating protein. These type-II ribosome inactivating proteins consist of (a) catalytic A-chain and (b) carbohydrate binding lectin B-chain (encoded by the same gene) (Hartley and Lord 2004). There is an absence of lectin domain in type-I ribosome inactivating proteins. The toxicity of type-II ribosome inactivating proteins is partially due to the capacity of the lectin binding to the cell surfaces and helps in mediating the arrival of the ribose inactivating protein into the cell (Olsnes et al. 1974).

Because of the absence of the lectin domain, the values of LD_{50} of type-I ribose inactivating proteins are mostly 1000 times higher as compared to the type-II ribose inactivating proteins in the entire animal model (mouse) (Barbieri et al. 1993).

Furthermore, type-I ribose inactivating proteins are observed in numerous plant material that are of edible nature such as cereal grains (wheat, barley, etc.) (Motto and Lupotto 2004), beetroot, leaves of spinach, asparagus, etc. (Barbieri et al. 2006). Thus, curcin is a major barrier toward feeding animals via processing of seed meal.

15.4.1.2 Phorbol-Esters

There are certain varieties of seeds of jatropha, in the country Mexico, that are edible apart from the toxic nature of the jatropha seeds which are usually employed by the population of local communities after cooking (Verma et al. 2012). By performing a scrutinization, the studies of edible seeds and nonedible seeds of jatropha have disclosed that there is an absence of phorbol esters in the varieties of edible seeds of jatropha (Makkar et al. 1998). It has been investigated that providing heat treatment to the edible seeds of jatropha makes it suitable for its usage as a foodstuff for rats with no unfortunate effects (Makkar and Becker 1999), even though more research work is required to test/check the fitness of a meal of jatropha as a feed for animals, which has the potential to prosper. Till date, there has been a discovery of the structures of 6 phorbol-esters by using the characterization technique, nuclear magnetic resonance (NMR) (Haas et al. 2002). Phorbol-esters are commonly present in the seeds or in the latex (King et al. 2009).

Phorbol-esters are equivalents of diacylglycerol which activates numerous protein kinase C (PKC) that show similar functionality with them (Zhang et al. 1995). This protein kinase C is the regulator of various processes that occur inside the cell. Protein kinase C gets activated for only a short time duration due to transient half-life of the diacylglycerol in the cell whereas when activating protein kinase C from phorbol-esters, it has been analyzed that the protein kinase C shows activation duration for a prolonged time and thus results in various biological activities (Griner and Kazanietz 2007).

Phorbol-esters are severely toxic in nature, and oils which include phorbol-esters are called as purgatives (Gandhi et al. 1995). Another toxic effect of phorbol-esters is that they cause irritation to the skin. It has been studied that phorbol-esters have a significant contribution toward numerous types of cancers and thus are named as co-carcinogens/tumor promoters. They are called co-carcinogens since they don't lead to the formation of the tumor by itself but result in the elevated levels of the risks of the tumor during co-exposure to a particular carcinogen (Griner and Kazanietz 2007).

Most studies, by seeing effects in mice, have suggested that phorbol-esters that promote tumor have used phorbol 13-myristate 12-acetate (PMA) from *Croton tiglium* which promotes tumor in both jatropha oil and its associated phorbol-esters (Horiuchi et al. 2010; Hirota et al. 2004). It has been reported that the effect of the phorbol-esters is also on the lytic cycle of the Epstein–Barr virus (latent) (MacNeil et al. 2003). Precisely, it is challenging in order to quantify the risks which are blended while handling the substances from the jatropha, but there is comparatively extended exposure toward crushed seeds or oil and thus increased risks. Thus, proper measures are needed to avoid such exposures (Gminski and Hecker 1998). If using varieties of jatropha which lacks phorbol-esters for feeding animals turns up as a promising source of income along with the elimination of the risks that are conjoined with the extended exposure of phorbol esters. Researchers are doing intense studies toward the fate of the phorbol-esters in the environment along with its impacts on the ecology of the soil prior to study on the varieties of jatropha in order to use them as a fertilizer for the meal (King et al. 2009).

15.4.2 Attributes of Jatropha *plant*

There are definitely some strengths and weaknesses, advantages and disadvantages of the biofuel produced from jatropha biodiesel. Jatropha is one such renewable crop with biological origin with proper maintenance of carbon cycle (closed) and hence an environmentally friendly fuel (Fig. 15.3). It is possible to curb the problem of soil erosion by promoting the use of barren lands via plantation of jatropha. Also, from the

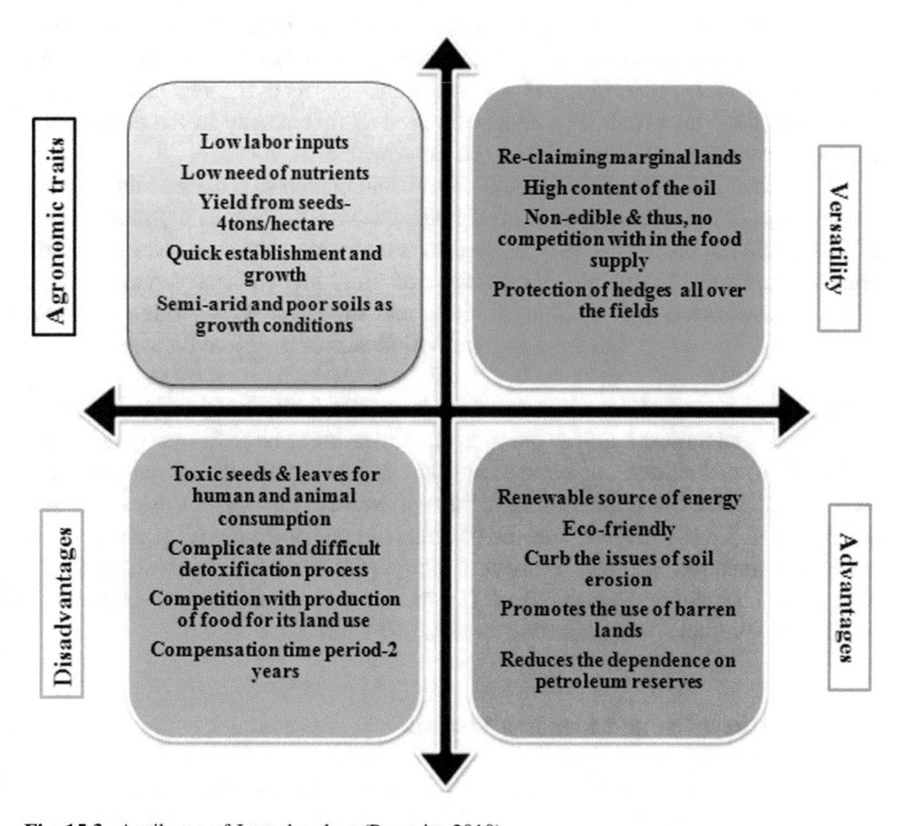

Fig. 15.3 Attributes of Jatropha plant (Parawira 2010)

jatropha seeds, there is a high amount of production of oil (2 tons per hectare per year) (Comprehensive Jatropha Report). Biofuels from the jatropha help in diminishing the dependence of a country on petroleum reserves which are imported from some other country (Mandal et al. 2011). Apart from advantages, the oil from jatropha seeds and its biofuel have numerous drawbacks. The time period for compensation is beyond 2 years. The production of the byproducts during the extraction of the oil in the shape of a cake is unfit for feeding animals because of the toxicity. Additionally, there are still no such developments of technologies for the production of biofuels from the viewpoint of commercialization. Furthermore, because of the production at a small scale, the cost of production of biofuels from jatropha is still acutely high (Datta and Mandal 2014).

15.4.3 Processing Methods

For the production of crude oil, vegetable oils and animal fats are being pressed which consists of (i) free fatty acids, (ii) phospholipids, (iii) sterols, (iv) water, (v) odorants, and some other types of impurities (Openshaw 2000a). Due to the presence of the aforementioned substances and other characteristics such as excessive viscous nature, low volatile properties, and polyunsaturated aspect of the vegetable oils, one cannot use them as a fuel; they cannot be used as fuel exactly in the compressed engines (Srivastava and Prasad 2000) (Banapurmath et al. 2008).

The content of the unsaturated fatty acids in conjugation with oleic acid following linoleic acid in the oil produced from the jatropha seed is 72%. The viscous nature of this produced oil is less as compared to the already published data of certain common and reliable oils at a temperature of 30°C like soybean having a value of 31cSt, cottonseed with a value of 36cSt, and sunflower with a value of 43cSt, respectively. This has pointed out the appropriateness of the oil to be used as a fuel (Akintayo 2004).

In order to surmount the aforementioned issue, that is the use of vegetable oils as such, there is an utmost requirement of the modification to be performed chemically in thought to make these oils equivalent to the features of the diesel obtained from fossil fuels. There are numerous techniques to process the oils but the significant techniques to process oils into fuel are (i) direct use and blending, (ii) pyrolysis, (iii) microemulsification, and (iv) transesterification (Demirbaş 2000), (Nwafor 2003). Researchers are doing extensive work in order to enhance the quality, output, and profit of the biofuels from vegetable oils (Parawira 2010).

15.4.3.1 Direct Use and Blending Method

Various demonstrations to run engines by using biofuels have been given like the working of an engine by using oil produced from peanuts with 100% efficiency, maintenance of the total power of the engine with no modifications of a mixture which consist of 10% vegetable oil has been used in the pre-combustion chamber of the engine (Agarwal 2007), the performance of the engine is checked by using a mixture of degummed soybean oil and diesel fuel (1:2) resulting in avoiding the thickening and gel nature of the lubricating oil, not similar to the 1:1 ratio, etc. (Parawira 2010).

Researchers have observed that blend oil from the seeds of jatropha (50%) can be used in diesel engines with no significant difficulties during operation. But it is a requirement to study further about the long-term effects on the engines (Pramanik 2003). Besides, the direct use and blend form of the vegetable oils are usually considered as an unsuitable and arduous approach to make use of them in the diesel engines (direct and indirect). The apparent issues can be (i) high viscosity, (ii) presence of acids in their composition, (iii) amount of free fatty acids, (iv) generation of gums during oxidation process, (v) occurrence of polymerization during storing and combustion process, (vi) deposition of carbon, (vii) occurrence of thick and gel-like properties of lubricating oils, etc. (Agarwal 2007; Meher et al. 2006).

15.4.3.2 Microemulsion Method

Picking the approach of microemulsion through solvents like methanol, ethanol, etc. helps in resolving the issue of the high viscous nature of vegetable oils (Agarwal 2007). The basic definition of microemulsion is dispersion of colloidal equilibrium of the microstructures of an optically active fluid of isotropic nature (1–150 nm) that are usually formed willingly from two immiscible liquids or one or more amphiphiles whether ionic or non-ionic (Ma and Hanna 1999). The constituents of biofuel from the microemulsion approach consist of (i) diesel fuel, (ii) vegetable oil, (iii) alcohol, (iv) surfactants, and (v) cetane. Alcohols like methanol, ethanol, etc. are used to lower the viscosity of the oil. Higher alcohols are usually engaged as surfactants whereas nitrates of alkyl are engaged to improve the function of cetane (Parawira 2010).

Microemulsions are also helpful in improving the spray characteristics via explosive vaporization of the components with low boiling point present in the micelles. This method reduces the viscosity by increasing the cetane number and endows better spray features in the biofuel. Withal, using engines with biofuels that opt the approach of microemulsion creates issues such as sticking of the injector needle, deposition of carbon, incomplete combustion process when used in a continuous manner (Parawira 2010).

15.4.3.3 Pyrolysis Method

The pyrolysis method deals with the transformation approach via oxygen or heat when the catalyst is present which cleaves the bonds and forms numerous small molecules. Performing pyrolysis of vegetable oils produces biofuel with the generation of the alkanes, alkenes, alkadienes, aromatic acids, carboxylic acids, etc. in different proportions (Ma and Hanna 1999).

The cost of the equipment to be used in the pyrolysis method is high for producing biofuels in developing countries. Also, by removing oxygen during the process results in providing no advantage to the environment regarding the use of oxygenated fuel (Ma and Hanna 1999). Another significant demerit of the pyrolysis method is the requirement of equipment called separate distillatory for segregating different fractions. Additionally, the product formed at the end of this method consists of sulfur and thus is a less environmentally friendly method (Ranganathan et al. 2008).

15.4.3.4 Transesterification Method

This method is considered as the best method for the production of biofuels (Fig. 15.4). Transesterification is defined as a set of reactions where fat/oil (triglycerides-TAGs) in the presence of an alcohol is converted into alkyl esters of fatty acids (methyl esters and ethyl esters which are an outstanding alternative of biofuel and glycerol) as shown in the reaction mentioned below (Parawira 2010):

It is a process used majorly in industries by heating extra alcohol present in the vegetable oils following various conditions under which the reaction will take place where an inorganic catalyst is also present. The nature of the reaction is reversible and hence surplus amount of alcohol (methanol, ethanol, propanol, and butanol) is used in shifting the equilibrium toward the product side. The reactions are generally catalyzed via acid, base, or an enzyme to enhance the rate and yield of the reactions. The process of transesterification when alkali (sodium hydroxide—NaOH, potassium hydroxide—KOH, carbonates) is used as a catalyst and is faster as compared to the transesterification when acid is used and thus is in use significantly at commercial levels (Ranganathan et al. 2008; Ma and Hanna 1999; Agarwal and Agarwal 2007). The products of the transesterification process are an amalgamation of esters, glycerol, alcohol, catalyst and triglycerides, diglycerides, and monoglycerides which are segregated later in the downstream process (Ma and Hanna 1999).

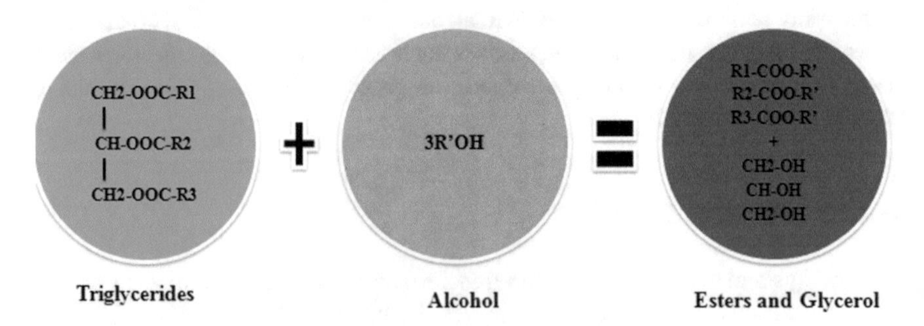

Fig. 15.4 Reaction showing the transesterification process (Parawira 2010)

Through the transesterification process, it has been observed that there is a severe change in the viscosity of the vegetable oil. The component having high viscosity and glycerol is taken out and thus the low viscosity product. After production of the biofuel via the transesterification process, there is a reduction in the flashpoint with the improvement of the cetane number. The yield is affected by various factors such as (i) moisture, (ii) free fatty acids (FFA), (iii) duration of the reaction, (iv) temperature of the reaction, (v) type of catalyst, and (vi) ratio of alcohol to oil (Parawira 2010).

15.4.4 Problems in Association with Environment

This life cycle approach (LCA) of jatropha and its impact on the environment is discussed in this particular section. In the life cycle approach (LCA), all the initial data and the products of every stage of the full cycle of the production of the biofuel are cataloged and calculated and after that, the effects are correlated with a standard or associated system. Here in this study, the standard system produces the same quantity of energy which is based on fossil fuels as a source of energy (Bernesson et al. 2006; Lettens et al. 2003). The study which consists of the life cycle approach of the biofuel from the sectors like agriculture and forests are constrained and usually focuses on the equilibrium of the energy and the repercussions of global warming (Moghbelli et al. 2007) whereas numerous groups are there where the impact is visibly available. The significance of the land use is one such group which is included not very often but flow behavior of a land, water supply, vegetation type, biodiversity are certain properties which are significant enough for the feasibility and renewability of the systems used during the production process which usually occupies a massive part of the land (Mattsson et al. 2000). By analyzing the significance of the land use, the will be a proper understanding of the renewable nature of vegetable oil or biofuel from the process employed for the production of biofuel (Achten et al. 2008).

15.4.4.1 The Equilibrium of Energy

It has been published that the life cycle of the equilibrium of energy of the production of the biofuel from the variety of jatropha is affirmative. Figure 15.5 deals with the initial data of energy to be entered after the allocation of the need for the total energy of an entire operation for various end- and byproducts (Achten et al. 2008).

This distribution is a proportional allotment of the input energy amid the products which are laid on the content of the energy of such products. From Fig. 15.5, it can be seen that the allotment of the input energy is surrounded by the jatropha methyl ester (JME) which is the end product, glycerin which is the byproduct of transesterification process, and seed cake which is the by-product during the extraction of oil. Other byproducts such as wood, husks of the fruit are not mentioned in this very allotment system due to their use in an energy-efficient way which is not a usual practice. The allotment usage is rationalized only if there is an efficient use of the byproducts into

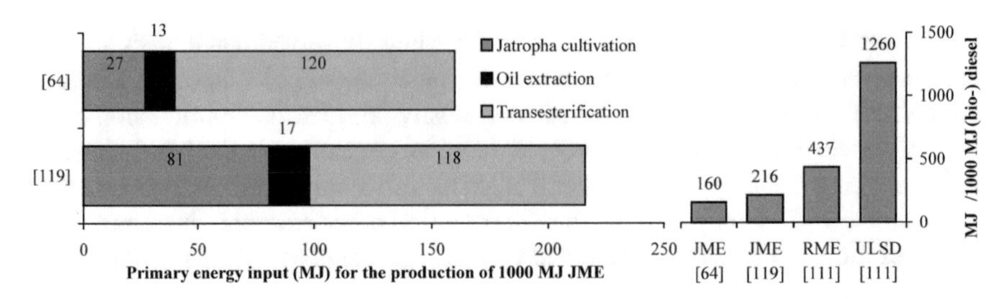

Fig. 15.5 Life cycle of the energy equilibrium of biofuel production (ULSD: Ultra-low sulfur diesel, RME: Rapeseed methyl ester, JME: Jatropha methyl ester) (Achten et al. 2008)

which the input energy is distributed. If in an event none of the byproducts is being employed, then there will be a slight positive value of the energy equilibrium, that is an input of 886 MJ for 1000 MJ jatropha methyl ester (JME) output or sometimes even negative whereas if all of the byproducts which include wood and husks of the fruit are employed, then there will be an effective utilization of this total input of 886 MJ which results in a total output of 17235 MJ, and hence the allotment of input energy would be 160 MJ/1000 MJ JME (Gmünder et al. 2012).

Figure 15.5 depicts the energy equilibrium of high and low input for the cultivation of jatropha. There is a clear difference between the two intensities of the cultivation that have been applied. Considering the intensity system for the cultivation of jatropha with low value, the primary input of energy in totality is 17% whereas it is 38% in case of the intensity system with high values. The two major practices for the cultivation process via energy are irrigation and fertilizer practices. Irrigation practices account for 46% whereas the fertilizer practices account for 45% of the input energy (total) (Gmünder et al. 2012). From Fig. 15.5, it can be inferred that the transesterification method consumes a huge amount of energy whereas the extraction stage has 78% contribution of the life cycle primary energy need (total) in both cases (Fig. 15.5).

From the aforementioned results, one can assume that the life cycle of energy equilibrium of biofuel is usually positive. The positivity, in fact, depends on how efficiently the byproducts of the system are being used. Since the transesterification process is considered as a big contributor toward the allotment of the energy for the end products of the biofuels (Fig. 15.5), crude oil used as an end product will undoubtedly enhance the equilibrium of energy. But, the combustion of the engine of clean jatropha oil is less effective in terms of energy (Bej 2002) with some issues associated with the engine (Meher et al. 2006).

15.4.4.2 Promising Repercussions on Global Warming

Two life cycle approaches has been exercised: (a) biofuel production from jatropha and (b) biofuel production from fossil fuels. Both these exercises have resulted in a positive outcome regarding the production of biofuel to reduce greenhouse gas

(GHG) emissions. The bulky greenhouse gas contributors are irrigation (26%), fertilizer (30%), and transesterification (24% and 70% depending on the intensity of the applied) (Gmünder et al. 2012).

Scientists have observed that (Gmünder et al. 2012) 90% of the total life cycle of greenhouse gas (GHG) emissions is because of the end use of the products. They have calculated that the potential of the global warming potential on the production and on the use of biofuel from jatropha is 23% of the potential of global warming from fossil fuels. Generally, this impact can be assumed as a positive notion when compared to fossil fuels. The extensive cultivation stage and the transesterification method will elevate the greenhouse gas need of the production method. This elevation will only be marginal when the overall impact of the life cycle on the potential of global warming is concerned because the significant contribution is of the end use of the products (90% of the total) (Gmünder et al. 2012). Withal, these two life cycle approaches did not mention the emissions of nitrous oxide (N_2O) because of the fertilization of nitrogen. The nitrous oxide global warming potential is 296-folds higher than an equivalent mass of carbon dioxide (Crutzen et al. 2008). According to the report of IPCC, the release of nitrous oxide is equivalent to 1% of the input of nitrogen from the mineral fertilizer or from the nitrogen fixed biologically. Considering the usage of the byproducts, the concept of greenhouse gas equilibrium and energy equilibrium is important (Achten et al. 2008).

15.4.4.3 The Significance of Land Use

Within the life cycle approach, methodologies to assess the impact of land use have been discussed a lot, but there is a general agreement regarding the fact that the impact of the use and occupation of the land use on soil and on the local biodiversity needs to be assessed properly (Lindeijer et al. 2002). But no such assessment of problems like these has been addressed for jatropha till date. It has been noticed that the structure of the oil from jatropha can be enhanced (Ogunwole et al. 2007), thus preventing the erosion of soil and sequestration of carbon (Achten et al. 2008).

It is worthy enough that the impact of land use occupation is dependent massively on the applied system for cultivation system and the intensity of the system. The extensive application of machinery and fertilizers are considered to be the significant forces to create a negative effect. For jatropha, being an exotic species, the impact of the change in the land use on biodiversity will be negative, but in majority, this is dependent majorly on both replacement of land use with jatropha and how the cultivation of jatropha will be done (Geertsma et al. 2009). Jatropha is an invasive species in South Africa and is considered as a weed in Australia (Low and Booth 2007). But so far, no such research has been carried out aiming at the quantification of the allelopathic effects of jatropha on local vegetation. Thus, land use (original), cultivation system, and intensity applied to the cultivation process are the most significant and determining characteristics (Achten et al. 2008).

15.4.5 Biodiesel Production via Transesterification Reaction Using Heterogeneous Catalysts

15.4.5.1 Bifunctional Catalyst for Biodiesel Production from *Jatropha* Oil

Considering the past few decades, the supply and security of energy is a significant problem all over the globe. There is a possibility in the advancements of the technology and escalation in the growth of the economy of the country by combusting fossil fuels which endows energy to the country. But the combustion of fossil fuels has an adverse impact on the environment since the combustion process results in the emission of greenhouse gases (GHGs) and certain other types of air pollutants (Demirbaş 2010; Lim and Keat Teong 2010). Also, there has been rapid depletion in the fossil fuel resources because of the increase in the demand of energy and hence there is an utmost requirement of alternative source of energy which favors the concept of renewability and sustainability. Developing biofuel is nowadays considered as an alternative for fuels obtained from fossil fuel resources and is also attracting global recognition (22.5 billion liters in the year 2012) (F. Rabiah Nizah et al. 2014). The advantages of biofuel (biodiesel here) are meritorious because of (a) it is biodegradable in nature, (b) eco-friendly option as compared to the fuel from fossil fuels, and (c) bio-renewable characteristics (Semwal et al. 2011; Wan Omar and Nor Aishah 2011). Biodiesel also named as fatty acid methyl ester (FAME) is generated (Fig. 15.6) through the transesterification method in which vegetable oils/animal fats are employed along with alcohol (of short chains) and later catalyzed by using acids or bases, respectively. This particular process includes three sequential reactions of reversible nature in which the triglycerides are transformed to diglycerides following monoglyceride conversion and end with the final transformation into glycerin (Umer et al. 2011).

Frequently used homogeneous catalysts in the transesterification process are sodium hydroxide (NaOH), potassium hydroxide (KOH), sulfuric acid (H_2SO_4), and phosphoric acid (H_3PO_4) (Berchmans et al. 2013). But, these homogeneous catalysts have certain demerits such as (i) requirement of high cost for the production of biodiesel as there is a demand for proper washing and purification of the product formed at the end of the reaction (Brito et al. 2008) and (ii) possible consequences of corrosion of equipment. On the other side, the employing heterogeneous catalysts can help in minimizing the issues affiliated with homogeneous catalysts. This is possible due to the easy separation of the heterogeneous catalyst from the liquid products which also provides high activity, high selectivity, and extended shelf-life of the catalyst (Zabeti et al. 2009). The only significant feature of a heterogeneous catalyst apart from the disadvantages is the reusable nature thus rendering it eco-friendly (Rabiah Nizah et al. 2014).

For producing biodiesel from the oil with a high content of free fatty acids (FFA) like the oil from Jatropha species, a two-step process is employed (acid catalyst

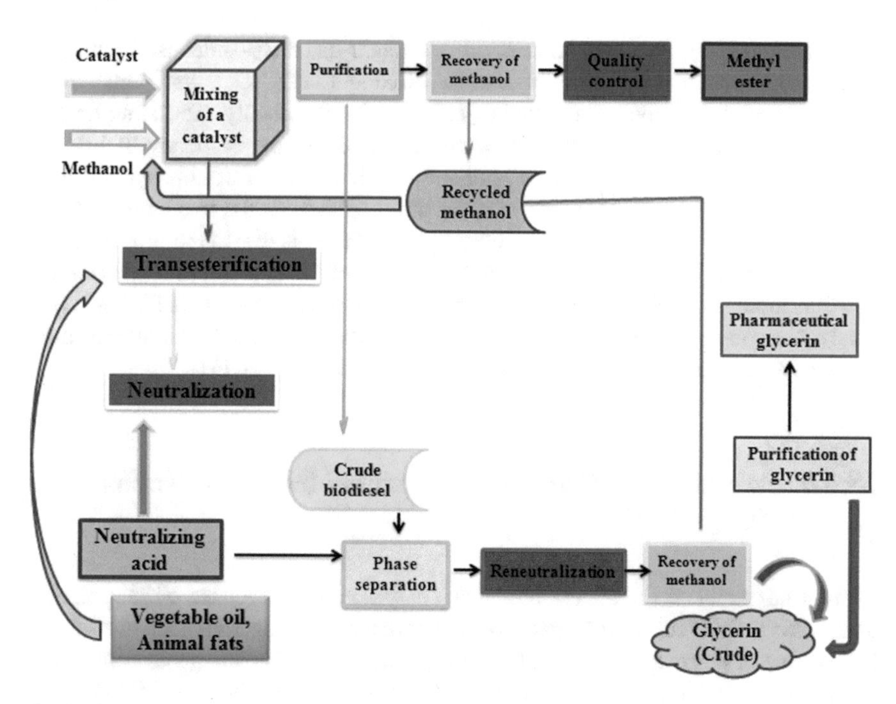

Fig. 15.6 Layout of biodiesel production (Folaranmi 2012)

and basic catalyst) (Choudhury et al. 2013; Ngo et al. 2008). Besides, this particular method needs numerous reactions along with washing and separation steps. To blow away this flaw, a one-step approach is required where there is simultaneous esterification and transesterification of the oil that takes place by using the catalyst which consists of acid and basic properties, respectively (Choudhury et al. 2014; Omota et al. 2003). Considering the above notion, researchers have employed Lanthanum oxide (La_2O_3) for the production of biodiesel; lanthanum oxide has a feature of both acid and basic sites and thus helps with simultaneous esterification and transesterification reactions. Lanthanum oxide was frequently employed as an aid to improve the activity of catalysts like magnesium oxide (MgO), zinc oxide (ZnO), calcium oxide (CaO), aluminum oxide (Al_2O_3), zirconium oxide (ZrO_2), and activated carbon (Russbueldt and Hoelderich 2010; Yan et al. 2009, 2010).

This research work deals with the development of an advanced type of bifunctional catalyst that is bismuth (III) oxide-lanthanum oxide (Bi_2O_3–La_2O_3) for performing simultaneous esterification and transesterification reactions. Researchers have scrutinized the best parameters of the reaction like the optimum temperature of the reaction (150 °C for 4 h), amount of catalyst (2 wt%) needed, and the molar ratio of methanol and oil (15:1) for transesterification in the best possible way and to discover the balance of bismuth (III) oxide-lanthanum oxide (Bi_2O_3–La_2O_3) catalyst. The synthesis of such catalysts was through an impregnation process. The characterization studies of the catalyst were done by techniques like X-ray diffraction (XRD), Nitrogen adsorption (Brunauer–Emmett–Teller-BET), temperature programmed desorption of carbon dioxide (TPD-CO_2), and temperature programmed desorption of ammonia (TPD-NH_3), specifically. The simultaneous esterification and transesterification were performed in the company of bismuth (III) oxide (1–7 wt%) and altered lanthanum oxide catalyst at atmospheric pressure. Analyzing the best condition to carry out the reaction, the maximum value of the biodiesel was 93%. The high activity of the catalyst is due to the proper dispersion of bismuth (III) oxide (Bi_2O_3) on the support of the lanthanum oxide (La_2O_3) and thus increases in the surface area of the bifunctional catalyst. This Bi_2O_3–La_2O_3 catalyst has managed 87% transformation of the fatty acid methyl ester after its three times of reuse in a consecutive manner (Rabiah Nizah et al. 2014).

15.4.6 Mixed Oxide Catalysts for Methanolysis of Jatropha *Oil*

Biofuel has grabbed the attention of many research areas because of the potential of an alternative fuel with respect to the apprehensions regarding the availability of reserves of petroleum (restorable nature) and the effect of the gases coming out from the fuels which are based on the fossil fuels on the environment (Muniyappa et al. 1996). Biodiesel is an amalgamation of fatty acid methyl esters (FAMEs) which acquire the same physicochemical and fuel characteristics as that of the fuel produced from the petroleum resources. Biodiesel, a clean fuel, is a derivative of vegetable oils/animal fats, is biodegradable, renewable with low amount of sulfur, less toxicity, low volatility, better transportation and storage features, a high cetane number, and a proper balance of atmospheric carbon dioxide for the production of biodiesel (Ngamcharussrivichai et al. 2008; Sharma and Singh 2009). Methods like microemulsification, pyrolysis, transesterification, and transesterification with methanol (alcohol) are considered as significant methods employed widely to treat the problems like high viscous nature of the oils obtained from the plant species (Sharma and Singh 2009). There are various applications of the byproduct of the transesterification method that is glycerol (Fig. 15.6) in the sectors such as food, cosmetics, pharmacy, etc. (Vicente et al. 2004).

In the recent past, researchers have been showing keen interest in calcium oxide (CaO) because of its merits it has from the economy point of view. A group of researchers (Kawashima et al. 2009) has observed that the activity of calcium oxide was so high that it produced 90% of biodiesel from rapeseed oil via the transesterification method along with reflux of methanol within a time duration of 3 h. Another group of researchers has published that (Chintareddy et al. 2006) there was a high activity of nanocrystalline calcium oxide that is 99% transformation of oil into biodiesel during the transesterification process of poultry fat and soybean oil at room temperature. Additionally, a team of researchers (Liu et al. 2008) has studied the impact of moisture on the catalytic activity of calcium oxide whereas a certain scientist has reported that during the transesterification process, there is a chance that soluble substance may leach away from calcium oxide. A scientist has published that calcium oxide dissolves in methanol, a little, and thus results in the formation of a suspension in the reaction blend because of its weak mechanical strength. The particular suspension creates issues during the separation of a catalyst from the products of biodiesel (Gryglewicz 1999). Hence, in order to support calcium oxide over the carriers, enhancing the stability of the calcium oxide catalysts is a must (Taufiq-Yap et al. 2011).

In this very particular example, the research work deals with the study of the behaviors of various solid catalysts of binary calcium oxides such as calcium-magnesium oxides (Ca–Mg), calcium-zinc oxides (Ca-Zn), etc. in the methanolysis (lysis of methanol) of oil from Jatropha curcas (JCO) for producing biodiesel. The synthesis of the required precursors was done by employing the co-precipitation method, and following the calcination of the precursors the metal oxides were generated (calcium with magnesium (CaMgO) and calcium with zinc (CaZnO)). Various techniques have been employed like X-ray diffraction (XRD), temperature programmed desorption of carbon dioxide (TPD-CO_2), scanning electron microscopy (SEM), nitrogen adsorption (Brunauer–Emmett–Teller-BET), etc. to determine the structural features, textural properties, and basic characteristics of the catalysts (Taufiq-Yap et al. 2011).

For the preparation of the catalysts, 2 M aqueous solution of metal nitrates such as calcium nitrate tetrahydrate ($Ca (NO_3)_2.4H_2O$), magnesium nitrate hexahydrate ($Mg (NO_3)_2.6H_2O$), zinc nitrate hexahydrate ($Zn (NO_3)_2.6H_2O$), etc. were added at a slow pace into the basic solution (aqueous) which consists of sodium carbonate (Na_2CO_3) and sodium hydroxide (NaOH) at a pH with a constant range of 8–9. The next step was continuous and intense stirring at a temperature of 338 K (64.85 °C) overnight. The final step was performed via filtration and washing (deionized water) of solids. The precursors formed were then dried overnight using an oven. After drying overnight, the precursors were then calcined at a temperature of 1073 K (799.85 °C) and 1173 K (899.85 °C) for 6 h. The precursor formed at 1073 K temperature was named as CaMgO (calcium with magnesium oxide catalyst) and the precursor formed at a temperature of 1173 K was named as CaZnO (calcium with zinc oxide catalyst). Both catalysts, CaMgO (calcium with magnesium oxide catalyst) and CaZnO (calcium with zinc oxide catalyst) exhibited a very high activity as calcium oxide with an easy separation from the product formed. It has been observed that the oxide catalyst of calcium with magnesium (CaMgO) was much more active as compared to the oxide catalyst of calcium with zinc (CaZnO) during the transesterification process of oil from jatropha in the presence of methanol. In the presence of appropriate conditions such as 338 K temperature, one-fourth of 4 wt% amount of catalyst, one-fourth of 15 as the molar ratio of methanol and oil, one-fourth of the 6 h as the time duration of the reaction, etc. under which the transesterification reaction was carried out, there has been more than 80% of the transformation of oil when CaMgO (calcium with magnesium oxide catalyst) and CaZnO (calcium with zinc oxide catalyst) were used. Although the high activity of calcium oxide has reduced the transformation of oil after reusing it consecutively for four cycles, the quality of CaMgO (calcium with magnesium oxide catalyst) and CaZnO (calcium with zinc oxide catalyst) during transformation decreases a little after the sixth cycle (Taufiq-Yap et al. 2011).

15.5 Investments in Biofuel and Their Cost of Production

Considering the recent status of the technologies prevailing all over the globe, second-generation biofuels emerge with a very soaring cost of capital investment (five folds compared to starchy ethanol plants). Generally, the estimated cost of capital investments for the production of ethanol from cellulose lies within the range of dollar 1.06 to 1.48 per liter of ethanol annually (Wright and Brown 2007). Recently, the associated costs of the operation lie in between dollar 0.35–0.45 per liter relying on the feedstocks and equivalent technologies. Possible advancements regarding technologies converting the biofuel help in reducing the requirement of capital investments from dollar 0.95 to dollar 1.27 per liter of the capacity of the ethanol (annually) along with the reduction of the costs of the operation to dollar 0.11–0.25 per liter of ethanol (Hamelinck et al. 2005). In order to accomplish such reductions in the costs of capital investment and cost of operation, bulky and precarious investments are required for the technological revolutions. It has also been assumed that the costs associated with the production of second-generation biofuels will be panoramic. With advancements in the field of technologies, it has been anticipated that there will be a reduction in the costs of biofuels to dollar 0.30–0.40 per liter by the year 2020 (International Energy Agency (IEA) 2005). Also, the contribution of the feedstock is roughly around 32–52% of the total costs associated with the production in all the studies whereas first-generation biofuels contribute only 55–70% (International Energy Agency (IEA), 2008).

The production costs of the biofuel from jatropha have been estimated around dollar 0.44 to 2.87 per liter with respect to developing countries. It has also been noticed that the yields and the costs of feedstock for the production of jatropha vary (Peters and Thielmann 2008). There are certain researchers who believed that there will be a decline in the costs because of the advancements in the production at a large scale and extraction of oil which in turn enhances the quality of the process along with the exploitation of the economies (Carriquiry et al. 2011), and some believe that shareholders must remain extra cautious regarding projections associated with the costs as costs may increase in operations working at a large scale (Peters and Thielmann 2008).

Among various studies, it has been noticed that biofuel is a costly form to produce energy in comparison with the energy obtained from fossil fuels. For example, the cost of ethanol (biofuel) from cellulose is 1.1–2.9 folds (per unit of energy) more as compared to the prices of gasoline. There is one optimistic postulate that the estimated cost of the biofuels either from jatropha or from algal oil will reside at the same levels as that of diesel but these low costs are not associated with the production at a large scale, especially to the biofuels generated from algae with an estimation of approximately 100 folds more costly as compared to diesel. Consequently, advancements are mandatory and important for the development of biofuels in the industries (Carriquiry et al. 2011).

15.6 Challenges and Policies

Myriad policies are usually affecting the markets associated with biofuels. It is comparatively hard to keep a track of biofuels such as their production process, consumption, nature of trade in the environment where it is very uncertain about any advancement in the policies relating to the biofuels. This is because the fresh policies providing information that energy obtained from biofuels is a costly affair in comparison to fossil fuels. Also, there is a dearth knowledge about the commercialization of the. The new policies are authorized very quickly by numerous countries along with the modification of the legislation made in the past (Carriquiry et al. 2011).

It has been reported by the Renewable Energy Policy Network for the 21st Century (REN21) that there is a total of 73 countries including developing nations that have goals of bioenergy during early 2009 (REN21, 2009). With a minimum of 23 countries, mandates are there which state to blend biofuels into fossil fuels for transportation. Numerous policies will provide incentives for the supply and employment of biofuels (first and second generation) to a similar extent, despite the prevailing circumstances of the costs of the production of biofuels and the associated benefits such as the net decrease in the amount of carbon.

There is a specific mandate by the US Energy Independence and Security Act (EISA) of 2007 regarding modified biofuels to be as a part of the second Renewable Fuels Standard (RFS2). The bottom side of that very particular market has a target value of 79.5 billion liters by the year 2022, out of which 60.5 billion liters are restrained for the biofuels generated from cellulose (Carriquiry et al. 2011). But, the US Environmental Protection Agency (US EPA) is significant enough to show their potential to prepare an annual estimate of the production of ethanol (biofuel) from cellulose to disclaim the part of ethanol (from cellulose) from the mandate of RFS2 if in case the capacity is not sufficient enough. For instance, in the month of December 2010, US Environmental Protection Agency (US EPA) surfaced the requirements for the year 2011 which consists of 25 million liters of ethanol produced from cellulose in lieu of the mandate of US Energy Independence and Security Act (EISA), that is 946 million liters (Carriquiry et al. 2011).

The significant hurdle which biofuels are facing is the economy. When a comparison is made with respect to the production costs, the production of biofuels is a costly affair as that of fossil fuels (Rajagopal and Zilberman, 2007). Policy interference will help a bit to speed up the conversion from first-generation biofuels to the commercial levels along with the consumption of second-generation biofuels. Additionally, it is a critical point that policies are modified in a way that will reinforce the production and development of meritorious biofuels and avert the production of unacceptable biofuels (International Energy Agency (IEA), 2008).

A significant policy will be regarding the cost of reduction of feedstocks, cost of the plant, conversion rates, yields, etc. For the production of the biofuels from cellulose, the cost associated with the conversion plays an important role whereas in first-generation biofuels (biofuels produced from algal species or jatropha) feedstock plays a significant role, and the collection of raw materials such as residues of agricultural practices are much more expensive. Issues of first-generation biofuels such as emission of greenhouse gases (GHG) have aggravated the urgent situation to shift to second-generation biofuels which results in the maintenance of equilibrium energy, more decrease in the emissions of greenhouse gases (GHGs), and less significant competition in prime lands having food crops. Whereas energy crops still compete for prime lands having food crops and thus envisaged that either by employing soils of low quality (jatropha varieties) or by endowing biomass that can be utilized per unit of the land such as switchgrass, and hence less or decrease the influence on the quality of soils. Residues of agricultural and forest practices, oil from microalgal species have resulted in less for prime land (Carriquiry et al. 2011).

Policy instruments like tax credits or tax exemptions provide incentives to the pathways employed for the production of biofuels with respect to the already established aims. For instance, with an objective to diminish the emissions of greenhouse gases (GHGs), soaring incentives like subsidies can be employed on biofuels with an increased level of reduction in the emissions of greenhouse gases. Additional policies like advancement in the source of income of small or poor farmers in developing nations is by giving incentives and acquiring the required raw materials from them. For example, the United States endows a tax credit in bulk for second-generation biofuels as compared to the ethanol produced from the corn. In Brazil, biofuel plants that acquire some feedstock from the farms of families along with some other needs can demand a social seal that entitles them to enjoy the benefits of tax provided by the government (Gordon 2008, (Carriquiry et al. 2011).

15.7 Conclusion and Future Scope

Jatropha *(Jatropha curcas* L.) is a versatile species with various aspects and appreciably promising features. Traditionally, the plant is employed with an objective in the domain of medicines, whereas it is also being employed to prevent and manage the erosion of soil. Nowadays, it is used as a resource for the production of biofuels. This plant species provides promising potential to cope with the emissions of greenhouse gases (GHGs), creates a chance of employment in rural areas, and is a significant source of renewable energy. The jatropha oil is considered as a worthy product as it is possible to convert oil into biofuel which is nowadays considered as an alternative source of energy with respect to the energy obtained from fossil fuels, petroleum reserves, etc. and also it is environmentally sound. It is grabbing the attention of a lot of researchers due to its good traits from an agriculture point of view, a high amount of oil from the seeds of jatropha, etc. There is an utmost requirement of extensive research on the analysis of the life cycle approach (LCA) for the production of biofuel from jatropha at small and industrial scales especially in developing countries where there is a production of jatropha in bulk amounts. Numerous production techniques to transform the oil into biofuel like pyrolysis, transesterification, blending, etc. have been known but the transesterification method to transform vegetable oils into biofuel is presently the most favorable method. This method makes use of catalysts (acid or base) for producing biodiesel and also has high conversion rates and reaction rates. This very biotechnological pathway is eco-friendly as well a proper usage of the byproducts (glycerol) in other sectors like food industry, pharmaceutical industry, etc. The characteristics of the biofuel produced from jatropha and that of the biofuel from the fossil fuels, petroleum reserves, etc. have been compared and thus complied according to the standards of America and Europe. Presently, the government over the globe is promoting the employment of biofuels. Considering the developing nations such as India, meeting the demands of energy in the coming future is an utmost concern and thus it has been implemented that that is necessary to use as an option of alternative fuel (biofuel from bio-renewable resources). For biofuel production, there is a requirement to cultivate seaweeds over a bulk scale; detail knowledge about the extraction process is mandatory to meet the demands of energy in an economical way. Boosting the use of biofuel as an alternative option requires acceptance from all over the globe along with the development of effective engines steered by biofuels. As a consequence, issues such as strengthening the economy of all the nations worldwide, mollifying the changes in the climate, improving the quality of the environment, and the production of biofuel from renewable resources will play a significant role and help in coping with such issues in a sustainable way providing benefits to both industries and environment.

Acknowledgements The author thankfully acknowledges Prof. Sanjeev Puri for believing in me and giving me this opportunity to explore and gain knowledge and excel.

References

Achten WMJ, Verchot L, Franken YJ, Mathijs E, Singh VP, Aerts R, Muys B (2008) *Jatropha* biodiesel production and use. Biomass Bioenerg 32:1063–1084. https://doi.org/10.1016/j.biombioe.2008.03.003

Agarwal AK (2007) Biofuels (alcohols and biodiesel) applications as fuels for internal combustion engines. Prog Energy Combust Sci 33:233–271. https://doi.org/10.1016/j.pecs.2006.08.003

Agarwal D, Agarwal A (2007) Performance and emissions characteristics of Jatropha oil (preheated and blends) in a direct injection compression ignition engine. Appl Therm Eng 27:2314–2323. https://doi.org/10.1016/j.applthermaleng.2007.01.009

Akintayo ET (2004) Characteristics and composition of *Parkia biglobbossa* and *Jatropha curcas* oils and cakes. Bioresour Technol 92:307–310. https://doi.org/10.1016/s0960-8524(03)00197-4

Banapurmath N, Tewari P, Hosmath R (2008) Performance and emission characteristics of a DI compression ignition engine operated on Honge, Jatropha and sesame oil methyl esters. Renew Energ 33:1982–1988. https://doi.org/10.1016/j.renene.2007.11.012

Barbieri L, Battelli MG, Stirpe F (1993) Ribosome-inactivating proteins from plants. Biochim Biophys Acta 1154:237–282. https://doi.org/10.1016/0304-4157(93)90002-6

Barbieri L, Polito L, Bolognesi A, Ciani M, Pelosi E, Farini V, Jha AK, Sharma N, Vivanco JM, Chambery A, Parente A, Stirpe F (2006) Ribosome-inactivating proteins in edible plants and purification and characterization of a new ribosome-inactivating protein from Cucurbita moschata. Biochim Biophys Acta 1760:783–792. https://doi.org/10.1016/j.bbagen.2006.01.002

Bej SK (2002) Performance evaluation of hydroprocessing catalysts: A review of experimental techniques. Energy Fuels 16:774–784. https://doi.org/10.1021/ef0102541

Berchmans HJ, Morishita K, Takarada T (2013) Kinetic study of hydroxide-catalyzed methanolysis of Jatropha curcas–waste food oil mixture for biodiesel production. Fuel 104:46–52. https://doi.org/10.1016/j.fuel.2010.01.017

Bernesson S, Nilsson D, Hansson PA (2006) A limited LCA comparing large- and small-scale production of ethanol for heavy engines under Swedish conditions. Biomass Bioenerg 30:46–57. https://doi.org/10.1016/j.biombioe.2005.10.002

Brito YC, Mello VM, Macedo CCS, Meneghetti MR, Suarez PAZ, Meneghetti SMP (2008) Fatty acid methyl esters preparation in the presence of maltolate and n-butoxide Ti(IV) and Zr (IV) complexes. App Catal A: Gen 351:24–28. https://doi.org/10.1016/j.apcata.2008.08.024

Carriquiry MA, Du X, Timilsina GR (2011) Second generation biofuels: Economics and policies. Energy Policy 39:4222–4234. https://doi.org/10.1016/j.enpol.2011.04.036

Carvalho CR, Clarindo WR, Praça MM, Araújo FS, Carels N (2008) Genome size, base composition and karyotype of *Jatropha curcas* L. An important biofuel plant. Plant Sci 174:613–617. https://doi.org/10.1016/j.plantsci.2008.03.010

Chintareddy VR, Oshel R, Verkade JG (2006) Room-temperature conversion of soybean oil and poultry fat to biodiesel catalyzed by nanocrystalline calcium oxides. Energ Fuel 20. https://doi.org/10.1021/ef050435d

Choudhury HA, Goswami PP, Malani RS, Moholkar VS (2014) Ultrasonic biodiesel synthesis from crude Jatropha curcas oil with heterogeneous base catalyst: Mechanistic insight and statistical optimization. Ultrason Sonochem 21:1050–1064. https://doi.org/10.1016/j.ultsonch.2013.10.023

Choudhury HA, Malani RS, Moholkar VS (2013) Acid catalyzed biodiesel synthesis from Jatropha oil: Mechanistic aspects of ultrasonic intensification. Chem Eng J 231:262–272. https://doi.org/10.1016/j.cej.2013.06.107

Crutzen PJ, Mosier AR, Smith KA, Winiwarter W (2008) N_2O release from agro-biofuel production negates global warming reduction by replacing fossil fuels. Atmos Chem Phys 8:389–395. https://doi.org/10.5194/acp-8-389-2008

Datta A, Mandal B (2014) Use of jatropha biodiesel as a future sustainable fuel 1. https://doi.org/10.1080/23317000.2014.930723

Demirbaş A (2000) Conversion of biomass using glycerin to liquid fuel for blending gasoline as alternative engine fuel. Energy Convers Manag 41:1741–1748. https://doi.org/10.1016/s0196-8904(00)00015-7

Demirbas A (2008) Biofuels sources, biofuel policy, biofuel economy and global biofuel projections. Energy Convers Manag 49:2106–2116. https://doi.org/10.1016/j.enconman.2008.02.020

Demirbaş A (2010) Biodiesel for future transportation energy needs. Energ Source Part A: Recovery, Utilization, and Environmental effects 32:1490–1508. https://doi.org/10.1080/15567030903078335

Escobar JC, Lora ES, Venturini OJ, Yáñez EE, Castillo EF, Almazan O (2009) Biofuels: environment, technology and food security. Renew Sust Energ Rev 13:1275–1287. https://doi.org/10.1016/j.rser.2008.08.014

F. Rabiah Nizah M, Taufiq-Yap YH, Rashid U, Teo SH, Nur ZA, Islam A (2014) Production of biodiesel from non-edible *Jatropha curcas* oil via transesterification using Bi_2O_3–La_2O_3 catalyst 88:1257–1262. https://doi.org/10.1016/j.enconman.2014.02.072

Fairless D (2007) Biofuel: The little shrub that could-maybe. Nature 449:652–655. https://doi.org/10.1038/449652a

Foidl N, Foidl G, Sanchez M, Mittelbach M, Hackel S (1996) *Jatropha curcas* L. as a source for the production of biofuel in Nicaragua. Bioresour Technol 58:77–82. https://doi.org/10.1016/S0960-8524(96)00111-3

Folaranmi J (2012) Production of biodiesel (B100) from jatropha oil using sodium hydroxide as catalyst. J Pet Eng 2013. https://doi.org/10.1155/2013/956479
Francis G, Edinger R, Becker K (2005) A concept for simultaneous wasteland reclamation, fuel production, and socio-economic development in degraded areas in India: Need, potential and perspectives of jatropha plantations. Nat Resour Forum 29:12–24. https://doi.org/10.1111/j.1477-8947.2005.00109.x
Gandhi VM, Cherian KM, Mulky MJ (1995) Toxicological studies on ratanjyot oil. Food Chem Toxicol 33:39–42
Gaurav N, Sivasankari S, Kiran GS, Ninawe A, Selvin J (2017) Utilization of bioresources for sustainable biofuels: A Review. Renew Sust Energ Review 73:205–214. https://doi.org/10.1016/j.rser.2017.01.070
Geertsma R, Wilschut L, Kauffman H (2009) Baseline review of the upper Tana. Kenya, Green Water Credits report
Ghosh SK (2016) Biomass and bio-waste supply chain sustainability for bio-energy and bio-fuel production. Procedia Environ Sci 31:31–39. https://doi.org/10.1016/j.proenv.2016.02.005
Gminski R, Hecker E (1998) Predictive and preventive toxicology of innovative industrial crops of the spurge (Euphorbiaceae) family vol 23. https://doi.org/10.1179/030801898789764615
Gmünder S, Singh R, Pfister S, Adheloya A, Zah R (2012) Environmental impacts of *Jatropha curcas* biodiesel in India. J Biomed Biotechnol 2012:623070. https://doi.org/10.1155/2012/623070
Griner EM, Kazanietz MG (2007) Protein kinase C and other diacylglycerol effectors in cancer. Nat Rev Cancer 7:281–294. https://doi.org/10.1038/nrc2110
Gryglewicz S (1999) Rapeseed oil methyl esters preparation using heterogeneous catalysts. Bioresour Technol 70:249–253. https://doi.org/10.1016/S0960-8524(99)00042-5
Haas W, Sterk H, Mittelbach M (2002) Novel 12-deoxy-16-hydroxyphorbol diesters isolated from the seed oil of *Jatropha curcas*. J Nat Prod 65:1434–1440
Hamelinck CN, Gv Hooijdonk, Faaij APC (2005) Ethanol from lignocellulosic biomass: techno-economic performance in short-, middle- and long-term. Biomass Bioenerg 28:384–410. https://doi.org/10.1016/j.biombioe.2004.09.002
Hartley MR, Lord JM (2004) Genetics of ribosome-inactivating proteins. Mini Rev Med Chem 4:487–492
Hassan MH, Kalam MA (2013) An overview of biofuel as a renewable energy source: development and challenges. Procedia Eng 56:39–53. https://doi.org/10.1016/j.proeng.2013.03.087
Haykiri-Acma H, Yaman S (2010) Interaction between biomass and different rank coals during co-pyrolysis. Renew Energ 35:288–292. https://doi.org/10.1016/j.renene.2009.08.001
Heller J, Engels J (1996) Physic nut. Jatropha curcas L, The International Plant Genetic Resource Institute (IPGRI)
Juan L, Yu C, Ying X, Fang YAN, Lin T, Fang C (2002) Cloning and expression of curcin, a ribosome-inactivating protein from the seeds of *Jatropha curcas*. Acta Bot Sin 45
Kang Q, Appels L, Tan T, Dewil R (2014) Bioethanol from lignocellulosic biomass: current findings determine research priorities. ScientificWorldJournal 2014:298153. https://doi.org/10.1155/2014/298153
Kawashima A, Matsubara K, Honda K (2009) Acceleration of catalytic activity of calcium oxide for biodiesel production. Bioresour Technol 100:696–700. https://doi.org/10.1016/j.biortech.2008.06.049
Kim S, Dale BE (2005) Life cycle assessment of various cropping systems utilized for producing biofuels: Bioethanol and biodiesel. Biomass Bioenerg 29:426–439. https://doi.org/10.1016/j.biombioe.2005.06.004
King AJ, He W, Cuevas JA, Freudenberger M, Ramiaramanana D, Graham IA (2009) Potential of Jatropha curcas as a source of renewable oil and animal feed. J Exp Bot 60:2897–2905. https://doi.org/10.1093/jxb/erp025
Kour D, Rana KL, Yadav N, Yadav AN, Rastegari AA, Singh C et al. (2019) Technologies for Biofuel Production: Current Development, Challenges, and Future Prospects. In: Rastegari AA, Yadav

AN, Gupta A (eds) Prospects of Renewable Bioprocessing in Future Energy Systems. Springer International Publishing, Cham, pp 1–50. https://doi.org/10.1007/978-3-030-14463-0_1

Kumar S, Sharma S, Thakur S, Mishra T, Negi P, Mishra S et al (2019) Bioprospecting of microbes for biohydrogen production: Current status and future challenges. In: Molina G, Gupta VK, Singh BN, Gathergood N (eds) Bioprocessing for Biomolecules Production. Wiley, USA, pp 443–471

Lettens S, Muys B, Ceulemans R, Moons E, Garcia J, Coppin P (2003) Energy budget and greenhouse gas balance evaluation of sustainable coppice systems for electricity production. Biomass Bioenerg 24:179–197. https://doi.org/10.1016/S0961-9534(02)00104-6

Lim S, Keat Teong L (2010) Recent trends, opportunities and challenges of biodiesel in Malaysia: An overview. Renew Sust Energ Review 14:938–954. https://doi.org/10.1016/j.rser.2009.10.027

Lindeijer E, Müller-Wenk R, Steen B (2002) Impact assessment of resources and land use

Liu X, He H, Wang Y, Zhu S, Piao X (2008) Transesterification of soybean oil to biodiesel using CaO as a solid base catalyst. Fuel 87:216–221. https://doi.org/10.1016/j.fuel.2007.04.013

Low T, Booth C (2007) The Weedy Truth about Biofuels. Invasive Species Council, Melbourne

Ma F, Hanna MA (1999) Biodiesel production: a review1Journal Series #12109, Agricultural Research Division, Institute of Agriculture and Natural Resources, University of Nebraska-Lincoln. 1. Bioresour Technol 70:1–15. https://doi.org/10.1016/S0960-8524(99)00025-5

MacNeil A, Sumba OP, Lutzke ML, Moormann A, Rochford R (2003) Activation of the Epstein-Barr virus lytic cycle by the latex of the plant *Euphorbia tirucalli*. Br J Cancer 88:1566–1569. https://doi.org/10.1038/sj.bjc.6600929

Makkar HP, Becker K (1999) Nutritional studies on rats and fish (carp *Cyprinus carpio*) fed diets containing unheated and heated *Jatropha curcas* meal of a non-toxic provenance. Plant Foods Hum Nutr 53:183–192

Makkar HPS, Aderibigbe AO, Becker K (1998) Comparative evaluation of non-toxic and toxic varieties of *Jatropha curcas* for chemical composition, digestibility, protein degradability and toxic factors. Food Chem 62:207–215. https://doi.org/10.1016/S0308-8146(97)00183-0

Mandal B, Palit S, Kumar Chowdhuri A, Kumar Mandal B (2011) Environmental impact of using biodiesel as fuel in transportation: A review. Int J of Globaal Warm 3. https://doi.org/10.1504/ijgw.2011.043421

Mattsson B, Cederberg C, Blix L (2000) Agricultural land use in life cycle assessment (LCA): Case studies of three vegetable oil crops. J Cleaner Prod 8:283–292. https://doi.org/10.1016/s0959-6526(00)00027-5

Meher LC, Vidya Sagar D, Naik SN (2006) Technical aspects of biodiesel production by transesterification—A review. Renew Sust Energ Review 10:248–268. https://doi.org/10.1016/j.rser.2004.09.002

Menezes RG, Rao NG, Karanth SS, Kamath A, Manipady S, Pillay VV (2006) *Jatropha curcas* poisoning. Indian J Pediatr 73:634; author reply 635

Moghbelli H, Halvaei A, Langari R (2007) New generation of passenger vehicles: FCV or HEV? Industrial Technology, IEEE Conference. https://doi.org/10.1109/icit.2006.372392

Motto M, Lupotto E (2004) The genetics and properties of cereal ribosome-inactivating proteins. Mini Rev Med Chem 4:493–503

Muniyappa PR, Brammer SC, Noureddini H (1996) Improved conversion of plant oils and animal fats into biodiesel and co-product. Bioresour Technol 56:19–24. https://doi.org/10.1016/0960-8524(95)00178-6

Ngamcharussrivichai C, Totarat P, Bunyakiat K (2008) Ca and Zn mixed oxide as a heterogeneous base catalyst for transesterification of palm kernel oil. Appl Catal A: Gen 341:77–85. https://doi.org/10.1016/j.apcata.2008.02.020

Ngo H, A. Zafiropoulos N, A. Foglia T, Samulski E, Lin W (2008) Efficient two-step synthesis of biodiesel from greases. Energ Fuel 22:626–634. https://doi.org/10.1021/ef700343b

Nwafor OMI (2003) The effect of elevated fuel inlet temperature on performance of diesel engine running on neat vegetable oil at constant speed conditions. Renew Energ 28:171–181. https://doi.org/10.1016/s0960-1481(02)00032-0

Ogunwole J, Patolia JS, Chaudhary D, Ghosh A, Chikara J (2007) Improvement of the quality of a degraded entisol with *Jatropha curcas* L. under Indian semi-arid conditions. 26

Olsnes S, Refsnes K, Pihl A (1974) Mechanism of action of the toxic lectins abrin and ricin. Nature 249:627–631. https://doi.org/10.1038/249627a0

Omota F, Dimian AC, Bliek A (2003) Fatty acid esterification by reactive distillation: Part 2—kinetics-based design for sulphated zirconia catalysts. Chem Eng Sci 58:3175–3185. https://doi.org/10.1016/S0009-2509(03)00154-4

Openshaw K (2000) A review of *Jatropha curcas*: An oil plant of unfulfilled promise. Biomass Bioenerg 19:1–15. https://doi.org/10.1016/s0961-9534(00)00019-2

Pandey VC, Singh K, Singh JS, Kumar A, Singh B, Singh RP (2012) *Jatropha curcas*: A potential biofuel plant for sustainable environmental development. Renew Sust Energ Review 16:2870–2883. https://doi.org/10.1016/j.rser.2012.02.004

Parawira W (2010) Biodiesel production from *Jatropha curcas*: A review. Sci Res Essays 5:1796–1808

Peters J, Thielmann S (2008) Promoting biofuels: Implications for developing countries. Energy Policy 36:1538–1544. https://doi.org/10.1016/j.enpol.2008.01.013

Pramanik K (2003) Properties and use of *Jatropha Curcas* oil and diesel fuel blends in compression ignition engine. Renew Energ 28:239–248. https://doi.org/10.1016/s0960-1481(02)00027-7

Pratap Singh S, Kumar Anand R, Vinay V, Sen D (2015) *Jatropha curcas*: A renewable biodiesel plant. International Advanced Research Journal in Science, Engineering and Technology (IARJSET) https://doi.org/10.17148/iarjsetp1

Qin W, Ming-Xing H, Ying X, Xin-Shen Z, Fang C (2005) Expression of a ribosome inactivating protein (curcin 2) in Jatropha curcas is induced by stress. J Biosci 30:351–357

Raajendiran A, Rajesh R, Rajesh E, Rajendran R, Jeyachandran S (2008) Production of bio-ethanol from cellulosic cotton waste through microbial extracellular enzymatic hydrolysis and fermentation.Electron J Environ Agric Food Chem

Ragauskas AJ, Williams CK, Davison BH, Britovsek G, Cairney J, Eckert CA, et al (2006) The path forward for biofuels and biomaterials. Science 311:484–489. https://doi.org/10.1126/science.1114736

Raja SA, Smart DSR, Lee CLR (2011) Biodiesel production from jatropha oil and its characterization. Res J Chem Sci 1

Ranganathan SV, Narasimhan SL, Muthukumar K (2008) An overview of enzymatic production of biodiesel. Bioresour Technol 99:3975–3981. https://doi.org/10.1016/j.biortech.2007.04.060

Rastegari AA, Yadav AN, Yadav N (2020) New and future developments in microbial biotechnology and bioengineering: trends of microbial biotechnology for sustainable agriculture and biomedicine systems: diversity and functional perspectives. Elsevier, Amsterdam

Rastegari AA, Yadav AN, Gupta A (2019) Prospects of Renewable Bioprocessing in Future Energy Systems. Springer International Publishing, Cham

Reddy L (2013) Potential bioresources as future sources of biofuels production: An overview. Biofuel Technologies 223-258. https://doi.org/10.1007/978-3-642-34519-7_9

Russbueldt BME, Hoelderich WF (2010) New rare earth oxide catalysts for the transesterification of triglycerides with methanol resulting in biodiesel and pure glycerol. J Catal 271:290–304. https://doi.org/10.1016/j.jcat.2010.02.005

Semwal S, Arora A, Badoni R, Tuli D (2011) Biodiesel production using heterogeneous catalysts vol 102

Sharma YC, Singh B (2009) Development of biodiesel: Current scenario. Renew Sust Energ Review 13:1646–1651. https://doi.org/10.1016/j.rser.2008.08.009

Srivastava A, Prasad R (2000) Triglycerides-based diesel fuels. Renewable and Sustainable Energ Review 4:111–133. https://doi.org/10.1016/S1364-0321(99)00013-1

Taherzadeh M, Karimi K (2007a) Acid-based hydrolysis processes for ethanol from lignocellulosic materials: A review. BioResources 2

Taherzadeh M, Karimi K (2007b) Enzyme-based hydrolysis processes for ethanol from lignocellulosic materials: a review. BioResources 2

Taufiq-Yap YH, Lee HV, Hussein MZ, Yunus R (2011) Calcium-based mixed oxide catalysts for methanolysis of Jatropha curcas oil to biodiesel. Biomass Bioenerg 35:827–834. https://doi.org/10.1016/j.biombioe.2010.11.011

Umer R, Rehman HA, Irshad H, Muhammad I, Haider MS (2011) Muskmelon (*Cucumis melo*) seed oil: a potential non-food oil source for biodiesel production. Energy (Oxford) 36:5632–5639. https://doi.org/10.1016/j.energy.2011.07.004

Verma S, Gupta A, Kushwaha P, Khare V, Srivastava S, Rawat AKS (2012) Phytochemical Evaluation and Antioxidant Study of Jatropha curcas Seeds. Pharmacog J 4:50–54. https://doi.org/10.5530/pj.2012.29.8

Vicente G, Martínez M, Aracil J (2004) Integrated biodiesel production: a comparison of different homogeneous catalysts systems. Bioresour Technol 92:297–305. doi:https://doi.org/10.1016/j.biortech.2003.08.014

Wan Omar WNN, Nor Aishah SA (2011) Optimization of heterogeneous biodiesel production from waste cooking palm oil via response surface methodology. Biomass Bioenerg 35:1329–1338. https://doi.org/10.1016/j.biombioe.2010.12.049

Weldu Y, Wondimagegnehu G (2015) Assessing the energy production and GHG emissions mitigation potential of biomass resources for Alberta. J Clean Prod 112:4257–4264. https://doi.org/10.1016/j.jclepro.2015.08.118

Wright M, Brown R (2007) Comparative economics of biorefineries based on the biochemical and thermochemical platforms. Biofuels, Bioprod 1:49–56. https://doi.org/10.1002/bbb.8

Yadav AN, Rastegari AA, Yadav N (2020) Microbiomes of Extreme Environments: Biodiversity and Biotechnological Applications. CRC Press, Taylor & Francis, Boca Raton, USA

Yan S, Kim M, Mohan S, Salley SO, Ng KYS (2010) Effects of preparative parameters on the structure and performance of Ca–La metal oxide catalysts for oil transesterification. Appl Cataly A: Gen 373:104–111. https://doi.org/10.1016/j.apcata.2009.11.001

Yan S, Salley SO, Simon Ng KY (2009) Simultaneous transesterification and esterification of unrefined or waste oils over ZnO-La_2O_3 catalysts. Appl Catal A: Gen 353:203–212. https://doi.org/10.1016/j.apcata.2008.10.053

Zabeti M, Wan Daud WMA, Aroua MK (2009) Activity of solid catalysts for biodiesel production: A review. Fuel Process Technol 90:770–777. https://doi.org/10.1016/j.fuproc.2009.03.010

Zhang G, Kazanietz MG, Blumberg PM, Hurley JH (1995) Crystal structure of the cys2 activator-binding domain of protein kinase C delta in complex with phorbol ester. Cell 81:917–924

Chapter 16
Biofuel Production: Global Scenario and Future Challenges

Manoj Kumar Mahapatra and Arvind Kumar

Abstract Biofuels have evolved as the fuel of the future. The technology hungry era has resulted in pollution escalation and fossil fuel depletion like never before. The scientific quest for sustainable fuel options finally culminated with the biofuels. The biomasses for biofuel production are the best candidates for carbon fixation. The biomass mediated carbon capture and storage/utilisation is an excellent natural route for reduction of carbon footprint. The substrates for biofuels are food crops, non-food crops, and microalgae. The food-fuel issues have enforced the biofuel producers and policymakers to give up the practice of food crop usage. As of now biofuel production from lignocellulosic and microalgae is the only remaining viable option. Biofuel production methods are necessarily less energy-consuming ones as compared to their fossil fuel counterparts. Fermentation, transesterification, and hydrothermal liquefaction are some of the biofuel production methods. Biofuels like bioethanol, biodiesel, and biobutanol are getting used in transportation sector as flex-fuels. The current blending percentages for flex-fuels are as high as 20%. The physicochemical properties of biofuels mandate engine retrofitting for biofuel usage. Biofuel sustainability is an utmost requirement for its social and economic acceptance. Biofuels productions must not be done at the cost of reduction in food supplies. Simultaneously biofuel production process should not put adverse effects on land quality, water reservoirs, and biodiversity. The biofuel policies all over the globe are more or less the same. They necessarily enforce biofuel blending into petroleum fuels. The producers are provided with lucrative fiscal supports by governments. Tax exemptions, frequent regulation of raw materials prices, and guarantee for biomass sales for over a decade are now attracting many farmers to initiate energy cropping. All these efforts from different stakeholders are gradually transforming the global energy sector scenario.

M. K. Mahapatra (✉) · A. Kumar
Environmental Pollution Abatement Laboratory, Department of Chemical Engineering, National Institute of Technology, Rourkela, India
e-mail: manojmahapatra2010@gmail.com

A. N. Yadav et al. (eds.), *Biofuels Production – Sustainability and Advances in Microbial Bioresources*, Biofuel and Biorefinery Technologies 11,
https://doi.org/10.1007/978-3-030-53933-7_16

16.1 Introduction

Energy is a quintessential need for survival. For living beings, it can be in the form of food, and for non-living entities, as fuel. The energy became the most significant necessity for developmental activities all around the globe. Till date, the best possible sources of energy are the conventional fossil fuels. However, the insecurity of a steady-state supply at par with the asking demand, varying prices and last but not the least the considerable adverse environmental impacts have forced us to look beyond fossil fuels for better alternatives (Brito and Martins 2017). The usages of fossil fuels have led to drastic climatic changes, loss of biodiversity, alterations in the quality of ecosystem and rapid exhaustion of fuel reserves (Kumar et al. 2019). These serious issues have rung the alarm to tackle them in the shortest period possible. These concerns have opened the doors for the usage of renewable energy sources and alternative ways for the production of energy and fuels (Renewable Energy Directive 2009; Correa et al. 2019).

Currently, the most debatable topic is how to cater the increasing energy demands while either doing no harm to the environment or even reversing the harmful effects (Heard et al. 2017). Nowadays, about 80% of the global energy demands are getting fulfilled by fossil fuels (coal, petroleum and natural gases) which amount to 5.8×10^{11} GJ as of 2016. The transportation sector alone has claimed about ~60% share out of those 80% energy demands (Correa et al. 2019; Joshia et al. 2017). Alternatives to conventional fuels must be renewable and sustainable. The biofuels have emerged as the sustainable fuel source having capabilities like reducing the greenhouse gas emissions, thereby simultaneously improving environmental health (Rauda et al. 2019). The amount of CO_2 emitted during the combustion is the amount of CO_2 assimilated by the biomass during its growth period (Kour et al. 2019). That is how carbon footprint is balanced via biofuels (Mahapatra and Kumar 2019). Biofuels are gaining lots of attention in the current times because of their extensive number of pros over minimal cons. Some of the striking pros and cons are mentioned in Table 16.1.

Table 16.1 Advantages and disadvantages of biofuels (Mahapatra and Kumar 2019)

Advantages of biofuels	Disadvantages of biofuels
The usage of biofuels can reduce the biofuel dependency to a large extent	Biofuel production can lead to food scarcity when the feedstocks are used as raw materials
Balance in the ecosystem is maintained since the emissions of greenhouse gases are controlled	
Reduction of waste handling issues since biomass wastes are used as the raw materials for biofuel production	The genetically engineered microbes and biomasses used for biofuel production can pose a threat to the ecosystem balance unless they are handled effectively
Biofuel generation process provides employment opportunities, which eventually helps in socio-economic development	

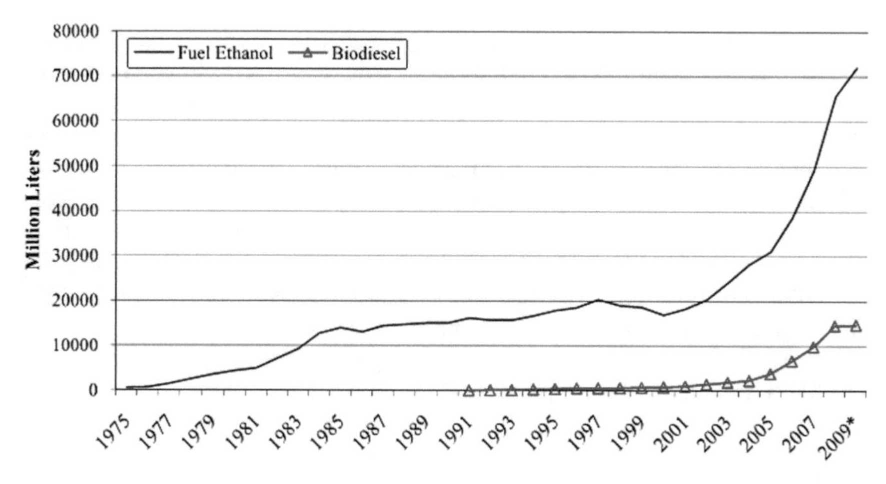

Fig. 16.1 The pattern of worldwide Ethanol and biodiesel production on an annual basis during 1975–2009 (Sorda et al. 2010)

The governmental interventions in the larger biofuel producing countries have acted as catalysts to enhance biofuel production. The US (global leader in bioethanol production) government provides financial incentives to the biofuel manufacturers. France and Germany have brought the mandates of biodiesel blending with conventional petroleum fuels to use them as transport fuels. This government supports resulted in an escalation of bioethanol production from 16.9 to 72.0 billion litres, while that of biodiesel has grown from 0.8 to 14.7 billion litres during the last decade between 2000 and 2009 (Sorda et al. 2010). The pattern of ethanol and biodiesel productions during 1975–2009 is represented as a plot in Fig. 16.1.

16.2 The Biofuels

The reports for the year 2014 of International Energy Agency, have predicted growth in energy demands of about 37% by 2040. However, the limited and fast exhausting fossil fuel reserves would not be able to meet the requirements alone by then. Hence the scientists are trying their best to come up with alternative fuel sources from biomasses, i.e. the biofuels. Although a lot of production methods are invented the commercial-scale productions at large scale are still to be untapped (Schiermeier et al. 2008).

The developed countries use biofuels for the transportation sector. Whereas, the developing nations are targeting in multifold such as maintaining the climatic harmony, creating employment opportunities at large, and last but not the least in the restoration of wastelands to their original state via biomass-energy plantation (Joshia et al. 2017). The global climate change issue is currently posing as one of the primary driving forces for emphasising biofuel production. Recently the International Energy

Agency has proposed a 2 °C scenario (2DS), which says that the global CO_2 emissions by the year 2060 should be cut down by a mark of 70% with reference to the 2014 level. The principal sectors which contribute to the CO_2 emissions at large are the transportation sector and electricity generation sector, with the former one alone providing to 23% of total emissions all across the globe. Both of these sectors are utilising fossil fuels for their operation (Yabe et al. 2012; Ahlgren et al. 2017). Based on the mobility model outcomes in accordance to the 2DS approach released by IEA, by the year 2060, the quota of biofuel usage in the transportation and electricity generation sectors should be at least 30.7% and 27%, respectively (Ho et al. 2014). The surge of biofuel usage in various areas is considered as favourable approach owing to their either carbon neutral or carbon negative based on their contribution to the CO_2 concentration (Naik et al. 2010).

16.3 The Generations of Biofuels

The biofuels are categorised into four different generations based on the raw materials used and are as follows,

- First generation (1G) biofuels
- Second generation (2G) biofuels
- Third generation (3G) biofuels
- Fourth generation (4G) biofuels

16.3.1 First Generation (1G) Biofuels

The first-generation biofuels are the biofuels which use edible biomasses as raw materials. Those raw materials can either be starch such as potato, wheat, barley, and corn or the sugars obtained from sugarcane and sugar beet (Alalwan et al. 2019). The first generation of biofuels is easy to produce because of the structural simplicity of biomasses. They hold their promises of cutting down fossil fuel usages considerably while lowering the atmospheric CO_2 concentration, which is consumed by the biomass during their growth (Rodionova et al. 2017). Some of the examples of first-generation biofuels are bioethanol, biodiesel, and bio ether, etc. Several criteria are to be looked upon before clearing up the food crops to be used as raw materials for the first-generation biofuel.

- Chemical composition of biomass
- Direct or indirect competition with the food crops
- Emission of harmful gases if any
- The extent of pesticide and assorted toxic chemical usage
- Cost of biomass transport and storage
- Employment opportunity creation

16.3.2 Second Generation (2G) Biofuels

This generation of biofuels follows an enhanced sustainable approach for their production. The carbon footprint from the utilisation of these 2G biofuels is either neutral or negative. Based on the raw material cost, the 2G biofuels is a cheaper option. Since most of the feedstocks are lignocellulosic biomasses, which are found as agriculture and forestry wastes (Trabelsi et al. 2018). The 2G biofuels include bioethanol, biodiesel, biobutanol, and acetone. Although acetone is a solvent, it gets produced as an outcome of ABE fermentation process (Alalwan et al. 2019).

The raw materials for 2G biofuels are necessarily lignocellulosic biomasses. Which are composed of lignin, cellulose, and hemicellulose (Ravindran and Jaiswal 2016). Figure 16.2 provides the complex chemical structures of these compounds. These compounds are consists of repeating cyclic units with varying functional groups. Cellulose is a homopolymeric carbohydrate consisting of hexose sugar (D-glucose) as the monomers connected by the β-1, 4-glycosidic linkages, and it constitutes the rigid primary cell wall of plants. The hemicellulose is a complex carbohydrate consisting of both hexose and pentose sugars. The glucose, mannose, and galactose are hexose sugars and the xylose, arabinose, and rhamnose are the pentose sugars found in hemicellulose. Apart from sugars, certain uronic acids like 4-o-methylglucuronic, D-glucuronic and D-galacturonic acids are also found in the strands of hemicellulose. The linkages between monomers are β-1, 4-glycosidic linkages and β-1, 3-glycosidic linkages. The lignin is the non-carbohydrate polymer where the monomers are derived from aromatic alcohols. The monomers of lignin are syringyl group, guaiacyl group, and p-hydroxyphenyl group, which are derived from sinapyl alcohol, coniferyl alcohol, and p-coumaryl alcohol, respectively (Mahapatra and Kumar 2019; Sadeek et al. 2015).

16.3.3 Third Generation (3G) Biofuels

In the third generation biofuels, the microbes like microalgae, bacteria, yeast, and fungi are the feedstocks. Amidst different type of microbes, the microalgae are the most promising one and are responsible for biodiesel production. The microalgae can be autotroph, heterotroph, and mixotrophs; they can exist in both fresh as well as marine waters. The predominant use of microalgae over other microbes is favoured because microalgae impart higher growth and biomass productivity tendencies and can accumulate lipids in the range of 20–77% (Bajracharya et al. 2017; Chelf et al. 1993). The residual biomass of microalgae is used for producing biomethane, bio-oil, bioethanol, and biohydrogen via separate biorefinery processes (Packer et al. 2016).

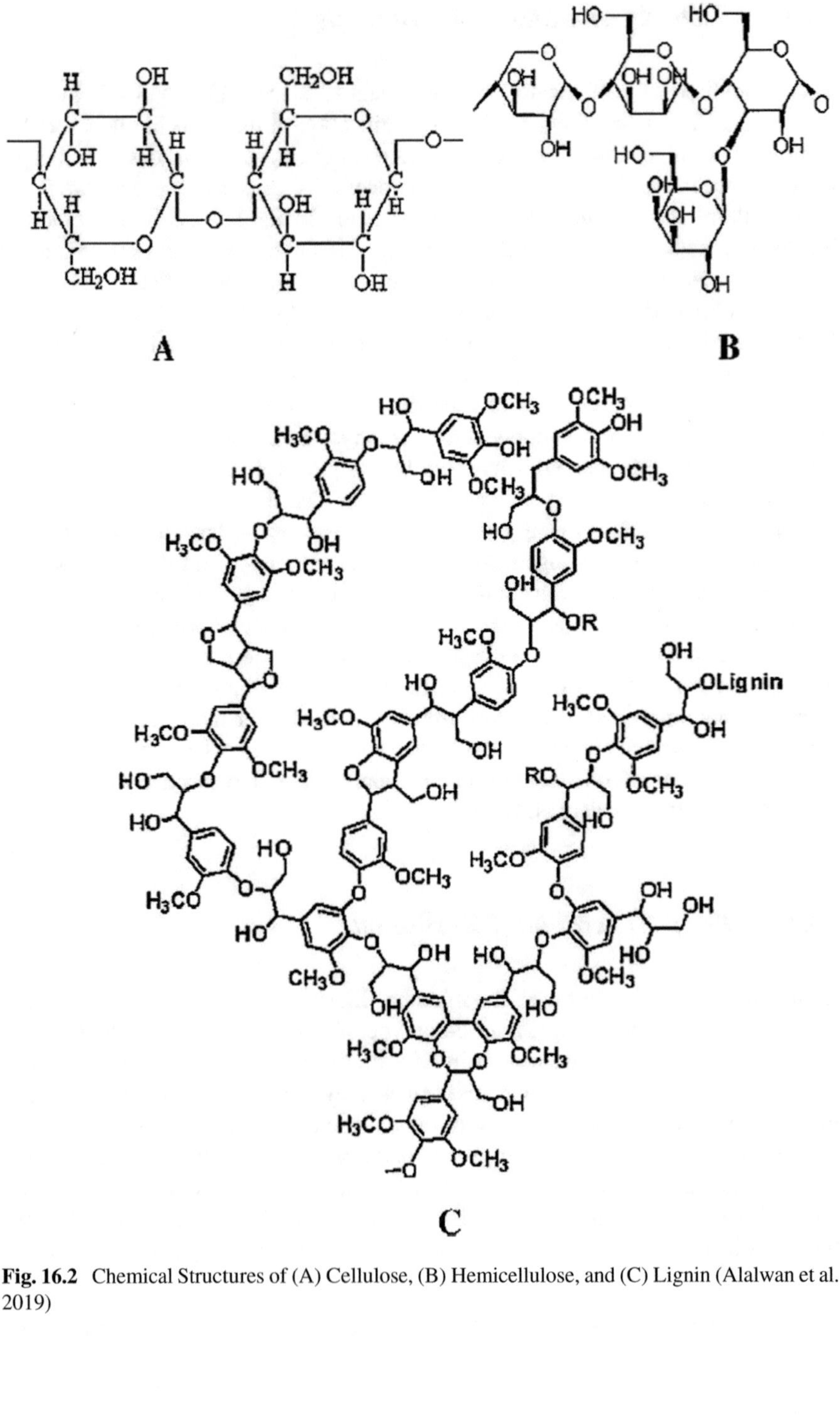

A

B

C

Fig. 16.2 Chemical Structures of (A) Cellulose, (B) Hemicellulose, and (C) Lignin (Alalwan et al. 2019)

16.3.3.1 Carbon Capture and Storage/Utilisation Approach of Microalgae During Biofuel Production

The microalgae can be autotrophs, heterotrophs, or even mixotrophs based on their mode of carbon source utilisation. The inorganic carbon such as CO_2 gets trapped by autotrophic microalgae via photosynthesis to obtain nutrient and energy. While, the heterotrophic microalgae use organic carbon as their nutrient and energy source. The mixotrophic ones as their name suggests utilise both the inorganic and organic carbon as their nutrient and energy sources. The microalgae use atmospheric CO_2 for its metabolism, as a result of which carbon concentration in the atmosphere gets reduced (Leong et al. 2018).

The Paris agreement of 2015 by the international community has decided to initiate the CO_2 mitigation steps with utmost priority. However, the mitigation if made by creating additional forestation will face the issues of land and food crisis (Dooley and Christoff 2018). Hence a newer concept of BECCS/U, i.e. bioenergy with carbon capture and storage/utilisation was considered to be more fruitful for the same purpose. Among different generations of fuels, the 3G biofuels derived from microalgae is the best candidate for the implementation of BECCS/U approach owing to the characteristic features of microalgae (Williamson 2016). The microalgae have several advantages for both the biofuel and CO_2 mitigation purposes (Choi et al. 2019) which are described as follows,

- Higher efficiency of photosynthesis and rapid growth ensures shorter harvest time.
- Microalgae can grow with wastewater, which in turn can serve as a mode of wastewater treatment and do not stress the clean water resources.
- Non-fertile and barren lands are the first choice places for creating algal ponds, hence no problem with cultivable landmass.
- As compared to their forestry equivalent microalgae need much lesser landmass to fulfil the desired objective.
- Microalgae do not need any additional chemicals in the form of fertiliser and pesticides, thus maintains the natural integrity of the ecosystem.
- The algal mass is tolerant to SO_x and NO_x; hence, the CO_2 laden flue gas stream can be fed to the biomass as the carbon source and mitigating the CO_2 simultaneously.

The microalgae biomasses are capable of capturing about 55–65% anthropogenic CO_2 emission from the atmosphere (Farrelly et al. 2013). CO_2 plays a crucial role in the microalgal photosynthesis process, and the dependency of microalgae on CO_2 can be better understood from the composition of carbon in the dried biomass of microalgae which ranges as 36%–65% (Chae et al. 2006). The low concentration of CO_2 in the atmosphere as high as 380 ppm and its poor solubility in water keeps the microalgae deprived of a continuous carbon source (McGinn et al. 2011). The compressed CO_2 supply will again add up the cost of biofuel production to as high as 41% (Grima et al. 2003). The only solution to this problem is the construction of in situ microalgal ponds at the emission sites of elevated CO_2 concentration.

Moreover, due to their tolerance to SO_x and NO_x, the CO_2 abundant flue gases can be fed to them as a carbon source (Choi et al. 2019).

Gonçalves et al. (2016) have reported about the involvement carbonic anhydrase and RuBisCo (Ribulose-1, 5- biphosphate carboxylase oxygenase) for CO_2 fixation in microalgae. Moreover, microalgae can utilise the bicarbonate and gaseous CO_2 as a carbon source. At the pH range of 6.5–10 for microalgae production media, the bicarbonate is the first choice as a carbon source. The flue gas (has a higher CO_2 concentration in the range of 0.03-0.05% as compared to the standard air), when fed as the carbon source, has resulted in a higher yield of the biomass (Gonçalves et al. 2016). Choi et al. (2017) have reported that the CO_2 when dissolved in media acts as a buffer to bring up the pH value to a favourable range for microalgae growth resulting in enhanced yield (Choi et al. 2017).

The chloroplast is the factory for the biosynthesis of lipid in microalgae. The CO_2 initially gets fixed as the endogenous source for Acetyl-CoA, and later becomes the carbon in the fatty acid chain of the lipid (De Bhowmick et al. 2015). Apart from *Chlorella sp.*, a few other species such as *Ostreococcus tauri*, *Phaeodactylum tricornutum*, *Nannochloropsis sp.*, and *Chlamydomonas reinhardtii* have shown promising capabilities for lipid biosynthesis (Zienkiewicz et al. 2016). The various 3G biofuels produced from microalgae via different biorefinery processes are represented in Fig. 16.3.

16.3.4 Fourth Generation (4G) Biofuels

The fourth generation biofuels are the advanced versions of 3G biofuels. Unlike the 3G biofuels, the 4G ones are produced from genetically modified microorganisms for better yield, and to avoid any inhibitory action from solvents. The widely used sources for 4G biofuels are microalgae, yeast, fungi, and cyanobacteria, etc. Additionally, thermochemical techniques such as gasification and pyrolysis in the range of 400–600 °Care also used for biofuel production (Azizi et al. 2018). The intention of using thermochemical conversion routes is to improve hydrocarbon yield and to reduce carbon emissions. However, the 4G biofuels are still in the developmental stage and need a lot of research inputs before hitting the commercial production market (Sikarwar et al. 2017).

16.3.4.1 Health and Environmental Issues Related to 4G Biofuels

The 4G biofuels use genetically modified (GM) microalgae as the feedstock for biofuel production. The GM microalgae are prepared for rapid growth and withstand adverse environmental conditions, as a result of which they impart the threat of replacing the native microalgae from the ecosystem. The absence of native microalgae in the ecosystem results in crashing of the biodiversity since GM microalgae are unable to provide natural qualities for biodiversity maintenance.

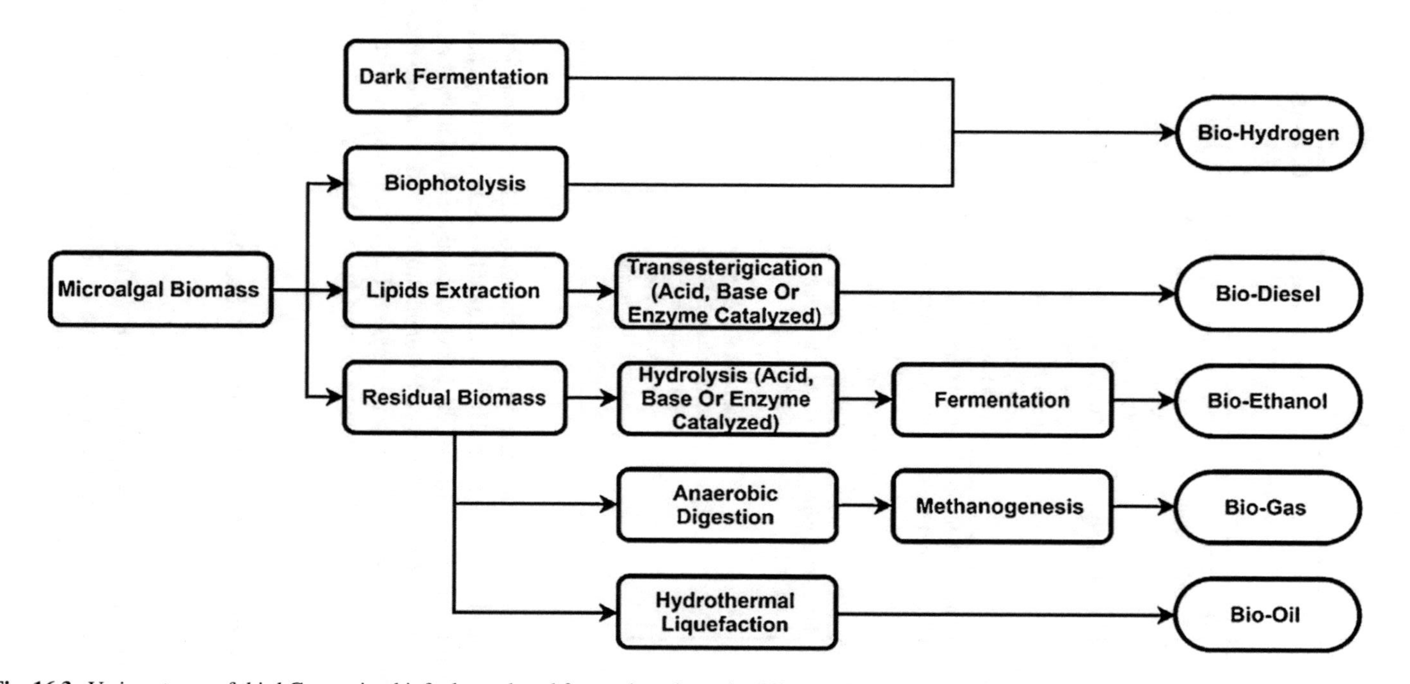

Fig. 16.3 Various types of third Generation biofuels produced from microalgae via different biorefinery processes (Alalwan et al. 2019)

Table 16.2 Health and environmental issues associated with the GM microalgae

Affected area	Type of risk	Brief description of the risk	References
Health hazard	Allergy	Affected areas are the GI tract, respiratory tract, and skin	(Genitsaris et al. 2011)
Health hazard	Resistance to antibiotic drugs	The medical treatments become prolonged, might end up with impaired immunity	(Wright et al. 2013)
Health hazard	Pathogenicity	The toxic residues are good candidates for imparting pathogenicity, toxicity, and carcinogenicity	(Menetrez 2012)
Environmental hazards	Alteration of the natural ecosystem	Depletion of nutrients in the eco-system leading to a sharp decline in biodiversity of flora and fauna	(Tucker and Zilinskas 2006)
Environmental hazards	Horizontal gene transfer	Very acute chances of mutation resulting in further deterioration of the natural ecosystem and toxins released from these species are deadly for other flora and fauna, naturally habituated in the ecosystem	(Raybould 2010)

The primary environmental concerns regarding the uncontrolled exploitation of GM microalgae are the alteration of natural habitats, toxicity, and horizontal gene transfer which pose the threat of mutation of the native microalgae (Abdullah et al. 2019). Apart from imparting adverse effects on the environment, GM microalgae are also responsible for health hazards in human beings. A list of health and environmental issues associated with the GM microalgae is provided in Table 16.2.

16.4 Understanding the Key Terms Like Advanced Biofuels and Drop-in Fuels

Specific interesting terminologies like advanced biofuels and drop-in fuels are being used in recent times and are capable of creating a significant amount of confusion in the course of understanding the biofuel concepts.

Advanced biofuels are a synonym to the 2G (second generation) biofuels, which uses non-food crops and their residues as well as waste materials such as animal fat, spent vegetable oil, greases, as the sources for biofuel generation. The biorefinery techniques for advanced biofuels are as follows;

- Hydrolysis of the raw materials and subsequent fermentation
- Thermochemical conversion route like pyrolysis
- Alcoholic fermentation of the syngas.
- Transesterification

The advanced biofuels are of two types, namely drop-in biofuels and the biobutanol. The drop-in biofuels are biodiesel and bioethanol produced either from lipids or lignocellulosic materials. These biofuels are good candidates to replace the conventional fossil fuels in IC engines with negligible revampments (Araújo et al. 2017). The hydrotreated biodiesel production has several advantages over the diesel produced by the transesterification process. The benefits range from zero enhancement of NO_x released to the atmosphere, no residue formation in engine cavity, enhanced life of engine oil, no sulphur contents (zero SO_x emission), and higher cetane number (Department of Energy 2019).

Biobutanol is another type of advanced biofuels produced via the fermentation route. However, unlike ethanol fermentation, biobutanol production employs ABE (acetone-butanol-ethanol) fermentation pathway. As the name suggests, the outcome is not only biofuels rather solvents are too produced. In the entire arena of biofuels, biobutanol is the only biofuel which can directly replace the gasoline in IC engines without any revampment. The physicochemical properties like higher energy density and lower vapour pressure of biobutanol have made it stand out as the ready to use biofuel in gasoline-powered IC engines (Bharathiraja et al. 2017).

16.5 The Carbon Sequestration Approach with the Biofuels

Carbon sequestration is defined as the method for capturing the carbon either from the atmosphere or from the effluent streams and securely store them rendering the atmospheric carbon levels within the acceptable limits (Jain et al. 2012). The carbon in the atmosphere is found in its oxide forms such as CO and CO_2, which are the causative agents of acid rain, global warming via greenhouse gas (GHG) effect, and poor quality of breathable air. The primary sources of carbon into the atmosphere are the combustion of fuels. The fossil fuels are the necessary evils for the overall global development but at the cost of heightened environmental pollution (Mathews 2008).

The most eco-friendly way of carbon capture from the atmosphere is photosynthesis by biomass. In recent times with the concept of biofuels, hope has been kindled for the carbon capture while generating biofuel in the course. Moreover, by taking into account the amount of carbon accumulation and subsequent releases from and to the atmosphere three distinct concepts came to existence namely,

Table 16.3 Summary of Carbon capture approaches via various fuels (Mathews 2008)

Type of approach	Carbon input to the atmosphere	Carbon capture from the atmosphere	Additional contribution from fossil fuels for product processing	Examples
Carbon positive	Yes	No	Yes	Fossil fuels
Carbon neutral	Yes	Yes	Yes, but in lesser quantities, as compared to the fossil fuels	2G biofuels
Carbon negative	Yes	Yes	Yes, the requirement is in a very minimal amount	3G biofuels

- Carbon positive approach
- Carbon-neutral approach
- Carbon-negative approach

Carbon positive approach is associated with fossil fuels which are the primary causatives for the enhancement of carbon footprint in the atmosphere. Hence they are also called as the carbon positive fuels (Mathews 2008).

The 2G biofuels are also synonymous as carbon-neutral fuels. Since during the growth of biomass a large amount of carbon gets captured, which is nearly equal to the amount of carbon that gets released while burning the biofuel. However, during transportation, and processing (tedious task) due to the expense of a certain amount of fossil fuels, sometimes these biofuels end up as the carbon positive fuels (Mathews 2008).

The biofuels from microalgae or the 3G biofuels are termed as the carbon-negative biofuels. The microalgae have excellent potential in capturing the CO_2 from the atmosphere in large amounts as compared to the forest biomasses. Harvesting and subsequent processing methods are not energy-intensive, resulting in lesser energy consumption for the intended purpose. The spent biomass after oil extraction can be carbonised, and resultant biochar is fed to the soil to produce a natural carbon sequestering agent while improving the soil quality (Mathews 2008; Rakshit et al. 2012). These three different approaches of carbon capture are enlisted in Table 16.3.

16.6 The Proven Biofuels Which Will Be the Key Players in Future

During the process of biofuel production, apart from fuels, solvents of industrial importance like acetone, glycerine, and ether are also get produced. However, here we will be discussing the key biofuels which are going to change the image of the fuel energy sector in upcoming times. Based on their production methods and properties three different types of biofuels qualify to gain such status and are as follows;

- Biodiesel
- Bioethanol
- Biobutanol

16.6.1 Biodiesel

Biodiesel is a renewable, eco-friendly, and non-toxic alternative to the petroleum diesel fuel. Due to the sustainable nature, biodiesel found its usage as mainstream fuel in transport sector via road, water bodies (with the ships), and aerial route (with aeroplanes). As of now, biodiesel is being used as blended fuel in the engines. Leading producers of biodiesel like Brazil, EU, and the US are targeting to enhance the blending percentage of biodiesel as 20, 10, and 25 by the year 2020 (Mofijur et al. 2016). The US Naval force is planning to use alternative fuels (primarily with biodiesel) in place of conventional fossil fuels to fulfil 50% of its energy requirement (Mabus 2010). Moreover, the production and usage of biodiesel in a sustainable manner is dependent on certain factors (Ntaribi and Paul 2019; Fazal et al. 2011) namely;

- The availability of better quality feedstocks
- Advanced processing technologies
- Physicochemical properties of biodiesel
- Engine compatibility
- Quantity of production and pricing
- Government regulations and financial support

16.6.1.1 The Sources of Biodiesel

The biodiesel is generally produced from two sources, namely; oil-rich food crops, and oil-rich non-food crops. These non-food crops are also called as the energy crops since their cultivation is aimed at biofuel production only. Some of the examples of both categories of sources for biodiesel production are as follows, rapeseed, linseed, rice bran, soybean, sunflower, corn, castor, coconut are a few among the most popular oil-rich food crops from which the 1G biodiesel can be produced, among the non-food crops, animal extracts, and wastes like *Jatropha curcas*, cottonseed, rubber seed, neem seed, apricot seed, desert date, jojoba, *Pongamia glabra*, *Pistacia chinensis*, *Moringa oleifera*, *Shorea robusta*, microalgae, fish oil, leather pre-fleshings, and the waste cooking oils are some of the notable ones for the production of 2G and 3G biodiesels (Alalwan et al. 2019; Karmakar et al. 2010).

16.6.1.2 Different Production Methods of Biodiesel

The various methods for biodiesel production are; blending of oils, micro-emulsification, pyrolysis, and the transesterification (Alalwan et al. 2019).

Blending of oils is a necessary step to reduce the viscosity. The vegetable oils can be preheated to make the blending easier since preheating the vegetable oils result in a reduction of viscosity and atomisation. As reported by Adams et al. (1983), the blending ratio of vegetable oil to diesel should be 1:2, to run the engine without any major modifications (Ghazali et al. 2015; Adams et al. 1983).

Micro-emulsification is another efficient method for biodiesel production. The microemulsions are nano molecules in the range of 1–150 nm in terms of their size and are composed of the oil phase, aqueous phase, and surfactant phase. With the use of butanol, hexanol, and octanol as the aqueous phase in the microemulsion, the desired viscosity limits can be met. A microemulsion with the components and their ratios as soybean oil: methanol: 2-octanol: cetane improver (surfactant), 52.7: 13.3: 33.3: 1 has successfully passed the significant 200 h EMA (engine manufacturers association) test for alternative fuels (Ghazali et al. 2015).

The pyrolysis is a thermochemical conversion process in the absence of oxygen. It is implemented as the alternative to catalytic cracking in the biodiesel production process. The raw materials for pyrolysis with the aim for optimisation of biodiesel are vegetable oils, lignocellulosic biomasses, animal fats, other oil-rich biological wastes, and the FAME (fatty acid methyl ester) (Ghazali et al. 2015). The biodiesel is produced using oils obtained from food crops via *transesterification* or the alcoholysis process. During the transesterification process, one alcohol in the ester is replaced by another desired alcohol in the presence of an alkali catalyst. The replacement of alcohol group from the ester results in a reduction of viscosity (Ghazali et al. 2015).

16.6.1.3 Technical Advantages and Disadvantages of Biodiesel

The molecular chemistry of biodiesel imparts several technical advantages over petroleum diesel; some of such benefits (Gopinath et al. 2010; Knothe et al. 2003; Qi et al. Qi et al. 2009) are enlisted here,

- The long-chain molecules, oxygenated moieties, give *enhanced lubricity* to the biodiesel.
- The long fatty acid chains of biodiesel help in intermolecular sliding, and thereby impart in *non-corrosiveness of the engine cavity.*
- The longer fatty acid chains and saturated molecules render *higher cetane number* to the biodiesel, which in turn result in *complete combustion, smoother engine performance, negligible carbonation in engine head, better fuel efficiency* with *lesser emissions.*

The chemistry of biodiesel gives many significant advantages for biodiesel. Unfortunately, it is the chemistry again which render some of the notable drawbacks to the biodiesel. Some of such disadvantages (Jakeria et al. 2014; Fazal et al. 2018; Tyson 2001) are enlisted below,

- *High viscosity*, this result in deposition of unwanted matter in the engine head, especially around the piston ring, thereby results in improper combustion.
- *Poor oxidation stability* brings in the clogging and sludging of fuel injectors and filters, crankcase, combustion chamber, which results in the below-par performance of the engine.
- *Higher corrosiveness* leads to the leakage in the fuel lines and simultaneous breakage of the seal. Thereby the maintenance costs go high.
- *Short storage period* forces the users to go for produce and use approach. Since when biodiesel is stored for longer durations that can result in loss of calorific value due to moisture accumulation. Moreover, the ester molecules get reverse-engineered via hydrolytic reactions to alcohol and free fatty acids, leaving the biodiesel with only option to discard them, instead of using the same.

16.6.2 Bioethanol

Bioethanol is produced by the fermentation of sugars obtained from the biomasses. This fermentation process is the most common type of fermentation process which commonly employs the yeast (*Saccharomyces cerevisiae*) as the fermenting agent (Rastegari et al. 2019). Although three main types of microbes exist those can bring out the ethanol fermentation, and are yeast (*Saccharomyces spp.*), mold (mycelium), and bacteria (*Zymomonas spp.*) (Yusoff et al. 2015). The industrial-scale production of bioethanol dates back to the year 1894 in France and Germany. Brazil pioneered to use bioethanol as a transport fuel in the year 1925. However, later due to high production costs bioethanol was not considered as a desirable option for any applications. During the oil crisis of 1970 and acute environmental pollutions have renewed the bioethanol production exclusively for its use as a transportation fuel (Alalwan et al. 2019).

Bioethanol can be used either in pure form or in the blended form with gasoline for Flex-fuel vehicle (FFV) fuel. Currently, bioethanol is blended at a low volume with gasoline at 10% v/v ratio and has obtained the brand name as E10. Bioethanol also acts as a precursor to the ethyl tertiary butyl ether (ETBE), which is mixed with the gasoline to enhance the oxygen content for pollution control purposes (Norkobilov et al. 2017). Like any other biofuels, bioethanol also helps in curbing CO_2 from the atmosphere by cutting down the amount of fossil fuel usage and capturing CO_2 via biomass during their growth (Li et al. 2017).

The USA leads the global bioethanol market scenario, contributing 47% of the total bioethanol production all across the globe (Balat and Balat 2009). It is estimated that the amount of bioethanol production has reached a value of 93 billion litres as of 2014 (Li et al. 2017). However, excessive usage of food crops for bioethanol production has raised the debate on the food crisis. It was estimated that the extent of food crops used for bioethanol production could have fed 200 million people. Hence the 1G bioethanol option is a strict no type production method (Rulli et al. 2016).

The bioethanol production from lignocellulosic biomasses is a tedious task as compared to the production from the food crops. The lignocellulosic biomasses must be hydrolysed before the fermentation process to release the sugars. 2G Bioethanol can also be produced via a thermochemical route called gasification followed by either of fermentation or enzyme catalysed reaction. The difficulties associated with the 2G bioethanol production are (Vyas et al. 2018) as follows,

- Lack of a remarkably efficient pretreatment method for effective release of sugar from biomasses.
- Release of undesired sugars along with important ones hampers the fermentation process efficacy.
- Presence of oligomers of sugars instead of simple monomers makes the implementation of genetically engineered microbes a must-have thing.
- Release of undesired by-products during hydrolysis leads to a reduction of bioethanol yield to a significant amount.
- Transportation and storage of biomasses at times pose as a costly affair.

As far as the 3G bioethanol is concerned, it is the most preferred bioethanol production process. Since 3G bioethanol, unlike its 1G counterpart, is not creating food scarcity issues, on the other hand, the biomass handling process is easier as compared to that of 2G bioethanol production. In the course of 3G bioethanol production, the algal carbohydrates such as starch and carbohydrates are subjected to hydrolysis to yield monomeric sugars and are subjected to fermentation, subsequently (Alalwan et al. 2019). Moreover, the yield of 3G bioethanol is very high. John et al. (2011) have reported that the Algenol Biofuels Inc. (current alias ALGENOL), a Florida-based company have achieved a production rate of 6000 gallons of ethanol per acre per year. This staggering amount of bioethanol is ~ 15 times higher than those of 2G biofuels (John et al. 2011). Bioethanol has several advantages and disadvantages as well, and those affect its use as a transportation fuel (Rastegari et al. 2020). On the pros side, it has very high octane number of 108, and with higher heat of vaporisation and wider flammability limits, together all these enhance the chances of bioethanol to be used as a reliable transport fuel. On the contrary bioethanol has some significant characteristic flaws associated with it, such as low energy density, corrosive nature, low vapour pressure, and water miscibility, these result in lower calorific value, faster wear and tear of engine parts, problem with cold start of the engine, and improper ignition due to the presence of moisture, respectively (Alalwan et al. 2019). Figure 16.4 represents the production flow sheet of different generations of bioethanol.

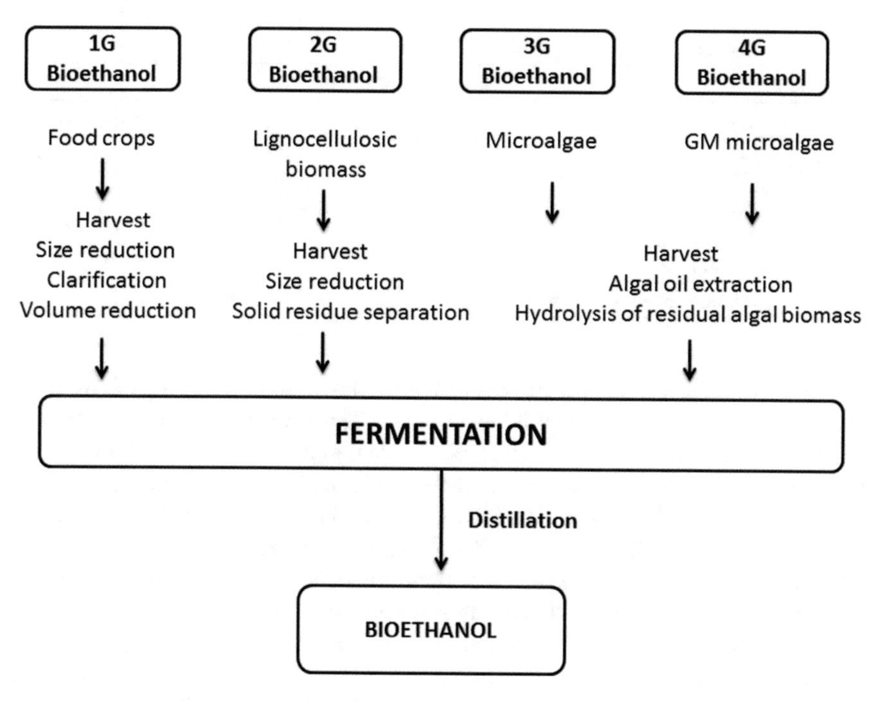

Fig. 16.4 The production flow sheet of different generations of bioethanol (Alalwan et al. 2019)

16.6.3 Biobutanol

Biobutanol has evolved as one of the most promising biofuels to replace the gasoline as the fuel without retrofitting the engine assembly. The mode of biobutanol synthesis is bacterial fermentation called ABE fermentation mediated by *Clostridium spp.* The raw materials for biobutanol synthesis are food crops and the lignocellulosic biomasses. Butanol is an excellent industrial chemical and has several applications (Mahapatra and Kumar 2017) as follows.

- Butanol is used as a solvent in rubber industries.
- It is also used as quick-drying lacquer for imparting smooth surface finish in the dye industry and printing presses.
- In the pharmaceutical industry, the butanol is used as an extractant for drugs, vitamins, and hormones.
- Butanol is used as a supplement in domestic and industrial cleaners.
- In thin-layer chromatography, butanol is used as eluent.
- It is used as a de-icing agent for gasoline-driven engines.
- Butanol is used as the precursor for the production of acrylic esters, glycol ethers, butyl acetate, butyl amines.

Table 16.4 The fuel properties of gasoline and butanol

Properties	Gasoline	Butanol	References
Energy density (MJ/L)	32.0	29.2	(Mahapatra and Kumar 2017)
Air to Fuel ratio	14.6	11.2	(Mahapatra and Kumar 2017)
Heat of vaporisation (MJ/Kg)	0.36	0.43	(Mahapatra and Kumar 2017)
Research octane testing value	91–99	96	(Mahapatra and Kumar 2017)
Motor octane testing value	81–89	78	(Mahapatra and Kumar 2017)
Rate of evaporation or reid value (psi)	8–15	0.33	(Alalwan et al. 2019)

The fuel properties, in comparison with gasoline listed in Table 16.4 depicts the ability of butanol to be used as IC engine fuel. The industrial synthesis of biobutanol dates back to 1912–1914 via ABE fermentation using molasses and cereal grains employing the Weizmann's organism (*Clostridium acetobutylicum*) (Jones and Woods 1986). The synthesis of biobutanol in the laboratory was first reported in 1861 by Louis Pasteur (Durre 1998). Apart from Weizmann's organism certain other native *Clostridium sp.* such as *Clostridium beijerinckii, Clostridium saccharoperbutylacetonicum, Clostridium saccharoacetobutylicum, Clostridium aurantibutyricum, Clostridium pasteurianum, Clostridium sporogenes, Clostridium cadaveris, and Clostridium tetanomorphum* are capable of butanol synthesis following the ABE fermentation route (Kumar and Gayen 2011). Among all these species of *Clostridium*, the *C. acetobutylicum, C. beijerinckii, C. saccharoperbutylacetonicum*, and *C. saccharoacetobutylicum* are the ones for higher biobutanol yield via ABE fermentation (Keis et al. 2001). The ABE fermentation process is strictly anaerobic, and the different products such as acetone, butanol, and ethanol are produced in the ratio 3:6:1 (Alalwan et al. 2019).

16.6.3.1 Biomasses for the Biobutanol Synthesis

Biobutanol production uses two categories of biomasses for their production, namely the food crops for 1G biobutanol production and lignocellulosic biomasses for 2G biobutanol productions. The food crops for biobutanol synthesis are sugarcane, sugar beet, wheat, rice, soybean oil, sunflower, and palm oil. However, utilisation of the food crops for biobutanol production eventually leads to the food scarcity and faces severe criticism with the title of price hikers in the food v/s energy debate (Kumari and Singh 2018). The 2G biobutanol, on the other hand, uses lignocellulosic biomasses such as rice straw, rice hulls, wheat straw, corn cobs, corn Stover, cane bagasse as the raw materials. These agricultural lignocellulosic wastes alone amount to 40 tons per hectare. Frequently, these wastes are either burnt as the most natural way of volume reduction, which is not a sustainable option or used as forage for the farm animals or as organic manures.

Apart from agricultural wastes, the forestry wastes are too good candidates for biobutanol production (Srirangan et al. 2012). The lignocellulosic residues are

Table 16.5 Inhibitor produced by various pretreatment processes (Baral and Shah 2014)

Biomass pretreatment process	Inhibitors produced
Acid hydrolysis	Furfural, hydroxyl methyl furfural, acetic acid, and phenolics
Alkali hydrolysis	Soluble salts (extremely difficult to separate them from the broth)
Steam explosion	Furfural, acetic acid, formic acid, and phenolics

the best candidates for the production of cost-effective 2G biobutanol. Although the lignocellulosic biomasses are significantly cheaper in their worth, their chemical composition is capable of bringing out a significant reduction in the efficiency of the fermentation process in terms of product yield. The pretreatment step produces certain toxic chemicals called 'inhibitors' which are capable of ceasing the metabolism of microbes leading to low yield (Baral and Shah 2014). Table 16.5 enlists the inhibitors produced by various pretreatment methods. Figure 16.5 represents the schematics of biobutanol production pathway from different biomasses.

The separation of inhibitors from fermentation media is called detoxification. A variety of detoxification methods are available which can be used based on the nature of hydrolysate. Mahapatra and Kumar (2019) have reported that the alkali treatment, LLE (liquid-liquid-extraction), membrane filtration, adsorption, microbial degradation, and enzymatic catalysis are some of the noteworthy detoxification methods for hydrolysate inhibitors. The details are reported elsewhere by the investigators (Mahapatra and Kumar 2019).

16.6.3.2 A Brief Discussion About the ABE Fermentation Process

The ABE fermentation is the second-largest industrial process after ethanol fermentation process. However, this fermentation process has seen its ups and downs over time. Before the inception of petrochemical solvents, butanol and acetone produced via ABE fermentation process during the early part of the twentieth century were in demand. Until the late twentieth century and early twenty-first century that dormant scenario prevailed, but with the invention of butanol's biofuel potential, the so-called fermentation process again raised to its previous glory. ABE fermentation yields acetone-butanol-ethanol simultaneously in the ratio 3:6:1. Like any other chemical/biochemical processes, the ABE fermentation can be carried out by any of the three modes, such as; batch, fed-batch, and continuous. The continuous method of fermentation has many advantages to make the process more favourable and efficient. A single inoculum batch is sufficient to carry out the process for a prolonged duration. Limited sterilisation and microbial inoculation steps enhance the productivity and process economy (Baral and Shah 2014).

The entire ABE fermentation is divided into two categories, namely, acidogenesis phase and solventogenesis phase. The microbes are in their log (exponential)

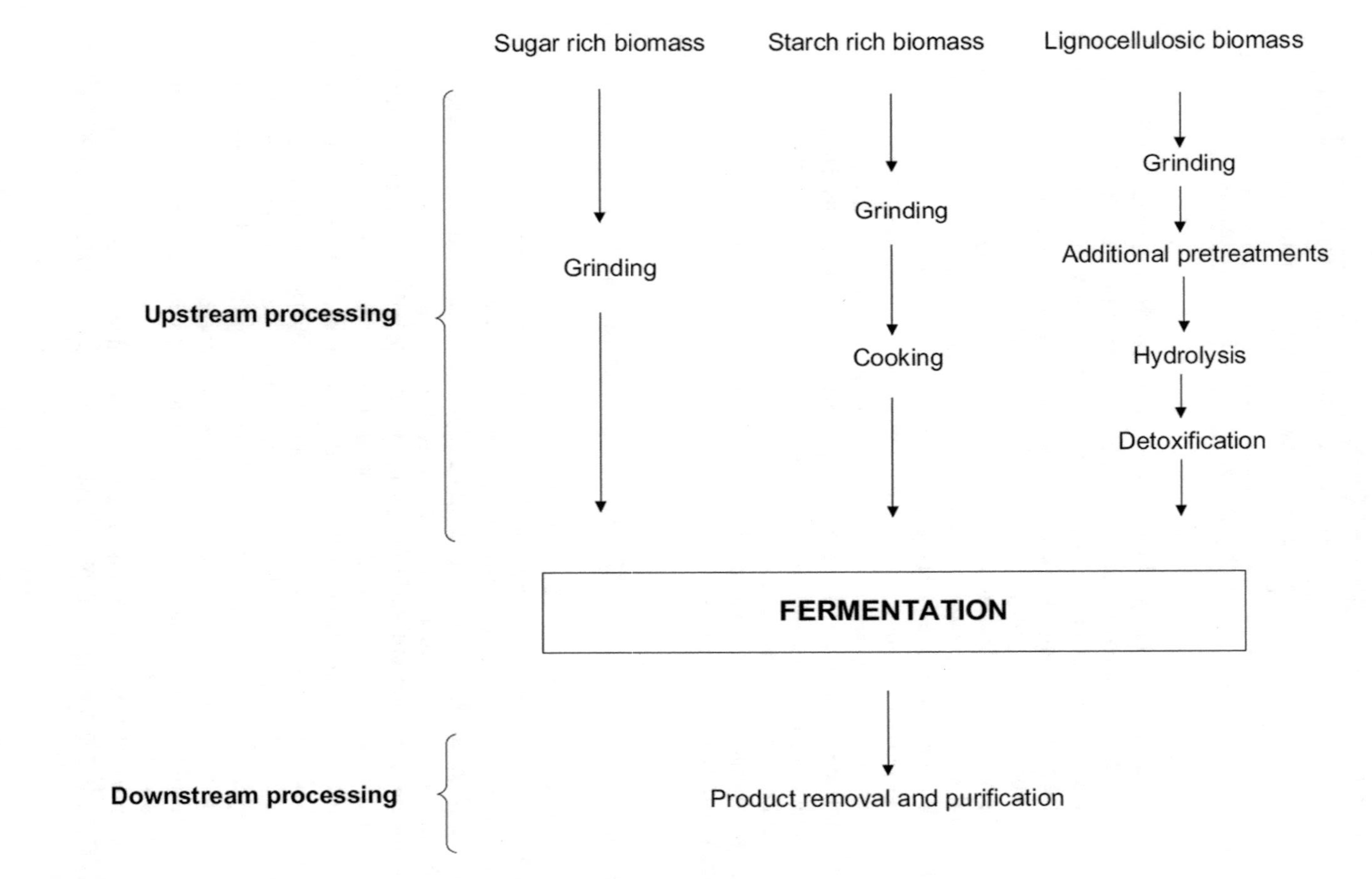

Fig. 16.5 Schematic representation of biobutanol production pathway from various biomasses (García et al. 2011)

phase of growth acidogenesis part. The synthesis of acids has led to the fermentation media pH declinement to ~4.5. The Glycolysis produces pyruvate from glucose, which eventually yields Acetyl-CoA. The Acetyl-CoA is the prime precursor for the synthesis of acetate, butyrate, ethanol, butanol, and acetone in an anaerobic mode. During the solventogenesis phase, which kicks in immediately after the acidogenesis phase, the acid production ceases due to low media pH. During this phase, the microbes have attained a stationary phase of their growth cycle. The acetaldehyde, acetate, and butyrate are depleted to yield ethanol, acetone, and butanol, respectively (Kumar and Gayen 2011).

The fermentation efficiency is evaluated based on the ABE yield and ABE productivity, respectively (Jin et al. 2019). The ABE yield and productivity can be assessed using the formula given by Eq. 16.1 and 16.2, respectively.

$$\text{Yield of ABE} = \frac{\text{g/L of Total ABE}}{\text{g/L of Total sugar utilised}} \tag{16.1}$$

$$\text{Productivity of ABE (g/L/h)} = \frac{\text{g/L of Total ABE}}{\text{h Duation of fermentation}} \tag{16.2}$$

Figure 16.6 depicts the pictorial representation of the biochemical pathway of ABE fermentation involving all the enzymes and intermediate.

16.7 Sustainability Parameters for Biofuels

Sustainable biofuel is the one which has satisfied all the parameters for evaluation. A few essential parameters are,

- The conflict between Food and fuel
- The Emission potentials
- Issues with the land, water, and biodiversity
- Performance of the biofuel

16.7.1 The Conflict Between Food and Fuel

The dispute arises with the food commodity price hikes. These conflicts were fueled by the fact that during the period of food price hikes, the biofuel production intensities were too at their peak. In G20 summits of 2008 and 2011, the agendas were focused primarily on the food prices, and biofuels were the ones to take the blame. The conclusions were, firstly the biofuels are leading to food price hikes and eventually, poor people will be affected at large. Secondly, the energy cropping is rendering the food croplands unusable for their intended purpose, forcing them to displacement (De Gorter and Drabik 2015; Tomei and Helliwell 2016). However, a thorough analysis

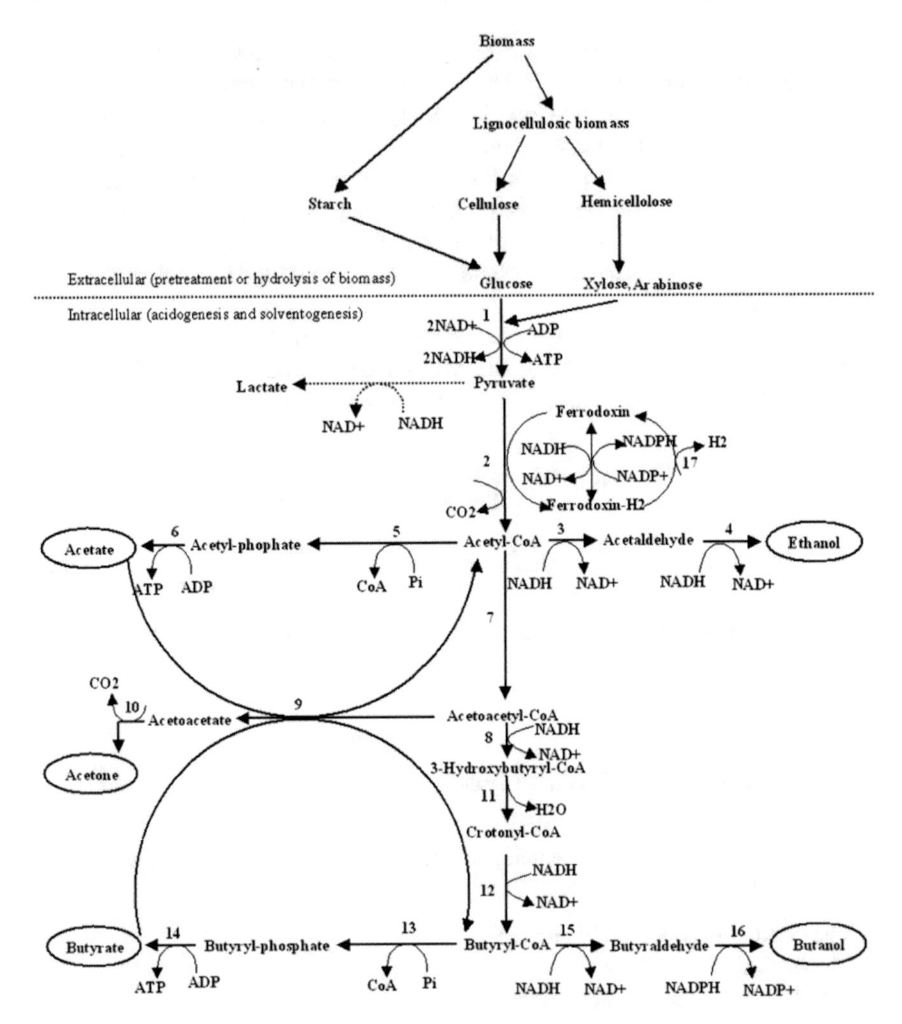

Fig. 16.6 Biochemical pathway of ABE fermentation (Kumar and Gayen 2011). The enzymes involved are the numbers depicts enzymes involved and are as follows (1) Enzymes of glycolysis process (2) Pyruvate ferredoxinoxidoreductase (3) Acetaldehyde dehydrogenase (4) Ethanol dehydrogenase (5) Phosphate acetyltransferase (phosphotransacettylase) (6) Acetate kinase (7) Thiolase (acetyl-CoA acetyltransferase) (8) 3-hydroxybutyryl-CoA dehydrogenase (9) Acetoacetyl-CoA: acetate/butyrate:CoA-transferase (10) Acetoacetate decarboxylase (11) Crotonase (12) Butyryl-CoA dehydrogenase (13) Phosphate butyltransferase (phosphotransbutyrylase) (14) Butyrate kinase (15) Butyraldehyde dehydrogenase (16) Butanol dehydrogenase

of the issue has concluded that apart from biofuel production, certain other factors such as transportation fuel price hikes, unpredictable weather conditions, and stock market performance at large play the key role for the conflict between food and fuel (Araújo et al. 2017).

16.7.2 The Emission Potentials

The staggering rise in global pollution has restrained the energy sectors to operate within the specified limits, and if that option is not possible, then switch over to a more eco-friendly option. Due to the detrimental effects of fossil fuel combustion, now the focus is on the biofuels. Several scientists have reported that biofuels are altogether an eco-friendly approach and are capable of reducing GHG emissions by 60–94% as compared to their fossil fuel counterparts (Highina et al. 2014). The life cycle assessment studies of the biofuels on climatic effects have revealed that they are astoundingly higher GHG emission contributors as compared to petroleum fuels (Searchinger et al. 2008). Extensive investigations on the biofuel sustainability on emission front have surfaced the involvement of inferior technologies behind heightened GHG release by biofuels (Ji and Long 2016). Xue et al. (2011) have reported an interesting fact about the biofuels, which can produce mixed outcomes in terms of sustainability. E.g. biodiesel is capable of cutting down particulate emissions by 88%, whereas, the same biodiesel releases additional amounts of NO_x as compared to the petroleum diesel bringing harmful effects on the ecosystem (Xue et al. 2011).

16.7.3 Issues with the Land, Water, and Biodiversity

Due to continuous population growth, there will be an ever-increasing demand for food supply. Fischer et al. (2002) have reported that in the last five decades, the amount of cultivable land area has increased by 12% equivalent to 159 million hectares (Mha). Apart from land area increment, extensive usage of pesticide and fertilisers, doubling the irrigation altogether resulted in an increase in food by a factor of 2.5–3 times (Fischer et al. 2002). Doornbosch and Steenblik (2007) have reported that only 5% of the total landmass of the globe is available for energy crop cultivation by the year 2050. Moreover, from that 5%, approximately 63% of the landmass is technically suitable for energy cropping (Doornbosch and Steenblik 2007). The availability of a meagre amount of land for energy crop cultivation discards the debate that energy crop cultivation is leading to the reduction of food croplands (Popp et al. 2014; Borras and Franco 2012). It is estimated that 80% expansion of cultivable landmass possibilities is there in the African, South American continents, and Central America (Araújo et al. 2017).

Biofuel production will stress out the freshwater resources since currently about 70% of freshwater is used for agriculture purposes (Fischer et al. 2002). The water quantity and quality as well will be affected by biofuel production. The runoff stream containing high concentrations of pesticides and chemical fertilisers will render the water resources unusable. Such contaminated areas will gradually become a dead zone (Solomon and Bailis 2014). However, the biosorption process comes as a rescue option to mitigate the contamination problem. Bransby et al. (1998) have reported about the prevention of nitrogen contamination to the water bodies via biosorption using switchgrass (Bransby et al. 1998).

The deforestation has devastating effects on biodiversity. This act of disruption brings imbalance to the ecosystem and pushes many species towards extinction (Cowie et al. 2016). The life cycle assessment of biofuels for biodiversity disruption is still in the rudimentary state. The assessment parameters are biodiversity damage potential and account of lost endemic species should be made more stringent (De Baan et al. 2013).

16.7.4 Performance of the Biofuel

Biofuels have a significantly higher octane rating, which indicates their capacity to withstand the pre-ignition compression process. However, the low energy density results in lesser fuel mileage as compared to their petroleum counterparts, bioethanol has shown a reduction in mileage by 25%–30% as compared to gasoline. The blending percentages like 20–40% (medium blending) the energy penalty gets lessened (Theiss et al. 2016). The biodiesel, on the other hand, is a superior fuel as compared to the bioethanol. Biodiesel, when used as sole fuel, has shown a reduction in hydrocarbons, particulate matter and CO, and NO_x by 70%, 50%, and 10%, respectively. Apart from the above reductions, the SO_x concentration also got reduced remarkably as compared to petroleum diesel. The lifecycle assessment of bioethanol and biodiesel shows carbon emissions as 2–69 kg CO_2-eq/GJ and 20–49 kg CO_2-eq/GJ, respectively, indicating that the biodiesel has a better sustainable fuel performance than bioethanol (Araújo et al. 2017).

16.8 The Biofuel Policies

In recent times the biofuel productions are undergoing an exponential escalation. However, to keep the ecological, economic, and social balances intact, while undertaking a sustainable route for biofuel production, specific guidelines are to be followed. These mandatory guidelines are also synonymous to policies made by the governing authorities (a group of politicians, bureaucrats, and scientists). This portion of the chapter will be dedicated to the various strategies for biofuels in action in some countries.

16.8.1 Canadian Biofuel Policies

The Environmental Protection Act Bill C-33 of Canada has mandated the biofuel blending content by 5% in gasoline and 2% in diesel fuel and heating oil by 2010 and 2012, respectively. These blending mandates have set a target of bioethanol and biodiesel productions counting to 1.9 billion litres and 520 million litres, respectively by the year 2012 (Sorda et al. 2010). While the bioethanol produced was strictly 1G, i.e. from the cereal grains, the biodiesel, on the other hand, got produced from animal fat hence qualifying to 2G category. Interestingly the biofuel production has positively boosted the Canadian economy by the mark of 2 billion Canadian dollars (Sorda et al. 2010).

16.8.2 *The United States Policies on Biofuels*

The updated version of Renewable fuel standard (RFS2) came into action on July 2010, according to which a staggeringly high amount of 36 billion gallons of biofuels to be used as transport fuels by the year 2022. The RFS2 has given preferences to 2G biofuels to curb the food-fuel issues, by mandating the cellulosic biofuels (2G biofuels) surge from 0.1 billion gallons to 21 billion gallons in a span of little over a decade. The Environmental protection agency (EPA) supervises RFS, which has a projection of 36 billion gallons of biofuel usage by 2022, against that of 9 billion gallons as of 2008 (Bramcourt 2016). RFS also mandates implementation of advanced technologies which will account for a reduction in the GHG emissions by 50% during lifecycle assessment. Moreover, the biofuels must practise the act of not surfacing the food crisis and land usage issues (Sorda et al. 2010; Araújo et al. 2017).

The biofuel usage needs retrofitting in the engine. Owing to the frequent evaluation and amendments on biofuel policies, all the gasoline-driven vehicles in the US, produced since 1970 are capable of using E10 as fuel. Biofuels due to their lesser energy densities are prone to give low mileage. It was challenging to meet the guidelines of Corporate Average Fuel Efficiency Requirements (CAFE) due to the low mileage received from biofuels. But the Alternative Motor Fuels Act (AMFA) enacted in 1988 in the US has resolved the problem (Koplow 2006).

16.8.3 *Argentinian Policies on Biofuels*

The biofuel blending enforcements came up in February 2007 have mandated a 5% biofuel blend composition in gasoline and diesel fuels. Interestingly the quality requirement policies came up after the blending enforcement, i.e. in November 2008 for bioethanol and February 2010 for the biodiesel.

The Argentinian biofuel manufacturers are more focused on producing biofuels for use in Argentina only. The background reasons are firstly stringent technical requirements of the biofuel importing countries, which the Argentinian biofuels are unable to comply. Secondly, the tax incentives and financial benefits for biofuel manufacturers who are providing their products for use in Argentina only. The government of Argentina has given assurances for the purchase of biofuels from the manufacturers in the country for 15 years with reimbursements for taxes and depreciation costs (Sorda et al. 2010).

16.8.4 Colombian Biofuel Policies

The government of Colombia had mandated E10 fuel usage in all the cities with a population of above 500,000 in the year based on law 963 (Sorda et al. 2010). This blending enforcement had resulted in an increment of 75% E10 fuel sale from the gasoline fuel market of 100% by the year 2009 (Sorda et al. 2010). The bioethanol in Colombia is obtained from sugarcane, and the government regulates the bioethanol prices based on international sugar prices.

As far as biodiesel is concerned the resolution 1289 of the year 2005 had mandated 5% biodiesel blend by 2008, with a projection of increment to 20% by 2012. Palm oil is the raw material used in Colombia. The government provides tax exemptions for palm oil production and any crop-based oil productions intended to use them as biodiesel. The automobile manufacturers were given stringent directives to make the vehicles capable of running on E85 flex-fuel after 2012, and on 100% biofuel by 2016 (Sorda et al. 2010).

16.8.5 Biofuel Policies of Brazil

The biofuel policies of Brazil are the most developed ones, and they are in existence since the 1970s. In the year 1975 National Alcohol Program 'Proàlcool' was introduced by the government, which had the focus of bioethanol production from sugarcane (Walter and Cortez 1999). The current scenario of biofuel blending in Brazil state that bioethanol blending has come up to 27% and that of biodiesel is 10% (Brazil Biofuels Annual 2016). Regional subsidy plans by the government help in balancing benefits from the energy crop cultivation in the underdeveloped region at par with those of developed regions. However, subsidies were not given in 2015, owing to the financial crisis in the country (Harto et al. 2010). From the vehicular point of view, tax incentives were provided to the flex-fuel run vehicle, with no such benefits for vehicles running on pure petroleum fuels. To attract small farmers and family farm producers of vegetable oil, the National Biodiesel Production Program (PNPB) launched in 2004 compelled suppliers to buy raw materials from them (Araújo 2017).

The Proàlcool program has played an important role in Brazil's economy by providing 3.6 million jobs and 3.5% of the GDP. The aggressive regulations for biofuel production have led Brazil to gain the most price-competitive biofuel status with a price tag of 0.23 US dollars per litre of bioethanol (de Almeida et al. 2008; Sorda et al. 2010).

16.8.6 Biofuel Policies of the EU

In 2009 the European Union energy and climate change package (CCP) had formulated regulations for the use of biofuels in the transportation sector. The CCP regulations have enforced 20% energy quota to be fulfilled by the renewable energies in 2020 (Sorda et al. 2010). The EU Directive 2009/28/EC on renewable energies has specified that the GHG emissions must be reduced by 35% at a minimum in their lifecycle. Apart from GHG emission norms, land management, social and economic compliances are also to be taken care of. To mitigate the food v/s fuel issues, the European Union in 2015 have enacted a cap for maximum 7% contribution will be from 1G biofuels till 2020. Beyond that only non-food crop-based biofuels will be used (EU Biofuels Annual 2016).

16.8.7 Biofuel Policies of China

China's bioenergy policies are very strategic and yet plausible enough to fulfil the goals of maximal usage of biofuels with the solutions to the crucial issues simultaneously. China has already discontinued the subsidies on the 1G biofuel production and utilisation, which began in 2000. China had targeted to produce 4 million tons of bioethanol and 1 million tons of biodiesel by 2015 as a part of their 12th fifth-year plan. China has planned 15% of total energy requirements to be fulfilled by biofuels with a mandate of minimum 10% by 2020 (Araújo et al. 2017; Lane 2016).

16.8.8 The Indian Biofuel Policies

The Indian National Policy on Biofuel has approved for 20% blending of bioethanol and biodiesel into gasoline and petroleum diesel, respectively by 2017. The same policy also enforced non-edible oil crop cultivation in wastelands for biodiesel production. To attract the farmers for the biodiesel cropping government of India has guaranteed the revision of minimum support price (MSP), and minimum purchase price (MPP) for the bioethanol and biodiesel over the time (Altenburg et al. 2009).

Another policy, named as Ethanol Blended Petrol (EBP), came into act in 2003 had mandated the 5% bioethanol blending requirement into gasoline in four union territories and nine states. The E10 implementation plans in 2008 got delayed due to a fluctuation in the supply of sugar molasses. The National Mission on Biodiesel which was begun in 2003 with targets of Jatropha cultivation on 11.2 million hectares (Mha) of wasteland and a 10% blending target by the year 2012. However, this mission had faced failure when the production cost was found to be surpassing the purchase price (Sorda et al. 2010).

Unlike other countries in India, financial support from the government is almost non-existent. E.g. the central government have exempted central excise tax of (4%) for biodiesel production, but the state governments have refused to do so. Amidst all these shortcomings government of India is providing subsidised loans to the sugar mills, which are setting up ethanol production units alongside (Sorda et al. 2010).

16.9 Conclusions and Future Prospect

Biofuel is turning out to be a global phenomenon with petroleum fuel reserve depletion and escalation of environmental pollution. The industrial-scale production of biofuels is in existence for over a century. The industrial-scale bioethanol and biobutanol began in 1894, and 1912, respectively. Although the biofuels have been in production since long, their applications as an energy source are untapped only in recent times. Biofuels are categorised into different generations based on the substrates used for their production. Due to the food scarcity issues, the 1G biofuel is now getting axed from production all over the globe.

The 2G biofuels although are very promising in terms of economy and yield, but their complex structures are making the production process a tedious affair. Although the 3G and 4G biofuels from are in the research and developmental stage, many scientists have already reported that the microalgae will eventually become the answer to a sustainable biofuel production approach. The biofuels are capable of carbon capture as compared to their petroleum counterparts since the biomass during its growth assimilates a significant amount of CO_2 from the atmosphere. It is estimated that biofuel usage follows either carbon neutral or carbon-negative approach during their life cycle. Biofuels are mostly ending up as transportation fuels, and three potential candidates such as biodiesel, bioethanol, and biobutanol have shown their abilities to cater to the need.

The effectiveness of any biofuel is evaluated based on specific sustainability parameters such as food-fuel issues, emission and engine performance, effects on land, water, and biodiversity. It is noteworthy to mention that none of the biofuels in use till date have scored a 100% on the sustainable efficiency front. To maintain a smooth operation for production, and utilisation of biofuels, the biofuel policies exist in every biofuel producing nations. Currently, the biofuel policies all over the globe have mandated biofuel blends to petroleum fuels in the range of 5–25%. Financial incentives like tax exemption, subsidised loans, and monetary rewards for following the sustainable approach of production are offered to the biofuel manufacturers. The biofuel production is still at a nascent stage and is in dire need of further inputs from cutting edge technologies, more stringent regulations for production and usage. The future of biofuels is definitely bright and they will definitely change the scenario of energy sector in near future. At last we can say public awareness has to play the greatest role in overall acceptance of biofuels sector and subsequently their usage at larger scales.

Acknowledgments The authors want to convey their gratitude to Springer Nature for providing an opportunity to become a part of the book series. The authors would like to extend their heartfelt thanks to all the editors of the book Dr. Ajar Nath Yadav, Dr. Ali Asghar Rastegari, Dr. Neelam Yadav, and Dr. Rajeeva Gaur. The authors are also very much thankful to the Ministry of human resource development (MHRD), Government of India and the Director of NIT, Rourkela for providing necessary facilities while penning this book chapter.

References

Abdullah B, Muhammad SA, Shokravi Z et al (2019) Fourth generation biofuel: a review on risks and mitigation strategies. Renew Sustain Energy Rev 107:37–50

Adams C, Peters JF, Rand MC et al (1983) Investigation of soybean oil as a diesel fuel extender: endurance tests. J Am Oil Chem Soc 60(8):1574–1579

Ahlgren E, Hagberg MB, Grahn M (2017) Transport biofuels in global energy -economy modelling – a review of comprehensive energy systems assessment approaches. GCB Bioenergy 9:1168–1180

Alalwan HA, Alminshid AH, Aljaafari HAS (2019) Promising evolution of biofuel generations. Sub Rev Renew Energy Focus 28:127–139

Altenburg T, Dietz H, Hahl M (2009) Biodiesel in India—value chain organisation and policy options for rural development. https://www.mysciencework.com/publication/download/biodiesel-in-india-value-chain-organisation-and-policy-options-for-rural-development/b434b63ad591c508c14331477eb621d7. Accessed 28 Sep 2019

Araújo K (2017) Low carbon energy transitions: turning points in national policy and innovation. Oxford University Press, New York

Araújo K, Mahajan D, Kerr R, da Silva M (2017) Global biofuels at the crossroads: an overview of technical, policy, and investment complexities in the sustainability of biofuel development. Agriculture 7(4):1–22

Azizi K, Moraveji MK, Najafabadi HA (2018) A review on bio-fuel production from microalgal biomass by using pyrolysis method. Renew Sustain Energy Rev 82:3046–3059

Bajracharya S, Vanbroekhoven K et al (2017) Bioelectrochemical conversion of CO_2 to chemicals: CO_2 as a next generation feedstock for electricity-driven bioproduction in batch and continuous modes. Faraday Discuss 202:433–449

Balat M, Balat H (2009) Recent trends in global production and utilization of bio-ethanol fuel. Appl Energy 86:2273–2282

Baral NR, Shah A (2014) Microbial inhibitors: formation and effects on acetone-butanol-ethanol fermentation of lignocellulosic biomass. Appl Microbiol Biotechnol 98:9151–9172

Bharathiraja B, Jayamuthunagai J, Sudharsana T et al (2017) Biobutanol – An impending biofuel for future: A review on upstream and downstream processing tecniques. Renew Sustain Energy Rev 68:788–807

Borras S, Franco J (2012) Global land grabbing and trajectories of agrarian change: a preliminary analysis. J Agrar Chang 12:34–59

Bramcourt K (2016) The renewable fuel standard (R43325). In: Congressional research service. Available via DIALOG. https://nationalaglawcenter.org/wp-content/uploads//assets/crs/R43325.pdf. Accessed 28 Sep 2019

Bransby D, McLaughlin S, Parrish D (1998) A review of carbon and nitrogen balances in swithgrass grown for energy. Biomass Bioenergy 14:379–384

Brazil Biofuels Annual (2016) U.S. Department of agriculture (USDA), Washington DC. https://apps.fas.usda.gov/newgainapi/api/report/downloadreportbyfilename?filename=Biofuels%20Annual_Sao%20Paulo%20ATO_Brazil_8-12-2016.pdf. Accessed 28 Sep 2019

Brito M, Martins F (2017) Life cycle assessment of butanol production. Fuel 208:476–482

Chae S, Hwang E, Shin HS (2006) Single cell protein production of Euglena gracilis and carbon dioxide fixation in an innovative photo-bioreactor. Bioresour Technol 97(2):322–329
Chelf P, Brown LM, Wyman CE (1993) Aquatic biomass resources and carbon dioxide trapping. Biomass Bioenergy 4:175–183
Choi YY, Joun JM, Lee J et al (2017) Development of large-scale and economic pH control system for outdoor cultivation of microalgae Haematococcus pluvialis using industrial flue gas. Bioresour Technol 244:1235–1244
Choi YY, Patel AK, Hong ME et al (2019) Microalgae bioenergy with carbon capture and storage (BECCS): An emerging sustainable bioprocess for reduced CO_2 emission and biofuel production. Bioresour Technol Rep 7(100270):1–14
Correa DF, Beyerb HL et al (2019) Towards the implementation of sustainable biofuel production systems. Renew Sust Energ Rev 107:250–263
Cowie A, Soimakallio S, Brandao M (2016) Environmental risks and opportunities of biofuels. In: Bouthillier YL, Cowie A, Martin P, McLeod-Kilmurray H (eds) The law and policy of biofuels, 1st edn. Edward Elgar, Cheltenham, UK, pp 3–29
de Almeida EF, Bomtempo JV, de Souza e Silva CM (2008) The performance of Brazilian biofuels: an economic, environmental and social analysis. https://www.itf-oecd.org/sites/default/files/docs/discussionpaper5.pdf. Accessed 27 Sep 2019
De Baan LD, Alkemade R, Koellner T (2013) Land use impacts on biodiversity in LCA: a global approach. Int J Life Cycle Assess 18:1216–1230
De Bhowmick G, Koduru L, Sen R (2015) Metabolic pathway engineering towards enhancing microalgal lipid biosynthesis for biofuel application-a review. Ren Sus Energ Rev 50:1239–1253
De Gorter H, Drabik D (2015) The economics of biofuel policies. Palgrave, New York
Department of Energy (DOE). Alternative fuels data center http://www.afdc.energy.gov/fuels/emerging.html. Accessed 22 Sept 2019
Directive 2009/28/EC of the European parliament and of the council on the promotion of the use of energy from renewable sources (2009) EU. http://data.europa.eu/eli/dir/2009/28/oj. Accessed 21 Sept 2019
Dooley K, Christoff P et al (2018) Co-producing climate policy and negative emissions: trade-offs for sustainable land-use. Glob Sustain 1(3):1–10
Doornbosch R, Steenblik R (2007) Biofuels: is the cure worse the curse? https://www.oecd.org/sd-roundtable/39411732.pdf. Accessed 26 Sep 2019
Durre P (1998) New insights an novel developments in clostridial acetone/butanol/ isopropane fermentation. Appl Microbiol Biotechnol 49:639–648
EU Biofuels Annual (2016) U.S. Department of agriculture (USDA), Washington DC. https://apps.fas.usda.gov/newgainapi/api/report/downloadreportbyfilename?filename=Biofuels%20Annual_The%20Hague_EU-28_6-29-2016.pdf. Accessed 27 Sep 2019
Farrelly DJ, Everard CD, Fagan CC et al (2013) Carbon sequestration and the role of biological carbon mitigation: a review. Ren Sus Energ Rev 21:712–727
Fazal MA, Haseeb ASMA, Masjuki HH (2011) Biodiesel feasibility study: an evaluation of material compatibility; performance; emission and engine durability. Renew Sust Energ Rev 15:1314–1324
Fazal MA, Suhaila NR, Haseeb ASMA et al (2018) Sustainability of additive doped biodiesel: analysis of its aggressiveness toward metal corrosion. J Clean Prod 181:508–516
Fischer G, van Velthuizen H, Shah M et al (2002) Global agro-ecological assessment for agriculture in the 21st century. http://pure.iiasa.ac.at/id/eprint/6667/1/RR-02-002.pdf. Accessed 28 Sep 2019
García V, Päkkilä J, Ojamo H et al (2011) Challenges in biobutanol production: How to improve the efficiency? Renew Sust Energ Rev 15:964–980
Genitsaris S, Kormas KA, Moustaka-Gouni M (2011) Airborne algae and cyanobacteria: occurrence and related health effcts. Front Biosci 3:772–787
Ghazali WNMW, Mamat R, Masjuki HH et al (2015) Effects of biodiesel from different feeds tocks on engine performance and emissions: a review. Renew Sust Energ Rev 51:585–602

Gonçalves AL, Rodrigues CM, Pires JC et al (2016) The effect of increasing CO_2 concentrations on its capture, biomass production and wastewater bioremediation by microalgae and cyanobacteria. Algal Res 14:127–136

Gopinath A, Puhan S, Nagarajan G (2010) Effect of biodiesel structural configuration on its ignition quality. Energy Environ 1:295–306

Grima EM, Belarbi E-H, Fernández FA et al (2003) Recovery of microalgal biomass and metabolites: process options and economics. Biotechnol Adv 20(7–8):491–515

Harto C, Meyers R, Williams E (2010) Life cycle water use of low carbon transport fuels. Energy Policy 38:4933–4944

Heard B, Brook B, Wigley T, Bradshaw C (2017) Burden of proof: a comprehensive review of the feasibility of 100% renewable-electricity systems. Renew Sustain Energy Rev 76:1122–1133

Highina B, Bugaje I, Umar B (2014) A review of second generation biofuel: a comparison of its carbon footprints. Eur J Eng Technol 2:117–125

Ho DP, Ngo HH, Guo W (2014) A mini review on renewable sources for biofuel. Bioresour Technol 169:742–749

Jain R, Urban L, Balbach H, Diana Webb M (2012) Handbook of environmental engineering assessment. Butterworth-Heinemann, Oxford, UK

Jakeria MR, Fazal MA, Haseeb ASMA (2014) Influence of different factors on the stability of biodiesel: a review. Renew Sust Energ Rev 30:154–163

Ji X, Long X (2016) A review of the ecological and socioeconomic effects of biofuel and energy policy recommendations. Renew Sustain Energy Rev 61:41–52

Jin Q, Qureshi N, Wang H et al (2019) Acetone-butanol-ethanol (ABE) fermentation of soluble and hydrolysed sugars in apple pomace by Clostridium beijerinckii P260. Fuel 244:536–544

John RP, Anisha G, Nampoothiri KM et al (2011) Micro and macroalgal biomass: a renewable source for bioethanol. Bioresour Technol 102:186–193

Jones DT, Woods DR (1986) Acetone–butanol fermentation revisited. Microbiol Rev 50(4):484–524

Joshia G, Pandey JK, Rana S, Rawat DS (2017) Challenges and opportunities for the application of biofuel. Renew Sust Energ Rev 79:850–866

Karmakar A, Karmakar S, Mukherjee S (2010) Properties of various plants and animals feedstocks for biodiesel production. Bioresour Technol 101:7201–7210

Keis S, Shaheen R, Jones TD (2001) Emended descriptions of *Clostridium acetobutylicum* and *Clostridium beijerinckii*, and descriptions of *Clostridium saccharoperbutylacetonicum* sp. nov. and *Clostridium saccharobutylicum* sp. nov. Int J Syst Evol Microbiol 51:2095–2103

Knothe G, Matheaus AC, RyanIii TW (2003) Cetane numbers of branched and straightchain fatty esters determined in an ignition quality tester. Fuel 82:971–975

Koplow D (2006) Biofuels—at what cost? Government support for ethanol and biodiesel in the United States. In: The global studies initiative, of the international institute for sustainable development. Available via DIALOG. https://www.iisd.org/gsi/sites/default/files/brochure_-_us_report.pdf. Accessed 28 Sep 2019

Kour D, Rana KL, Yadav N, Yadav AN, Rastegari AA, Singh C et al. (2019) Technologies for biofuel production: current development, challenges, and future prospects. In: Rastegari AA, Yadav AN, Gupta A (eds) Prospects of renewable bioprocessing in future energy systems. Springer International Publishing, Cham, pp 1–50. https://doi.org/10.1007/978-3-030-14463-0_1

Kumar M, Gayen K (2011) Developments in biobutanol production: new insights. Appl Energy 88:1999–2012

Kumar S, Sharma S, Thakur S, Mishra T, Negi P, Mishra S et al (2019) Bioprospecting of microbes for biohydrogen production: Current status and future challenges. In: Molina G, Gupta VK, Singh BN, Gathergood N (eds) Bioprocessing for Biomolecules Production. Wiley, USA, pp 443–471

Kumari D, Singh R (2018) Pretreatment of lignocellulosic wastes for biofuel production: a critical review. Renew Sust Energ Rev 90:877–891

Lane J (2016) Biofuels mandates around the world: 2016. Biofuels Digest. http://www.biofuelsdigest.com/bdigest/2016/01/03/biofuels-mandates-around-the-world-2016/. Accessed 10 Sep 2019

Leong W-H, Lim J-W et al (2018) Third generation biofuels: A nutritional perspective in enhancing microbial lipid production. Renew Sustain Energy Rev 91:950–961

Li Z, Wang D, Shi Y-C (2017) Effects of nitrogen source on ethanol production in very high gravity fermentation of corn starch. J Taiwan Inst Chem Eng 70:229–235

Mabus R (2010) Department of the navy's energy program for security and independence. Diane Publishing Co., Darby, PA

Mahapatra MK, Kumar A (2019) Fermentation of oil extraction: bioethanol, acetone and butanol production. In: Rastegari AA, Yadav AN, Gupta A (eds) Prospects of renewable bioprocessing in future energy systems, 1st edn. Springer Nature, Switzerland

Mahapatra MK, Kumar A (2017) A short review on biobutanol, a second generation biofuel production from lignocellulosic biomass. J Clean Energy Technol 5(1):27–30

Mathews JA (2008) Carbon-negative biofuels. Energy Policy 36:940–945

McGinn PJ, Dickinson KE, Bhatti S et al (2011) Integration of microalgae cultivation with industrial waste remediation for biofuel and bioenergy production: opportunities and limitations. Photosyn Res 109(1–3):231–247

Menetrez MY (2012) An overview of algae biofuel production and potential environmental impact. Environ Sci Technol 46(13):7073–7085

Mofijur M, Rasul MG, Hyde J et al (2016) Role of biofuel and their binary (diesel –biodiesel) and ternary (ethanol –biodiesel –diesel) blends on internal combustion engines emission reduction. Renew Sust Energ Rev 53:265–278

Naik S, Goud VV et al (2010) Production of first and second generation biofuel: a comprehensive review. Renew Sustain Energy Rev 14(2):578–597

Norkobilov A, Gorri D, Ortiz I (2017) Process flowsheet analysis of pervaporation-based hybrid processes in the production of ethyl tert-butyl ether. J Chem Technol Biotechnol 92:1167–1177

Ntaribi T, Paul DI (2019) The economic feasibility of Jatropha cultivation for biodiesel production in Rwanda: a case study of Kirehe district. Energy Sustain Dev 50:27–37

Packer MA, Harris GC, Adams SL (2016) Food and feed applications of algae, In: Bux F, Chisti Y (Eds) Algae biotechnology: products and processes, 1st edn. Springer International Publishing Switzerland

Popp J, Lakner Z, Harangi-Rakos M et al (2014) The effect of bioenergy expansion: food, energy and environment. Renew Sustain Energy Rev 32:559–578

Qi HD, Geng LM, Chen H et al (2009) Combustion and performance evaluation of a diesel engine fueled with biodiesel produced from soybean crude oil. Renew Energy 34:2706–2713

Rakshit R, Patra AK, Das A (2012) Biochar applicatin in soils mitiate climate change through carbon sequestratin. Int J Bio-Resour Stress Manage 3(1):079–083

Rastegari AA, Yadav AN, Yadav N (2020) New and future developments in microbial biotechnology and bioengineering: trends of microbial biotechnology for sustainable agriculture and biomedicine systems: diversity and functional perspectives. Elsevier, Amsterdam

Rastegari AA, Yadav AN, Gupta A (2019) Prospects of renewable bioprocessing in future energy systems. Springer International Publishing, Cham

Rauda M, Kikas T, SippulaO Shurpali NJ (2019) Potentials and challenges in lignocellulosic biofuel production technology. Renew Sust Energ Rev 111:44–56

Ravindran R, Jaiswal AK (2016) A comprehensive review on pre-treatment strategy for lignocellulosic food industry waste: challenges and opportunities. Bioresour Technol 199:92–102

Raybould A (2010) The bucket and the searchlight: formulating and testing risk hypotheses about the weediness and invasiveness potential of transgenic crops. Environ Biosaf Res 9:123–133

Rodionova M, Poudyal R, Tiwari I et al (2017) Biofuel production challenges and opportunities. Int J Hydrogen Energy 42:8450–8461

Rulli MC, Bellomi D, Cazzoli A et al (2016) The water-land-food nexus of firstgeneration biofuels. https://www.nature.com/articles/srep22521.pdf. Accessed 26 Sep 2019

Sadeek SA, Negm NA et al (2015) Metal adsorption by agricultural biosorbents: adsorption isotherm, kinetic and biosorbents chemical structures. Int J Biol Macromol 81:400–409

Schiermeier Q, Tollefson J et al (2008) Energy alternatives: electricity without carbon. Nat News 454(7206):816–823
Searchinger T, Heimlich R, Houghton RA et al (2008) Use of U.S. croplands for biofuels increases greenhouse gases through emissions from land-use change. Science 319:1238–1240
Sikarwar VS, Zhao M, Fennell PS et al (2017) Progress in biofuel production from gasification. Prog Energy Combust Sci 61:189–248
Solomon B, Bailis R (eds) (2014) Sustainable development of biofuels in Latin America and the caribbean. Springer, New York
Sorda G, Banse M, Kemfert C (2010) An overview of biofuel policies across the world. Energy Policy 38:6977–6988
Srirangan K, Akawi L, Moo-Young M et al (2012) Towards sustainable production of clean energy carriers from biomass resources. Appl Energy 100:172–186
Theiss T, Alleman T, Brooker A et al (2016) Summary of high-octane, mid-level ethanol blends study. https://info.ornl.gov/sites/publications/Files/Pub109556.pdf. Accessed 26 Sep 2019
Tomei J, Helliwell R (2016) Food versus fuel? Going beyond biofuels. Land Use Policy 56:320–326
Trabelsi ABH, Zaafouri K, Baghdadi W et al (2018) Second generation biofuels production from waste cooking oil via pyrolysis process. Renew Energy 126:888–896
Tucker JB, Zilinskas RA (2006) The promise and perils of synthetic biology. New Atlantis 12:25–45
Tyson KS (2001) Biodiesel handling and use guidelines, technical report NREL/TP-580-30004. https://p2infohouse.org/ref/40/39041.pdf. Accessed 25 Sep 2019
Vyas P, Kumar A, Singh S (2018) Biomass breakdown: a review on pretreatment, instrumentations and methods. Front Biosci (Elite Ed) 10:155–174
Walter A, Cortez L (1999) An historical overview of the Brazilian Bioethanol Program. Renew Energy Dev 11(1):1–4
Williamson P (2016) Emissions reduction: scrutinize CO_2 removal methods. Nature 530:153–155
Wright O, Stan G-B, Ellis T (2013) Building-in biosafety for synthetic biology. Microbiology 159:1221–1235
Xue J, Grift T, Hansen A (2011) Effect of biodiesel on engine performance and emissions. Renew Sustain Energy Rev 15:1098–1116
Yabe K, Shinoda Y, Seki T et al (2012) Market penetration speed and effects on CO_2 reduction of electric vehicles and plug-in hybrid electric vehicles in Japan. Energy Policy 45:529–540
Yusoff MNAM, Zulkifli NWM, Masum BM et al (2015) Feasibility of bioethanol and biobutanol as transportation fuel in spark-ignition engine: a review RSC Adv. 5:100184–100211
Zienkiewicz K, Du Z-Y, Ma W et al (2016) Stress-induced neutral lipid biosynthesis in microalgae—molecular, cellular and physiological insights. Mol Cell Biol Lipids, Biochim Biophys Acta 1861(9):1269–1281

Chapter 17
Advances in Microbial Bioresources for Sustainable Biofuels Production: Current Research and Future Challenges

Tanvir Kaur, Rubee Devi, Divjot Kour, Neelam Yadav, Shiv Prasad, Anoop Singh, Puneet Negi, and Ajar Nath Yadav

Abstract Long ago, fossil fuel likes coal, oil, and natural gases have been exploited massively to run various engines, automobiles, and other purposes. Presently, also demand for fossils is increasing with the increasing population, and its global reserves are depleting speedily and will disappear in the future. Apart from the depletion, fossil fuel use also has some destructive effects on environment as it releases huge amounts of carbon dioxide (CO_2) and some other pollutants in the atmosphere when burned. Considering all the fossil fuels deleterious factors, a substitute has been searched and named biofuel. Biofuels are the liquid or gaseous fuels (bioethanol, biomethanol, biodiesel, biohydrogen) which are a renewable source and also environmentally friendly. Various bioresources are being utilized for biofuels' production such as agriculture byproducts, food processing wastes or lignocellulosic waste, animal and poultry wastes, and microbial biomass. Microbial bioresources are the most significant resource for the production of biofuel as they can be achieved in less time and can be cultivated using CO_2 which provides greenhouse gas alleviation

T. Kaur · R. Devi · D. Kour · A. N. Yadav (✉)
Department of Biotechnology, Dr. KSG Akal College of Agriculture, Eternal University, Baru Sahib, Sirmour, Himachal Pradesh, India
e-mail: ajarbiotech@gmail.com; ajar@eternaluniversity.edu.in

N. Yadav
Gopi Nath P.G. College, Veer Bahadur Singh Purvanchal University, Ghazipur, Uttar Pradesh, India

P. Negi
Department of Physics, Akal College of Basic Sciences, Eternal University, Baru Sahib, Sirmour, Himachal Pradesh, India

S. Prasad
Centre for Environment Science & Climate Resilient Agriculture, Indian Agricultural Research Institute, New Delhi, India

A. Singh
Department of Scientific and Industrial Research, Ministry of Science and Technology, Government of India, New Mehrauli Road, New Delhi, India

A. N. Yadav et al. (eds.), *Biofuels Production – Sustainability and Advances in Microbial Bioresources*, Biofuel and Biorefinery Technologies 11,
https://doi.org/10.1007/978-3-030-53933-7_17

benefits. Diverse microbes have the ability to produce biofuels like algae, bacteria, cyanobacteria, and fungi by using different methods that have been discussed in this chapter.

This book contains current knowledge about functional annotation of microbial bioresources for sustainable biofuels' production. The book covers the current knowledge of recent advancement of bioresources which provides state-of-the-art information in the area of biofuel and biorefinery technologies, broadly involving microbial-based innovations and applications. This book is directed toward presenting drawbacks in existing processes and technologies in current biofuels options. The book will be highly useful to the faculty, researchers, and students associated with microbiology, biotechnology, agriculture, molecular biology, environmental biology, and related subjects. Currently, demand for energy in the universe is met by the fossil fuel combustions like coal, oil, and natural gases, which is used to run various engines. The enormous exploitation of fossil fuels has limited the quantities of fossil fuels in global reservoirs that will be depleted in the coming years. In addition, fossil fuel usage has resulted in the production of CO_2, and other pollutants in the atmosphere in large amounts that have harmful effects on the environment (Hong 2012). So, looking up to the increasing fossil fuel demand and need of sustainable environment, alternative biomaterials have been searched against non-renewable source of energy known as biofuels. Biofuel is produced from the biomass of plant and human waste which is a liquid or gaseous form of fuels. Bioethanol, biomethanol, biodiesel, and biohydrogen are the variety of biofuels that can be synthesized from the biomass which can be used as fuel for vehicles, engines, and cells for electricity (Kour et al. 2019; Yadav et al. 2019a). From ancient time to present time, there are advancements in biofuels which are categorized in different generations from first to fourth (Fig. 17.1). Apart from the easy production of these bio-derived compounds, it has several advantages like it is environmentally friendly, cheap in cost, and represents a carbon dioxide cycle in combustion; they are biodegradable and contribute to sustainability and benefits of the consumers, economy, and environment (Puppan 2002).

The biofuel production can be achieved by using different types of bioresources, which has been used like agriculture byproducts, food processing wastes or lignocellulosic waste, animal and poultry wastes, and microbial biomass. Microorganisms, the producer of different biofuels, easily broke cellulose into simple sugars, that is, glucose or pentose, and convert the sugar into biofuel, that is, long-chain fatty acids derived from fats, oils, or lipids (Dunlop et al. 2010). Bacteria, cyanobacteria, algae, and fungi are the different microbes that have been exercised for the production of biofuels (Rastegari et al. 2019; Yadav et al. 2020).

In the present scenario, there is a necessity to explore the green resources of energy because of the various environmental issues, and hydrogen (H_2) energy is one of them. Regarding the same, scientific communities are involved extensively to develop several H_2 production techniques that include thermochemical, electrochemical, and

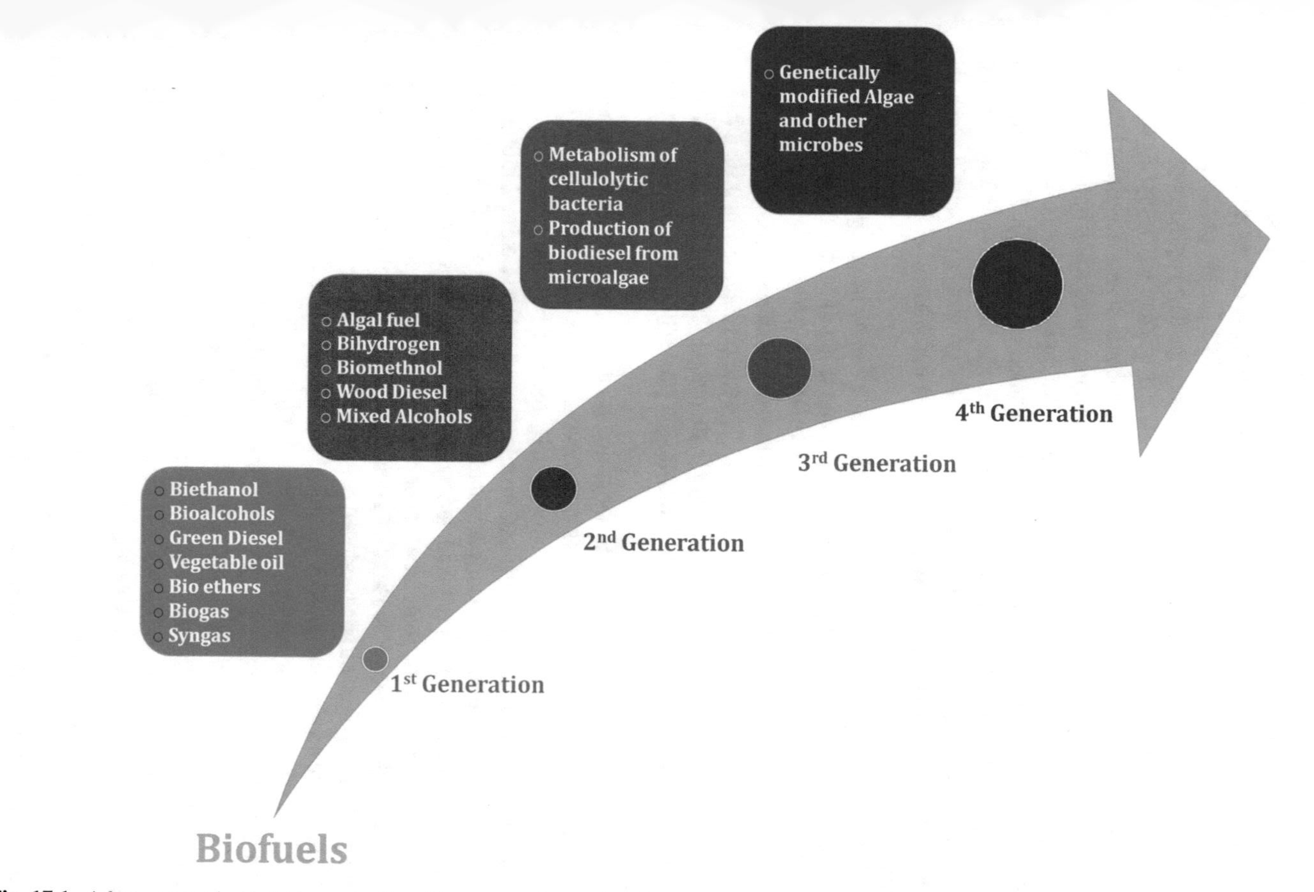

Fig. 17.1 Advancement in biofuels production: First generation to fourth generation Adapted with permission from Kour et al. (2019)

biological techniques. Moreover, as being low energy demanding and eco-friendly nature, biological techniques are found to be much beneficial than thermochemical and electrochemical processes for H_2 production.

Since the last few years, prominence is being observed on the H_2 gas production through electrohydrogenesis or bioelectrochemical process. Among the bioelectrochemical systems, microbial electrolysis cell (MEC) is one in which bacterial metabolism and electrochemistry jointly are responsible for H_2 production. MEC technology generates H_2 through microbially decomposing organic materials on applying a threshold electric current (Khan et al. 2017; Liu et al. 2005; Logan et al. 2008; Meda et al. 2015). In MECs, exoelectrogenic bacteria or anode aspiring bacteria oxidize the organic compounds and generate electrons, protons, and carbon dioxide. Bacteria extracellularly transfer the electron to the anode in anaerobic condition and release protons in the solution (Montpart i Planell et al. 2014). Thereafter, H_2 produces as electrons combine with the free protons in the solution after traveling through a wire to the cathode. However, to initiate the process, a threshold potential (>0.2 V) is applied to the electrodes at neutral pH value (Khan et al. 2017; Liu et al. 2005).

MEC has a great potential for sustainable and clean hydrogen production from wastewaters and biomass. The H_2 production rate is significantly higher (~100%) in MECs as a comparison to the fermentation process and water electrolysis (Khan et al. 2017; Wang et al. 2013). Moreover, the threshold external voltage required for conventional water electrolysis is (>1.6–1.8 V), which is large enough than (>0.2–0.8 V) required in MECs at neutral pH (Kadier et al. 2016; Khan et al. 2017). Figure 17.2 shows the schematic of biohydrogen production by a double-chamber MEC.

The performance of MECs can be measured on the basis of various parameters. Basically, in the MEC chamber, a substrate material such as acetate is being present in the biomass and wastewater that decomposes by the bacteria. The different parameters of this substrate material are utilized in order to measure the Coulombic efficiency of MEC. The Coulombic efficiency is calculated as $C_{Ef} = \frac{C_{To}}{C_{Th}} \times 100\%$, where C_{To} is the total Coulombs which is calculated by integrating the current over time. C_{Th} is the theoretical amount of coulombs that can be produced from substrate, calculated as $C_{Th} = \frac{FaS_cV}{M}$, where F is Faraday's constant (96485 C/mol electrons), a is the number of moles of electrons produced per mol of substrate ($a = 8$ for acetate), S_c is the substrate concentration, V is the liquid volume, and M is the molecular weight of substrate ($M = 82$ for acetate) (Liu et al. 2005).

Further, the overall hydrogen recovery is calculated as $R_{(H_2)} = C_{Ef}\, R_{Cat}$. Here, R_{Cat} is the cathodic hydrogen recovery which is calculated as $R_{Cat} = \frac{n_{H_2}}{n_{To}}$, where n_{H_2} is the total moles of hydrogen produced and $n_{To} = \frac{C_{To}}{2F}$ is the moles of hydrogen that could be produced from the measured current. Hydrogen yield $Y_{(H_2)}$ is calculated as $Y_{(H_2)} = \frac{n_{H_2}}{n_s}$, where n_s is substrate removal calculated on the basis of chemical oxygen demand. The hydrogen production rate $Q_{(H_2)}$ is measured in the unit of $m^3 \cdot d^{-1} \cdot m^{-3}$, which is based on the measured daily hydrogen production normalized to the reactor volume (Association et al. 1920).

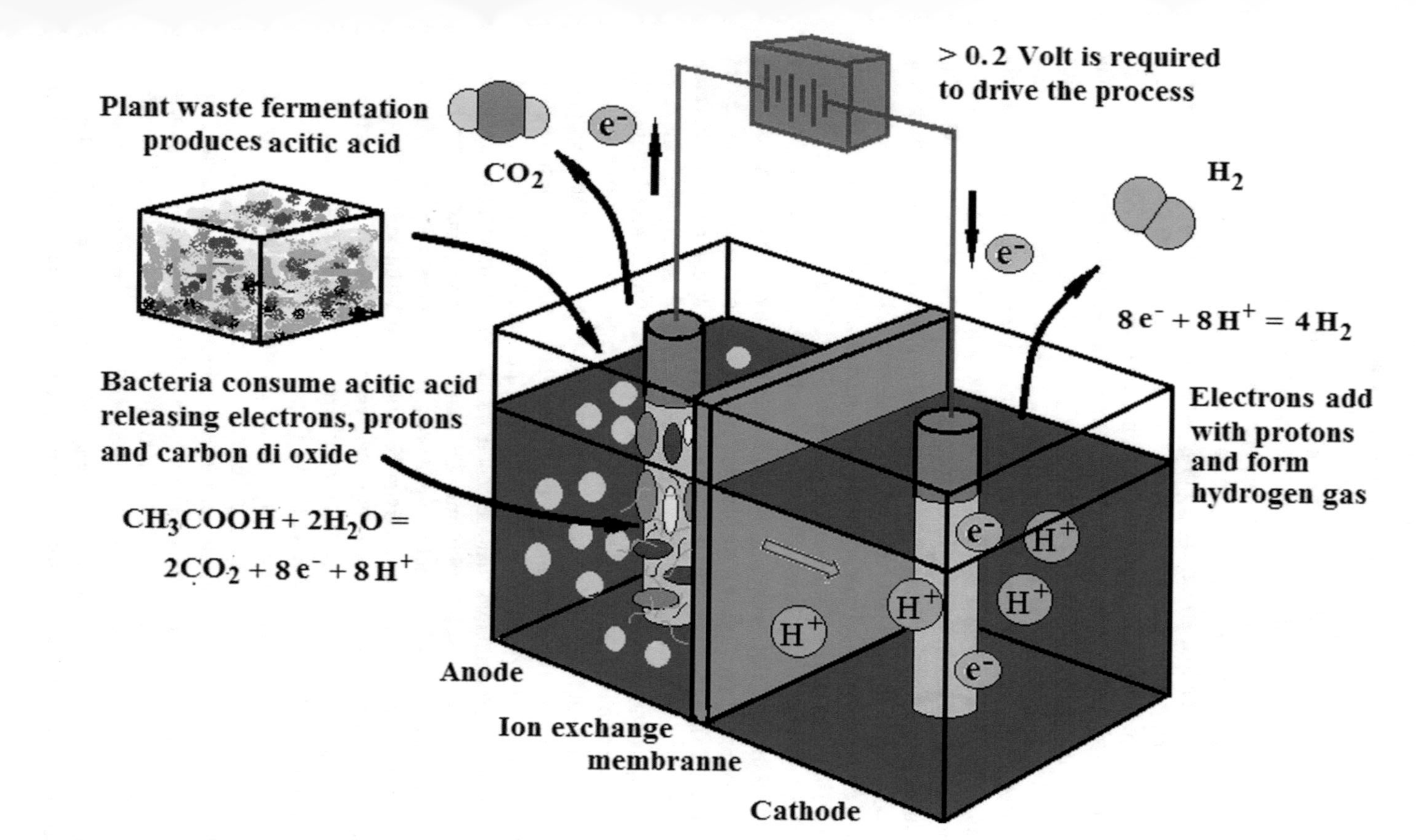

Fig. 17.2 Biohydrogen production by double-chamber microbial electrolysis cell Adapted with permission from Kour et al. (2019)

Biofuels are a liquid or gaseous type of biomass-processed fuels that can substitute fossil fuels for diesel, petrol, and other transportation fuels; it is used for running various automotive powers and mechanical machineries. Biofuels are derivable from various sources including bio-waste, forest biomass, and agricultural crop. There are various forms of biofuels like biogas, biodiesel, and bioethanol (Peskett et al. 2007). Biofuels are organic primary and secondary biomass-derived fuels which may be used for combustion or other technology to generate thermal energy. This concept includes multipurpose plantation, grown energy crop, and agricultural and non-agricultural products. Bioethanol is an alcohol-derived biological derivable from a wide range of sugar or starch including sugarcane crop, coconut, and cassava (Fig. 17.3). The local ethanol has been produced from corn, guinea, maize, millet, and other starchy substrates. Local production of ethanol from palm wine is yet another growing livelihood in southwestern and southern Nigeria. They help to minimize waste from cassava crop, which is prevalent phenomenon during gluts. It will also give priority to this essential crop which was a long neglected important crop. Biodiesel is a renewable fuel which increases global acceptance. It can be achieved by blending with the standard diesel. Higher yield of biodiesel is produced in Europe from crop feedstock such as soybean and rapeseed, while Asian countries are also investigating the less desirable biodiesel feedstock from non-edible seed oil palm and groundnut based on the estimates for commodity production at the FAO2007 according to world ranking (Sarin et al. 2007). Non-edible oil production has also a high potential which is suitable for biodiesel manufacturing. In recent time, hydrogen gas is one of the most important clean biofuel resources and raw materials (Fan et al. 2004). Many scientists have been focusing that biohydrogen production by carbonate fermentation large consideration has been paid that production of biohydrogen through the acidogenic phages of anaerobic process from wastewater treatment. Valdez-Vazquez et al. (2005) had determined the hydrogen production from paper mill wastes using consortium of solid-state fermentation and anaerobic degustation. One of the methods of biohydrogen production fermentation type of ethanol is the known best method of fermentation. Biohydrogen has been categorized into dark fermentative, biophotolysis, and photofermentations that are beneficial for the environment safety as they don't liberate CO_2 during combustion. The most biological processes are biohydrogen production showing that hydrogen produced from fermentation has become more favorable condition due to advantages like high sustainability, low energy required, and high hydrogen production rate. System production efficiency and cost-effectiveness could be improved. Glucose is the most preferred carbon source for the production of biohydrogen (Muri et al. 2016).

The biobutanol production economies in a fermentation broth are largely relying on the efficiency of bioconversion and the purity of the product. As a typical biofuel, biobutanol produced by biomass fermentation is critical for the production of renewable source of energy, due to the oil supply shortage and the demand for protection of environmental (Kumar et al. 2019; Yadav 2019). Biobutanol should become an attractive, inexpensive, and renewable fuel as petroleum oil contributes to costly fuel due to reduced oil supplies and rising atmospheric greenhouse gases (Tigunova et al. 2013). Low butanol titer, feedstock availability, and product inhibition are the main

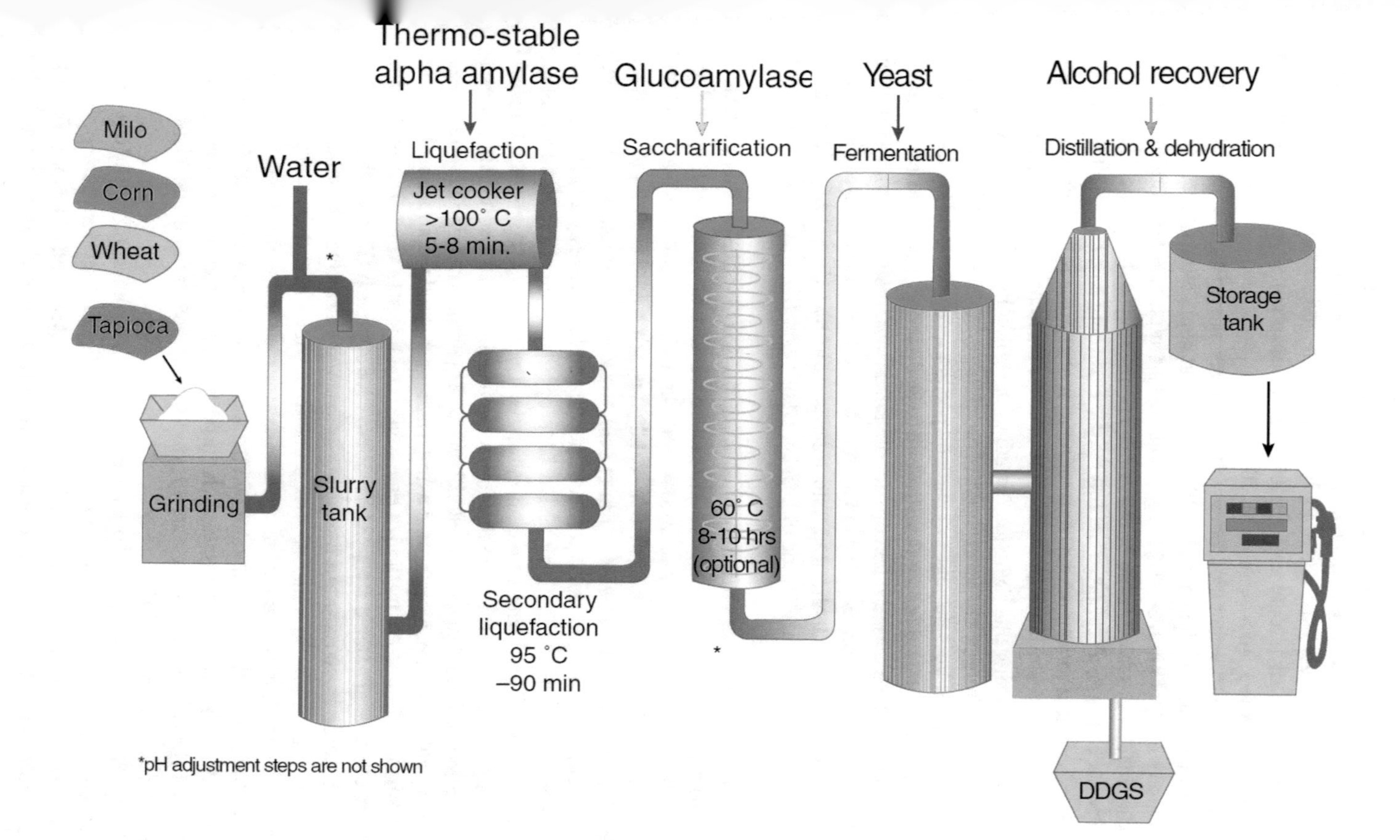

Fig. 17.3 Steps in the process of making bioethanol Adapted with permission from Schubert (2006)

challenges in butanol development. Biofuels like biobutanol have been classified into the first and second generations. Biofuel is based on the use of feedstock. The first-generation biofuel uses materials like sugarcane and cereal grains, while the second generation of biofuel uses lignocellulosic materials (e.g., agricultural and forest wastes) as feedstocks (Liu et al. 2014).

A method of generating biofuels involves the dewatering of substantially intact algal cells to create an algal biomass. The sequential application of algal biomass is solvent sets to algal biomass and the sequential separation of solid biomass fraction from liquid fractions to achieve a liquid fraction containing neutral lipids. The process also involves the esterification of neutral lipids, separating a water-miscible fraction of glycerine from a water-immiscible fraction of fuel esters, carotenoids, and omega-3 fatty acid. Biofuel or ingredients of fuels or other chemicals may be generated by living organisms using this invention including but not limited to oils, vegetable oil, hydrocarbon, hydrogen, biodiesel, lipids, fats, butane, methane, methanol, ethanol, and alcohols. To address the abovementioned problem, vegetable oil is used as such; there is an almost needed for the modification to be done chemically in order to make such oils equal to the diesel characteristics obtained from fossil fuels (Marchetti et al. 2007). There are numerous procedures available for biofuel production like pyrolysis method, micro-emulsification, and transesterification (Fig. 17.4). Scientist is doing extensive researches for the improvement of vegetable oils, biofuel efficiency, production, and benefits.

The most widely used reaction is to turn oils into biodiesel; this process is called transesterification. In this method, reaction is performed between triglycerides, which are obtained from oily feedstock and short-chain alcohol (usually ethanol and methanol alcohol). In this reaction, catalyst is being used in order to insert the alcohol alkyl group into the chain of fatty acid, to form ester of fatty acid (biodiesel) and glycerol molecules as aside product (Stephen and Periyasamy 2018). There are four major types of catalysts that have been applied to algal biodiesel including lipase, acid, alkali, and heterogeneous catalysts. Two steps have involved biodiesel processing; it requires separate extraction of oil, usually with the solvent or a solvent combination and subsequent transesterification. Algal lipid extraction for transesterification method has been performed using a hexane method to extract neutral lipids and methanol: chloroform method of extracting total lipids. In situ transesterification is known as a single-step concept id, an alternative approach (Daroch et al. 2013).

This process is exercised in industries by heating extra alcohol available in the vegetable oils following under several conditions where inorganic catalyst is also present; the reaction would take place. The essence of the reaction is the reversible, and thus surplus amount of alcohol (methanol, ethanol, propanol, and butanol) is used to move the equilibrium toward the product side. These reactions are usually catalyzed through acid, base, or an enzyme to enhance the rate and yield of the reaction (Selvaraj et al. 2019).

The pyrolysis method has been used for the transformation approach by oxygen or heat which cleaves the bond and numerous forms of smaller molecules. Pyrolysis is the performing vegetable oil produced biofuel in various amounts with the carboxylic acid, aromatic acids, alkadienes, alkenes, and alkanes production (Demirbaş 2000).

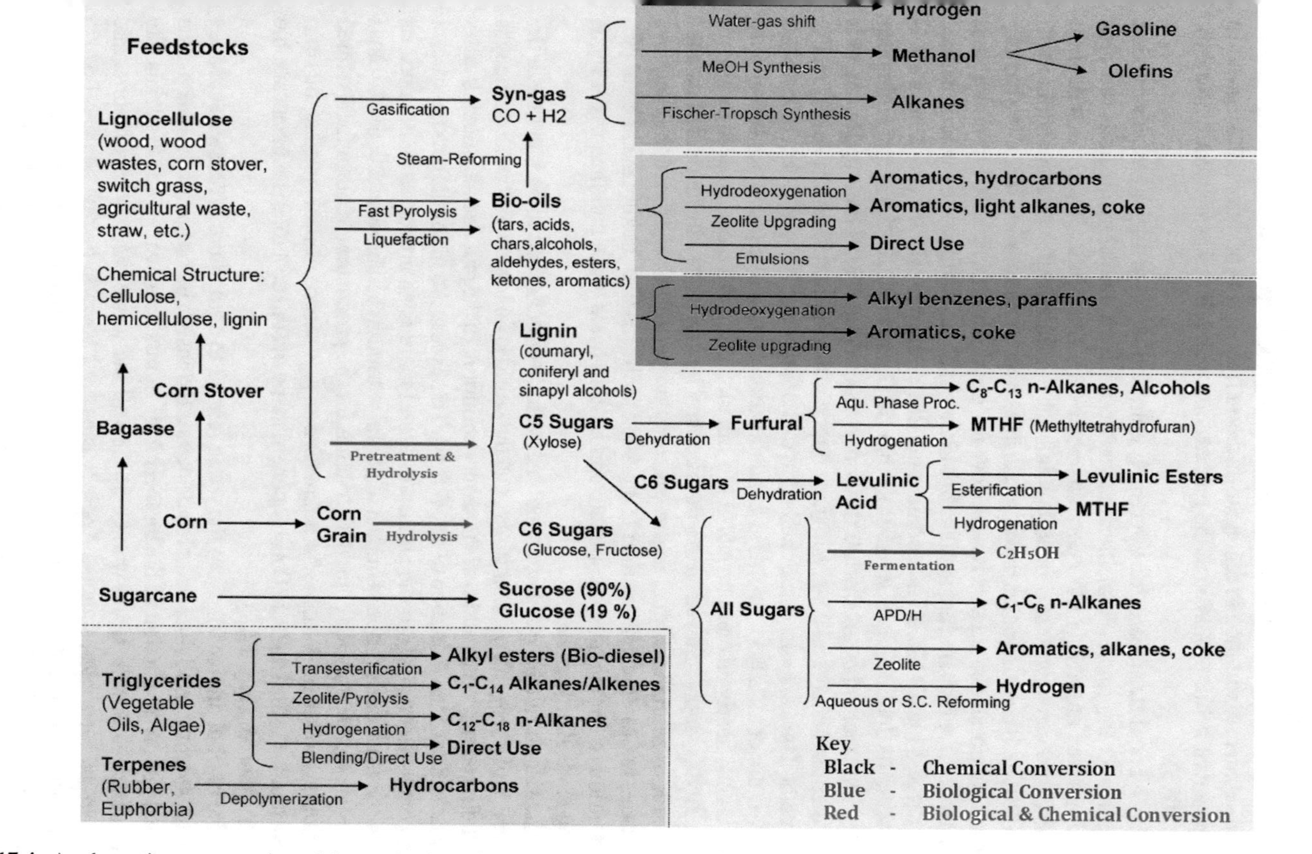

Fig. 17.4 A schematic representation of the production of liquid fuels from lignocellulosic biomass Adapted with permission from Huber et al. (2006)

For the developing countries, the cost of the equipment to be used in the pyrolysis process is higher for biofuel development. In addition, the depletion of oxygen during the cycle results in no environmental gain with respect to the use of oxygenated fuel. Another important demerit of the pyrolysis process is needed for the equipment called separate distillation to segregate different fractions. In addition, substances are composed of sulfur produced at the process end which is less eco-friendly (Strezov et al. 2015).

The method developed for upgrading commodity fossil fuel by emulsion fuel formation is the micro-emulsification method. Micro-emulsion methods are also beneficial for improving the spray characteristics through explosive vaporization of the components with low boiling points present in micelles (Hasannuddin et al. 2016, 2018). It can also be used to upgrade flues from other sources, including bio-oil. Bio-oil micro-emulsification is a process for combining the whole bio-oil without assistance of surfactant into another liquid fuel such as biodiesel and other mixture. This process blends bio-oil into commodity diesel and removes the problems associated with the application of stand-alone bio-oil (Leng et al. 2015; Lif and Holmberg 2006). This method is an effective method for commodity fossil fuels upgrading as it has potential benefits as it helps in the reduction of pollutant emission. Additionally, this method is also a successful method to make full use of bio-oil (Chiaramonti et al. 2003; Ikura et al. 2003). As micro-emulsification technology combines bio-oil into fossil fuels, nearly all components of bio-oil, including water, can be used as biofuel tools. In this, approaches decrease the viscosity by increasing the amount of catanes and endows with the biofuel improved spray properties. Using the biofuel engines produces micro-emulsion method that caused problems like sticking the injector pin, deposition of carbon, and incomplete combustion cycle while continuously using it (Leng et al. 2018).

The rise of human population and industrialization has led to an increase in the demands of energy all over the globe, and the world is facing two foremost challenges including energy crisis and environmental pollution. In the past decades, energy crisis occurred due to the reduction of fossil fuels. The excess use of fossil fuels has caused high emissions of CO_2 to the atmosphere, and there is an urgent need to reduce its emission to avoid destructive impact of global warming (Milano et al. 2016). All these challenges have attracted a greater attention for the production of clean fuels termed as biofuels which are suitable for alternative energy source. Biofuel production from bioresources is thus an essential component in the overall development of sustainable energy sources (Fig. 17.5).

The biofuel production from the waste related to agriculture is renewable, easily and abundantly available, and cheap source. Extensive investigation has been even carried out on production of the biofuels from agricultural waste. The materials like crop residues, sawdust, and wood chips are all the waste based on agriculture, and they are efficient bioresources. Apple wastes are known to be rich source of acetic acid, citric acid, glucose, malic acid, succinate, and sucrose, and their utilization for the production of biohydrogen is known to be a promising strategy (Lu et al. 2016). The waste of coffee generated during the transformation and processing of coffee beans from fruit is also an essential agricultural product (Karmee 2018; Mussatto

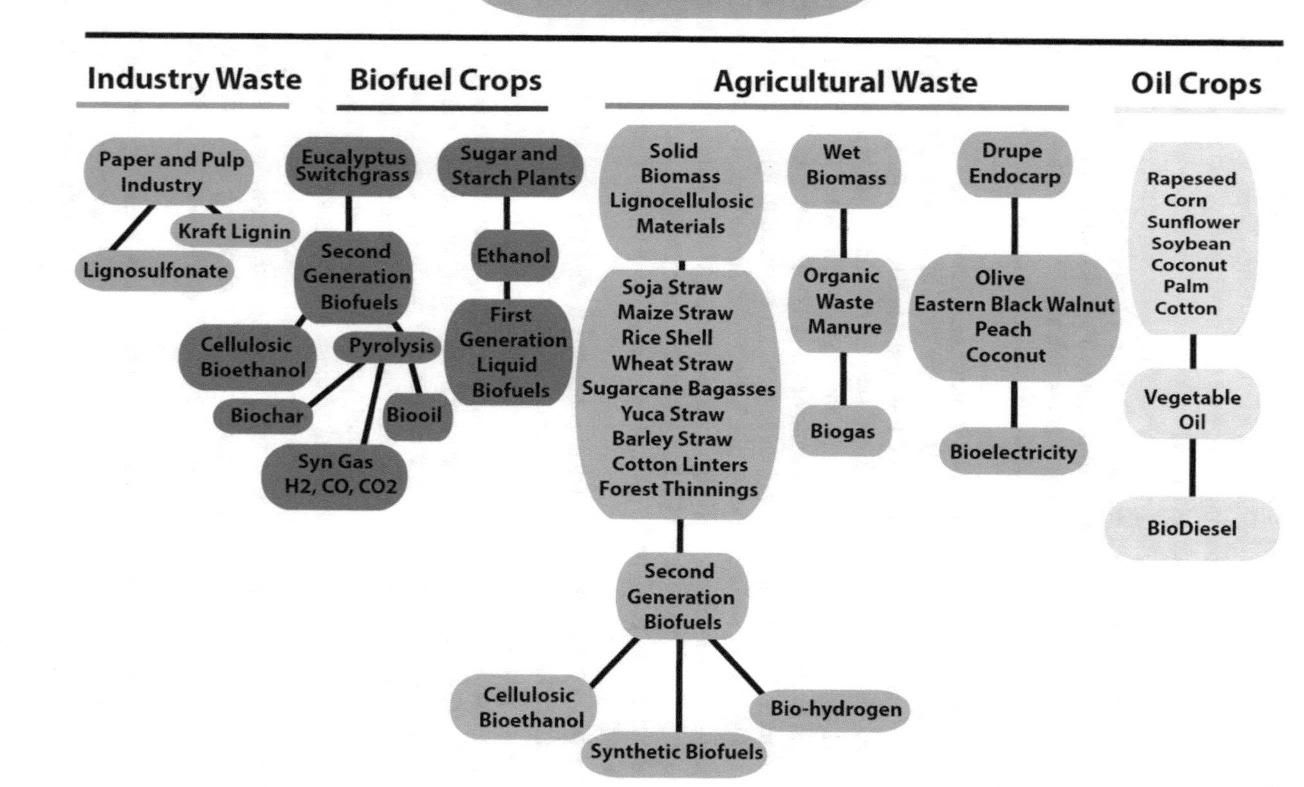

Fig. 17.5 Biomass feedstocks and their utilization in the production of biofuels, bioenergy, and bioproducts Adapted from Welker et al. (2015)

et al. 2011). The spent coffee grounds are among the valuable bioresources that could be converted into biofuels and are even gaining attraction from the point of view of sustainable waste management policy (Campos-Vega et al. 2015). The other agricultural waste sugar cane bagasse and corncobs can also be used for the production of alcohol components (Reno et al. 2011). In some studies, bran, straw, and husks of rice have been tested for the production of methanol (Nakagawa et al. 2007). The corncobs and sugar cane bagasse have been proved to contain biomethanol (Shamsul et al. 2014). Cassava peel, bagasse, leaves, rhizome, and stem are another valuable agricultural bioresource that can be exploited for biofuel production. Cassava peel can be utilized for the production of bioethanol and biogas as it is rich in carbohydrates, lipids, and proteins. Cassava stem is rich source of cellulose, hemicellulose, and lignin and is used for the production of bioethanol, bio-oil, and biogas (Sivamani et al. 2018).

Food waste is one of the most problematic organic solid wastes, accounting for 15–63% of total municipal solid wastes worldwide (Yun et al. 2018). Thus, its management is very significant; otherwise, it releases odor and leachate during collection and transportation due to its high moisture content, and volatile solids at the same time have high energy content. Thus, using food waste as a resource for biofuel production could be an ideal strategy (Breunig et al. 2017). Food wastes like mixed food wastes, meat, rice, wheat, and vegetable peelings are used to make liquid biofuels (Pham et al. 2015), whereas wastes from bakery are being to hydrolyze bi-enzymatic systems in order to obtain crude hydrolysate that consists of lipids, carbohydrates, amino acids, and phosphates. Carbohydrate and lipid portions obtained during the process are further used for the production of bioethanol and biodiesel, respectively (Karmee 2016).

The poultry industry growth has generated a massive amount of waste, for instance, poultry meal and poultry litter resulting from production facilities and processing plants and it needs to be managed properly. Poultry litter consists of poultry manure, feathers, and lignocellulosic bedding material mixture (Chan et al. 2008). Alternatively, the unavoidable parts in the poultry meat processing like poultry meal consist of ground, rendered, clean parts of poultry carcasses and bones, and offal, and undeveloped eggs are also the poultry wastes (Kantarli et al. 2019). All these wastes consist of abundance of enzymes, proteins, and lipids (Lasekan et al. 2013; Onwosi et al. 2020). The mishandling of these wastes can firstly lead to potential raw material loss and secondly to biological, economic, environmental, and industrial problems (Ashayerizadeh et al. 2017). Feathers are known to be very valuable feedstock for the production of biogas because of high protein content (Mézes et al. 2015). Wastes from the animal husbandry and slaughterhouses have also been used for biogas generation (Afazeli et al. 2014).

Feedstock of algal biomass is the most attractive and promising feedstock for the production of biofuel. Microalgae possess many desirable traits of a renewable energy resource such as they are fast growing, can be cultivated using carbon dioxide which provides greenhouse gas alleviation benefits, have high oil content, do not compete for arable lands and potable water, can use growth nutrients such as phosphorus and nitrogen from waste streams, and do not require complex treatment methods (Lee et al. 2015; Sawayama et al. 1995; Voloshin et al. 2016). Microalgae accumulate high amounts of carbohydrates and lipids. The carbohydrates present in the biomass of microalgae are the chief source of cellulose which is present in the cell wall, and the starch that is without lignin and low hemicellulose contents present in the plastids can be readily converted into fermentable sugars (Chen et al. 2013) via microbial fermentation (Wang et al. 2011). The high content of oil in microalgae can be converted into biodiesel (Ahmad et al. 2012). The content of lipid of some species, for instance, *Botryococcus braunii,* exceeds 80% of the dry weight (Lee et al. 2015); *Dunaliella* and *Chlorella* are known to have lipid contents as great as 50% of the dry weight.

Microorganisms have advantages in various fields such as food, agriculture, and environment (Kour et al. 2020). It is has been reported for producing different biofuels including biodiesel, bioethanol, biomethanol, and many more. Numerous bacteria, namely, *Clostridium* sp., *Escherichia coli*, *Thermoanaerobacter* sp., *Lactobacillus* species, and *Rhodococcus* sp. has been reported for producing biofuels (Rastegari et al. 2020; Yadav et al. 2019b). Algae, particularly green microalgae which are unicellular, have been used for a long time as a potential renewable fuel source (Panjiar et al. 2017). It has a potential to produce biomass of significant quantities and then plant crops for the biofuel production. *Chlamydomonas reinhardtii*, *Scenedesmus obliquus*, *Botryococcus braunii*, *Chlorella* sp., and *Scenedesmus* sp. are the few algae species that have been reported for producing different biofuels (Pittman et al. 2011). Cyanobacteria are an advantageous organism that has several applications in the various industries as its cell grows faster with very simple nutrient requirements (like sunlight, water, and CO_2) and have ability for transforming naturally. This organism has also been reported for producing various biofuels. *Synechocystis* sp., *S. elongates* (Machado and Atsumi 2012), *Cyanobacterium aponinum,* and *Phormidium* sp. (Karatay and Dönmez 2011) are some reported cyanobacteria for producing biofuels. Fungi, the member of the taxonomic form eukarya, have also been reported for producing biofuels. *Mucor circinelloides*, *Cunninghamella japonica*, *Rhizopus oryzae*, *Clostridium acetobutylicum*, *Aspergillus awamori* are few reported fungi (Subhash and Mohan 2014; Vicente et al. 2009; Yadav et al. 2018).

Presently, mankind is dealing with the crucial issues like increased fuel prices, environmental pollution, and climate change due to the fossil fuel reduction and emissions of CO_2 in the atmosphere by the excess fossil fuel utilization. To these all the problems biofuel has is the one alternative which environment amiable and sustainable that is produced by various biomasses of agriculture, food and poultry waste, and microbial biomass. Microbial bioresources are the most significant feedstock for the production of biofuel because of their advantages like fast growing, using CO_2 for cultivation that provides greenhouse gas alleviation benefits, have high oil content, do not compete for arable lands, potable water, can use growth nutrients such as phosphorus and nitrogen from waste streams, and do not require complex treatment methods. So, biofuel produced using microbes is one of the sustainable methods and it can easily meet the growing requirement of biofuels.

References

Afazeli H, Jafari A, Rafiee S, Nosrati M (2014) An investigation of biogas production potential from livestock and slaughterhouse wastes. Renew Sustain Energy Rev 34:380–386
Ahmad A, Yasin NM, Derek C, Lim J (2012) Crossflow microfiltration of microalgae biomass for biofuel production. Desalination 302:65–70
Ashayerizadeh O, Dastar B, Samadi F, Khomeiri M, Yamchi A, Zerehdaran S (2017) Study on the chemical and microbial composition and probiotic characteristics of dominant lactic acid bacteria in fermented poultry slaughterhouse waste. Waste Manage 65:178–185
Association APH, Association AWW, Federation WPC, Federation WE (1920) Standard methods for the examination of water and wastewater. American Public Health Association
Breunig HM, Jin L, Robinson A, Scown CD (2017) Bioenergy potential from food waste in California. Environ Sci Technol 51:1120–1128
Campos-Vega R, Loarca-Pina G, Vergara-Castañeda HA, Oomah BD (2015) Spent coffee grounds: A review on current research and future prospects. Trends Food Sci Technol 45:24–36
Chan K, Van Zwieten L, Meszaros I, Downie A, Joseph S (2008) Using poultry litter biochars as soil amendments. Soil Research 46:437–444
Chen C-Y, Zhao X-Q, Yen H-W, Ho S-H, Cheng C-L, Lee D-J et al (2013) Microalgae-based carbohydrates for biofuel production. Biochem Eng J 78:1–10
Chiaramonti D, Bonini M, Fratini E, Tondi G, Gartner K, Bridgwater A et al (2003) Development of emulsions from biomass pyrolysis liquid and diesel and their use in engines—Part 1: emulsion production. Biomass Bioenerg 25:85–99
Daroch M, Geng S, Wang G (2013) Recent advances in liquid biofuel production from algal feedstocks. Appl Energy 102:1371–1381
Demirbaş A (2000) Mechanisms of liquefaction and pyrolysis reactions of biomass. Energy Convers Manag 41:633–646
Dunlop MJ, Keasling JD, Mukhopadhyay A (2010) A model for improving microbial biofuel production using a synthetic feedback loop. Syst Synth Biol 4:95–104
Fan Y, Li C, Lay J-J, Hou H, Zhang G (2004) Optimization of initial substrate and pH levels for germination of sporing hydrogen-producing anaerobes in cow dung compost. Biores Technol 91:189–193
Hasannuddin A, Wira J, Sarah S, Ahmad M, Aizam S, Aiman M et al (2016) Durability studies of single cylinder diesel engine running on emulsion fuel. Energy 94:557–568
Hasannuddin A, Yahya W, Sarah S, Ithnin A, Syahrullail S, Sugeng D et al (2018) Performance, emissions and carbon deposit characteristics of diesel engine operating on emulsion fuel. Energy 142:496–506
Hong J (2012) Uncertainty propagation in life cycle assessment of biodiesel versus diesel: global warming and non-renewable energy. Biores Technol 113:3–7
Huber GW, Iborra S, Corma A (2006) Synthesis of transportation fuels from biomass: chemistry, catalysts, and engineering. Chem Rev 106:4044–4098
Ikura M, Stanciulescu M, Hogan E (2003) Emulsification of pyrolysis derived bio-oil in diesel fuel. Biomass Bioenerg 24:221–232
Kadier A, Simayi Y, Abdeshahian P, Azman NF, Chandrasekhar K, Kalil MS (2016) A comprehensive review of microbial electrolysis cells (MEC) reactor designs and configurations for sustainable hydrogen gas production. Alexandria Eng J 55:427–443. https://doi.org/10.1016/j.aej.2015.10.008
Kantarli IC, Stefanidis SD, Kalogiannis KG, Lappas AA (2019) Utilisation of poultry industry wastes for liquid biofuel production via thermal and catalytic fast pyrolysis. Waste Manage Res 37:157–167
Karatay SE, Dönmez G (2011) Microbial oil production from thermophile cyanobacteria for biodiesel production. Appl Energy 88:3632–3635
Karmee SK (2016) Liquid biofuels from food waste: current trends, prospect and limitation. Renew Sustain Energy Rev 53:945–953

Karmee SK (2018) A spent coffee grounds based biorefinery for the production of biofuels, biopolymers, antioxidants and biocomposites. Waste Manage 72:240–254

Khan M, Nizami A, Rehan M, Ouda O, Sultana S, Ismail I et al (2017) Microbial electrolysis cells for hydrogen production and urban wastewater treatment: a case study of Saudi Arabia. Appl Energy 185:410–420

Kour D, Rana KL, Yadav AN, Yadav N, Kumar M, Kumar V et al (2020) Microbial biofertilizers: Bioresources and eco-friendly technologies for agricultural and environmental sustainability. Biocatal Agric Biotechnol 23:101487. https://doi.org/10.1016/j.bcab.2019.101487

Kour D, Rana KL, Yadav N, Yadav AN, Rastegari AA, Singh C et al. (2019) Technologies for Biofuel Production: Current Development, Challenges, and Future Prospects. In: Rastegari AA, Yadav AN, Gupta A (eds) Prospects of renewable bioprocessing in future energy systems. Springer International Publishing, Cham, pp 1–50. https://doi.org/10.1007/978-3-030-14463-0_1

Kumar M, Saxena R, Rai PK, Tomar RS, Yadav N, Rana KL et al (2019) Genetic diversity of methylotrophic yeast and their impact on environments. In: Yadav AN, Singh S, Mishra S, Gupta A (eds) Recent advancement in white biotechnology through fungi: Volume 3: Perspective for Sustainable Environments. Springer International Publishing, Cham, pp 53–71. https://doi.org/10.1007/978-3-030-25506-0_3

Lasekan A, Bakar FA, Hashim D (2013) Potential of chicken by-products as sources of useful biological resources. Waste Manage 33:552–565

Lee OK, Seong DH, Lee CG, Lee EY (2015) Sustainable production of liquid biofuels from renewable microalgae biomass. J Ind Eng Chem 29:24–31

Leng L, Li H, Yuan X, Zhou W, Huang H (2018) Bio-oil upgrading by emulsification/microemulsification: a review. Energy 161:214–232

Leng L, Yuan X, Zeng G, Chen X, Wang H, Li H et al (2015) Rhamnolipid based glycerol-in-diesel microemulsion fuel: formation and characterization. Fuel 147:76–81

Lif A, Holmberg K (2006) Water-in-diesel emulsions and related systems. Adv Coll Interface Sci 123:231–239

Liu G, Wei W, Jin W (2014) Pervaporation membranes for biobutanol production. ACS Sustain Chem Eng 2:546–560

Liu H, Grot S, Logan BE (2005) Electrochemically assisted microbial production of hydrogen from acetate. Environ Sci Technol 39:4317–4320

Logan BE, Call D, Cheng S, Hamelers HV, Sleutels TH, Jeremiasse AW et al (2008) Microbial electrolysis cells for high yield hydrogen gas production from organic matter. Environ Sci Technol 42:8630–8640

Lu C, Zhang Z, Ge X, Wang Y, Zhou X, You X et al. (2016) Bio-hydrogen production from apple waste by photosynthetic bacteria HAU-M1. Int J Hydrogen Energy 41:13399–13407

Machado IM, Atsumi S (2012) Cyanobacterial biofuel production. J Biotechnol 162:50–56

Marchetti J, Miguel V, Errazu A (2007) Possible methods for biodiesel production. Renew Sustain Energy Rev 11:1300–1311

Meda U, Rakesh S, Raj M (2015) Bio-hydrogen production in microbial electrolysis cell using waste water from sugar industry. Int J Eng Sci Res Technol 4:452–458

Mézes L, Nagy A, Gálya B, János T (2015) Poultry feather wastes recycling possibility as soil nutrient. Eur J Soil Sci 4:244–252

Milano J, Ong HC, Masjuki H, Chong W, Lam MK, Loh PK et al (2016) Microalgae biofuels as an alternative to fossil fuel for power generation. Renew Sustain Energy Rev 58:180–197

Montpart i Planell N, Guisasola i Canudas A, Baeza Labat JA (2014) Hydrogen production from wastewater in single chamber microbial electrolysis cells: studies towards its scaling-up

Muri P, Črnivec IGO, Djinović P, Pintar A (2016) Biohydrogen production from simple carbohydrates with optimization of operating parameters. Acta Chim Slov 63:154–164

Mussatto SI, Machado EM, Martins S, Teixeira JA (2011) Production, composition, and application of coffee and its industrial residues. Food Bioprocess Technol 4:661

Nakagawa H, Harada T, Ichinose T, Takeno K, Matsumoto S, Kobayashi M et al (2007) Biomethanol production and CO2 emission reduction from forage grasses, trees, and crop residues. Jpn Agr Res Q: JARQ 41:173–180

Onwosi CO, Igbokwe VC, Odimba JN, Nwagu TN (2020) Anaerobic bioconversion of poultry industry-derived wastes for the production of biofuels and other value-added products. In: Biovalorisation of wastes to renewable chemicals and biofuels. Elsevier, pp 113–131

Panjiar N, Mishra S, Yadav AN, Verma P (2017) Functional foods from cyanobacteria: an emerging source for functional food products of pharmaceutical importance. In: Gupta VK, Treichel H, Shapaval VO, Oliveira LAd, Tuohy MG (eds) Microbial functional foods and nutraceuticals. John Wiley & Sons, USA, pp 21–37. https://doi.org/10.1002/9781119048961.ch2

Peskett L, Slater R, Stevens C, Dufey A (2007) Biofuels, agriculture and poverty reduction. Nat Resour Perspect 107:1–6

Pham TPT, Kaushik R, Parshetti GK, Mahmood R, Balasubramanian R (2015) Food waste-to-energy conversion technologies: current status and future directions. Waste Manage 38:399–408

Pittman JK, Dean AP, Osundeko O (2011) The potential of sustainable algal biofuel production using wastewater resources. Biores Technol 102:17–25

Puppan D (2002) Environmental evaluation of biofuels. Periodica Polytechnica Soc Manage Sci 10:95–116

Rastegari AA, Yadav AN, Yadav N (2020) New and future developments in microbial biotechnology and bioengineering: Trends of microbial biotechnology for sustainable agriculture and biomedicine systems: diversity and functional perspectives. Elsevier, Amsterdam

Rastegari AA, Yadav AN, Gupta A (2019) Prospects of renewable bioprocessing in future energy systems. Springer International Publishing, Cham

Reno M, Lora E, Palacio J, Venturini O, Buchgeister J, Almazan O (2011) A LCA (life cycle assessment) of the methanol production from sugarcane bagasse. Energy 36:3716e3726

Sarin R, Sharma M, Sinharay S, Malhotra RK (2007) Jatropha–palm biodiesel blends: an optimum mix for Asia. Fuel 86:1365–1371

Sawayama S, Inoue S, Dote Y, Yokoyama S-Y (1995) CO_2 fixation and oil production through microalga. Energy Convers Manag 36:729–731

Schubert C (2006) Can biofuels finally take center stage? Nat Biotechnol 24:777

Selvaraj R, Praveenkumar R, Moorthy IG (2019) A comprehensive review of biodiesel production methods from various feedstocks. Biofuels 10:325–333

Shamsul N, Kamarudin SK, Rahman NA, Kofli NT (2014) An overview on the production of bio-methanol as potential renewable energy. Renew Sustain Energy Rev 33:578–588

Sivamani S, Chandrasekaran AP, Balajii M, Shanmugaprakash M, Hosseini-Bandegharaei A, Baskar R (2018) Evaluation of the potential of cassava-based residues for biofuels production. Rev Environ Sci Bio/Technol 17:553–570

Stephen JL, Periyasamy B (2018) Innovative developments in biofuels production from organic waste materials: a review. Fuel 214:623–633

Strezov V, Evans T, Kan T, Strezov V, Evans T (2015) Lignocellu-118. Losic biomass pyrolysis: a review of product properties and effects o f pyrolysis parameters and effects of pyrolysis parameters. Renew Sustain Energy Rev 57:1126–1140

Subhash GV, Mohan SV (2014) Lipid accumulation for biodiesel production by oleaginous fungus Aspergillus awamori: influence of critical factors. Fuel 116:509–515

Tigunova O, Shulga S, Blume YB (2013) Biobutanol as an alternative type of fuel. Cytol Gen 47:366–382

Valdez-Vazquez I, Sparling R, Risbey D, Rinderknecht-Seijas N, Poggi-Varaldo HM (2005) Hydrogen generation via anaerobic fermentation of paper mill wastes. Biores Technol 96:1907–1913

Vicente G, Bautista LF, Rodríguez R, Gutiérrez FJ, Sádaba I, Ruiz-Vázquez RM et al (2009) Biodiesel production from biomass of an oleaginous fungus. Biochem Eng J 48:22–27

Voloshin RA, Rodionova MV, Zharmukhamedov SK, Veziroglu TN, Allakhverdiev SI (2016) Biofuel production from plant and algal biomass. Int J Hydrogen Energy 41:17257–17273

Wang J, Xu S, Xiao B, Xu M, Yang L, Liu S et al (2013) Influence of catalyst and temperature on gasification performance of pig compost for hydrogen-rich gas production. Int J Hydrogen Energy 38:14200–14207

Wang X, Liu X, Wang G (2011) Two-stage Hydrolysis of Invasive Algal Feedstock for Ethanol Fermentation F. J Integr Plant Biol 53:246–252

Welker C, Balasubramanian V, Petti C, Rai K, DeBolt S, Mendu V (2015) Engineering plant biomass lignin content and composition for biofuels and bioproducts. Energies 8:7654–7676

Yadav AN (2019) Fungal white biotechnology: conclusion and future prospects. In: Yadav AN, Singh S, Mishra S, Gupta A (eds) Recent Advancement in white biotechnology through fungi: volume 3: perspective for sustainable environments. Springer International Publishing, Cham, pp 491–498. https://doi.org/10.1007/978-3-030-25506-0_20

Yadav AN, Kour D, Rana KL, Yadav N, Singh B, Chauhan VS et al. (2019a) Metabolic engineering to synthetic biology of secondary metabolites production. In: Gupta VK, Pandey A (eds) New and future developments in microbial biotechnology and bioengineering. Elsevier, Amsterdam, pp 279–320. https://doi.org/10.1016/B978-0-444-63504-4.00020-7

Yadav AN, Singh S, Mishra S, Gupta A (2019b) Recent advancement in white biotechnology through fungi. volume 2: perspective for value-added products and environments. Springer International Publishing, Cham

Yadav AN, Rastegari AA, Yadav N (2020) Microbiomes of extreme environments: biodiversity and biotechnological applications. CRC Press, Taylor & Francis, Boca Raton, USA

Yadav AN, Verma P, Kumar V, Sangwan P, Mishra S, Panjiar N et al. (2018) Biodiversity of the genus *Penicillium* in different habitats. In: Gupta VK, Rodriguez-Couto S (eds) New and future developments in microbial biotechnology and bioengineering, *Penicillium* system properties and applications. Elsevier, Amsterdam, pp 3–18. https://doi.org/10.1016/b978-0-444-63501-3.000 01-6

Yun Y-M, Lee M-K, Im S-W, Marone A, Trably E, Shin S-R et al (2018) Biohydrogen production from food waste: current status, limitations, and future perspectives. Bioresour Technol 248:79–87

Correction to: Bioprospecting of Microorganisms for Biofuel Production

Sonali Bhardwaj, Sachin Kumar, and Richa Arora

Correction to:
Chapter 2 in: A. N. Yadav et al. (eds.),
***Biofuels Production – Sustainability and Advances in Microbial Bioresources*,**
Biofuel and Biorefinery Technologies 11,
https://doi.org/10.1007/978-3-030-53933-7_2

In the original version of the chapter, the following correction has been incorporated:

The authors' names have been changed from "Richa Arora Sonali and Sachin Kumar" to "Sonali Bhardwaj, Sachin Kumar and Richa Arora".

The updated version of this chapter can be found at
https://doi.org/10.1007/978-3-030-53933-7_2

A. N. Yadav et al. (eds.), *Biofuels Production – Sustainability and Advances in Microbial Bioresources*, Biofuel and Biorefinery Technologies 11,
https://doi.org/10.1007/978-3-030-53933-7_18

Printed in the United States
by Baker & Taylor Publisher Services